Path Integrals in Physics
Volume II
Quantum Field Theory, Statistical Physics and other Modern Applications

Path Integrals in Physics

Volume II
Quantum Field Theory, Statistical Physics and other Modern Applications

M Chaichian

Department of Physics, University of Helsinki
and
Helsinki Institute of Physics, Finland

and

A Demichev

Institute of Nuclear Physics, Moscow State University, Russia

I₀P

Institute of Physics Publishing
Bristol and Philadelphia

© IOP Publishing Ltd 2001

All rights reserved. No part of this publication may be reproduced, stored in a retrieval system or transmitted in any form or by any means, electronic, mechanical, photocopying, recording or otherwise, without the prior permission of the publisher. Multiple copying is permitted in accordance with the terms of licences issued by the Copyright Licensing Agency under the terms of its agreement with the Committee of Vice-Chancellors and Principals.

British Library Cataloguing-in-Publication Data

A catalogue record for this book is available from the British Library.

ISBN 0 7503 0801 X (Vol. I)
       0 7503 0802 8 (Vol. II)
       0 7503 0713 7 (2 Vol. set)

Library of Congress Cataloging-in-Publication Data are available

Commissioning Editor: James Revill
Production Editor: Simon Laurenson
Production Control: Sarah Plenty
Cover Design: Victoria Le Billon
Marketing Executive: Colin Fenton

Published by Institute of Physics Publishing, wholly owned by The Institute of Physics, London

Institute of Physics Publishing, Dirac House, Temple Back, Bristol BS1 6BE, UK

US Office: Institute of Physics Publishing, The Public Ledger Building, Suite 1035, 150 South Independence Mall West, Philadelphia, PA 19106, USA

Typeset in TEX using the IOP Bookmaker Macros
Printed in the UK by Bookcraft, Midsomer Norton, Bath

Contents

	Preface to volume II		ix
3	**Quantum field theory: the path-integral approach**		**1**
	3.1	Path-integral formulation of the simplest quantum field theories	2
		3.1.1 Systems with an infinite number of degrees of freedom and quantum field theory	2
		3.1.2 Path-integral representation for transition amplitudes in quantum field theories	14
		3.1.3 Spinor fields: quantization via path integrals over Grassmann variables	21
		3.1.4 Perturbation expansion in quantum field theory in the path-integral approach	22
		3.1.5 Generating functionals for Green functions and an introduction to functional methods in quantum field theory	27
		3.1.6 Problems	38
	3.2	Path-integral quantization of gauge-field theories	49
		3.2.1 Gauge-invariant Lagrangians	50
		3.2.2 Constrained Hamiltonian systems and their path-integral quantization	54
		3.2.3 Yang–Mills fields: constrained systems with an infinite number of degrees of freedom	60
		3.2.4 Path-integral quantization of Yang–Mills theories	64
		3.2.5 Covariant generating functional in the Yang–Mills theory	67
		3.2.6 Covariant perturbation theory for Yang–Mills models	73
		3.2.7 Higher-order perturbation theory and a sketch of the renormalization procedure for Yang–Mills theories	80
		3.2.8 Spontaneous symmetry-breaking of gauge invariance and a brief look at the standard model of particle interactions	88
		3.2.9 Problems	98
	3.3	Non-perturbative methods for the analysis of quantum field models in the path-integral approach	101
		3.3.1 Rearrangements and partial summations of perturbation expansions: the $1/N$-expansion and separate integration over high and low frequency modes	101
		3.3.2 Semiclassical approximation in quantum field theory and extended objects (solitons)	110
		3.3.3 Semiclassical approximation and quantum tunneling (instantons)	120
		3.3.4 Path-integral calculation of quantum anomalies	130
		3.3.5 Path-integral solution of the polaron problem	137
		3.3.6 Problems	144
	3.4	Path integrals in the theory of gravitation, cosmology and string theory: advanced applications of path integrals	149

	3.4.1	Path-integral quantization of a gravitational field in an asymptotically flat spacetime and the corresponding perturbation theory	149
	3.4.2	Path integrals in spatially homogeneous cosmological models	154
	3.4.3	Path-integral calculation of the topology-change transitions in $(2+1)$-dimensional gravity	160
	3.4.4	Hawking's path-integral derivation of the partition function for black holes	166
	3.4.5	Path integrals for relativistic point particles and in the string theory	174
	3.4.6	Quantum field theory on non-commutative spacetimes and path integrals	185

4 Path integrals in statistical physics — 194

- 4.1 Basic concepts of statistical physics — 195
- 4.2 Path integrals in classical statistical mechanics — 200
- 4.3 Path integrals for indistinguishable particles in quantum mechanics — 205
  - 4.3.1 Permutations and transition amplitudes — 206
  - 4.3.2 Path-integral formalism for coupled identical oscillators — 210
  - 4.3.3 Path integrals and parastatistics — 216
  - 4.3.4 Problems — 221
- 4.4 Field theory at non-zero temperature — 223
  - 4.4.1 Non-relativistic field theory at non-zero temperature and the diagram technique — 223
  - 4.4.2 Euclidean-time relativistic field theory at non-zero temperature — 226
  - 4.4.3 Real-time formulation of field theory at non-zero temperature — 233
  - 4.4.4 Path integrals in the theory of critical phenomena — 238
  - 4.4.5 Quantum field theory at finite energy — 245
  - 4.4.6 Problems — 252
- 4.5 Superfluidity, superconductivity, non-equilibrium quantum statistics and the path-integral technique — 257
  - 4.5.1 Perturbation theory for superfluid Bose systems — 258
  - 4.5.2 Perturbation theory for superconducting Fermi systems — 261
  - 4.5.3 Non-equilibrium quantum statistics and the process of condensation of an ideal Bose gas — 263
  - 4.5.4 Problems — 277
- 4.6 Non-equilibrium statistical physics in the path-integral formalism and stochastic quantization — 280
  - 4.6.1 A zero-dimensional model: calculation of usual integrals by the method of 'stochastic quantization' — 281
  - 4.6.2 Real-time quantum mechanics within the stochastic quantization scheme — 284
  - 4.6.3 Stochastic quantization of field theories — 288
  - 4.6.4 Problems — 293
- 4.7 Path-integral formalism and lattice systems — 295
  - 4.7.1 Ising model as an example of genuine discrete physical systems — 296
  - 4.7.2 Lattice gauge theory — 302
  - 4.7.3 Problems — 308

Supplements — 311

- I Finite-dimensional Gaussian integrals — 311
- II Table of some exactly solved Wiener path integrals — 313
- III Feynman rules — 316
- IV Short glossary of selected notions from the theory of Lie groups and algebras — 316

V	Some basic facts about differential Riemann geometry	325
VI	Supersymmetry in quantum mechanics	329

Bibliography   **332**

Index   **337**

Preface to volume II

In the second volume of this book (chapters 3 and 4) we proceed to discuss path-integral applications for the study of systems with an infinite number of degrees of freedom. An appropriate description of such systems requires the use of second quantization, and hence, field theoretical methods. The starting point will be the quantum-mechanical phase-space path integrals studied in volume I, which we suitably generalize for the quantization of field theories.

One of the central topics of chapter 3 is the formulation of the celebrated Feynman diagram technique for the perturbation expansion in the case of field theories with constraints (gauge-field theories), which describe all the fundamental interactions in elementary particle physics. However, the important applications of path integrals in quantum field theory go far beyond just a convenient derivation of the perturbation theory rules. We shall consider, in this volume, various modern non-perturbative methods for calculations in field theory, such as variational methods, the description of topologically non-trivial field configurations, the quantization of extended objects (solitons and instantons), the $1/N$-expansion and the calculation of quantum anomalies. In addition, the last section of chapter 3 contains elements of some advanced and currently developing applications of the path-integral technique in the theory of quantum gravity, cosmology, black holes and in string theory.

For a successful reading of the main part of chapter 3, it is helpful to have some acquaintance with a standard course of quantum field theory, at least at a very elementary level. However, some parts (e.g., quantization of extended objects, applications in gravitation and string theories) are necessarily more fragmentary and presented without much detail. Therefore, their complete understanding can be achieved only by rather experienced readers or by further consultation of the literature to which we refer. At the same time, we have tried to present the material in such a form that even those readers not fully prepared for this part could get an idea about these modern and fascinating applications of path integration.

As we stressed in volume I, one of the most attractive features of the path-integral approach is its universality. This means it can be applied without crucial modifications to statistical (both classical and quantum) systems. We discuss how to incorporate the statistical properties into the path-integral formalism for the study of many-particle systems in chapter 4. Besides the basic principles of path-integral calculations for systems of indistinguishable particles, chapter 4 contains a discussion of various problems in modern statistical physics (such as the analysis of critical phenomena, calculations in field theory at non-zero temperature or at fixed energy, as well as the study of non-equilibrium systems and the phenomena of superfluidity and superconductivity). Therefore, to be tractable in a single book, these examples contain some simplifications and the material is presented in a more fragmentary style in comparison with chapters 1 and 2 (volume I). Nevertheless, we have again tried to make the text as self-contained as possible, so that all the crucial points are covered. The reader will find references to the appropriate literature for further details.

Masud Chaichian, Andrei Demichev
Helsinki, Moscow
December 2000

Chapter 3

Quantum field theory: the path-integral approach

So far, we have been discussing systems containing only one or, at most, a few particles. However, the method of path integrals readily generalizes to systems with many and even an arbitrary number of degrees of freedom. Thus in this chapter we shall consider one more infinite limit related to path integrals and discuss applications of the latter to systems with an *infinite* number of degrees of freedom. In other words, we shall derive path-integral representations for different objects in *quantum field theory* (QFT). Of course, this is nothing other than quantum mechanics for systems with an arbitrary or non-conserved number of excitations (particles or quasiparticles). Therefore, the starting point for us is the quantum-mechanical phase-space path integrals studied in chapter 2. In most practical applications in QFT, these path integrals can be reduced to the Feynman path integrals over the corresponding *configuration spaces* by integrating over momenta. This is especially important for relativistic theories where this transition allows us to keep relativistic invariance of all expressions *explicitly*.

Apparently, the most important result of path-integral applications in QFT is the formulation of the celebrated *Feynman rules* for perturbation expansion in QFT with constraints, i.e. in *gauge-field theories* which describe all the fundamental interactions of elementary particles. In fact, Feynman derived his important rules (Feynman 1948, 1950) (in *quantum electrodynamics* (QED)) just using the path-integral approach! Later, these rules (graphically expressed in terms of *Feynman diagrams*) were rederived in terms of the standard operator approach. But the appearance of more complicated *non-Abelian gauge-field theories* (which describe weak, strong and gravitational interactions) again brought much attention to the path-integral method which had proved to be much more suitable in this case than the operator approach, because the latter faces considerable combinatorial and other technical problems in the derivation of the Feynman rules. In fact, it is this success that attracted wide attention to the path-integral formalism in QFT and in quantum mechanics in general.

Further development of the path-integral formalism in QFT has led to results far beyond the convenient derivation of perturbation theory rules. In particular, it has resulted in various non-perturbative approximations for calculations in field theoretical models, variational methods, the description of topologically non-trivial field configurations, the discovery of the so-called *BRST (Becchi–Rouet–Stora–Tyutin) symmetry* in gauge QFT, clarification of the relation between quantization and the theory of stochastic processes, the most natural formulation of string theory which is believed to be the most realistic candidate for a 'theory of everything', etc.

In the first section of this chapter, we consider path-integral quantization of the simplest field theories, including scalar and spinor fields. We derive the path-integral expression for the generating functional of the Green functions and develop the perturbation theory for their calculation. In section 3.2, after an introduction to the quantization of quantum-mechanical systems with constraints, we proceed to the path-integral description of gauge theories. We derive the covariant generating functional and covariant perturbation expansion for Yang–Mills theories with exact and spontaneously broken gauge symmetry, including the realistic *standard model* of electroweak interactions and *quantum chromodynamics* (QCD), which is the gauge theory of strong interactions.

In section 3.3, we present non-perturbative methods and results in QFT based on the path-integral approach. They include $1/N$-expansion, separate integration over different Fourier modes (with appropriate approximations for different frequency ranges), semiclassical, in particular *instanton*, calculations and the quantization of extended objects (*solitons*), the analysis and calculation of *quantum anomalies* in the framework of the path integral and the Feynman variational method in non-relativistic field theory (on the example of the so-called *polaron problem*).

Section 3.4 contains some advanced applications of path-integral techniques in the theory of quantum gravity, cosmology, black holes and string theory. Reading this section requires knowledge of the basic facts and notions from Einstein's general relativity and the differential geometry of Riemann manifolds (some of these are collected in supplement V).

We must stress that, although we intended to make the text as self-contained as possible, this chapter by no means can be considered as a comprehensive introduction to such a versatile subject as QFT. We mostly consider those aspects of the theory which have their natural and simple description in terms of path integrals. Other important topics can be found in the extensive literature on the subject (see e.g., Wentzel (1949), Bogoliubov and Shirkov (1959), Schweber (1961), Bjorken and Drell (1965), Itzykson and Zuber (1980), Chaichian and Nelipa (1984), Greiner and Reinhardt (1989), Peskin and Schroeder (1995) and Weinberg (1995, 1996, 2000)).

3.1 Path-integral formulation of the simplest quantum field theories

After a short exposition of the postulates and main facts from conventional field theory, we present the path-integral formulation of the simplest models: a single scalar field and a fermionic field. The latter requires path integration over the Grassmann variables considered at the end of chapter 2. Then we consider the perturbation expansion and generating functional for these simple theories which serve as introductory examples for the study of the realistic models presented in the next section.

3.1.1 Systems with an infinite number of degrees of freedom and quantum field theory

There are various formulations of quantum field theory, differing in the form of presentation of the basic quantities, namely transition amplitudes. In the operator approach, the transition amplitudes are expressed as the vacuum expectation value of an appropriate product of particle creation and annihilation operators. These operators obey certain commutation relations (generalization of the standard canonical commutation relations to a system with an infinite number of degrees of freedom). Another formulation is based on expressing the transition amplitudes in terms of path integrals over the fields. In studying the gauge fields, the path-integral formalism has proven to be the most convenient. However, for an easier understanding of the subject we shall start by considering unconstrained fields and then proceed to gauge-field theories (i.e. field theories with constraints).

Let us consider, as a starting example, a single scalar field. From the viewpoint of Hamiltonian dynamics, a field is a system with an infinitely large number of degrees of freedom, for the field is characterized by a generalized coordinate $\varphi(x)$ and a generalized momentum $\pi(x)$ at each space point $x \in \mathbb{R}^d$.

It is worth making the following remark. If we were intending to provide an introduction to the very subject of quantum field theory, it would be pedagogically more reasonable to start from non-relativistic many-body problems and the corresponding non-relativistic quantum field

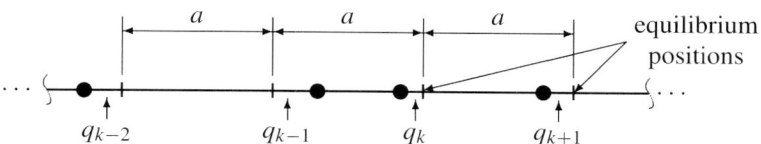

Figure 3.1. Vibrating chain of coupled oscillators; the distances between the equilibrium positions of the particles are equal to some fixed value a, the displacements of the particles from the equilibrium positions are the dynamical variables and are denoted by q_k ($k = 1, \ldots, K$).

theories, as they are the closest generalization of one (or at most a few) particle problems in quantum mechanics. However, the area of the most fruitful applications of non-relativistic field theories is the physics of quantum statistical systems, in general with non-zero temperature. Path integrals for statistical systems have some peculiarities (in particular, the corresponding trajectories may have a rather specific meaning, one which is quite different from that in quantum mechanics). Therefore, we postpone discussion of such systems until the next chapter and now proceed to consider path-integral formulation of quantum field theories at zero temperature which finds its main application in the description of the *relativistic* quantum mechanics of elementary particles. In this chapter, we shall encounter only one example of a non-relativistic field theoretical model which describes the behaviour of an electron inside a crystal (the so-called *polaron problem*).

◇ **Quantum fields as an infinite number of degrees of freedom limit of systems of coupled oscillators**

In order to approach the consideration of systems with an infinite number of degrees of freedom (quantum fields) we start from a chain of K coupled oscillators with equal masses and frequencies, in the framework of ordinary quantum mechanics (see figure 3.1).

The Hamiltonian of such a system has the form

$$H = \sum_{k=1}^{K} \tfrac{1}{2}[p_k^2 + \Omega^2(q_k - q_{k+1})^2 + \Omega_0^2 q_k^2] \tag{3.1.1}$$

where p_k, q_k ($k = 1, \ldots, K$) are the canonical variables (momentum and position) of the kth oscillator and the equations of motion read:

$$\dot{q}_k = p_k$$
$$\dot{p}_k = \Omega^2(q_{k+1} + q_{k-1} - 2q_k) - \Omega_0^2 q_k \tag{3.1.2}$$

or, written only in terms of coordinates,

$$\ddot{q}_k = \Omega^2(q_{k+1} + q_{k-1} - 2q_k) - \Omega_0^2 q_k. \tag{3.1.3}$$

The frequency Ω_0 defines the potential energy of an oscillator due to a shift from its equilibrium position and the frequency Ω defines the interaction of an oscillator with its neighbours. Since we shall use this model as a starting point for the introduction of quantum fields, a concrete value of the particle masses in (3.1.1) is not important and for convenience we have put it equal to unity (cf (2.1.42)). Besides, as is usual in relativistic quantum field theory, we use units such that $\hbar = 1$.

The equations of motion must be accompanied by some boundary conditions. Since we are going to pass later to systems in infinite volumes (of infinite sizes), the actual form of the boundary conditions should not have a crucial influence on the behaviour of the systems. Therefore, we can choose them freely and the most convenient one is the periodic condition:

$$q_{k+K} = q_k. \tag{3.1.4}$$

After the quantization, the canonical variables become operators with the following canonical commutation relations:

$$[\widehat{q}_k, \widehat{p}_l] = i\delta_{kl}$$
$$[\widehat{q}_k, \widehat{q}_l] = [\widehat{p}_k, \widehat{p}_l] = 0 \qquad k, l = 1, \ldots, K. \tag{3.1.5}$$

In order to find the eigenvalues of the corresponding quantum Hamiltonian

$$\widehat{H} = \sum_{k=1}^{K} \tfrac{1}{2}[\widehat{p}_k^2 + \Omega^2(\widehat{q}_k - \widehat{q}_{k+1})^2 + \Omega_0^2 \widehat{q}_k^2] \tag{3.1.6}$$

it is helpful to introduce new variables (the so-called *normal coordinates*) $\widehat{Q}_r$, $\widehat{P}_r$ via the discrete Fourier transform:

$$\widehat{q}_k = \frac{1}{\sqrt{K}} \sum_{r=-K/2+1}^{K/2} \widehat{Q}_r e^{i2\pi rk/K}$$
$$\widehat{p}_k = \frac{1}{\sqrt{K}} \sum_{r=-K/2+1}^{K/2} \widehat{P}_r e^{-i2\pi rk/K} \tag{3.1.7}$$

with the analogous commutation relations

$$[\widehat{Q}_r, \widehat{P}_s] = i\delta_{rs}$$
$$[\widehat{Q}_r, \widehat{Q}_s] = [\widehat{P}_r, \widehat{P}_s] = 0 \tag{3.1.8}$$

where r and s are integers from the interval $[-K/2+1, K/2]$. It is easy to verify that the normal coordinates also satisfy the periodic conditions: $\widehat{Q}_{-K/2} = \widehat{Q}_{K/2}$ and $\widehat{P}_{-K/2} = \widehat{P}_{K/2}$, so that we again have $2N$ independent variables (as in the case of $\widehat{q}_k$, $\widehat{p}_k$). This restriction, as well as the range of the summations in (3.1.7), follows from the periodic boundary conditions (3.1.5). Since q_k, p_k are Hermitian operators, the new operators satisfy the conditions

$$\widehat{Q}_k^\dagger = \widehat{Q}_{-k} \qquad \widehat{P}_k^\dagger = \widehat{P}_{-k}. \tag{3.1.9}$$

The Kronecker symbol δ_{ln} can be represented as the sum

$$\sum_{k=1}^{K} e^{i2\pi k(l-n)/K} = K\delta_{ln}. \tag{3.1.10}$$

This is an analog of the integral representation (1.1.22) for the δ-function, adapted to the discrete finite lattice with a periodic boundary condition. Using this formula, we can invert the transformation (3.1.7) of the dynamical variables:

$$\widehat{Q}_r = \frac{1}{\sqrt{K}} \sum_{k=1}^{K} \widehat{q}_k e^{-i2\pi rk/K}$$
$$\widehat{P}_r = \frac{1}{\sqrt{K}} \sum_{k=1}^{K} \widehat{p}_k e^{i2\pi rk/K}. \quad (3.1.11)$$

In the normal coordinates Q_r, P_r the Hamiltonian (3.1.6) takes the simpler form

$$\widehat{H} = \tfrac{1}{2} \sum_{r=-K/2+1}^{K/2} [\widehat{P}_r \widehat{P}_r^\dagger + \omega_r^2 \widehat{Q}_r \widehat{Q}_r^\dagger] \quad (3.1.12)$$

$$\omega_r^2 \equiv \Omega^2 \left(2 \sin \frac{2\pi r}{K}\right) + \Omega_0^2. \quad (3.1.13)$$

Thus, in the normal coordinates we have K non-interacting oscillators and it is natural to introduce the creation and annihilation operators (cf (2.1.47), taking into account that Q_r, P_r now are not Hermitian operators):

$$\widehat{a}_r = \frac{1}{\sqrt{\omega_r}}(\omega_r \widehat{Q}_r + i\widehat{P}_r^\dagger)$$
$$\widehat{a}_r^\dagger = \frac{1}{\sqrt{\omega_r}}(\omega_r \widehat{Q}_r^\dagger - i\widehat{P}_r) \quad (3.1.14)$$

(note that $\widehat{a}_{-r} \neq \widehat{a}_r^\dagger$). The commutation relations for $\widehat{a}_r$, $\widehat{a}_r^\dagger$ are derived from (3.1.8) with the expected result:

$$[\widehat{a}_r, \widehat{a}_s^\dagger] = \delta_{rs}$$
$$[\widehat{a}_r, \widehat{a}_s] = [\widehat{a}_r^\dagger, \widehat{a}_s^\dagger] = 0. \quad (3.1.15)$$

In terms of these operators, the Hamiltonian (3.1.12) reads as

$$\widehat{H} = \sum_{r=-K/2+1}^{K/2} \omega_r (\widehat{a}_r^\dagger \widehat{a}_r + \tfrac{1}{2}). \quad (3.1.16)$$

Eigenstates of the Hamiltonian written in the latter form can be constructed in the standard way: the state

$$|n_{-K/2+1}, n_{-K/2+2}, \ldots, n_{K/2}\rangle = \prod_{r=-K/2+1}^{K/2} \frac{1}{\sqrt{n_r!}} (\widehat{a}_r^\dagger)^{n_r} |0\rangle \quad (3.1.17)$$

is the Hamiltonian eigenstate with energy (eigenvalue)

$$E = E_0 + \sum_r n_r \omega_r. \quad (3.1.18)$$

The state $|0\rangle$ in (3.1.17) has the lowest energy:
$$E_0 = \sum_r \frac{\omega_r}{2} \tag{3.1.19}$$

and is defined by the conditions
$$\hat{a}_r|0\rangle = 0 \qquad r = -K/2 + 1, \ldots, K/2. \tag{3.1.20}$$

Let us consider the continuous limit for a chain of coupled oscillators $K \to \infty$, $a \to 0$, with a finite value of the product $aK \equiv L$. Technically, this corresponds to the following substitutions:
$$q_k \longrightarrow \frac{q(x)}{\sqrt{a}} \qquad \sum_k \longrightarrow \frac{1}{a}\int_0^L dx \qquad \Omega \longrightarrow \frac{v}{a} \tag{3.1.21}$$

and Hamiltonian (3.1.1) takes the following form in the limit
$$H = \int_0^L dx \, \frac{1}{2}\left[p^2(x,t) + v^2\left(\frac{\partial q}{\partial x}\right)^2 + \Omega_0^2 q^2(x,t)\right]. \tag{3.1.22}$$

Now the degrees of freedom of the system are 'numbered' by the continuous variable x. However, for a finite length L, the normal coordinates Q_r, P_r are still countable:
$$q(x) = \frac{1}{\sqrt{L}} \sum_{r=-\infty}^{\infty} e^{i2\pi r/L} Q_r$$
$$p(x) = \frac{1}{\sqrt{L}} \sum_{r=-\infty}^{\infty} e^{i2\pi r/L} P_r \tag{3.1.23}$$

though the index r is now an arbitrary unbounded integer. The quantum Hamiltonian can be cast again into the form (3.1.12) or (3.1.16):
$$\hat{H} = \tfrac{1}{2} \sum_{r=-\infty}^{\infty} (\hat{P}_r \hat{P}_r^\dagger + \omega_r^2 \hat{Q}_r \hat{Q}_r^\dagger)$$
$$= \sum_{r=-\infty}^{\infty} \omega_r (\hat{a}_r^\dagger \hat{a}_r + \tfrac{1}{2}) \tag{3.1.24}$$
$$\omega_r^2 = v^2 k^2 + \Omega_0^2 \qquad k \equiv \frac{2\pi r}{L} \tag{3.1.25}$$

with the only difference begin that the sums run over all integers. The eigenstates and eigenvalues of this Hamiltonian are given by (3.1.17)–(3.1.20). The essentially new feature of this system with an *infinite* number of degrees of freedom (i.e. after the transition $K \to \infty$) is that the energy (3.1.19) of the lowest eigenstate $|0\rangle$ becomes *infinite*. We can circumvent this difficulty by redefining the Hamiltonian as follows:
$$\hat{H} \longrightarrow \hat{H} - E_0 = \tfrac{1}{2} \sum_{r=-\infty}^{\infty} \omega_r \hat{a}_r^\dagger \hat{a}_r \tag{3.1.26}$$

i.e. counting the energy with respect to the lowest state $|0\rangle$. This is the simplest example of the so-called renormalizations in quantum field theory.

All the considerations outlined here can easily be generalized to higher-dimensional lattices and corresponding higher-dimensional spaces in the continuous limit. In the latter case, the dynamical variables depend on (are labeled by) d-dimensional vectors:

$$\hat{q}(x,t) \longrightarrow \hat{\varphi}(x,t) \qquad \hat{p}(x,t) \longrightarrow \hat{\pi}(x,t) \qquad x \in \mathbb{R}^d \qquad (3.1.27)$$

so that we have arrived in this way at the notion of the *quantum field in the* $(d+1)$-*dimensional spacetime*. Note that the straightforward generalization of the coupled oscillator model previously considered in the one-dimensional space leads to the *vector fields* $\hat{\varphi}(x,t), \hat{\pi}(x,t)$ because the displacements and momenta of oscillators in d-dimensional spaces are described by vectors. However, if we assume that for some reason the displacements are confined to *one* direction, we obtain the physically important case of *scalar* quantum fields $\hat{\varphi}(x,t), \hat{\pi}(x,t)$.

Hamiltonians for quantum fields in higher-dimensional spaces are the direct generalizations of those for the one-dimensional case (cf (3.1.22)). In particular, for the most realistic three-dimensional space, we have

$$\hat{H} = \tfrac{1}{2} \int d^3r \, [\hat{\pi}^2(r,t) + v^2(\nabla\hat{\varphi}(r,t))^2 + \Omega_0^2 \hat{\varphi}^2(r,t)]. \qquad (3.1.28)$$

The operators of the quantum field $\hat{\varphi}(r,t)$ and the corresponding momentum $\hat{\pi}(r,t)$ satisfy the canonical commutation relations at equal times:

$$[\hat{\varphi}(r,t), \hat{\pi}(r',t)] = i\delta^3(r-r')$$
$$[\hat{\varphi}(r,t), \hat{\varphi}(r',t)] = [\hat{\pi}(r,t), \hat{\pi}(r',t)] = 0. \qquad (3.1.29)$$

The three-dimensional periodic boundary conditions require the following equalities:

$$\varphi(x+L,y,z,t) = \varphi(x,y+L,z,t) = \varphi(x,y,z+L,t) = \varphi(x,y,z,t) \qquad (3.1.30)$$

and the corresponding Fourier transform,

$$\hat{\varphi}(r,t) = \frac{1}{L^{3/2}} \sum_{k_x,k_y,k_z=-\infty}^{\infty} (2\omega_k)^{-1/2} [e^{i(k \cdot r - \omega_k t)} \hat{a}_k + e^{-i(k \cdot r - \omega_k t)} \hat{a}_k^\dagger] \qquad (3.1.31)$$

$$k_{x,y,z} = \frac{2\pi l_{x,y,z}}{L} \qquad (3.1.32)$$

$$\omega_k^2 = v^2 k^2 + \Omega_0^2 \qquad (3.1.33)$$

allows us once again to convert (3.1.28) into the Hamiltonian for an infinite set of independent oscillators:

$$\hat{H} = \sum_{l_{x,y,z}=-\infty}^{\infty} \omega_k (\hat{a}_k^\dagger \hat{a}_k + \tfrac{1}{2}) \qquad (3.1.34)$$

with $\hat{a}_k^\dagger, \hat{a}_k$ being the creation and annihilation operators subjected to the following commutation relations:

$$[\hat{a}_k, \hat{a}_{k'}^\dagger] = \delta_{kk'}$$
$$[\hat{a}_k, \hat{a}_{k'}] = [\hat{a}_k^\dagger, \hat{a}_{k'}^\dagger] = 0. \qquad (3.1.35)$$

The eigenstates and eigenvalues of this Hamiltonian again have the form (3.1.17)–(3.1.20), so that the energy of a state $|n_{k_1}, n_{k_2}, \ldots\rangle$ is completely defined by the set $\{n_{k_i}\}$ of *occupation numbers* $\{n_k\} = n_{k_1}, n_{k_2}, \ldots$ (i.e. by powers of the creation operators on the right-hand side of (3.1.17)):

$$E_{\{n_k\}} - E_0 = \sum_i n_{k_i} \omega_{k_i}. \qquad (3.1.36)$$

Note that, if we put

$$v = c \qquad \Omega_0 = mc^2 \qquad (3.1.37)$$

where c is the speed of light and m is the mass of a particle, equation (3.1.33) exactly coincides with the relativistic relation between energy, mass and momentum k of a particle. Hence, expression (3.1.36) for the energy of the quantum field $\varphi(x, t)$ can be interpreted as the sum of energies of the set (defined by the occupation numbers $\{n_{k_i}\}$) of free relativistic particles. To simplify the formulae, we shall, in what follows, put the speed of light equal to unity, $c = 1$; the latter can be achieved by an appropriate choice of units of measurement.

- Thus, we have obtained a remarkable result: a quantum field with the Hamiltonian (3.1.28) (or (3.1.22)) and the choice of parameters as in (3.1.37) is equivalent to a system of an arbitrary number of free relativistic particles. According to the commutation relations (3.1.35), these particles obey Bose–Einstein statistics.

We have already mentioned the specific problems of quantum systems with an infinite number of degrees of freedom, that is, the appearance of divergent expressions. One example is the energy of 'zero oscillations' (3.1.19), which diverges for an infinite number of oscillators. Another example is the expression for 'zero fluctuations' of the field $\varphi(t, r)$, in other words for the dispersion of the field in the lowest energy state:

$$(\mathbb{D}_{|0\rangle} \hat{\varphi})^2 \equiv \langle 0 | \hat{\varphi}^2 | 0 \rangle$$
$$= \frac{1}{(2\pi)^3} \int d^3 k \frac{1}{2\omega_k} = \frac{1}{(2\pi)^3} \int d^3 k \frac{1}{2\sqrt{k^2 + m^2}} \to \infty. \qquad (3.1.38)$$

The reason for the infinite value of the fluctuation is related to the fact that $\hat{\varphi}$, acting on an arbitrary state with finite energy, gives a state with an infinite norm. Thus $\hat{\varphi}$ does not belong to well-defined operators in the Hilbert space of states of the Hamiltonian under consideration. Another way to express this fact is to say that $\hat{\varphi}$ is an operator-valued distribution (generalized function). To construct a well-defined operator, we have to smear $\hat{\varphi}$ with an appropriate test function, e.g., to consider the quantity

$$\bar{\varphi}_\lambda \stackrel{\text{def}}{\equiv} \frac{1}{(2\pi \lambda^2)^{3/2}} \int d^3 r \, e^{-r^2/(2\lambda^2)} \hat{\varphi}(t, r) \qquad (3.1.39)$$

which can be interpreted as an average value of the field in the volume λ^3 around the point r. The reader may check that the dispersion of $\bar{\varphi}_\lambda$ is finite:

$$\langle 0 | \bar{\varphi}_\lambda^2 | 0 \rangle \approx \frac{1}{\lambda^3 \sqrt{\lambda^{-2} + m^2}} \qquad (3.1.40)$$

(problem 3.1.1, page 38). The last expression shows that the smaller the volume λ^3 is, the stronger the fluctuations of the field are. This fact, of course, is in full correspondence with the quantum-mechanical uncertainty principle.

◇ **Relativistic invariance of field theories and Minkowski space**

To reveal explicitly the relativistic symmetry of the system described by the Hamiltonian (3.1.28) with the parameters (3.1.37), we should pass to the Lagrangian formalism:

$$H[\pi(\mathbf{r},t), \varphi(\mathbf{r},t)] \longrightarrow L[\dot{\varphi}(\mathbf{r},t), \varphi(\mathbf{r},t)]$$

where L is the classical Lagrangian defined by the classical Hamiltonian H via the Legendre transformation:

$$L[\dot{\varphi}(\mathbf{r},t), \varphi(\mathbf{r},t)] = \int d^3r\, \pi(\mathbf{r},t)\dot{\varphi}(\mathbf{r},t) - H[\pi(\mathbf{r},t), \varphi(\mathbf{r},t)]. \qquad (3.1.41)$$

The momentum π on the right-hand side of (3.1.41) is assumed to be expressed through $\dot{\varphi}$, φ with the help of the Hamiltonian equation of motion. In our case,

$$\dot{\varphi} = \{\varphi, H\} = \pi \qquad (3.1.42)$$

(recall that $\{\cdot,\cdot\}$ is the Poisson bracket). Thus, the Lagrangian for the scalar field reads as

$$L(t) = \int d^3r\, \tfrac{1}{2}[\dot{\varphi}^2(\mathbf{r},t) - (\nabla\varphi(\mathbf{r},t))^2 - m^2\varphi^2(\mathbf{r},t)]. \qquad (3.1.43)$$

To demonstrate the invariance of the Lagrangian (3.1.43) with respect to transformations forming relativistic kinematic groups, i.e. the *Lorentz* or *Poincaré* groups, it is helpful to pass to four-dimensional notation. Let us introduce the four-dimensional *Minkowski* space with the coordinates:

$$x^\mu \stackrel{\mathrm{def}}{\equiv} \{t, \mathbf{r}\} \qquad \mu = 0, 1, 2, 3 \qquad (3.1.44)$$

i.e.

$$x^0 = t \qquad x^i = r_i \qquad i = 1, 2, 3,$$

and the metric tensor

$$g_{\mu\nu} = \mathrm{diag}\{1, -1, -1, -1\} \qquad (3.1.45)$$

which defines the scalar product of vectors in the Minkowski space:

$$xy \equiv x^\mu y_\mu \stackrel{\mathrm{def}}{\equiv} x^\mu g_{\mu\nu} y^\nu$$

(repeating indices are assumed to be summed over). In particular, the squared vector in the Minkowski space reads as

$$x^2 \equiv (x^\mu)^2 = x^\mu g_{\mu\nu} x^\nu = (x^0)^2 - (x^1)^2 - (x^2)^2 - (x^3)^2$$
$$= t^2 - \mathbf{r}^2 = t^2 - r_1^2 - r_2^2 - r_3^2 \qquad (3.1.46)$$

or, for the infinitesimally small vector dx^μ,

$$(dx^\mu)^2 = dx^\mu g_{\mu\nu} dx^\nu = (dt)^2 - (d\mathbf{r})^2. \qquad (3.1.47)$$

In the literature on relativistic field theory, it is common to drop boldface type for four-dimensional vectors and we shall follow this custom. If the vector indices $\mu, \nu, \ldots$ take, in some expressions,

only spacelike values 1, 2, 3, we shall denote them by Latin letters $l, k, \ldots$ and use the following shorthand notation:
$$A_l B_l = \sum_{l=1}^{3} A_l B_l$$
where A_l, B_l are the spacelike components of some four-dimensional vectors $A_\mu = \{A_0, A_l\}$, $B_\nu = \{B_0, B_l\}$.

The Minkowski metric tensor $g_{\mu\nu}$ is invariant with respect to the transformations defined by the pseudo-orthogonal 4×4 matrices $\Lambda^\mu{}_\nu$ from the Lie group $SO(1,3)$, called the *Lorentz group*:
$$\Lambda^\rho{}_\mu g_{\rho\sigma} \Lambda^\sigma{}_\nu = g_{\mu\nu}. \tag{3.1.48}$$

This means that any scalar product in the Minkowski space is invariant with respect to the Lorentz transformations. Moreover, the scalar products of vectors (recall that the latter are expressed through the *differences* in the coordinates of two points) are also invariant with respect to the four-dimensional translations forming the Abelian (commutative) group T_4. In particular, the reader can easily verify that $dx^\mu g_{\mu\nu} dx^\nu$ and $(\partial/\partial x^\mu) g^{\mu\nu} (\partial/\partial x^\nu)$, where $g^{\mu\nu}$ denotes the inverse matrix
$$g^{\mu\rho} g_{\rho\nu} = \delta^\mu{}_\nu$$
are invariant with respect to both Lorentz 'rotations' as well as translations and, hence, with respect to the complete *Poincaré group* $SO(1,3) \mathbin{\textcircled{s}} T_4$.

We shall not go further into the details of relativistic kinematics, referring the reader to, e.g., Novozhilov (1975), Chaichian and Hagedorn (1998) or any textbook on quantum field theory (in particular, those mentioned at the very beginning of this chapter).

To restore full equivalence between the time and space coordinates, it is useful to introduce the Lagrangian density. This is nothing other than the integrand of (3.1.43), which, in four-dimensional notation, takes the form
$$\mathcal{L}_0(\dot\varphi, \varphi) = \tfrac{1}{2}[g^{\mu\nu} \partial_\mu \varphi(x) \partial_\nu \varphi(x) - m^2 \varphi^2(x)]. \tag{3.1.49}$$

The action for a scalar relativistic field and for the entire time line $-\infty < x^0 < \infty$ can now be written as follows:
$$S_0[\varphi] = \int_{\mathbb{R}^4} d^4x \, \mathcal{L}_0(\dot\varphi, \varphi). \tag{3.1.50}$$

Taking into account the fact that the integration measure $d^4x = dx^0 dx^1 dx^2 dx^3$ is invariant with respect to the pseudo-orthogonal Lorentz transformations as well as with respect to translations, we can readily check that action (3.1.50) is indeed Poincaré invariant.

For a finite-time interval, $t_0 < t \equiv x^0 < t_f$, the action reads as
$$S[\varphi] = \int_{t_0}^{t_f} dx^0 \, L \equiv \int_{t_0}^{t_f} dx^0 \int_{\mathbb{R}^3} dx^1 dx^2 dx^3 \, \mathcal{L}(\dot\varphi, \varphi). \tag{3.1.51}$$

The equation of motion can now be derived from the extremality of action (3.1.51): $\delta S = 0$, together with the boundary conditions that variations of the field at times t_0 and t_f vanish: $\delta\varphi(t_0) = \delta\varphi(t_f) = 0$, which result in the Euler–Lagrange equation
$$\frac{\partial}{\partial t} \frac{\delta L}{\delta \dot\varphi} = \frac{\delta L}{\delta \varphi} \tag{3.1.52}$$

or
$$\frac{\partial}{\partial t}\frac{\partial \mathcal{L}}{\partial \dot{\varphi}} = \frac{\partial \mathcal{L}}{\partial \varphi} - \nabla \frac{\partial \mathcal{L}}{\partial \nabla \varphi}. \qquad (3.1.53)$$

For a free scalar field with the Lagrangian density (3.1.49), the Euler–Lagrange equation is equivalent to the so-called *Klein–Gordon equation*:

$$(\Box + m^2)\varphi(x) = 0 \qquad (3.1.54)$$

where

$$\Box \stackrel{\text{def}}{\equiv} g^{\mu\nu}\partial_\mu \partial_\nu \equiv \frac{\partial^2}{\partial t^2} - \nabla^2. \qquad (3.1.55)$$

In order to describe *interacting* particles, we have to add, to the Lagrangian density (3.1.49), higher powers of the field $\varphi(x)$:

$$\mathcal{L}(\partial_\mu \varphi, \varphi) = \tfrac{1}{2}[g^{\mu\nu}\partial_\mu \varphi(x)\partial^\mu \varphi(x) - m^2\psi^2(x)] - V(\varphi(x)). \qquad (3.1.56)$$

Here the function $V(\varphi(x))$ describes a field self-interaction. The equation of motion for φ now becomes

$$(\Box + m^2)\varphi(x) = -\frac{\partial V(\varphi)}{\partial \varphi}. \qquad (3.1.57)$$

Most often, we shall consider a self-interaction of the form

$$V(\varphi(x)) = \frac{g}{4!}\varphi^4(x) \qquad (3.1.58)$$

where $g \in \mathbb{R}$ is called the *coupling constant*. In systems described by the Lagrangian (3.1.57) and expressions similar to it (i.e. with interaction terms), particles (field excitations) can arise and disappear, so that the total number of particles is *not a conserved quantity*. This is a characteristic property of relativistic particle theory. Vice versa, it is clear that a system with an *arbitrary* number of particles definitely requires, for its description, a formalism with an infinite number of degrees of freedom, i.e. the quantum field theory.

◇ **Lagrangian for spin-$\tfrac{1}{2}$ field, Dirac equation and operator quantization**

Many well-established types of particle in nature (for example, electrons, positrons, quarks, neutrinos) have half-integer spin $J = \tfrac{1}{2}$ and obey Fermi statistics. Systems of such particles are described by *spinor (fermion) quantum fields* satisfying *canonical anticommutation relations* (see any textbook on quantum field theory, e.g., Bogoliubov and Shirkov (1959), Bjorken and Drell (1965) and Itzykson and Zuber (1980)).

A system of free relativistic spin-$\tfrac{1}{2}$ fermions is described by a four-component complex field $\psi_\alpha(x)$, $\alpha = 1, \ldots, 4$ and has the Lagrangian density

$$\mathcal{L}(x) = \bar{\psi}(x)(i\slashed{\partial} - m)\psi(x) \equiv \sum_{\alpha,\beta=1}^{4} \bar{\psi}_\alpha(x)(i\slashed{\partial}_{\alpha\beta} - m\delta_{\alpha\beta})\psi_\beta(x) \qquad (3.1.59)$$

where we have introduced the standard notation: for any four-dimensional vector A_μ, the quantity $\slashed{A}$ means

$$\slashed{A} \stackrel{\text{def}}{\equiv} \gamma^\mu A_\mu = g_{\mu\nu}\gamma^\mu A^\nu = \gamma^0 A^0 - \boldsymbol{\gamma} \cdot \boldsymbol{A} \qquad (3.1.60)$$

and γ^μ, $\mu = 0, 1, 2, 3$ are the *Dirac matrices* satisfying the defining relations

$$\gamma^\mu \gamma^\nu + \gamma^\nu \gamma^\mu = 2g^{\mu\nu} \mathbb{I}_4 \qquad (3.1.61)$$

($\mathbb{I}_4$ is the 4×4 unit matrix). In particular, $\slashed{\partial} \equiv \gamma^\mu \partial_\mu$. One possible representation of the γ-matrices has the form

$$\gamma^0 = \begin{pmatrix} \mathbb{I}_2 & 0 \\ 0 & -\mathbb{I}_2 \end{pmatrix} \qquad \gamma^i = \begin{pmatrix} 0 & \sigma^i \\ -\sigma^i & 0 \end{pmatrix} \qquad i = 1, 2, 3. \qquad (3.1.62)$$

Here σ^i are the Pauli matrices and $\mathbb{I}_2$ is the 2×2 unit matrix. The *Dirac conjugate spinor* $\bar{\psi}(x)$ in (3.1.59) is defined as follows:

$$\bar{\psi}(x) \stackrel{\text{def}}{\equiv} \psi^\dagger(x)\gamma^0 \qquad \text{or} \qquad \bar{\psi}_\alpha(x) \stackrel{\text{def}}{\equiv} \sum_{\beta=1}^{4} \psi^\dagger_\beta(x)\gamma^0_{\beta\alpha}. \qquad (3.1.63)$$

Note that, as is usual in the literature on quantum field theory, we do not use special print for either the γ-matrices or the Pauli matrices (similarly to four-dimensional vectors).

The extremality condition for the action with the density (3.1.59) (Euler equation) gives the *Dirac equation* for a spin-$\frac{1}{2}$ field

$$(i\slashed{\partial} - m)\psi(x) = 0. \qquad (3.1.64)$$

The general form for the expansion of a solution of the Dirac equation (3.1.64) over plane waves is the following:

$$\psi(t, \boldsymbol{r}) = \frac{1}{(2\pi)^{3/2}} \sum_{i=1}^{2} \int d^3k \, [b^*_i(\boldsymbol{k})u_i(k)e^{ikx} + c_i(\boldsymbol{k})v_i(k)e^{-ikx}]$$

$$\psi^\dagger(t, \boldsymbol{r}) = \frac{1}{(2\pi)^{3/2}} \sum_{i=1}^{2} \int d^3k \, [b_i(\boldsymbol{k})u^\dagger_i(k)e^{-ikx} + c^*_i(\boldsymbol{k})v^\dagger_i(k)e^{ikx}] \qquad (3.1.65)$$

where $k^0 = \omega_k \equiv \sqrt{\boldsymbol{k}^2 + m^2}$ and $u_i(k)$, $v_i(k)$, $i = 1, 2$ comprise the complete set of orthonormal solutions of the Dirac equation (in the momentum representation):

$$(\slashed{k} - m)u_i(k)\big|_{k^0 = \sqrt{\boldsymbol{k}^2 + m^2}} = 0$$

$$(\slashed{k} + m)v_i(k)\big|_{k^0 = -\sqrt{\boldsymbol{k}^2 + m^2}} = 0$$

so that, in fact, u_i and v_i are only functions of three-dimensional momentum $\boldsymbol{k}$.

The orthogonality relations read as

$$\bar{v}_i(\boldsymbol{k})v_j(\boldsymbol{k}) \equiv \sum_{\alpha=1}^{n} (\bar{v}_i(\boldsymbol{k}))_\alpha (v_j(\boldsymbol{k}))_\alpha$$

$$= -\bar{u}_i(\boldsymbol{k})u_j(\boldsymbol{k}) = \frac{m}{\omega_k}\delta_{ij}$$

$$v^\dagger_i(\boldsymbol{k})v_j(\boldsymbol{k}) = u^\dagger_i(\boldsymbol{k})u_j(\boldsymbol{k}) = \delta_{ij} \qquad (3.1.66)$$

$$\bar{u}_i(\boldsymbol{k})v_j(\boldsymbol{k}) = u^\dagger_i(\boldsymbol{k})v_j(-\boldsymbol{k}) = 0$$

and the completeness relations are

$$\sum_i [(v_i(\bm{k}))_\alpha (\bar{v}_i(\bm{k}))_\beta - (u_i(\bm{k}))_\alpha (\bar{u}_i(\bm{k}))_\beta] = \frac{m}{\omega_k}\delta_{\alpha\beta}$$

$$\sum_i (v_i(\bm{k}))_\alpha (\bar{v}_i(\bm{k}))_\beta = \frac{1}{2\omega_k}(\slashed{k}_{\alpha\beta} + m\delta_{\alpha\beta}) \qquad (3.1.67)$$

$$\sum_i (u_i(\bm{k}))_\alpha (\bar{u}_i(\bm{k}))_\beta = \frac{1}{2\omega_k}(\slashed{k}_{\alpha\beta} - m\delta_{\alpha\beta}).$$

The quantization procedure converts the amplitudes b_i, b^*, c_i, c_i^* into creation and annihilation operators for fermionic particles. To take into account the Pauli principle, we have to impose *anticommutation relations*:

$$\begin{aligned}
\{\widehat{b}_i(k), \widehat{b}_j^\dagger(k')\} &= \delta_{ij}\delta^3(\bm{k}-\bm{k}') \\
\{\widehat{c}_i(k), \widehat{c}_j^\dagger(k')\} &= \delta_{ij}\delta^3(\bm{k}-\bm{k}') \\
\{\widehat{b}_i(k), \widehat{b}_j(k')\} &= \{\widehat{c}_i(k), \widehat{c}_j(k')\} = 0 \\
\{\widehat{b}_i^\dagger(k), \widehat{b}_j^\dagger(k')\} &= \{\widehat{c}_i^\dagger(k), \widehat{c}_j^\dagger(k')\} = 0 \\
\{\widehat{b}_i(k), \widehat{c}_j(k')\} &= \{\widehat{b}_i(k), \widehat{c}_j^\dagger(k')\} = 0 \\
\{\widehat{c}_i(k), \widehat{b}_j^\dagger(k')\} &= \{\widehat{c}_i^\dagger(k), \widehat{b}_j^\dagger(k')\} = 0.
\end{aligned} \qquad (3.1.68)$$

From these relations, it is easy to derive the equal-time anticommutation relations for the fields $\hat{\psi}, \hat{\psi}^\dagger$:

$$\begin{aligned}
\{\hat{\psi}_\alpha(t,\bm{r}), \hat{\psi}_\beta^\dagger(t,\bm{r}')\} &= \delta_{\alpha\beta}\delta^3(\bm{r}-\bm{r}') \\
\{\hat{\psi}_\alpha(t,\bm{r}), \hat{\psi}_\beta(t,\bm{r}')\} &= 0 \qquad \{\hat{\psi}_\alpha^\dagger(t,\bm{r}), \hat{\psi}_\beta^\dagger(t,\bm{r}')\} = 0.
\end{aligned} \qquad (3.1.69)$$

Note that the canonical momentum π_α conjugated to the field ψ_α with respect to the Lagrangian (3.1.59) is equal to $i\psi_\alpha^\dagger$:

$$\pi_\alpha \equiv \frac{\partial \mathcal{L}}{\partial \dot{\psi}_\alpha} = i\psi_\alpha^\dagger. \qquad (3.1.70)$$

Thus the commutation relations (3.1.69) are nothing other than the fermionic generalization of the canonical commutation relations for (generalized) coordinates and conjugate momenta. The corresponding Hamiltonian

$$H = \int d^3r \, (\pi\dot{\psi} - \mathcal{L})$$

($\mathcal{L}$ is the Lagrangian density (3.1.59)) in the quantum case can be written in terms of the fermionic creation and annihilation operators:

$$\widehat{H} = \sum_{i=1}^{2} \int d^3k \, \omega_k [\widehat{b}_i^\dagger(\bm{k})\widehat{b}_i(\bm{k}) - \widehat{c}_i(\bm{k})\widehat{c}_i^\dagger(\bm{k})]. \qquad (3.1.71)$$

In order to make this Hamiltonian operator positive definite, we can use the anticommutation relations (3.1.68) together with an infinite shift of the vacuum energy similarly to the bosonic case (cf (3.1.26)):

$$\widehat{H} \to \widehat{H} - E_0 = \sum_{i=1}^{2} \int d^3k\, \omega_k [\widehat{b}_i^\dagger(k)\widehat{b}_i(k) + \widehat{c}_i^\dagger(k)\widehat{c}_i(k)].$$

Sometimes this procedure of energy subtraction in the fermionic case is carried out by invoking the qualitative concept of a Dirac 'sea', i.e. we assume that all negative-energy states are occupied and real experimentally observable particles correspond to excitations of this background state of the whole fermion system (see e.g., Bjorken and Drell (1965)).

3.1.2 Path-integral representation for transition amplitudes in quantum field theories

In the operator approach, there exist different representations for the canonical field operators $\widehat{\varphi}(r)$ and $\widehat{\pi}(r)$ satisfying the commutation relations (3.1.29) or (3.1.69). For example, in the case of the scalar field theory and in the *coordinate representation*, the vectors of the corresponding Hilbert space of states are *functionals* $\Phi[\varphi(r)]$ of the field $\varphi(t, r)$ and the operator $\widehat{\varphi}$ is diagonal:

$$\widehat{\varphi}(r)\Phi[\varphi(r)] = \varphi(r)\Phi[\varphi(r)]$$
$$\widehat{\pi}(r)\Phi[\varphi(r)] = -\mathrm{i}\frac{\delta}{\delta\varphi(r)}\Phi[\varphi(r)].$$

Thus, when quantizing a field theory (in other words, a system with an infinite number of degrees of freedom), even in the operator approach, we have to deal anyway with functionals, so that an application of *path (functional) integrals* in this area is highly natural.

◇ Path integrals in scalar field theory

In fact, the introduction of quantum fields, presented in this section, as the limit of systems with a finite number of degrees of freedom (coupled oscillators), allows us to write immediately an expression for the corresponding transition amplitude (the quantum-mechanical propagator). Indeed, in the case of a field theory, the space coordinates $r = \{x^1, x^2, x^3\}$ label the different degrees of freedom. For the lattice approximation (with spacing a) in a finite volume L^3, from which we started in this section, we have simply a finite number K^3 of oscillators $q_k = \varphi(r_k)$ (cf (3.1.21) and (3.1.27); generally speaking, we have anharmonic oscillators because of the self-interaction term $V(\varphi)$ in (3.1.56)). Therefore, we can write the transition amplitude for the quantum fields as a direct generalization (infinite limit) of the path-integral representation for propagators of quantum-mechanical systems with a finite number of degrees of freedom obtained in chapter 2 (cf (2.2.9) and (2.2.21)):

$$\langle \varphi(t, r), t | \varphi_0(t_0, r), t_0 \rangle = \langle \varphi(r) | e^{-\mathrm{i}(t-t_0)\widehat{H}} | \varphi_0(r) \rangle$$
$$= \lim_{\substack{L \to \infty \\ a \to 0}} \lim_{\substack{K \to \infty \\ \varepsilon \to 0}} \lim_{N \to \infty} \prod_{k=1}^{K} \left\{ \prod_{j=1}^{N} \left[\int_{-\infty}^{\infty} d\varphi_j(r_k) \right] \prod_{j=1}^{N+1} \left[\int_{-\infty}^{\infty} \frac{d\pi_j(r_k)}{2\pi} \right] \right\}$$
$$\times \exp\{\mathrm{i}S_N(\pi_i(r_l), ; \varphi_s(r_l),)\} \qquad (3.1.72)$$

where the discrete-time and discrete-space approximated action S_N depends on all $\pi_i(r_l)$, $i = 1, \ldots, N+1$; $l = 1, \ldots, K$ and $\varphi_s(r_l)$, $s = 0, \ldots, N+1$; $l = 1, \ldots, K$ variables (N is the number of time slices and ε is the 'distance' between the time slices). In the case of Hamiltonian (3.1.28), the continuous limits in (3.1.72) correspond to the following path integral:

$$
\langle\varphi(t,\boldsymbol{r}),t|\varphi_0(t_0,\boldsymbol{r}),t_0\rangle = \int_{\mathcal{C}\{\varphi_0(\boldsymbol{r}),t_0;\varphi(\boldsymbol{r}),t\}} \mathcal{D}\varphi(\tau,\boldsymbol{r})\frac{\mathcal{D}\pi(\tau,\boldsymbol{r})}{2\pi}\exp\left\{i\int_{t_0}^{t}d\tau\int_{\mathbb{R}^3}d^3r\left[\pi(\tau,\boldsymbol{r})\partial_\tau\varphi(\tau,\boldsymbol{r})\right.\right.
$$
$$
\left.\left.-\frac{1}{2}\sum_{i=1}^{3}(\partial_i\varphi(\tau,\boldsymbol{r}))^2 - \frac{1}{2}m^2\varphi^2(\tau,\boldsymbol{r}) - V(\varphi(\tau,\boldsymbol{r}))\right]\right\}. \quad (3.1.73)
$$

Gaussian integration over the momenta $\pi(\tau,\boldsymbol{r})$ yields the Feynman path integral in the coordinate space:

$$
\langle\varphi(t,\boldsymbol{r}),t|\varphi_0(t_0,\boldsymbol{r}),t_0\rangle = \mathfrak{N}^{-1}\int_{\mathcal{C}\{\varphi_0(\boldsymbol{r}),t_0;\varphi(\boldsymbol{r}),t\}} \mathcal{D}_{d\tau}\varphi(\tau,\boldsymbol{r})\exp\left\{i\int_{t_0}^{t}dx^0\int_{\mathbb{R}^3}dx^1\,dx^2\,dx^3\,\mathcal{L}(\varphi)\right\} \quad (3.1.74)
$$

where the Lagrangian density $\mathcal{L}(\varphi)$ is defined in (3.1.56) and

$$
\mathfrak{N}^{-1} \stackrel{\text{def}}{\equiv} \int \mathcal{D}\pi(x)\exp\left\{-\frac{i}{2}\int dx\,\pi^2(x)\right\} \quad (3.1.75)
$$

is the normalization constant for expressing the transition amplitude via the Feynman (configuration) path integral.

Expression (3.1.74) is almost invariant with respect to relativistic Poincaré transformations. The only source of non-invariance is the restriction of the time integral in the exponent to the finite interval $[t_0, t]$. However, it is necessary to point out that elementary particle experimentalists do not measure probabilities directly related to amplitudes for transitions between eigenstates $|\varphi(t_0,\boldsymbol{r}),t_0\rangle$ and $|\varphi(t,\boldsymbol{r}),t\rangle$ of the quantum field $\hat{\varphi}(\tau,\boldsymbol{r})$, but rather probabilities related to S-matrix elements, i.e. to probability amplitudes for transitions between states which, at $t \to \pm\infty$, contain definite numbers of particles of various types. These are called 'in' and 'out', $|\alpha,\text{in}\rangle$ and $|\beta,\text{out}\rangle$, where α and β denote sets of quantum numbers characterizing momenta, spin z-components and types of particle (e.g., photons, leptons, etc). The *S-matrix operator* is defined as follows (cf (2.3.136)):

$$
\widehat{S} = \lim_{\substack{t\to+\infty \\ t'\to-\infty}} e^{it\widehat{H}_0}e^{-i(t-t')\widehat{H}}e^{-it'\widehat{H}_0} \quad (3.1.76)
$$

where $\widehat{H}_0$ is the free Hamiltonian (without the self-interaction term $V(\varphi)$). Physically, this operator describes the *scattering* of elementary particles, i.e. we assume that

- initially, the particles under consideration are far from each other and can be described by the free Hamiltonian (because the distance between particles is much larger than the radius of the action of the interaction forces);
- then the particles become closer and interact; and
- finally, the particles which have appeared as a result of the interaction again move far away from each other and behave like free particles.

The advantage of operator (3.1.76) is that its matrix elements prove to be explicitly relativistically invariant (see later).

◇ **The path integral in holomorphic representation**

The path-integral representation for transition amplitudes (3.1.74) is not particularly convenient for deriving the matrix elements of the scattering operator (3.1.76). Even the path-integral expression for the free evolution operator $e^{it\hat{H}_0}$ is rather cumbersome in this representation. The operator formulation of the field theory briefly presented earlier prompts a formalism based on the creation and annihilation operators and the corresponding path integration variables may appear to be much more suitable. Indeed, as we have learned, determining the eigenstates and eigenvalues of the free Hamiltonian is much simpler in terms of these operators. Similar reasons make path integrals constructed on the basis of creation and annihilation operators, i.e. for *normal symbols*, more convenient. Fortunately, we are quite ready for this construction due to our considerations in sections 2.3.1 and 2.3.3. In quantum field theory, such path integrals are called path integrals in a *holomorphic representation*.

We start from the continuous analog of the Fourier transform (3.1.31):

$$\varphi(r) = \frac{1}{(2\pi)^{3/2}} \int d^3r \frac{1}{\sqrt{2\omega_k}} (a^*(k)e^{-ik\cdot r} + a(k)e^{ik\cdot r}) \tag{3.1.77}$$

$$\pi(r) = \frac{i}{(2\pi)^{3/2}} \int d^3r \, i\sqrt{\frac{\omega_k}{2}} (a^*(k)e^{-ik\cdot r} - a(k)e^{ik\cdot r})$$

$$\omega_k = \sqrt{k^2 + m^2}. \tag{3.1.78}$$

The free-particle Hamiltonian (3.1.34) in the continuous limit has the form

$$H_0 = \int d^3k \, \omega_k a^*(k)a(k) \tag{3.1.79}$$

and is the continuous sum of an infinite number of oscillators. Here the variable k 'numbers' the oscillators and ω_k are their frequencies. The total Hamiltonian H for a field with self-interaction also contains the term

$$V[a^*, a] = \int_{\mathbb{R}^3} d^3r \, V(\varphi(r)). \tag{3.1.80}$$

The evolution operator is defined by its normal symbol $U(a^*(k), a(k); t, t_0)$, which is expressed through the path integral via the straightforward generalization of expression (2.3.103) for one oscillator:

$$U(a^*(k), a(k); t, t_0) = \int \mathcal{D}a^*(k, \tau) \mathcal{D}a(k, \tau)$$
$$\times \exp\left\{ \int d^3k \, [a^*(k, t)a(k, t) - a^*(k, t)a(k, t_0)] \right\}$$
$$\times \exp\left\{ \int_{t_0}^{t} d\tau \int d^3k \, [-a^*(k, \tau)\dot{a}(k, \tau) - i\omega_k a^*(k, \tau)a(k, \tau)] \right.$$
$$\left. - \int_{t_0}^{t} d\tau \, V[a^*, a] \right\} \tag{3.1.81}$$

where the boundary conditions are 'asymmetrical', as usual for normal symbols: we fix $a^*(k, t)$ at time t and $a(k, t_0)$ at t_0:

$$a^*(k, t) = a^*(k) \qquad a(k, t_0) = a(k). \tag{3.1.82}$$

The corresponding integral kernel can be immediately written down, using relation (2.3.57):

$$K_U(a^*(k), a(k); t, t_0) = \int \mathcal{D}a^*(k, \tau) \mathcal{D}a(k, \tau) \exp\left\{ \int d^3k \, a^*(k, t)a(k, t) \right\}$$
$$\times \exp\left\{ \int_{t_0}^{t} d\tau \int d^3k \, [-a^*(k, \tau)\dot{a}(k, \tau) - i\omega_k a^*(k, \tau)a(k, \tau)] \right.$$
$$\left. - i\int_{t_0}^{t} d\tau \, V[a^*, a] \right\}. \tag{3.1.83}$$

Representation (3.1.83) allows us to derive very easily the path-integral representation for the S-matrix. To this aim, we can use a nice property of kernels obtained from normal symbols. Namely, if some operator $\widehat{A}$ has a kernel $K_A(a^*(\boldsymbol{k}), a(\boldsymbol{k}))$, the operator

$$e^{i\widehat{H}_0 t} \widehat{A} e^{-i\widehat{H}_0 t_0}$$

has the kernel (cf (2.3.153))

$$K_A(a^*(\boldsymbol{k})e^{i\omega t}, a(\boldsymbol{k})e^{-i\omega t_0}). \qquad (3.1.84)$$

The latter substitution only has influence on the boundary conditions (3.1.82). Hence, the kernel of the S-matrix is obtained from (3.1.83) with the help of the infinite time limit:

$$K_S(a^*(\boldsymbol{k}), a(\boldsymbol{k})) = \lim_{\substack{t \to \infty \\ t_0 \to -\infty}} K_U(a^*(\boldsymbol{k}), a(\boldsymbol{k}); t, t_0)$$

$$= \lim_{\substack{t \to \infty \\ t_0 \to -\infty}} \int \mathcal{D}a^*(\boldsymbol{k}, \tau)\mathcal{D}a(\boldsymbol{k}, \tau) \exp\left\{\int d^3k\, a^*(\boldsymbol{k}, t)a(\boldsymbol{k}, t)\right\}$$

$$\times \exp\left\{\int_{t_0}^{t} d\tau \int d^3k\, [-a^*(\boldsymbol{k}, \tau)\dot{a}(\boldsymbol{k}, \tau) - i\omega_k a^*(\boldsymbol{k}, \tau)a(\boldsymbol{k}, \tau)]\right.$$

$$\left. - i\int_{t_0}^{t} d\tau\, V[a^*, a]\right\}$$

$$= \lim_{\substack{t \to \infty \\ t_0 \to -\infty}} \int \mathcal{D}a^*(\boldsymbol{k}, \tau)\mathcal{D}a(\boldsymbol{k}, \tau) \exp\left\{\int d^3k\, \frac{1}{2}[a^*(\boldsymbol{k}, t)a(\boldsymbol{k}, t) + a^*(\boldsymbol{k}, t_0)a(\boldsymbol{k}, t_0)]\right\}$$

$$\times \exp\left\{i\int_{t_0}^{t} d\tau \int d^3k \left[\frac{1}{2i}(\dot{a}^*(\boldsymbol{k}, \tau)a(\boldsymbol{k}, \tau) - a^*(\boldsymbol{k}, \tau)\dot{a}(\boldsymbol{k}, \tau))\right.\right.$$

$$\left.\left. - \omega_k a^*(\boldsymbol{k}, \tau)a(\boldsymbol{k}, \tau)\right] - i\int_{t_0}^{t} d\tau\, V[a^*, a]\right\} \qquad (3.1.85)$$

with the conditions

$$a^*(\boldsymbol{k}, t) = a^*(\boldsymbol{k})\exp\{i\omega_k t\} \qquad a(\boldsymbol{k}, t_0) = a(\boldsymbol{k})\exp\{-i\omega_k t_0\}. \qquad (3.1.86)$$

The last expression in (3.1.85), which has a more symmetrical form, has been derived using integration by parts in the exponent.

◇ S-matrix for a scalar field in the presence of an external source and generating functional for Green functions in quantum field theory

Let us calculate expression (3.1.85) for an infinite collection of oscillators in a *field of external forces*. In other words, we shall calculate the S-matrix for a scalar field φ in the presence of an external source, i.e. with a potential term of the form

$$V_J(\varphi) = -J(x)\varphi(x). \qquad (3.1.87)$$

In terms of the variables a^*, a, the functional $V[a^*(\boldsymbol{k}), a(\boldsymbol{k})]$ reads as

$$V_J[a^*(\boldsymbol{k}), a(\boldsymbol{k})] = \int d^3k\, (\widetilde{J}(t, \boldsymbol{k})a^*(\boldsymbol{k}) + \widetilde{J}^*(t, \boldsymbol{k})a(\boldsymbol{k})) \qquad (3.1.88)$$

where

$$\widetilde{J}(t, \boldsymbol{k}) = -\frac{1}{\sqrt{2k^0}}\left(\frac{1}{2\pi}\right)^{3/2} \int d^3r\, e^{-i\boldsymbol{k}\boldsymbol{r}} J(t, \boldsymbol{r}). \qquad (3.1.89)$$

The direct physical meaning of (3.1.85) with potential term (3.1.88) is the S-matrix for scattering of particles on the external *classical* source $J(x)$. However, as we have learned in chapter 1 (section 1.2.8), explicit calculation of such path integrals allows us to find (via functional differentiation) any correlation functions easily. In quantum field theory, a path integral with a linear potential term also plays the role of a generating functional and of a basic tool for the perturbation expansion.

Since path integral (3.1.85), (3.1.88) is quadratic the stationary-phase method gives an exact result for it (cf section 2.2.3). The extremality equations are as follows:

$$\dot{a}(\tau, \mathbf{k}) + i\omega(\mathbf{k})a(\tau, \mathbf{k}) + i\widetilde{J}(\tau, \mathbf{k}) = 0$$
$$\dot{a}^*(\tau, \mathbf{k}) - i\omega(\mathbf{k})a^*(\tau, \mathbf{k}) - i\widetilde{J}^*(\tau, \mathbf{k}) = 0 \qquad (3.1.90)$$
$$a^*(\mathbf{k}, t) = a^*(\mathbf{k})\exp\{i\omega_k t\} \qquad a(\mathbf{k}, t_0) = a(\mathbf{k})\exp\{-i\omega_k t_0\}$$

and the solution is found to be

$$a^*(\tau, \mathbf{k}) = a^*(\mathbf{k})e^{i\omega_k \tau} - ie^{i\omega_k \tau} \int_\tau^t ds\, e^{-i\omega_k s} \widetilde{J}^*(s, \mathbf{k})$$
$$a(\tau, \mathbf{k}) = a(\mathbf{k})e^{-i\omega_k \tau} - ie^{-i\omega_k \tau} \int_{t_0}^\tau ds\, e^{i\omega_k s} \widetilde{J}(s, \mathbf{k}). \qquad (3.1.91)$$

Substituting this solution into the exponent of (3.1.85), we obtain in the limit of the infinite-time interval an expression for the kernel of the S-matrix:

$$K_S[a^*, a; J^*, J] = \exp\left\{\int d^3k \left[a^*(\mathbf{k})a(\mathbf{k})\right.\right.$$
$$+ \frac{1}{(2\pi)^{3/2}\sqrt{2\omega_k}} \int_{-\infty}^\infty d\tau \int_{\mathbb{R}^3} d^3r\, J(t, \mathbf{r})(a^*(\mathbf{k})e^{i\omega_k \tau}e^{-i\mathbf{k}\mathbf{r}} + a(\mathbf{k})e^{-i\omega_k \tau}e^{i\mathbf{k}\mathbf{r}})$$
$$\left.\left. - \frac{1}{4\omega_k}\frac{1}{(2\pi)^3}\int_{-\infty}^\infty d\tau\, ds \int_{\mathbb{R}^3} d^3r\, d^3r'\, J(\tau, \mathbf{r})e^{i\mathbf{k}(\mathbf{r}-\mathbf{r}')}e^{-i\omega|\tau-s|}J(s, \mathbf{r}')\right]\right\}. \qquad (3.1.92)$$

This formula is a generalization of the corresponding expression for a single oscillator in an external field (recall that in problem 2.2.14, page 198, volume I, we calculated the transition amplitude (2.2.200) for one oscillator in an external field with fixed initial and final *positions* in contrast with (3.1.92), where the boundary conditions are imposed on the oscillator variables).

Note that we have put the fluctuation factor in (3.1.92) equal to unity. The reason for this is the definition of the phase-space integral as the ratio (2.3.77) and the fact that the *first* term in the exponent of (3.1.92) gives the correct kernel for the *unit* operator. On the other hand, let us rescale the frequencies and external source by a factor λ: $\omega_k \to \lambda\omega_k$, $J \to \lambda J$. Then, in the limit $\lambda \to 0$, the Hamiltonian (3.1.79), (3.1.87) vanishes, turning the S-matrix into the unit operator. Simultaneously, the second and third terms in the exponent of (3.1.92) become zero too, leaving the correct expression for the kernel of the unit operator. This proves that the fluctuation factor indeed equals unity (cf also the calculation of the scattering operator in non-relativistic quantum mechanics in chapter 2, equations (2.3.150) and (2.3.151)).

The first term in the exponent of (3.1.92) suggests that it is reasonable to pass to the normal symbol for the S-matrix. The relation (2.3.57) between a kernel and the normal symbol shows that the transition to the normal symbol for (3.1.92) just reduces to dropping the first term in the exponent. The essential advantage of writing the S-matrix as the normal symbol is that the remaining terms in (3.1.92) can be presented in an explicitly relativistically invariant form. Indeed, the term bilinear with respect to the source function can be rewritten with the help of the relativistic *causal Green function* $D_c(x)$ which plays

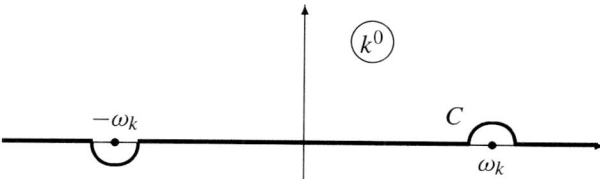

Figure 3.2. The contour C of integration in the complex plane k^0, in the representation (3.1.93) for the causal Green function D_c.

a crucial role in any formulation of quantum field theory (see e.g., Bogoliubov and Shirkov (1959) and Itzykson and Zuber (1980)):

$$D_c(x) \stackrel{\text{def}}{\equiv} -\frac{1}{(2\pi)^4} \int d^4k \, e^{-ikx} \frac{1}{k^2 - m^2 + i\varepsilon} \qquad (3.1.93)$$

$$= -\frac{1}{(2\pi)^4} \int \frac{dk^1 \, dk^2 \, dk^3}{2i\omega_k} e^{-ikr} e^{-i\omega_k |x^0|}. \qquad (3.1.94)$$

The rule of bypassing the singular points in the explicitly relativistic invariant expression (3.1.93) for D_c is defined by the infinitesimal addition $+i\varepsilon$ in the denominator of the integrand, as illustrated in figure 3.2. Recall that $D_c(x)$ is one of the Green functions of the Klein–Gordon equation

$$(\Box + m^2) D_c(x) = \delta^2(x). \qquad (3.1.95)$$

The term which is linear in the source $J(x)$ in (3.1.92) can be rewritten via the solution φ_0 of the *homogeneous* (with zero right-hand side) Klein–Gordon equation:

$$\varphi_0(x) \stackrel{\text{def}}{\equiv} \frac{1}{(2\pi)^{3/2}} \int d^4k \, \delta(k^2 - m^2) [a^*(\mathbf{k}) e^{ikx} + a(\mathbf{k}) e^{-ikx}]$$

$$= \frac{1}{(2\pi)^{3/2}} \int \frac{d^3k}{\sqrt{2k^0}} [a^*(\mathbf{k}) e^{i(k^0 x^0 - \mathbf{k}\mathbf{r})} + a(\mathbf{k}) e^{-i(k^0 x^0 - \mathbf{k}\mathbf{r})}]. \qquad (3.1.96)$$

The first line in (3.1.96) is explicitly Lorentz invariant. In the second line, it is implied that $k^0 = \omega_k = \sqrt{\mathbf{k}^2 + m^2}$ and thus

$$(\Box + m^2)\varphi_0 = 0. \qquad (3.1.97)$$

In terms of $\varphi_0(x)$ and $D_c(x)$, the normal symbol for the S-matrix takes the relativistically invariant form

$$S_0[\varphi_0; J] = \exp\left\{ i \int d^4x \, J(x) \varphi_0(x) + \int d^4x \, d^4x' \, J(x) D_c(x - x') J(x') \right\}. \qquad (3.1.98)$$

In this formula, we have substituted the pair of oscillator variables a^*, a by the single field φ_0 because, due to definition (3.1.96), they are in one-to-one correspondence. The subscript '0' of the S-matrix functional indicates that there is no self-interaction in the model under consideration (it exists only in interactions with an external source).

◇ Generating functional for Green functions in scalar field theory

If we put $\varphi_0 = 0$ or, equivalently, $a^*(t, \boldsymbol{k}) = a(t_0, \boldsymbol{k}) = 0$, the normal symbol for the S-matrix turns into the *generating functional for Green functions* $\mathcal{Z}[J(x)]$:

$$\mathcal{Z}[J(x)] \stackrel{\text{def}}{\equiv} S[\varphi_0; J]|_{\varphi_0=0} \qquad (3.1.99)$$

(this definition is valid for a field Hamiltonian with an arbitrary interaction term). In particular,

$$\mathcal{Z}_0[J(x)] \stackrel{\text{def}}{\equiv} S_0[\varphi_0; J]|_{\varphi_0=0} = \int \mathcal{D}\varphi(x) \exp\left\{i \int d^4x \, [\mathcal{L}_0(\varphi(x)) + J(x)\varphi(x)]\right\}$$

$$= \exp\left\{\int d^4x \, d^4x' \, J(x) D_c(x - x') J(x')\right\}. \qquad (3.1.100)$$

The direct physical meaning of $\mathcal{Z}[J(x)]$ is the transition amplitude from the vacuum state $|0\rangle$ to the same vacuum state in the presence of an external source $J(x)$. This is clear from the following arguments. Let us consider a normally ordered operator

$$\widehat{A}(\widehat{a}^\dagger(\boldsymbol{k}), \widehat{a}(\boldsymbol{k}')) = \sum_{m,n=0}^{\infty} \int d^3k \, d^3k' \, C_{mn}(\boldsymbol{k}, \boldsymbol{k}') (\widehat{a}^\dagger(\boldsymbol{k}))^m (\widehat{a}(\boldsymbol{k}'))^n$$

and the corresponding normal symbol

$$A(a^*(\boldsymbol{k}), a(\boldsymbol{k}')) = \sum_{m,n=0}^{\infty} \int d^3k \, d^3k' \, C_{mn}(\boldsymbol{k}, \boldsymbol{k}') (a^*(\boldsymbol{k}))^m (\widehat{a}(\boldsymbol{k}'))^n.$$

Then, using the definition of the vacuum state, i.e.

$$\widehat{a}(\boldsymbol{k})|0\rangle = 0 \qquad \text{for all } \boldsymbol{k}$$

we obtain

$$\langle 0|\widehat{A}(a^*(\boldsymbol{k}), a(\boldsymbol{k}'))|0\rangle = \int d^3k \, d^3k' \, C_{00}(\boldsymbol{k}, \boldsymbol{k}')$$

$$= A(a^*(\boldsymbol{k}), a(\boldsymbol{k}'))|_{a^*(\boldsymbol{k})=a(\boldsymbol{k}')=0}. \qquad (3.1.101)$$

Thus any normal symbol with a zero value for its arguments, and hence $\mathcal{Z}[J(x)]$, is equal to the vacuum expectation value of the corresponding operator.

The functional derivatives of the generating functional give the Green functions. In particular,

$$\frac{\delta}{\delta J(x)} \frac{\delta}{\delta J(x')} \mathcal{Z}_0[J] \bigg|_{J=0} = D_c(x - x') = \langle 0|\mathbf{T}(\hat{\varphi}(x)\hat{\varphi}(x'))\rangle.$$

Therefore, $\mathcal{Z}_0[J]$ generates the *causal* Green function, in other words, the vacuum expectations of the *time-ordered* products of the field operators. This observation can be expanded to the generating functional $\mathcal{Z}[J]$ for an action with an arbitrary interaction term and to an arbitrary Green function: $\mathcal{Z}[J]$ generates vacuum expectations

$$\frac{\delta}{\delta J(x_1)} \frac{\delta}{\delta J(x_2)} \cdots \frac{\delta}{\delta J(x_n)} \mathcal{Z}[J] \bigg|_{J=0} = \langle 0|\mathbf{T}(\hat{\varphi}(x_1)\hat{\varphi}(x_2) \cdots \hat{\varphi}(x_n))|0\rangle \qquad (3.1.102)$$

of *time-ordered* products of field operators. The general reason for this fact is that such a product provides an appropriate order for the field operators along the trajectories in the path integral. We suggest the reader verifies the correspondence between path integrals and vacuum expectations of *time-ordered* operator products explicitly (problem 3.1.2).

We shall return to the physical meaning of the generating functional $\mathcal{Z}[J(x)]$ and the technical merit of its use in section 3.1.5. However, before then, we should introduce the path-integral construction for fermionic fields which describe systems of an arbitrary number of spin-$\frac{1}{2}$ particles.

3.1.3 Spinor fields: quantization via path integrals over Grassmann variables

In the preceding subsection we presented the path-integral formalism for the quantization of scalar field theory. The latter describes systems of an arbitrary number of scalar, i.e. spin-0, particles obeying Bose–Einstein statistics. In this subsection, we present the path-integral approach to the quantization of spinor fields. The consideration of fermionic systems with a finite number of degrees of freedom in section 2.6 and the scalar field theory in the preceding subsection provide a good basis for generalization to systems with an arbitrary number of fermions.

◇ Path-integral quantization of spinor fields

As we have learned in section 2.6, the path-integral quantization of fermionic systems uses anticommuting (Grassmann) variables. For spinor fields these variables are the anticommuting counterparts b_i, b_i^*, c_i, c_i^* of the operators $\widehat{b}, \widehat{b}^\dagger, \widehat{c}, \widehat{c}^\dagger$ satisfying the commutation relations (3.1.68). The anticommutation relations for the former are obtained from those for the latter by putting all right-hand sides in (3.1.68) equal to zero.

Let us consider again, as in the bosonic case, the basic example of a spinor field interacting with the external sources $\eta(x), \bar{\eta}(x)$. Note that for fermionic systems, the external sources are also chosen to be *Grassmann* variables. The Hamiltonian of this system reads as

$$H[b^*, b; c^*, c] = \int d^3 \left[i\bar{\psi}(\boldsymbol{r}) \sum_{j=1}^{3} \gamma_j \partial_j \psi(\boldsymbol{r}) + m\bar{\psi}(\boldsymbol{r})\psi(\boldsymbol{r}) + \bar{\psi}(\boldsymbol{r})\eta(x) + \bar{\eta}(x)\psi(\boldsymbol{r}) \right]$$

$$= \sum_{i=1}^{2} \int d^3k \, [\omega_k(b_i^*(\boldsymbol{k})b_i(\boldsymbol{k}) + c_i^*(\boldsymbol{k})c_i(\boldsymbol{k}))$$
$$+ \xi_i^*(t, \boldsymbol{k})b_i(\boldsymbol{k}) + b_i^*(\boldsymbol{k})\xi_i(t, \boldsymbol{k}) + \zeta_i^*(t, \boldsymbol{k})c_i(\boldsymbol{k}) + c_i^*(\boldsymbol{k})\zeta(t, \boldsymbol{k})]. \quad (3.1.103)$$

In the latter expression, we have introduced the new sources

$$\xi_i(t, \boldsymbol{k}) = u_i^* \widetilde{\eta}(t, \boldsymbol{k}) \qquad \zeta(t, \boldsymbol{k}) = v_i^* \widetilde{\eta}(t, \boldsymbol{k})$$

where $\widetilde{\eta}(t, \boldsymbol{k})$ is the Fourier transform of $\eta(x)$:

$$\widetilde{\eta}(t, \boldsymbol{k}) = \frac{1}{(2\pi)^{3/2}} \int d^3r \, e^{ikr} \eta(t, \boldsymbol{r})$$

(we have used the orthonormality of the spinors u_i and v_i, $i = 1, 2$, cf (3.1.66)). Combining the results and methods of sections 3.1.1 and 2.6, we obtain the kernel of the *S*-matrix operator for the Hamiltonian (3.1.71):

$$K_S(b^*, c^*; b, c) = \lim_{\substack{t \to \infty \\ t_0 \to -\infty}} \int \mathcal{D}b_1^*(\tau, \boldsymbol{k}) \mathcal{D}b_1(\tau, \boldsymbol{k}) \mathcal{D}b_2^*(\tau, \boldsymbol{k}) \mathcal{D}b_2(\tau, \boldsymbol{k}) \mathcal{D}c_1^*(\tau, \boldsymbol{k}) \mathcal{D}c_1(\tau, \boldsymbol{k})$$

$$\times \mathcal{D}c_2^*(\tau, \boldsymbol{k})\mathcal{D}c_2(\tau, \boldsymbol{k}) \exp\left\{\sum_{i=1}^{2}\left[\frac{1}{2}\int d^3k\, (b_i^*(t,\boldsymbol{k})b_i(t,\boldsymbol{k}) + b_i^*(t_0,\boldsymbol{k})b_i(t_0,\boldsymbol{k})\right.\right.$$
$$+ c_i^*(t,\boldsymbol{k})c_i(t,\boldsymbol{k}) + c_i^*(t_0,\boldsymbol{k})c_i(t_0,\boldsymbol{k}))$$
$$+ \mathrm{i}\int_{t_0}^{t} d\tau\left[\int d^3k\, \frac{1}{2\mathrm{i}}(\dot{b}_i^*(\tau,\boldsymbol{k})b_i(\tau,\boldsymbol{k}) - b_i^*(\tau,\boldsymbol{k})\dot{b}_i(\tau,\boldsymbol{k})\right.$$
$$\left.\left.\left.+ \dot{c}_i^*(\tau,\boldsymbol{k})c_i(\tau,\boldsymbol{k}) - c_i^*(\tau,\boldsymbol{k})\dot{c}_i(\tau,\boldsymbol{k})) - H[b^*,b;c^*,c]\right]\right]\right\} \quad (3.1.104)$$

where we keep b_i, c_i fixed at the initial time t_0 and b_i^*, c_i^* at the time t:

$$b_i(t_0,\boldsymbol{k}) = b_i \qquad c_i(t_0,\boldsymbol{k}) = c_i$$
$$b_i^*(t,\boldsymbol{k}) = b_i^* \qquad c_i^*(t,\boldsymbol{k}) = c_i^*.$$

Calculating integral (3.1.104) by the stationary-phase method is quite similar to the case of the scalar field. To obtain an explicitly Lorentz-invariant expression, we have to pass again to the normal symbol for the S-matrix which reads as

$$S_0[b^*,c^*;b,c;\bar{\eta},\eta] = \exp\left\{\mathrm{i}\int dx\, dy\, \bar{\eta}(x)S_\mathrm{c}(x-y)\eta(y) + \mathrm{i}\int dx\, (\bar{\eta}(x)\psi_0(x) + \bar{\psi}_0(x)\eta(x))\right\} \quad (3.1.105)$$

where $S_\mathrm{c}(x-y)$ is the causal Green function of the Dirac equation

$$(S_\mathrm{c})_{\alpha\beta}(x-y) = -\mathrm{i}\frac{1}{(2\pi)^{3/2}}\sum_{i=1}^{2}\int d^3k\, \exp\{\mathrm{i}[\boldsymbol{k}\cdot(\boldsymbol{x}-\boldsymbol{y}) - k^0|x^0 - y^0|]\}$$
$$\times (v_{i\alpha}(\boldsymbol{k})v_{i\beta}^*(\boldsymbol{k}) + u_{i\alpha}(\boldsymbol{k})u_{i\beta}^*(\boldsymbol{k})).$$

Using the completeness (3.1.67) of the spinors u_i, v_i, we can write the spinor Green function in matrix form:

$$S_\mathrm{c}(x-y) = -\frac{1}{(2\pi)^4}\int d^4k\, \frac{\exp\{-\mathrm{i}k(x-y)\}}{\gamma_\mu k^\mu - m + \mathrm{i}\varepsilon}. \quad (3.1.106)$$

The field ψ_0 in (3.1.105) is expressed in terms of the variables b_i, b^*, c_i, c_i^* via the Fourier transform (3.1.65) (and, hence, $\psi_0(x)$ in expression (3.1.105) for the normal symbol of the S-matrix is the solution of the free Dirac equation).

Again, similarly to the case of a scalar field, the symbol (3.1.105) with zero field ψ (or, equivalently, with $b_i = b^* = c_i = c_i^* = 0$) turns into the generating functional for the Green functions of free spinor fields, that is for vacuum expectation values of time-ordered products of spinor field operators:

$$\mathcal{Z}_0[\bar{\eta},\eta] = S_0[b^*,c^*;b,c;\bar{\eta},\eta]|_{b^*=c^*=b=c=0}. \quad (3.1.107)$$

We shall consider this topic from a general point of view and in more detail in the next subsection.

3.1.4 Perturbation expansion in quantum field theory in the path-integral approach

Let us now consider the S-matrix kernel and the generating functional for Green functions in the case of an arbitrary potential $V(x)$. As we know, e.g., from consideration of the non-relativistic scattering operator (section 2.3.3), we may hope to calculate the path integrals for rather exceptional cases of non-trivial potentials (using some special methods, e.g., transformations analogous to those in section 2.5). In the general case, we are confined to using approximation methods and one of the most important among these is perturbation theory.

◇ **Formal calculation of the path integral for the *S*-matrix in scalar field theory with an arbitrary self-interaction and perturbation expansion**

In fact, the construction of the *S*-matrix (or the generating functionals) for linear potentials in the preceding subsection supplies us with all the necessary ingredients for developing the perturbation theory. The basic observation is based on the following obvious formula:

$$\varphi(x_1)\varphi(x_2)\cdots\varphi(x_n) = \frac{1}{i}\frac{\delta}{\delta J(x_1)}\frac{1}{i}\frac{\delta}{\delta J(x_2)}\cdots\frac{1}{i}\frac{\delta}{\delta J(x_n)}\exp\left\{i\int d^4x\,\varphi(x)J(x)\right\}\bigg|_{J=0}. \quad (3.1.108)$$

This formula allows us to write any functional $\Phi[\varphi(x)]$ in the form

$$\Phi[\varphi(x)] = \Phi\left[\frac{1}{i}\frac{\delta}{\delta J(x)}\right]\exp\left\{i\int d^4x\,\varphi(x)J(x)\right\}\bigg|_{J=0}. \quad (3.1.109)$$

In particular,

$$\exp\left\{-i\int d^4x\,V(\varphi(x))\right\} = \exp\left\{-i\int d^4x\,V\left(\frac{1}{i}\frac{\delta}{\delta J(x)}\right)\right\}\exp\left\{i\int d^4x\,\varphi(x)J(x)\right\}\bigg|_{J=0}. \quad (3.1.110)$$

The first exponential on the right-hand side of (3.1.110) is understood in the sense of the Taylor expansion, so that, in fact, we have an infinite series of variational operators with the raising power of the functional derivative. If the potential term $V(\varphi(x))$ contains a small parameter, we can restrict the expansion to the first few terms and calculate the functional with the desired accuracy.

Thus, to calculate the functional integral (3.1.85), defining the *S*-matrix kernel for an arbitrary potential, we first introduce an auxiliary additional potential term (3.1.87). Now we can substitute the functional $\exp\{-i\int d^4x\,V(\varphi(x))\}$ by the right-hand side of (3.1.110) and move the variational derivatives out of the path-integral sign. The rest of the path integral coincides with that for the linear potential which we have already calculated. Thus, using result (3.1.98), we can immediately write the formal expression for the normal symbol of the *S*-matrix describing the scattering of scalar particles with an arbitrary interaction:

$$S[a^*, a] \equiv S[\varphi_0] = \exp\left\{-i\int d^4x\,V\left(\frac{1}{i}\frac{\delta}{\delta J(x)}\right)\right\}$$
$$\times \exp\left\{i\int d^4x\,\varphi(x)J(x) + \frac{i}{2}\int dx\,dy\,J(x)D_c(x-y)J(y)\right\}\bigg|_{J=0} \quad (3.1.111)$$

(recall that φ_0 and a^*, a are in one-to-one correspondence, cf (3.1.96)). Expanding this functional as a power series in φ_0:

$$S[\varphi_0] = \sum_{n=0}^{\infty}\frac{1}{n!}\int d^4x_1\,d^4x_2\cdots d^4x_n\,S_n(x_1, x_2, \ldots, x_n)\varphi_0(x_1)\varphi_0(x_2)\cdots\varphi_0(x_n) \quad (3.1.112)$$

we obtain the so-called *coefficient functions* $S_n(x_1, x_2, \ldots, x_n)$ of the *S*-matrix. In the operator approach, these appear in the process of expanding the *S*-matrix in a series over normal products of free fields. Convolution of these coefficient functions with the initial $\psi_1(x_1), \psi_2(x_2), \ldots, \psi_l(x_l)$ and final $\psi_{l+1}(x_{l+1}), \psi_{l+2}(x_{l+2}), \ldots, \psi_n(x_n)$ wavefunctions of the particles participating in the scattering gives the corresponding probability amplitude:

$$\langle\psi_{l+1}, \psi_{l+2}, \ldots, \psi_n; \text{out}|\psi_1, \psi_2, \ldots, \psi_l; \text{in}\rangle = \int d^4x_1\,d^4x_2\cdots d^4x_n$$
$$\times \psi_1(x_1)\psi_2(x_2)\cdots\psi_l(x_l)S_n(x_1, x_2, \ldots, x_n)\psi_{l+1}(x_{l+1})\psi_{l+2}(x_{l+2})\cdots\psi_n(x_n) \quad (3.1.113)$$

(the labels 'in' and 'out' denote the states at $t \to -\infty$ and $t \to \infty$, respectively). From expansion (3.1.112) it is clear that the coefficient functions S_n can be obtained by formal differentiation of the functional $S[\varphi_0]$ over φ_0 (though we remember that φ_0 is the solution of the free Klein–Gordon equation, in other words, it belongs to the mass surface):

$$S_n(x_1, x_2, \ldots, x_n) = \frac{\delta}{\delta\varphi_0(x_1)} \frac{\delta}{\delta\varphi_0(x_2)} \cdots \frac{\delta}{\delta\varphi_0(x_n)} S[\varphi_0(x)]\bigg|_{\varphi_0=0}. \qquad (3.1.114)$$

Therefore, the functional $S[\varphi_0(x)]$ is called the *generating functional for S-matrix coefficient functions*.

We have called expression (3.1.111) 'formal' because we cannot calculate the action of the complete variational operator on the right-hand side explicitly and we have to use its Taylor expansion up to some power of the functional derivative, i.e. to use the *perturbation theory approximation*.

Example 3.1. Let us consider, as an example, a field theory defined by the Lagrangian density

$$\mathcal{L} = \frac{1}{2}[(\partial_\mu \varphi)^2 - m^2 \varphi^2] - \frac{g}{4!}\varphi^4 \qquad (3.1.115)$$

so that the self-interaction term has the form

$$V(\varphi) = \frac{g}{4!}\varphi^4. \qquad (3.1.116)$$

The constant g defines the strength of the scalar field self-interaction and is called the *coupling constant*. The factor $1/4!$ is introduced for further technical convenience. Because of the form of the potential term, the theory with Lagrangian (3.1.115) is called the φ^4-*interaction model* (or simply φ^4-*model*).

If the coupling constant is small enough, we can expand the first exponential on the right-hand side of (3.1.111) in a Taylor series and calculate the generating S-matrix functional up to the desired accuracy (the power of the coupling constant g). For example, up to second order, the S-matrix functional for the φ^4-model becomes

$$S[\varphi_0] = \left[1 - \mathrm{i}\frac{g}{4!}\int dx \left(\frac{\delta}{\delta J(x)}\right)^4 + \frac{1}{2}\left(\mathrm{i}\frac{g}{4!}\right)^2 \int dx\, dy \left(\frac{\delta}{\delta J(x)}\right)^4 \left(\frac{\delta}{\delta J(y)}\right)^4 + \cdots\right]$$
$$\times \exp\left\{\mathrm{i}\int d^4x\, \varphi_0(x)J(x) + \frac{\mathrm{i}}{2}\int dx\, dy\, J(x)D_\mathrm{c}(x-y)J(y)\right\}\bigg|_{J=0} \qquad (3.1.117)$$

so that in this approximation we have the following expression for $S[\varphi_0]$:

(i) In the zeroth-order perturbation theory, i.e. if we use only the first term (unit) in the square brackets on the right-hand side of (3.1.117), the S-matrix is trivial:

$$S^{(0)}[\varphi_0] = 1. \qquad (3.1.118)$$

This result is physically obvious: this zeroth-order approximation corresponds to a total neglect of the potential term and in the absence of any interaction, the S-matrix operator is equal to unity (cf (3.1.76)).

(ii) In the first order, differentiation gives

$$S^{(1)}[\varphi_0] = \frac{g}{4!}\left[-\mathrm{i}\int d^4x\, \varphi_0^4(x) + 6D_\mathrm{c}(0)\int d^4x\, \varphi_0^2(x) + 3\mathrm{i}D_\mathrm{c}^2(0)\int d^4x\right]. \qquad (3.1.119)$$

(iii) The second-order term has the form:

$$\begin{aligned}S^{(2)}[\varphi_0] = &-\frac{1}{2}\left(\frac{g}{4!}\right)^2 \int d^4x\, d^4y\, [\varphi_0^4(x)\varphi_0^4(y) + 16\mathrm{i}\varphi_0^3(x)\varphi_0^3(y)D_{\mathrm{c}}(x-y)\\&- 72\varphi_0^2(x)\varphi_0^2(y)D_{\mathrm{c}}^2(x-y) - \mathrm{i}96\varphi_0(x)\varphi_0(y)D_{\mathrm{c}}^3(x-y) + 24D_{\mathrm{c}}^4(x-y)\\&+ 12\mathrm{i}\varphi_0^2(x)\varphi_0^4(y)D_{\mathrm{c}}(0) - 6\varphi_0^4(y)D_{\mathrm{c}}^2(0) - 96\varphi_0(x)\varphi_0^3(y)D_{\mathrm{c}}(x-y)D_{\mathrm{c}}(0)\\&- 36\varphi_0^2(x)\varphi_0^2(y)D_{\mathrm{c}}^2(0) - 144\mathrm{i}\varphi_0^2(y)D_{\mathrm{c}}^2(x-y)D_{\mathrm{c}}(0) - 36\mathrm{i}\varphi_0^2(y)D_{\mathrm{c}}^3(0)\\&- 144\mathrm{i}\varphi_0(x)\varphi_0(y)D_{\mathrm{c}}(x-y)D_{\mathrm{c}}^2(0) + 72D_{\mathrm{c}}^2(x-y)D_{\mathrm{c}}^2(0)].\end{aligned} \quad (3.1.120)$$

⎯⎯ ○ ⎯⎯

◇ Remark on the renormalization of field theories

As can be seen from (3.1.119) and (3.1.120), the perturbation expansion terms contain the causal Green function at the zero value of the argument, $D_{\mathrm{c}}(0)$. This is an undefined (infinite) quantity:

$$D_{\mathrm{c}}(0) = -\frac{1}{(2\pi)^4}\int d^4k\, \frac{1}{k^2 - m^2 + \mathrm{i}\varepsilon} = \infty. \quad (3.1.121)$$

Thus, the perturbation expansion (3.1.117)–(3.1.120) requires a more thorough treatment and some improvement. We have already mentioned the problem of divergence in quantum field theory (cf (3.1.19) and (3.1.38)). In the case of the simple divergence of the 'zero-oscillation' energy E_0, the solution of this problem was quite obvious: we just redefined the background energy or, equivalently, the Hamiltonian: $\widehat{H} \to \widehat{H} - E_0$, counting only the difference which is physically observable and finite. Essentially, the same idea allows us to overcome the problem of divergence in the so-called *renormalizable field theories*, in general. The point is that divergent terms in quantum-mechanical amplitudes for a system described by some renormalizable field theory can be combined with the initial parameters of the corresponding Lagrangian (such as the masses and coupling constants). Sometimes, these initial parameters are called *bare masses and coupling constants*, while their combinations with divergent terms are called *renormalized parameters* of the theory. Since all physically measurable quantities only contain these combinations and not solely the bare parameters, we can claim that only the renormalized parameters correspond to the physical masses of the known particles and to their coupling constants. In other words, we substitute combinations of the bare parameters and corresponding divergent terms by *finite* values known from physical measurements. This procedure is called the *renormalization of a quantum field theory*. In order for this procedure to be mathematically meaningful, at intermediate steps we have to work with a *regularized theory*. This means that we have to consider the actual model as a limit of some other theory which does not contain the divergences. Then we carry out, at first, all the renormalization procedure for the regularized model and only at the final step do we take the limit corresponding to the initial field theoretical model. In fact, we have already dealt with an example of regularization: field theoretical systems in this chapter were introduced as the *limit* of systems with a finite number of degrees of freedom defined on a lattice (which, of course, have no divergence problems). Clearly, this is the most physically transparent regularization of field theories. Its obvious shortcoming is that it violates the essential symmetries of a field theory on continuous spacetime: rotational, Lorentz and translational. There exist many other regularization schemes preserving spacetime symmetries and adjusted to other specific invariance properties of field theoretical models, which we shall meet later in this book.

All the realistic field theoretical models possess the so-called *multiplicative renormalizability*. Roughly speaking, this property means that all the ultraviolet divergences can be absorbed into certain factors, related to the renormalization of the bare masses and coupling constants. For example, in the case of scalar φ^4-field theory we define:

$$\varphi = Z_\varphi^{-1/2}\varphi_B$$
$$g = Z_g^{-1} Z_\varphi^2 g_B \qquad (3.1.122)$$
$$m^2 = m_B^2 + \delta m^2$$

where φ_B, g_B, m_B are the bare field, coupling constant and mass, respectively and Z_φ, Z_g are the renormalization constants (depending on the regularization parameters). Then, after substituting the bare quantities with the renormalized ones according to (3.1.122), all coefficient functions of the S-matrix and all Green functions can be made finite (renormalized) by multiplying them by the appropriate power of the Z_φ-constant.

Since path-integral techniques do not bring anything essentially new to the regularization and renormalization procedure, we shall not describe it in detail, referring the reader to textbooks on quantum field theory (e.g., Bogoliubov and Shirkov (1959) and Itzykson and Zuber (1980)).

◇ Scattering amplitudes for scalar particles and first encounter with Feynman diagrams

Combining formulae (3.1.113) and (3.1.114), together with the perturbation expansion of the type (3.1.117), allows us to calculate scattering amplitudes. For example, for the scattering of two scalar particles ($2 \to 2$-scattering), we obtain in the second order of the perturbation expansion:

$$\langle \psi(p_3)\psi(p_4); \text{out}|\psi(p_1)\psi(p_2); \text{in}\rangle = \Bigg[-ig\psi(p_3)\psi(p_4)\psi(p_1)\psi(p_2)$$
$$- (ig)^2 \psi(p_3)\psi(p_4)\bigg(\int d^4k\, \widetilde{D}_c(k)\widetilde{D}_c(k - p_1 - p_2)\bigg)\psi(p_1)\psi(p_2)\Bigg]\delta(p_1 + p_2 - p_3 - p_4).$$
$$(3.1.123)$$

Here we assume that the particles have definite momenta in the initial and final states (this is a standard situation in scattering experiments) and that the terms in expansion (3.1.120) containing the divergent quantity $D_c(0)$ are removed by the renormalization procedure. $D_c(p)$ is the Fourier transform of the causal Green function; from (3.1.93), it is clear that

$$D_c(p) = \frac{1}{p^2 - m^2 + i\varepsilon}. \qquad (3.1.124)$$

We see that the integral in (3.1.123) is also logarithmically divergent and hence requires regularization.

The different terms in the perturbation theory expansion of the S-matrix or any other quantity can be graphically represented by the well-known Feynman diagrams. In these diagrams, each graphical element is in one-to-one correspondence with the building blocks of the perturbation expansion. In particular, for the φ^4-model, this correspondence is summarized in table 3.1. Using these graphical elements, the first- and second-order terms in the expansion (3.1.123) are depicted in figures 3.3(a) and (b), respectively.

We shall not carefully derive the Feynman rules for the construction of amplitudes from the diagrams (see the textbooks on quantum field theory). A short practical collection of these rules is presented in supplement III. Note that, while in the standard operator approach the graphical method of Feynman diagrams is extremely important because it considerably simplifies complicated combinatorial calculations, in the functional path-integral approach all calculations in perturbation theory are reduced to simple manipulations with derivatives. Therefore, generally speaking, we can quite comfortably work

Table 3.1. Correspondence rules for the φ^4-model.

Physical quantity	Mathematical expression	Diagram element
Wavefunction	$\psi(p)$	•—
Causal Green function (propagator)	$D_c = -\dfrac{1}{p^2 - m^2 + i\varepsilon}$	•—•
Interaction vertex	$-ig$	•

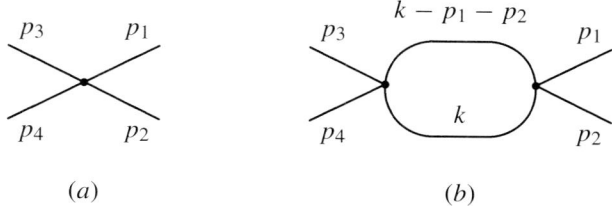

(a) (b)

Figure 3.3. Feynman diagrams for the φ^4-model in second-order perturbation theory.

directly with expressions of the type (3.1.117). However, the Feynman diagrams still prove to be very illustrative and a convenient accompanying tool.

All the formalism of perturbation theory, together with the graphical Feynman techniques, can be straightforwardly generalized to fields with higher spins, in particular to anticommuting spinor fields. A more convenient approach to field theoretical calculations, however, uses Green functions (vacuum expectations of chronologically ordered products of quantum fields). We shall proceed to discuss this technique in the next subsection.

3.1.5 Generating functionals for Green functions and an introduction to functional methods in quantum field theory

Green functions $G^{(n)}$ in quantum field theory, defined as vacuum expectations of chronological products of quantum fields

$$G^{(n)}(x_1, x_2, \ldots, x_n) \stackrel{\text{def}}{\equiv} \langle 0 | \mathbf{T}(\varphi(x_1)\varphi(x_2) \cdots \varphi(x_n)) | 0 \rangle \quad (3.1.125)$$

play an outstanding role in the field theoretical formalism. Their exceptional significance is explained by the fact that they are convenient for practical calculations and, at the same time, contain complete information about quantum systems. In particular, the spectrum of a system (including bound states) and corresponding wavefunctions can be extracted from the Green functions by studying their singularities. The relation of the Green functions to the scattering amplitudes is established by the so-called *reduction formula* within the Lehmann–Symanzik–Zimmermann formalism (Lehmann *et al* 1955) (see also, e.g., Bjorken and Drell (1965) and Weinberg (1995)). For a scalar field, the reduction formula reads as

$$\langle \psi_{p_1}, \ldots, \psi_{p_n}; \text{out} | \psi_{k_1}, \ldots, \psi_{k_n}; \text{in} \rangle = \left(\frac{i}{\sqrt{Z_\varphi}} \right)^{m+n} \prod_{i=1}^{m} \int d^4 x_i \prod_{j=1}^{m} \int d^4 y_j$$

$$\times \psi_{k_j}^*(y_j) \psi_{k_i}(x_i) (\Box_{x_i} + m^2)(\Box_{y_j} + m^2) \langle 0 | \mathbf{T}(\varphi(y_1) \cdots \varphi(y_n)\varphi(x_1) \cdots \varphi(x_m)) | 0 \rangle \quad (3.1.126)$$

where Z_φ is the field renormalization constant and we assumed for simplicity that all $p_i \neq k_j$ (in the general case, additional terms corresponding to zero scattering angle appear). Analogous formulae exist for any other fields (with higher spin).

Thus, we can concentrate our efforts on the study of different Green functions or their generating functional $\mathcal{Z}[J(x)]$ introduced in section 3.1.1. For an arbitrary interaction term, the generating functional is defined as in the free theory (cf (3.1.99)):

$$\mathcal{Z}[J(x)] \stackrel{\text{def}}{\equiv} S[\varphi_0; J]|_{\varphi_0=0}$$
$$= \int \mathcal{D}\varphi(x) \exp\left\{i \int d^4x \, [\mathcal{L}(\varphi(x)) + J(x)\varphi(x)]\right\}. \qquad (3.1.127)$$

This functional produces the Green functions (vacuum expectations) (3.1.125) for the field theory with arbitrary (self-) interaction terms defined by the Lagrangian density $\mathcal{L}(\varphi(x))$. We suggest to the reader, as a useful exercise, to derive the expression (3.1.127) directly from the path integral (3.1.74), without using the S-matrix functional (see problem 3.1.2, page 39).

◇ **Perturbation series for the Green functions**

Similarly to the case of the S-matrix symbol, we can represent the generating functional for field theories with an arbitrary interaction as an infinite series of variational operators acting on the explicit form of the generating functional for the free theory, in the presence of an external source:

$$\mathcal{Z}[J] = \exp\left\{-i \int d^4x \, V\left(\frac{1}{i}\frac{\delta}{\delta J(x)}\right)\right\} \mathcal{Z}_0[J(x)]. \qquad (3.1.128)$$

Recall that $\mathcal{Z}_0[J]$ is the generating functional for the free theory (i.e. for the Lagrangian with $V(\varphi(x)) = 0$) in the presence of an external source $J(x)$ and path integration of it gives the following explicit expression:

$$\mathcal{Z}_0[J] = \exp\left\{\frac{i}{2} \int d^4x \, d^4y \, J(x) D_c(x-y) J(y)\right\}. \qquad (3.1.129)$$

In problem 3.1.4, page 40 we suggest deriving the expression (3.1.129) for the free-field generating functional by another, perhaps the simplest, method of *square completion*.

Let us consider again the φ^4-model. Expanding the exponential in (3.1.128), we obtain a perturbation series for the Green function generating functional:

$$\mathcal{Z}[J] = \mathcal{Z}_0[J](1 + g\mathcal{Z}^{(1)}[J] + g^2\mathcal{Z}^{(2)}[J] + \cdots). \qquad (3.1.130)$$

Here $\mathcal{Z}^{(1)}, \mathcal{Z}^{(2)}$ are obtained by functional differentiation and have the form

$$\mathcal{Z}^{(1)}[J] = -\frac{i}{4!}\bigg[\int d^4x \, d^4y_1 \cdots d^4y_4 \, D_c(x-y_1) D_c(x-y_2)$$
$$\times D_c(x-y_3) D_c(x-y_4) J(y_1) J(y_2) J(y_3) J(y_4)$$
$$- i3! \int d^4x \, d^4y_1 \, d^4y_2 \, D_c(x-y_1) D_c(x-y_2) D_c(x-x) J(y_1) J(y_2)$$
$$+ 3! \int d^4x \, D_c^2(x-x)\bigg] \qquad (3.1.131)$$

$$\mathcal{Z}^{(2)}[J] = \frac{1}{2}(\mathcal{Z}^{(1)}[J])^2 + \frac{i}{2(3!)^2} \int d^4x_1 \, d^4x_2 \, d^4y_1 \cdots d^4y_6$$

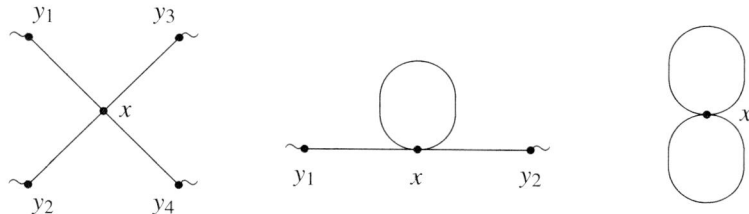

Figure 3.4. Graphical representation of the first-order contributions to the Green function generating functional in the φ^4-model.

$$\times D_c(x_1 - y_1)D_c(x_1 - y_2)D_c(x_1 - y_3)D_c(x_1 - x_2)D_c(x_2 - y_1)$$
$$\times D_c(x_2 - y_2)D_c(x_2 - y_3)J(y_1)J(y_2)J(y_3)J(y_4)J(y_5)J(y_6)$$
$$+ \frac{3}{2(4!)^2} \int d^4x_1 d^4x_2 d^4y_1 \cdots d^4y_4 \, D_c(x_1 - y_1)D_c(x_1 - y_2)D_c^2(x_1 - x_2)$$
$$\times D_c(x_2 - y_3)D_c(x_2 - y_4)J(y_1)J(y_2)J(y_3)J(y_4)$$
$$+ \frac{2}{24!} \int d^4x_1 d^4x_2 d^4y_1 \cdots d^4y_4 \, D_c(x_1 - y_1)D_c(x_1 - x_1)D_c(x_1 - x_2)D_c(x_2 - y_2)$$
$$\times D_c(x_2 - y_3)D_c(x_2 - y_4)J(y_1)J(y_2)J(y_3)J(y_4)$$
$$- \frac{i}{8} \int d^4x_1 d^4x_2 d^4y_1 d^4y_2 \, D_c(x_1 - y_1)D_c(x_1 - x_1)D_c(x_1 - x_2)$$
$$\times D_c(x_2 - x_2)D_c(x_2 - y_2)J(y_1)J(y_2)$$
$$- \frac{i}{8} \int d^4x_1 d^4x_2 d^4y_1 d^4y_2 \, D_c(x_1 - y_1)D_c^2(x_1 - x_2)D_c(x_2 - x_2)D_c(x_1 - y_2)J(y_1)J(y_2)$$
$$- \frac{i}{12} \int d^4x_1 d^4x_2 d^4y_1 d^4y_2 \, D_c(x_1 - y_1)D_c^3(x_1 - x_2)D_c(x_2 - y_2)J(y_1)J(y_2)$$
$$+ (J\text{-independent terms}). \tag{3.1.132}$$

Note that the most convenient way for field theory renormalization uses Green functions, i.e. we first renormalize the Green functions and then all the other physical quantities are expressed through them. Therefore we keep in (3.1.131) and (3.1.132) the divergent terms, assuming that they are suitably regularized. We also keep formal arguments of the type $(x - x)$ in the causal functions, though, of course, we should just write $D_c(0)$. We do this for easier comparison with the corresponding graphical representation in terms of Feynman diagrams. To construct these diagrams, we have to add a graphical element corresponding to the source $J(x)$; this shall be represented as follows: $J(x) = $ •~. The first-order contribution (3.1.131) is depicted in figure 3.4. The first term of the second-order contribution (3.1.132), being the power of the first-order contribution $\mathcal{Z}^{(1)}$, is represented by a *disconnected* graph (diagram). Since such terms are powers or products of terms which correspond to *connected* diagrams, we can reduce the study to the latter only. The second-order J-dependent terms represented by connected diagrams are depicted in figure 3.5 (we have dropped all J-independent terms because they vanish under the action of the functional derivatives and hence do not contribute to the Green functions).

For systematization and further use, let us recall some nomenclature from graph (diagram) theory:

- Diagrams containing pieces not connected by lines are called *disconnected diagrams*.
- If any vertex of a diagram can be reached from any other vertex by moving along the lines of the graph, the diagram is said to be *connected*.
- *One-particle irreducible* (OPI in abbreviation) *diagrams* cannot be converted into disconnected

30 Quantum field theory: the path-integral approach

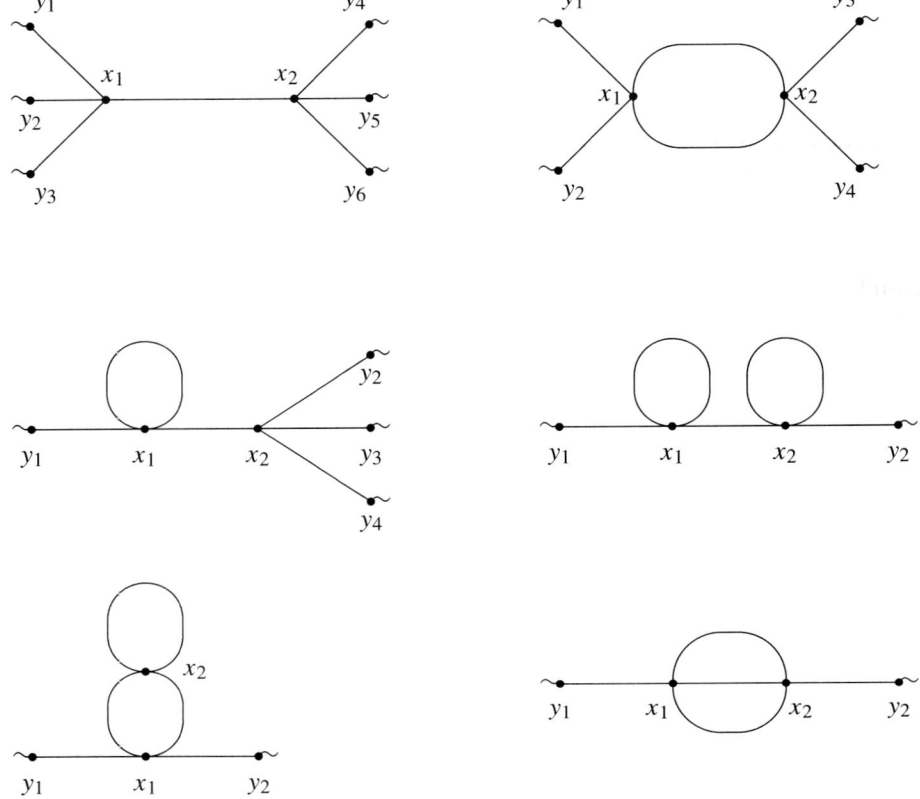

Figure 3.5. Connected diagrams contributing to the second-order contributions to the Green-function generating functional in the φ^4-model.

graphs by cutting just one internal line.

A similar terminology is applied to the actual Green functions:

- Parts of Green functions represented by connected and OPI diagrams are called *connected Green functions* W_n and *OPI Green functions* Γ_n, respectively.
- Truncated (amputated) connected Green functions $W^{(\text{tr})}$ are defined by the relation:

$$W_n(x_1,\ldots,x_n) = \int d^4y_1 \cdots d^4y_n\, W_2(x_1,y_1)\cdots W_2(x_n,y_n) W_n^{(\text{tr})}(y_1,\ldots,y_n). \qquad (3.1.133)$$

Here $W_2(x_i, y_i)$ are two-point connected Green functions (total propagator).

Some lower-point Green functions have special names:

- The zero-point connected Green functions W_0, which do not have external lines, are said to be *connected vacuum loops*.
- A connected one-point Green function (with one external line) is called a '*tadpole*'.

- The Green function $D_c^{(\text{tot})} \stackrel{\text{def}}{\equiv} W_2$ is called the *total propagator* to distinguish it from the *bare* propagator D_c (causal Green function of the free Klein–Gordon equation).

Different Green functions are obtained by differentiation of the expansion (3.1.131), (3.1.132). For example, the terms with four source factors contribute to the four-point Green function in first- and second-order perturbation theory.

◇ Generating functional for connected Green functions

It is helpful to construct the modified generating functional which only directly produces *connected Feynman diagrams*. It appears that the logarithm

$$W[J(x)] \stackrel{\text{def}}{\equiv} \ln \mathcal{Z}[J(x)] \qquad (3.1.134)$$

of the functional $\mathcal{Z}[J(x)]$ satisfies this requirement: the functional derivatives of $W[J(x)]$,

$$W_n(x_1, \ldots, x_n) \stackrel{\text{def}}{\equiv} \left. \frac{\delta^n W[J]}{\delta J(x_1) \cdots \delta J(x_n)} \right|_{J=0} \qquad (3.1.135)$$

correspond to *connected* Feynman diagrams and, according to the previous definition, are called *connected n-point Green functions* $W_n(x_1, \ldots, x_n)$. The heuristic proof of this fact goes as follows. Consider the obvious identity:

$$\begin{aligned} \ln \mathcal{Z}[J] &= \ln\{\mathcal{Z}_0[J](1 + \mathcal{Z}_0^{-1}[J](\mathcal{Z}[J] - \mathcal{Z}_0[J]))\} \\ &= \ln \mathcal{Z}_0[J] + \ln\{1 + (\mathcal{Z}[J]\mathcal{Z}_0^{-1}[J] - 1)\}. \end{aligned} \qquad (3.1.136)$$

Expansion (3.1.130) allows us to rewrite identity (3.1.136) as a series in the coupling constant g (perturbation expansion):

$$\begin{aligned} W[J] &= \ln \mathcal{Z}_0[J] + (g\mathcal{Z}^{(1)} + g^2 \mathcal{Z}^{(2)} + \cdots) - \tfrac{1}{2}(g\mathcal{Z}^{(1)} + g^2 \mathcal{Z}^{(2)} + \cdots)^2 + \cdots \\ &= \ln \mathcal{Z}_0[J] + g\mathcal{Z}^{(1)} + g^2(\mathcal{Z}^{(2)} - \tfrac{1}{2}(\mathcal{Z}^{(1)})^2) + \cdots. \end{aligned} \qquad (3.1.137)$$

Thus, the contribution of the disconnected diagrams $1/2(\mathcal{Z}^{(1)})^2$ to the $\mathcal{Z}^{(2)}$ (the first term in (3.1.132)) is indeed canceled out (by the second term in the parentheses) in the logarithm of the generating functional. This result can be generalized to all disconnected contributions (see e.g., Itzykson and Zuber (1980)). Note that the functional $W[J]$ is insensitive to normalization factors multiplying the functional $\mathcal{Z}[J]$ (an additive constant is absolutely inessential for generating functionals). This feature of $W[J]$ is convenient for heuristic simple methods of calculation of the corresponding path integrals (see, for example, problem 3.1.4, page 40).

◇ Variational equations for Green functions from path integrals

The so-called *Dyson–Schwinger equations* (Dyson 1949, Schwinger 1951) are exact relations between different Green functions. All these relations can be presented as one equation with variational derivatives for the generating functional $\mathcal{Z}[J]$. The simplest way of deriving it is to use the path-integral representation for $\mathcal{Z}[J]$ (Feynman and Hibbs 1965).

The key observation for this derivation is that the functional integration measure $\mathcal{D}\varphi(x)$ is invariant with respect to the translation

$$\varphi(x) \to \varphi(x) + f(x) \qquad (3.1.138)$$

where $f(x)$ is an arbitrary function well decreasing at infinity (so that it belongs to the class of functions to be integrated over in the path integral). This invariance of the measure (including the functional measure for fermion fields, i.e. for Grassmann variables) can be easily verified with the help of the time-sliced approximation for the path integrals. The translational invariance implies that

$$\int c\,\mathcal{D}\varphi(x)\exp\{i(S[\varphi]+J\varphi)\} = \int \mathcal{D}\varphi(x)\exp\{i(S[\varphi+f]+J(\varphi+f))\} \qquad (3.1.139)$$

where $S[\varphi]$ is an action for the field $\varphi(x)$ and we have adopted here the shorthand notations

$$J\varphi \equiv \int d^4x\, J(x)\varphi(x). \qquad (3.1.140)$$

In the infinitesimal form, equality (3.1.139) reads:

$$\int \mathcal{D}\varphi(x) \frac{\delta}{\delta\varphi(x)} \exp\{i(S[\varphi]+J\varphi)\} = 0 \qquad (3.1.141)$$

or, after differentiation of the exponential,

$$\int \mathcal{D}\varphi(x)\left[\frac{\delta S[\varphi]}{\delta\varphi(x)}+J(x)\right]\exp\{i(S[\varphi]+J\varphi)\} = 0. \qquad (3.1.142)$$

Using relation (3.1.108), we can rewrite (3.1.142) as the *Schwinger variational equation* for the Green-function generating functional:

$$\left[\frac{\delta S[\varphi]}{\delta\varphi(x)}\bigg|_{\varphi=-i\delta/\delta J}+J(x)\right]\mathcal{Z}[J] = 0. \qquad (3.1.143)$$

The Schwinger equation is homogeneous and, hence, defines $\mathcal{Z}[J]$ up to a factor. Therefore, it is convenient to substitute $\mathcal{Z}[J]$ with a functional $W[J] = \ln \mathcal{Z}[J]$ for connected Green functions. Then, expanding $W[J]$ in powers of $J(x)$ and equating, in equation (3.1.143), the coefficients with different powers of $J(x)$, we obtain the infinite chain (system) of the differential *Dyson–Schwinger equations* for n-point Green functions with increasing n.

Example 3.2. Let us consider as an example the simplest scalar φ^3-model with the action

$$S[\varphi] = \int d^4x \left(\frac{1}{2}\varphi(x)(\Box+m^2)\varphi(x) + \frac{g}{3!}\varphi^3(x)\right). \qquad (3.1.144)$$

The renormalization of this model requires the addition of a φ^4-vertex and hence, rigorously speaking, it is not self-consistent. However, this fact is not important for our formal functional manipulations and we choose this example as the simplest one to illustrate the general functional techniques.

The Schwinger equation for this model takes the form

$$\left[\int d^4x'\, K_{\text{KG}}(x,x')\frac{1}{i}\frac{\delta}{\delta J(x')} - \frac{g}{2}\frac{\delta^2}{\delta J(x)^2} + J(x)\right]\mathcal{Z}[J] = 0 \qquad (3.1.145)$$

where we have denoted by $K_{\text{KG}}(x,x')$ the integral kernel of the Klein–Gordon operator, i.e. for any function $f(x)$ from the domain of definition of the latter, $K_{\text{KG}}(x,x')$ satisfies the equality

$$(\Box+m^2)f(x) \equiv \int d^4x\, K_{\text{KG}}(x,x')f(x').$$

$W_n \Leftrightarrow$ [diagram: blob with n legs] $g \Leftrightarrow \bullet$ $D_c \Leftrightarrow$ [line]

Figure 3.6. Graphical notation for the elements of the Schwinger equation (3.1.148), after the expansion (3.1.149).

In particular,

$$\int d^4x'\, K_{KG}(x,x') D_c(x'-y) = \int d^4x'\, D_c(x-x') K_{KG}(x',y) = \delta(x-y). \tag{3.1.146}$$

Substituting in (3.1.145) $\mathcal{Z}[J] = \exp\{W[J]\}$, we obtain the equation for the generating functional $W[J]$ of connected Green functions:

$$\int d^4x'\, K_{KG}(x,x') \frac{1}{i} \frac{\delta W}{\delta J(x')} - \frac{g}{2}\left[\frac{\delta^2 W}{\delta J(x)^2} + \left(\frac{\delta W}{\delta J(x)}\right)^2\right] + J(x) = 0. \tag{3.1.147}$$

For an iterative solution of this equation, it is convenient, at first, to convolute it with the function $\int d^4x'\, J(x') D_c(x'-x)$:

$$\int d^4x'\, J(x') \frac{\delta W}{\delta J(x')} = \frac{ig}{2}\bigg[\int d^4x'\, d^4x\, J(x') D_c(x'-x) \frac{\delta^2 W}{\delta J(x)^2}$$
$$+ \int d^4x'\, d^4x\, J(x') D_c(x'-x) \left(\frac{\delta W}{\delta J(x)}\right)^2\bigg]$$
$$- \int d^4x'\, d^4x\, J(x') D_c(x'-x) J(x). \tag{3.1.148}$$

Expansion of the functional $W[J]$ in powers of the sources:

$$W[J] = \sum_{n=0}^{\infty} \int d^4x_1 \cdots d^4x_n\, \frac{i^n}{n!} W_n(x_1,\ldots,x_n) J(x_1) \cdots J(x_n) \tag{3.1.149}$$

converts the functional equation (3.1.148) into an infinite chain of differential equations, called the *Dyson–Schwinger equations*, for the connected Green functions $W_n(x_1,\ldots,x_n)$ (by equating the factors with equal powers of the source function $J(x)$). It is illustrative to represent this chain of equations graphically, using the Feynman-like notation depicted in figure 3.6. In this notation, the chain of equations for connected Green functions is presented in figure 3.7. To illustrate the general formalism, the first two equations from the chain are explicitly depicted in figure 3.8. Multiplying the second equation in figure 3.8 by K_{KG} from the right and by W_2^{-1} from the left, we obtain the equality for the so-called *mass operator* (or *proper energy*) Σ of a particle. The resulting equality for Σ has the graphical image as in figure 3.9. In the latter figure, the element

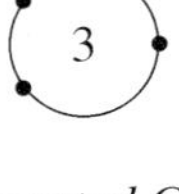

[diagram: blob labeled 3 with one leg] $=$ truncated Green function $W_3^{(tr)}$

denotes the three-point *truncated Green function* (cf definition (3.1.133)).

In the same way we can derive the Schwinger equation for any field theoretical model with a polynomial Lagrangian. The starting point is the translational invariance of the path-integral measure and hence a relation of the type (3.1.141) for all fields enters the model.

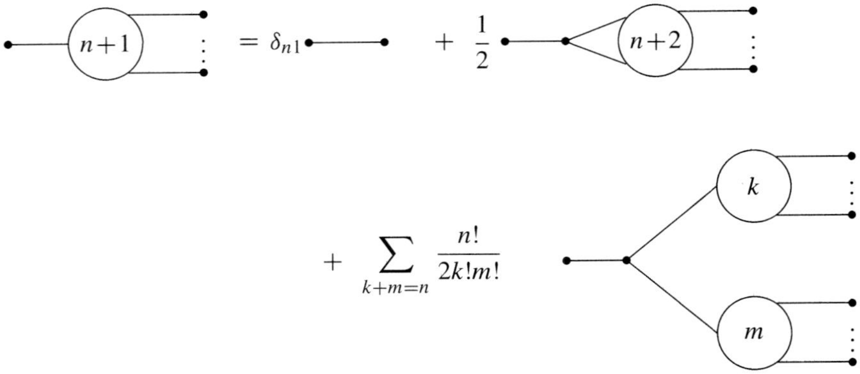

34   Quantum field theory: the path-integral approach

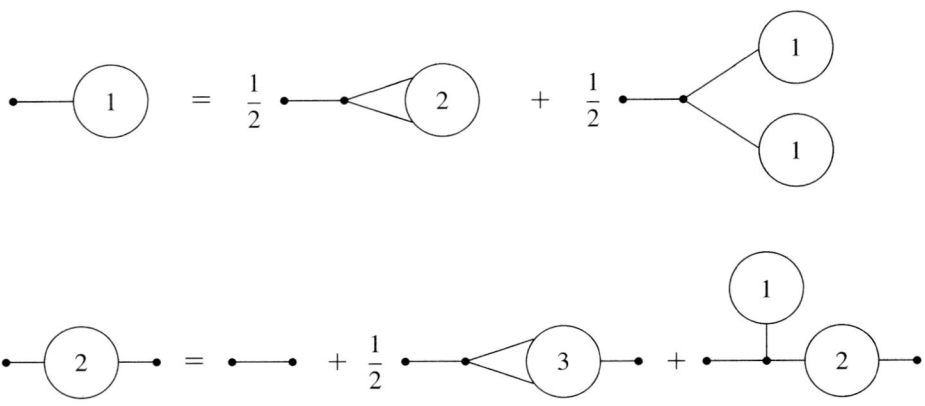

Figure 3.7. Graphical representation of the Dyson–Schwinger chain of equations for connected Green functions.

Figure 3.8. Graphical representation of the first two equations from the Dyson–Schwinger chain of equations for connected Green functions.

◇ Ward–Takahashi identities as the result of a special change of variables in the path integrals

We may consider more general transformations

$$\varphi \longrightarrow F(x, \{a\}; \varphi) \qquad (3.1.150)$$

of integration variables in path integrals which produce the generating functional for Green functions. Any type of transformation forms a group and we assume that transformations (3.1.150) form a *Lie group* with the set of parameters $\{a\}$ (for some basic notions from group theory, see supplement IV). If transformations (3.1.150) have a unit Jacobian, then by changing the integration variables, we again obtain the infinitesimal condition for measure invariance (cf (3.1.141)):

$$\int \mathcal{D}\varphi(x) \left(\int d^4y \, \delta_a\varphi(y) \frac{\delta S[\varphi]}{\delta\varphi(y)} \right) \exp\{S[\varphi]\} = 0 \qquad (3.1.151)$$

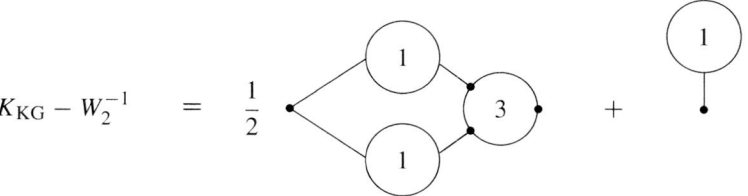

Figure 3.9. Equation for the mass operator Σ.

where
$$\delta_a \varphi \stackrel{\text{def}}{\equiv} \sum_i \left.\frac{\partial F(x, \{a\}; \varphi)}{\partial a^i}\right|_{a^i=0} da^i.$$

In the general situation, the equations obtained from (3.1.151) follow from the Schwinger equations and hence do not provide new information about the Green functions. Indeed, after the substitution of $\delta_a \varphi$ and $\delta S[\varphi]/\delta \varphi$ by variational operators as in (3.1.143), we obtain an equation for the generating functional $\mathcal{Z}[J]$ in the form

$$L_a L_{\text{Schw}} \mathcal{Z}[J] = 0 \qquad (3.1.152)$$

where L_a and L_{Schw} are variational operators, the latter being the ordinary Schwinger equation operator (the variational operator on the left-hand side of (3.1.143)). Thus, (3.1.152) follows from the Schwinger equation. However, in special cases, the combination $L_a L_{\text{Schw}}$ may prove to be a lower-order variational operator than just the Schwinger operator L_{Schw}. This happens if terms with *higher powers* of the field variable in the action functional $S[\varphi]$ are *invariant* with respect to the group of transformations (3.1.150). The resulting relations are called *Ward–Takahashi identities*. These identities are extremely important for the proof of the renormalizability of quantum gauge theories, including electrodynamics (*Abelian gauge theory*) and the Yang–Mills theory (*non-Abelian gauge theory*). In the latter case, the identities are called *generalized Ward–Takahashi* or *Slavnov–Taylor–Ward–Takahashi identities*. We shall discuss the (generalized) Ward–Takahashi identities in some detail in section 3.2.7 devoted to quantum gauge theories (where path integrals find one of their most important applications).

◇ **Generating functional for one-particle irreducible Green functions**

One-particle irreducible (OPI) Green functions play an important role in the renormalization of quantum field theories (especially those theories with gauge invariance) as well as in non-perturbative calculations (the so-called *effective action* method). Therefore, it is desirable to construct for them a generating functional. This aim is achieved via a *Legendre transformation* of the functional $W[J]$ generating connected Green functions:

$$W[J] \longrightarrow \Gamma[\phi]: \qquad \Gamma[\phi] = W[J(\phi)] - \int d^4x\, \phi(x) J(x) \qquad (3.1.153)$$

where the source $J(x)$ on the right-hand side is expressed through ϕ with the help of the equation

$$\phi(x) \stackrel{\text{def}}{\equiv} W_1(x) = \frac{\delta W[J]}{\delta J(x)}. \qquad (3.1.154)$$

Recall that the Legendre transformation relates Lagrangians and the corresponding Hamiltonians of physical systems, thus (3.1.153) is a formal analog of the transition from the Lagrangian to the

36   Quantum field theory: the path-integral approach

Hamiltonian formalism in classical mechanics. Note also that the direct physical meaning of the quantity $\phi(x)$ is the vacuum expectation of the quantum field $\hat{\varphi}$ in the presence of the external source $J(x)$:

$$\phi(x) = \langle 0|\hat{\varphi}(x)|0\rangle_J \qquad (3.1.155)$$

and thus it is the classical counterpart of the field operator. In the literature, W_1 is usually denoted by the same letter as the corresponding field operator. We have denoted it by the slightly modified character ϕ to avoid confusion with the path integration variable φ. The generating functional $\Gamma[\phi]$ for OPI Green functions is often called the *effective action* for the corresponding quantum field theory. The reason for this name is that in the lowest approximation, $\Gamma[\phi]$ exactly coincides with the classical action of the theory.

Differentiation of the equality (3.1.153) taking into account (3.1.154) yields the relation

$$\begin{aligned}\frac{\delta \Gamma}{\delta \phi(x)} &= \int d^4y \, \frac{\delta W}{\delta J(y)} \frac{\delta J(y)}{\delta \phi(x)} - \frac{\delta}{\delta \phi(x)} \int d^4x \, \phi(x) J(x) \\ &= -J(x)\end{aligned} \qquad (3.1.156)$$

which explicitly defines J (for known Γ) as a functional of ϕ.

The following simple chain of equalities shows that the second derivatives of $\Gamma[\phi]$ and $-W[J]$ are the kernels of the inverse operators:

$$\begin{aligned}\delta(x-y) &= \frac{\delta J(x)}{\delta J(y)} = \int d^4x' \, \frac{\delta J(x)}{\delta \phi(x')} \frac{\delta \phi(x')}{\delta J(y)} \\ &= -\int d^4x' \, \frac{\delta^2 \Gamma}{\delta \phi(x)\delta \phi(x')} \frac{\delta^2 W}{\delta J(x')\delta J(y)}.\end{aligned} \qquad (3.1.157)$$

Thus the two-point connected Green function is easily expressed through the corresponding two-point OPI Green function. Similarly, any higher connected Green functions can be expressed through OPI functions, the diagrams of the connected Green functions being constructed from OPI parts linked by lines in such a way that cutting any of these lines converts the diagrams to disconnected ones. Substituting the Legendre transformation into the Schwinger equation (3.1.143) allows us to present the functional equation directly in terms of OPI functions and to develop the corresponding iterative (approximate, perturbative) method of its solution.

As seen from our short discussion, once the Dyson–Schwinger equation (or Ward–Takahashi identity) is derived from the path integral for generating functionals, further functional manipulations (transitions to connected, OPI Green functions etc) have no *direct* relationship with the main object of this book, the path integral. Thus we shall not go further into a *general* consideration of functional methods in quantum field theory, referring the reader to the special literature (see, e.g., Itzykson and Zuber (1980) and Vasiliev (1998)). However, we shall meet important concrete applications of these methods combined with the path-integral formalism in the subsequent sections.

◇ **Generating functional and Feynman diagrams for the Yukawa model**

All the consideration in the present subsection can be straightforwardly generalized to the case of several fields and fields with higher spins including fermion fields. In the latter case, the only peculiarity is that after introduction of anticommuting integration variables and sources, as explained in section 3.1.3, we must care about the order of all factors and Grassmann derivatives. Instead of a general consideration of these generalizations, we shall discuss here an example illustrating the functional methods for a model with spinor fields, namely for the Yukawa-interaction model, while in the subsequent sections we shall consider concrete practically most important examples (types) of field theoretical models.

Path-integral formulation of the simplest quantum field theories

The *Yukawa model* contains a spin-$\frac{1}{2}$ field $\psi(x)$ and a scalar field $\varphi(x)$, with the following Lagrangian density:

$$\mathcal{L}_{\text{Yu}} = \mathcal{L}_0(\varphi) + \mathcal{L}_0(\bar{\psi}, \psi) + \mathcal{L}_{\text{int}}(\bar{\psi}, \psi, \varphi) \tag{3.1.158}$$

where $\mathcal{L}_0(\varphi)$ and $\mathcal{L}_0(\bar{\psi}, \psi)$ are the free Lagrangian densities (3.1.49) and (3.1.59) for scalar and spinor fields, respectively; $\mathcal{L}_{\text{int}}(\bar{\psi}, \psi, \varphi)$ defines an interaction of these fields, the so-called *Yukawa coupling*:

$$\mathcal{L}_{\text{int}}(\bar{\psi}, \psi, \varphi) = g\bar{\psi}(x)\psi(x)\varphi(x). \tag{3.1.159}$$

The path-integral representation for the generating functional reads as

$$\mathcal{Z}[\bar{\eta}, \eta, J] = \int \mathcal{D}\varphi\, \mathcal{D}\bar{\psi}\, \mathcal{D}\psi\, \exp\left\{ i \int d^4 x\, [\mathcal{L}_0(\varphi(x)) + \mathcal{L}_0(\bar{\psi}(x), \psi(x)) + \mathcal{L}_{\text{int}}(\bar{\psi}(x), \psi(x), \varphi(x)) \right.$$
$$\left. + J(x)\varphi(x) + \bar{\psi}(x)\eta(x) + \bar{\eta}(x)\psi(x)] \right\}. \tag{3.1.160}$$

Similarly to the case of the purely scalar theory, the starting point for developing the perturbation expansion is the representation of the generating functional (3.1.160) in the form

$$\mathcal{Z}[\bar{\eta}, \eta, J] = \exp\left\{ \mathcal{L}_{\text{int}}\left(\frac{1}{i}\frac{\delta}{\delta\eta}, \frac{1}{i}\frac{\delta}{\delta\bar{\eta}}, \frac{1}{i}\frac{\delta}{\delta J} \right) \right\} \mathcal{Z}_0[\bar{\eta}, \eta, J]. \tag{3.1.161}$$

The generating functional $\mathcal{Z}_0[\bar{\eta}, \eta, J]$ is simply given by the product of the generating functionals for the free scalar and free spinor theories:

$$\mathcal{Z}_0[\bar{\eta}, \eta, J] = \exp\left\{ -i \int d^4 x\, d^4 y\, \bar{\eta}(x) S_{\text{c}}(x-y) \eta(y) \right\} \exp\left\{ -\frac{i}{2} \int d^4 x\, d^4 y\, J(x) D_{\text{c}}(x-y) J(y) \right\}. \tag{3.1.162}$$

The expansion of (3.1.161) up to second order in the coupling constant g leads to

$$\mathcal{Z}[\bar{\eta}, \eta, J] = \left[1 - g i \int d^4 x\, \frac{\delta^3}{\delta\bar{\eta}(x)\delta\eta(x)\delta J(x)} \right.$$
$$\left. - \frac{g^2}{2} \int d^4 x\, \frac{\delta^3}{\delta\bar{\eta}(x)\delta\eta(x)\delta J(x)} \int d^4 y\, \frac{\delta^3}{\delta\bar{\eta}(y)\delta\eta(y)\delta J(y)} + \cdots \right] \mathcal{Z}_0[\bar{\eta}, \eta, J]. \tag{3.1.163}$$

Since the differentiation refers to different fields, the calculation for the Yukawa model is even simpler than for a scalar field with self-interaction. Nevertheless, the derivation of the terms in the second and higher orders is still quite laborious and it is reasonable to exploit again the Feynman graphical techniques.

The generating functional $\mathcal{Z}[\bar{\eta}, \eta, J]$ contains several fields and sources and therefore we need different graphical elements for them. The correspondence rules for the Yukawa model are summarized in table 3.2.

The differentiation in formula (3.1.163) gives a generating functional in the first $\mathcal{Z}^{(1)}[\bar{\eta}, \eta, J]$ and second $\mathcal{Z}^{(2)}[\bar{\eta}, \eta, J]$ order of perturbation theory. In the diagram representation, they are represented as follows:

$$\mathcal{Z}^{(1)}[\bar{\eta}, \eta, J] = \left[\quad \cdots\bigcirc \quad - \quad \longrightarrow\!\!\!|\!\!\!\longleftarrow \quad \right] \mathcal{Z}_0[\bar{\eta}, \eta, J] \tag{3.1.164}$$

Table 3.2. Correspondence rules for the Yukawa model.

Physical quantity	Mathematical expression	Diagram element
Propagators	$D_c(x-y)$	•----•
	$S_c(x-y)$	•——•
Interaction vertex	g	(vertex)
External sources	$J(x)$	∼•
	$\bar{\eta}(x)$	•⇐
	$\eta(x)$	•⇒

$$\mathcal{Z}^{(2)}[\bar{\eta}, \eta, J] = \Bigg[\text{(diagrams)} \Bigg] \mathcal{Z}_0[\bar{\eta}, \eta, J]. \quad (3.1.165)$$

Using table 3.2, we can easily translate this result back to the algebraic form. The last term in (3.1.165) corresponds to the disconnected part of the Green functions. A noticeable feature of theories which include fermions is that any fermion loop produces the factor (-1). In functional formalism, this is a consequence of the anticommutativity of the Grassmann variables corresponding to fermion fields and sources.

3.1.6 Problems

Problem 3.1.1. Derive expression (3.1.40) for the dispersion of the smeared field $\hat{\varphi}_\lambda$.

Hint. Using expression (3.1.31) for the quantum field in terms of independent oscillators we readily obtain

$$\langle 0|\bar{\hat{\varphi}}_\lambda^2|0\rangle = \frac{1}{(2\pi\lambda)^6} \int d^3k\, d^3r\, d^3r' \frac{1}{2\omega_k} \exp\{i\mathbf{k}\cdot(\mathbf{r}-\mathbf{r}')\} - (r^2+r'^2)/(2\lambda^2)$$

$$= \frac{1}{2(2\pi)^3}\int \frac{d^3k}{2\omega_k} e^{-k^2\lambda^2} \approx \frac{1}{\lambda^3\sqrt{\lambda^{-2}+m^2}}.$$

Problem 3.1.2. Prove that the path integral

$$\int \mathcal{D}\varphi(x)\varphi(x_1)\varphi(x_2) e^{iS}$$

gives the vacuum expectation

$$\langle 0|\mathbf{T}(\varphi(x_1)\varphi(x_2))|0\rangle$$

of the *time-ordered* product of the field operators.

Hint. As a hint, we shall consider the simplified case of a quantum-mechanical system with one degree of freedom. The field theoretical path integral is considered via a straightforward generalization: at first, we consider a system with N degrees of freedom on a lattice and then pass to the continuum limit (cf section 3.1.2).

We start from the matrix element

$$\langle a,t|\mathbf{T}(\widehat{A}^{(\mathrm{H})}(\widehat{a}^\dagger(t_1),\widehat{a}(t_1))\widehat{A}^{(\mathrm{H})}(\widehat{a}^\dagger(t_2),\widehat{a}(t_2)))|a_0,t_0\rangle$$

where $|a\rangle \equiv |\Upsilon_a\rangle$ represents the (non-normalized) coherent state (cf (2.3.44)) for the creation–annihilation operators $\widehat{a}^\dagger, \widehat{a}$; $|a,t\rangle = \exp\{it\widehat{H}\}|a\rangle$ and $\widehat{A}^{(\mathrm{H})}(\widehat{a}^\dagger(t),\widehat{a}(t))$ is some operator made of $\widehat{a}^\dagger(t), \widehat{a}(t)$, the latter being in the *Heisenberg* representation (this is indicated by the superscript '(H)'). We assume that $\widehat{A}^{(\mathrm{H})}(\widehat{a}^\dagger,\widehat{a})$ is written in the normal form. Then, we obtain:

$$\langle a,t|\mathbf{T}(\widehat{A}^{(\mathrm{H})}(\widehat{a}^\dagger(t_1),\widehat{a}(t_1))\widehat{A}^{(\mathrm{H})}(\widehat{a}^\dagger(t_2),\widehat{a}(t_2)))|a_0,t_0\rangle$$
$$= \langle a|e^{-i(t-t_1)\widehat{H}}\widehat{A}^{(\mathrm{S})}(\widehat{a}^\dagger,\widehat{a})e^{i(t-t_1)\widehat{H}}e^{-i(t-t_2)\widehat{H}}\widehat{A}^{(\mathrm{S})}(\widehat{a}^\dagger,\widehat{a})e^{i(t-t_2)\widehat{H}}e^{-i(t-t_0)\widehat{H}}|a_0\rangle$$
$$= \langle a|e^{-i(t-t_1)\widehat{H}}\widehat{A}^{(\mathrm{S})}(\widehat{a}^\dagger,\widehat{a})e^{-i(t_1-t_2)\widehat{H}}\widehat{A}^{(\mathrm{S})}(\widehat{a}^\dagger,\widehat{a})e^{-i(t_2-t_0)\widehat{H}}|a_0\rangle$$
$$= \int da^*_1\, da_1\, db^*_1\, db_1\, da^*_2\, da_2\, db^*_2\, da_2\, \langle a|e^{-i(t-t_1)\widehat{H}}|a_1\rangle e^{-a^*_1 a_1}$$
$$\times \langle a_1|\widehat{A}^{(\mathrm{S})}(\widehat{a}^\dagger,\widehat{a})|b_1\rangle e^{-b^*_1 b_1} \langle b_1|e^{-i(t_1-t_2)\widehat{H}}|a_2\rangle e^{-a^*_2 a_2}$$
$$\times \langle a_2|\widehat{A}^{(\mathrm{S})}(\widehat{a}^\dagger,\widehat{a})|b_2\rangle e^{-b^*_2 b_2} \langle b_2|e^{-i(t_2-t_0)\widehat{H}}|a_0\rangle.$$

Here $\widehat{A}^{(\mathrm{S})}(\widehat{a}^\dagger,\widehat{a})$ denotes the operator in the Schrödinger representation. Since $\widehat{A}^{(\mathrm{S})}(\widehat{a}^\dagger,\widehat{a})$ is written in the normal form, we have

$$\langle a_i|\widehat{A}^{(\mathrm{S})}(\widehat{a}^\dagger,\widehat{a})|a'_i\rangle = A^{(\mathrm{S})}(a^*_i,a'_i)e^{a^*_i a'_i}$$

and using the path integral for the evolution amplitudes in the holomorphic representation (cf (2.3.103)), we arrive at the required result

$$\langle a,t|\mathbf{T}(\widehat{A}^{(\mathrm{H})}(\widehat{a}^\dagger(t_1),\widehat{a}(t_1))\widehat{A}^{(\mathrm{H})}(\widehat{a}^\dagger(t_2),\widehat{a}(t_2)))|a_0,t_0\rangle$$
$$= \int \mathcal{D}a^*(\tau)\,\mathcal{D}a(\tau)\, A(a^*(t_1),a(t_1))A(a^*(t_2),a(t_2))$$
$$\times \exp\{[a^*(t)a(t) - a^*(t)a(t_0)]\}$$
$$\times \exp\left\{\int_{t_0}^{t} d\tau\,[-a^*(\tau)\dot{a}(\tau) - i\omega_k a^*(\tau)a(\tau)] - \int_{t_0}^{t} d\tau\, V[a^*,a]\right\}.$$

The case $t_2 > t_1$ gives essentially the same result. Finally, we use the fact that $|a=0\rangle = |0\rangle$, where $|0\rangle$ is the ground state ('vacuum vector'), see (2.3.42), (2.3.44) or (2.3.108) (cf also (3.1.99)): choosing

$a = a_0 = 0$, we convert the considered matrix element into the vacuum expectation value. This proves the required statement for a system with one degree of freedom. The generalization to an arbitrary number of degrees of freedom and to an arbitrary number of field operators in the time-ordered product is straightforward and gives the proof of formula (3.1.102).

Problem 3.1.3. Derive a relation between the vacuum expectation

$$G(x_1, x_2) \equiv \langle 0|\mathbf{T}(\hat{\varphi}(x_1)\hat{\varphi}(x_1))|0\rangle$$

of the time-ordered product of field operators (the field theoretical Green function) and the amplitude $\langle\varphi'(\boldsymbol{r}), t|\mathbf{T}(\hat{\varphi}(x_1)\hat{\varphi}(x_1))|\varphi(\boldsymbol{r}), t_0\rangle$ (here $x_i = (\boldsymbol{r}_i, t_i)$) in the *coordinate* representation.

Hint. Again, we present, as a hint, the main points of the derivation for a quantum-mechanical system with one degree of freedom. We have

$$\langle x, t|\mathbf{T}(\widehat{x}^{(H)}(t_1)\widehat{x}^{(H)}(t_2))|x_0, t_0\rangle = \sum_{m,n}\langle x, t|n\rangle\langle n|\mathbf{T}(\widehat{x}^{(H)}(t_1)\widehat{x}^{(H)}(t_2))|m\rangle\langle m|x_0, t_0\rangle$$

where $|n\rangle$ are the eigenvectors of the Hamiltonian under consideration: $\widehat{H}|n\rangle = E_n|n\rangle$, $\widehat{x}^{(H)}(t)$ is the position operator in the Heisenberg representation and $|x, t\rangle = \exp\{it\widehat{H}\}|x\rangle$. To understand the last relation, note that *without* the time-ordered product, the matrix element under consideration would coincide with the usual transition amplitude

$$\langle x, t|x_0, t_0\rangle = \langle x|\mathrm{e}^{-\mathrm{i}(t-t_0)\widehat{H}}|x_0\rangle.$$

Since we assume that E_0 is the lowest eigenvalue of $\widehat{H}$, the following limit gives the required relation:

$$\lim_{\substack{t\to\mathrm{i}\infty \\ t_0\to-\mathrm{i}\infty}} \langle x, t|\mathbf{T}(\widehat{x}^{(H)}(t_1)\widehat{x}^{(H)}(t_2))|x_0, t_0\rangle = \langle x|0\rangle\mathrm{e}^{-E_0|t|}\langle 0|x_0\rangle\mathrm{e}^{-E_0|t_0|}\langle 0|\mathbf{T}(\widehat{x}^{(H)}(t_1)\widehat{x}^{(H)}(t_2))|0\rangle$$

which can be rewritten in a more compact form:

$$\langle 0|\mathbf{T}(\widehat{x}^{(H)}(t_1)\widehat{x}^{(H)}(t_2))|0\rangle = \lim_{\substack{t\to\mathrm{i}\infty \\ t_0\to-\mathrm{i}\infty}} \frac{\langle x, t|\mathbf{T}(\widehat{x}^{(H)}(t_1)\widehat{x}^{(H)}(t_2))|x_0, t_0\rangle}{\langle x, t|x_0, t_0\rangle}. \qquad (3.1.166)$$

As in the solution of the preceding problem, we can show that

$$\langle x, t|\mathbf{T}(\widehat{x}^{(H)}(t_1)\widehat{x}^{(H)}(t_2))|x_0, t_0\rangle = \int_{\mathcal{C}\{x,t;x_0,t_0\}} \mathcal{D}x(\tau)\frac{\mathcal{D}p(\tau)}{2\pi}x(t_1)x(t_2)\mathrm{e}^{\mathrm{i}S}.$$

Of course, the transition amplitude in the denominator is also represented in terms of path integrals and this gives the path-integral expression for the vacuum expectation value.

Formula (3.1.99) and the solution of the preceding problem show that path integrals in the holomorphic representation are much more convenient for this aim.

Problem 3.1.4. Calculate path integral (3.1.127) (see also (3.1.100) and (3.1.129)) using the translational invariance $\varphi \to \varphi + \varphi_\mathrm{c}$ of the functional measure and carrying out a *square completion* in the exponent of the integrand.

Hint. We want to calculate the integral

$$\mathcal{Z}_0[J] = \int \mathcal{D}\varphi(x) \exp\left\{i \int d^4x \left[\tfrac{1}{2}(\partial_\mu\varphi)^2 + \tfrac{1}{2}m^2\varphi^2 + J(x)\varphi(x)\right]\right\}$$

$$= \int \mathcal{D}\varphi(x) \exp\left\{i \int d^4x \left[\tfrac{1}{2}\varphi(x)(-\partial_\mu^2 + m^2 - i\varepsilon)\varphi + J(x)\varphi(x)\right]\right\}.$$

Here we have introduced, in the exponent, the regularization term $-\varepsilon \int d^4x\, \varphi^2/2$ which provides the convergence of the integral. Let φ_c be the solution of the classical equation of motion:

$$(-\partial_\mu^2 + m^2 - i\varepsilon)\varphi_c(x) = -J(x)$$

that is,

$$\varphi_c(x) = -\int d^4y\, D_c(x-y)J(y)$$

where $D_c(x)$ is the causal Green function (3.1.93). Let us change the integration variables:

$$\varphi(x) \to \varphi'(x) = \varphi(x) - \varphi_c(x).$$

Then, we obtain

$$\mathcal{Z}_0 = \mathfrak{N}^{-1} \exp\left\{-i \int d^4x\, d^4y \left[\tfrac{1}{2}J(x) D_c(x-y) J(y)\right]\right\}$$

i.e. the required expression up to the factor $\mathfrak{N}^{-1}$ which does not depend on the external source

$$\mathfrak{N}^{-1} = \int \mathcal{D}\varphi' \exp\left\{i \int d^4x\, \varphi'(-\partial_\mu^2 + m^2 - i\varepsilon)\varphi'\right\}.$$

By the simple method of square completion, this factor cannot be calculated directly. Our more rigorous consideration in sections 3.1.2 and 3.1.3 shows that it must be put equal to unity. Note that the connected Green functions introduced in section 3.1.5 are insensitive to such a factor, so that for them this method is well suited. It is also worth mentioning that, while in this problem we have introduced the ε-term by hand (to improve convergence of the path integral), in section 3.1.2 we obtained the ε-prescription for the Green functions (see (3.1.93)) from the correctly chosen boundary conditions.

Problem 3.1.5. Using the path-integral representation for the generating functional $\mathcal{Z}[J]$ prove that $\delta \mathcal{Z}[J]/\delta J(x)$, for a field theory with the Lagrangian density $\mathcal{L}(\varphi) = \mathcal{L}_0(\varphi) + \mathcal{L}_{\text{int}}(\varphi)$, where $\mathcal{L}_0$ is the free particle part and $\mathcal{L}_{\text{int}}$ is the self-interaction Lagrangian, satisfies the equation

$$\frac{1}{i}(\Box + m^2)\frac{\delta \mathcal{Z}[J]}{\delta J(x)} - \mathcal{L}'_{\text{int}}\left(\frac{1}{i}\frac{\delta}{\delta J(x)}\right) \mathcal{Z}[J] = J(x) \mathcal{Z}[J]. \quad (3.1.167)$$

Here

$$\mathcal{L}'_{\text{int}}\left(\frac{1}{i}\frac{\delta}{\delta J(x)}\right) \equiv \left.\frac{\partial \mathcal{L}_{\text{int}}(\varphi)}{\partial \varphi}\right|_{\varphi = \delta/\delta J}.$$

This differential equation is obviously the quantum counterpart of the classical equation for a field φ with the Lagrangian $\mathcal{L}(\varphi)$. By direct substitution, show that a solution of equation (3.1.167) in free-field theory has the form (3.1.100) for the generating functional and that the general solution of (3.1.167) can be written as in (3.1.128), i.e.

$$\mathcal{Z}[J(x)] = \exp\left\{-i \int d^4x\, V\left(\frac{1}{i}\frac{\delta}{\delta J(x)}\right)\right\} \mathcal{Z}_0[J(x)]. \quad (3.1.168)$$

Hint. From the definition of $\mathcal{Z}[J]$, we have

$$\frac{1}{i}\frac{\delta \mathcal{Z}[J]}{\delta J(x)} = \int \mathcal{D}\varphi \, \varphi(x) e^{iS[\varphi,J]}. \qquad (3.1.169)$$

Now we use the simple identity

$$\frac{1}{i}\frac{\delta}{\delta \varphi(x)} e^{iS[\varphi,J]} = \frac{\delta S[\varphi,J]}{\delta \varphi(x)} e^{iS[\varphi,J]}$$
$$= (-(\Box + m^2)\varphi(x) + \mathcal{L}'_{\text{int}}(\varphi(x)) + J(x)) e^{iS[\varphi,J]} \qquad (3.1.170)$$

so that acting by the Klein–Gordon operator $(\Box + m^2)$ on both sides of (3.1.169), we obtain

$$(\Box + m^2)\frac{1}{i}\frac{\delta \mathcal{Z}[J]}{\delta J(x)} = \int \mathcal{D}\varphi \left(-\frac{1}{i}\frac{\delta}{\delta \varphi(x)} \mathcal{L}'_{\text{int}}(\varphi(x)) + J(x)\right) e^{iS[\varphi,J]}.$$

The first term in the integrand on the right-hand side is a total derivative and can only produce boundary terms which vanish (for a rigorous proof of this fact for the Feynman path integral we need, as usual, some regularization, e.g., a transition to imaginary time and, hence, to the Wiener path integral). The rest of the equality is equivalent to (3.1.167).

For the free-field theory, equation (3.1.167) can be rewritten as follows:

$$\frac{\delta \mathcal{Z}_0[J]}{\delta J(x)} = i(\Box + m^2)^{-1} J(x) \mathcal{Z}_0[J]$$
$$= -i \int d^4y \, D_c(x-y) J(y) \mathcal{Z}_0[J]$$

with the obvious solution (3.1.100).

The fact that (3.1.168) is the solution of (3.1.167) is proved by direct substitution. To compare the left- and right-hand sides of equation (3.1.167) after the substitution (3.1.168), we should move the operator

$$\exp\left\{-i \int d^4x \, V\left(\frac{1}{i}\frac{\delta}{\delta J(x)}\right)\right\}$$

in front of $J(x)$. This is achieved with the help of the operator identity

$$e^{-\widehat{B}} \widehat{A} e^{\widehat{B}} = \widehat{A} + [\widehat{A}, \widehat{B}] + \tfrac{1}{2}[[\widehat{A}, \widehat{B}], \widehat{B}] + \cdots \qquad (3.1.171)$$

and the commutator

$$\left[\int d^4y \, V\left(\frac{1}{i}\frac{\delta}{\delta J(y)}\right), J(x)\right] = -i\mathcal{L}'_{\text{int}}\left(\frac{1}{i}\frac{\delta}{\delta J(x)}\right).$$

Problem 3.1.6. Derive the graphical representation (a complete collection of Feynman diagrams together with the symmetry factors, i.e. the numerical factors with which a given diagram enters the expression for the Green function) for the two- and four-point Green functions in the first-order approximation of the perturbation theory for the scalar field theory with φ^4 self-interaction.

Hint. The expressions are obtained by using the genuine formula (3.1.128) or by the differentiation of (3.1.131). The result is:

$$G_2(x-y) = \underset{x \quad\quad y}{\bullet\!\!-\!\!\!-\!\!\bullet} + \frac{1}{2}g \;\; \underset{x \quad\quad y}{\bullet\!\!-\!\!\bigcirc\!\!-\!\!\bullet}$$

$$G_4(x_1, x_2, x_3, x_4) = \left(\cdots + || + \times \right)$$

$$+ \frac{1}{2}g \left(\cdots + \cdots + \triangleleft | + | \triangleright + \cdots + \cdots \right)$$

$$+ g \times$$

Thus, G_4 contains in addition to the free four-point function all possible propagators with a self-energy insertion and a single 'true' interaction graph.

Problem 3.1.7. Prove the following relations between the ordinary G_n and connected W_n Green functions in scalar field theory:

$$G_2(x - y) = W_2(x - y) \qquad (3.1.172)$$

$$G_4(x_1, x_2, x_3, x_4) = W_4(x_1, x_2, x_3, x_4) + W_2(x_1 - x_2)W_2(x_3 - x_4)$$
$$+ W_2(x_1 - x_3)W_2(x_2 - x_4) + W_2(x_1 - x_4)W_2(x_2 - x_3) \qquad (3.1.173)$$

provided that

$$G_1 = \left.\frac{\delta \mathcal{Z}[J]}{\delta J(x)}\right|_{J=0} = 0. \qquad (3.1.174)$$

The latter condition physically means that the vacuum expectation of the quantum field vanishes. If the quantum field can be expressed via creation and annihilation operators with the condition (3.1.20), the validity of condition (3.1.174) is obvious. Note, however, that in an important class of models with so-called *spontaneous symmetry-breaking* this condition is not fulfilled (see section 3.2.8).

Hint. Recalling that connected functions are generated by the logarithm of the generating functional for ordinary Green functions, we have

$$W_2(x - y) = \left.\frac{\delta^2 \ln \mathcal{Z}[J]}{\delta J(x)\delta J(y)}\right|_{J=0}$$
$$= \left. \left(-\frac{1}{\mathcal{Z}[J]^2} \frac{\delta \mathcal{Z}[J]}{\delta J(x)} \frac{\delta \mathcal{Z}[J]}{\delta J(y)} + \frac{1}{\mathcal{Z}[J]} \frac{\delta^2 \mathcal{Z}[J]}{\delta J(x)\delta J(y)} \right) \right|_{J=0}$$
$$= G_2(x - y). \qquad (3.1.175)$$

Here we have used (3.1.174) and the normalization $\mathcal{Z}[0] = 1$. Derivation of (3.1.173) is quite analogous.

Problem 3.1.8. Find the relation between connected and OPI three- and four-point Green functions.

Hint. Differentiation of the equality (3.1.157) with the help of the relation

$$\frac{\delta}{\delta J(x)} = \int d^4y \, \frac{\delta \phi(y)}{\delta J(x)} \frac{\delta}{\delta \phi(y)} \qquad (3.1.176)$$

yields

$$0 = \int d^4x \, \frac{\delta^3 W}{\delta J(x_3)\delta J(x_1)\delta J(x)} \frac{\delta^2 \Gamma}{\delta\phi(x)\delta\phi(x_2)}$$
$$+ \int d^4x'_1 \, d^4x'_3 \frac{\delta^2 W}{\delta J(x_1)\delta J(x'_1)} \frac{\delta^3 \Gamma}{\delta\phi(x'_1)\delta\phi(x'_2)\delta\phi(x'_3)} \frac{\delta\phi(x'_3)}{\delta J(x_3)}.$$

Again using (3.1.157), we obtain

$$\frac{\delta^3 W}{\delta J(x_1)\delta J(x_2)\delta J(x_3)} = \int d^4x'_1 \, d^4x'_2 \, d^4x'_3 \, \frac{\delta^2 W}{\delta J(x_1)\delta J(x'_1)} \frac{\delta^2 W}{\delta J(x_2)\delta J(x'_2)}$$
$$\times \frac{\delta^2 W}{\delta J(x_3)\delta J(x'_3)} \frac{\delta^3 \Gamma}{\delta\phi(x'_1)\delta\phi(x'_2)\delta\phi(x'_3)}. \qquad (3.1.177)$$

Graphically this relation is depicted as follows:

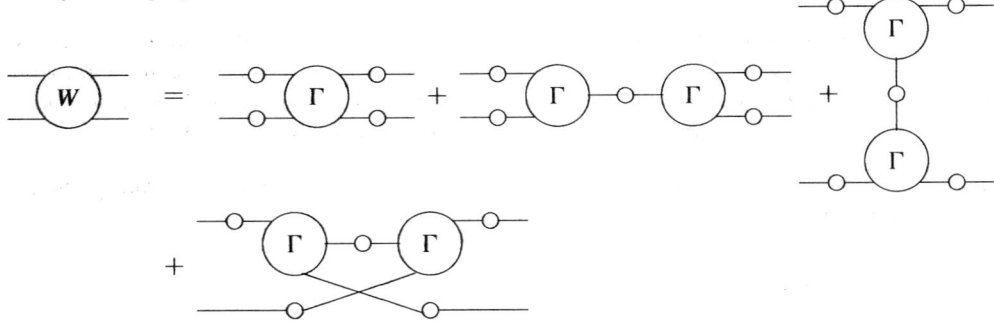

Here the lines with the circles denote the two-point Green function (recall that connected and ordinary *two-point* Green functions coincide, see (3.1.172)).

One more differentiation of (3.1.177) over $J(x)$ gives the relation for connected and OPI four-point Green functions. The calculation is rather cumbersome but straightforward. As a hint, we present this relation only in the graphical form:

Problem 3.1.9. Consider the anharmonic oscillator with the action

$$S = \int d\tau \left[\frac{m\dot{x}^2}{2} - \frac{m\omega^2 x^2}{2} + \alpha x^3 + \beta x^4 \right].$$

Of course, the corresponding non-Gaussian quantum-mechanical path integral cannot be written exactly. Using the relation between vacuum expectation values and the transition amplitudes as well as the explicit expression for the transition amplitudes of a harmonic oscillator in the presence of a time-dependent external force (driven oscillator), see (2.2.200), develop the perturbation theory expansion and Feynman diagram technique for this *quantum-mechanical* system. Calculate in the two-loop order the shift in the ground-state energy of the *harmonic* oscillator due to the presence of αx^3- and βx^4-terms and corrections to the two-point Green function $G_0(t_1, t_2) = \langle 0|x(t_1)x(t_2)|0\rangle$ of the harmonic oscillator ($|0\rangle$ is the ground state of the harmonic oscillator).

Do the same calculation also for the Euclidean version (i.e. after the analytic continuation to an imaginary (Euclidean) time $\tau_E = it$.

Hint. Calculations in the real-time formalism and with the Euclidean imaginary time are essentially the same. Since later (in section 3.3.3) we shall consider tunneling phenomena for the anharmonic oscillator in the Euclidean-time formalism (the Euclidean time is much more convenient for the study of tunneling phenomena), we give the hint for the imaginary-time version of the calculation.

The Euclidean action reads (we put, for brevity, $m = 1$):

$$S_E = \int d\tau \left[\frac{\dot{x}^2}{2} + \frac{\omega^2 x^2}{2} + \alpha x^3 + \beta x^4 \right]. \tag{3.1.178}$$

Expanding the path integral in powers of α and β, we can derive the Feynman rules for an anharmonic oscillator. Using the explicit form (1.2.262) of the generating functional for the harmonic oscillator with an external force (in the Euclidean time), and calculating

$$\lim_{\substack{\tau_0 \to -\infty \\ \tau \to \infty}} \frac{\delta}{\delta\eta(\tau)} \frac{\delta}{\delta\eta(\tau_0)} \mathcal{Z}[\eta; (x_\tau, \tau | x_0, \tau_0)]$$

(cf (3.1.166)), we obtain the free propagator (vacuum expectation for the time-ordered product of two coordinate operators in the Heisenberg representation):

$$G_0(\tau_1, \tau_2) = \langle 0|\hat{x}^{(H)}(\tau_1)\hat{x}^{(H)}(\tau_2)|0\rangle = \frac{1}{2\omega} \exp\{-\omega|\tau_1 - \tau_2|\}. \tag{3.1.179}$$

In addition to this, there are three- and four-point vertices with coupling constants α and β. To calculate an n-point Green function we have to sum over all diagrams with n external legs and integrate over the time variables corresponding to internal vertices.

The vacuum energy is given by the sum of all closed diagrams. At one-loop order, there is only one diagram, the free-particle loop diagram. At two-loop order, there are two $\mathcal{O}(\alpha^2)$ and one $\mathcal{O}(\beta)$ diagram:

$$\bigcirc\!\bigcirc \qquad \ominus \qquad \bigcirc\!\!-\!\!\bigcirc \qquad . \tag{3.1.180}$$

Calculating the diagrams is not difficult. Since the propagator is exponentially suppressed for large times, everything is finite (in contrast to the field theory case, see sections 3.1.4 and 3.2.7). Summing all the diagrams, we get

$$\langle 0|\exp(-H\tau)|0\rangle = \sqrt{\frac{\omega}{\pi}} \exp\left(-\frac{\omega\tau}{2}\right) \left[1 - \left(\frac{3\beta}{4\omega^2} - \frac{11\alpha^2}{8\omega^4}\right)\tau + \cdots \right]. \tag{3.1.181}$$

For small α^2 and β and τ not too large, we can exponentiate the result and read off the correction to the ground-state energy:

$$E_0 = \frac{\omega}{2} + \frac{3\beta}{4\omega^2} - \frac{11\alpha^2}{8\omega^4} + \cdots. \tag{3.1.182}$$

Of course, we could obtain the result using the ordinary Rayleigh–Schrödinger perturbation theory, but the method discussed here proves to be much more powerful when we come to non-perturbative effects (see section 3.3) and to the field theory.

Evaluating the first perturbative correction to the Green function corresponds to the diagrams:

$$ \text{(diagrams)} \tag{3.1.183} $$

Calculation gives

$$\Delta G_0(0,\tau) = \frac{9\alpha^2}{4\omega^6} + \frac{\alpha^2}{2\omega^6}e^{-2\omega\tau} + \frac{15\alpha^2}{4\omega^5}\tau e^{-\omega\tau} - \frac{3\beta}{2\omega^3}\tau e^{-\omega\tau}. \tag{3.1.184}$$

Comparing this result with the decomposition in terms of stationary states:

$$G(0,\tau) = \sum_{n=0}^{\infty} e^{-(E_n-E_0)\tau} |\langle 0|\widehat{x}^{(H)}|n\rangle|^2 \tag{3.1.185}$$

we can identify the first (time-independent) term with the square of the ground-state expectation value $\langle 0|\widehat{x}^{(H)}|0\rangle$ (which is non-zero due to the tadpole diagram). The second term comes from the excitation of two quanta, and the last two (with extra factors of τ) are the lowest-order 'mass renormalizations', or corrections to the zeroth-order gap between the ground and first excited states, $E_1 - E_0 = \omega$.

Problem 3.1.10. If we agreed to work *only* in the framework of the perturbation theory, we could use a formula of the type (3.1.111) or (3.1.128) as the starting *definition* of the corresponding path integral. More precisely, we could define (Faddeev and Slavnov 1980)

$$\int \mathcal{D}\varphi(x)\,\varphi(x_1)\cdots\varphi(x_n)\exp\left\{\frac{i}{2}\int d^4x\,d^4y\,\varphi(x)K(x-y)\varphi(y) + i\int d^4x\,\varphi(x)J(x)\right\}$$

$$\stackrel{\text{def}}{\equiv} (-i)^n \frac{\delta}{\delta J(x_1)}\cdots\frac{\delta}{\delta J(x_n)}\exp\left\{-\frac{i}{2}\int d^4x\,d^4y\,J(x)K^{-1}(x-y)J(y)\right\}. \tag{3.1.186}$$

Here $K(x-y)$ and $K^{-1}(x-y)$ are integral kernels of mutually inverse operators:

$$\int dx'\, K(x-x')K^{-1}(x'-y) = \int dx'\, K^{-1}(x-x')K(x'-y) = \delta(x-y). \tag{3.1.187}$$

Note that it is assumed that on the set of fields $\varphi(x)$ which are integrated over in (3.1.186) $K^{-1}(x-y)$ is uniquely defined. Recall that in (3.1.111) and (3.1.128) we have distinguished the causal Green function $D_c(x)$ as the inverse of the Klein–Gordon operator $(\Box + m^2)$ by imposing suitable asymptotic conditions for fields φ.

Show that the perturbative definition (3.1.186) of the path integral fulfils the main properties of any integral. In particular, prove that definition (3.1.186) is compatible with the rule of integration by parts:

$$\int \mathcal{D}\varphi(x)\left[\frac{\delta}{\delta\varphi(x')}\exp\left\{\frac{i}{2}\int d^4x\,d^4y\,\varphi(x)K(x-y)\varphi(y)\right\}\right]\exp\left\{i\int d^4x\,\varphi(x)J(x)\right\}$$

$$= \int \mathcal{D}\varphi(x)\exp\left\{\frac{i}{2}\int d^4x\,d^4y\,\varphi(x)K(x-y)\varphi(y)\right\}\frac{\delta}{\delta\varphi(x')}\exp\left\{i\int d^4x\,\varphi(x)J(x)\right\}. \tag{3.1.188}$$

Three subsequent problems concern other properties of the path integrals defined by (3.1.186).

Hint. The left-hand side of (3.1.188) can be easily calculated:

$$\int \mathcal{D}\varphi(x) \left[\frac{\delta}{\delta\varphi(x')} \exp\left\{\frac{i}{2}\int d^4x\, d^4y\, \varphi(x) K(x-y)\varphi(y)\right\}\right] \exp\left\{i\int d^4x\, \varphi(x) J(x)\right\}$$

$$= i\int \mathcal{D}\varphi(x) \left[\int d^4x''\, K(x'-x'')\varphi(x'')\right]$$

$$\times \exp\left\{\frac{i}{2}\int d^4x\, d^4y\, \varphi(x) K(x-y)\varphi(y) + i\int d^4x\, \varphi(x) J(x)\right\}$$

$$= i\int d^4x''\, K(x'-x'') \int \mathcal{D}\varphi(x)\, \varphi(x'')$$

$$\times \exp\left\{\frac{i}{2}\int d^4x\, d^4y\, \varphi(x) K(x-y)\varphi(y) + i\int d^4x\, \varphi(x) J(x)\right\}$$

$$= -iJ(x') \exp\left\{-\frac{i}{2}\int d^4x\, d^4y\, J(x) K^{-1}(x-y) J(y)\right\}. \qquad (3.1.189)$$

The last equality follows from definition (3.1.186). The right-hand side of (3.1.188) gives, clearly, the same result and this proves the required equality.

Problem 3.1.11. Show that a path integral over several fields in perturbation theory can be defined by formula (3.1.186) as an iterated integral; in other words, starting from (3.1.186), prove that

$$\int \mathcal{D}\varphi_1(x)\cdots\mathcal{D}\varphi_n(x) \exp\left\{\frac{i}{2}\sum_{i,j=1}^{n}\int d^4x\, d^4y\, \varphi_i(x) K_{ij}(x-y)\varphi_j(y) + i\sum_{i=1}^{n}\int d^4x\, \varphi_i(x) J_i(x)\right\}$$

$$= \exp\left\{-\frac{i}{2}\sum_{i,j=1}^{n}\int d^4x\, d^4y\, J_i(x) K_{ij}^{-1}(x-y) J_j(y)\right\}. \qquad (3.1.190)$$

Hint. Assume that (3.1.190) is correct for some integer n and directly show (using (3.1.186)) that it is then correct for $n+1$. This induction proves the statement.

Problem 3.1.12. Prove that definition (3.1.186) implies the relation

$$\int \mathcal{D}\varphi(x)\, F[\varphi]\left[\int \mathcal{D}\lambda(x)\, \det\left(\frac{\delta f(\varphi)}{\delta\varphi}\right) \exp\left\{i\int d^4x\, \lambda(x)(f(\varphi(x)) - g(x))\right\}\right] = F[\widetilde{\varphi}] \qquad (3.1.191)$$

where $F[\varphi]$ is an arbitrary functional, g is some new field variable and $\widetilde{\varphi}$ is a solution of the equation

$$f(\widetilde{\varphi}(x)) - g(x) = 0. \qquad (3.1.192)$$

This means that the integral on the left-hand side of (3.1.191) contains the δ-functional

$$\int \mathcal{D}\lambda(x)\, \exp\left\{i\int d^4x\, \lambda(x)(f(\varphi(x)) - g(x))\right\} = \delta(f(\varphi(x)) - g(x)). \qquad (3.1.193)$$

Hint. The function $f(\varphi(x))$ can be presented in the form

$$f(\varphi(x)) = c_0(x) + \varphi(x) + \widetilde{f}(\varphi(x)) \qquad (3.1.194)$$

where
$$\widetilde{f}(\varphi(x)) = \int d^4y\, c_1(x,y)\varphi(y) + \int d^4y\, d^4y'\, c_2(x,y,y')\varphi(y)\varphi(y') + \cdots.$$

For simplicity, we put the coefficient at $\varphi(x)$ in (3.1.194) equal to unity; the consideration is trivially generalized for an arbitrary value of the coefficient.

The determinant in the path integral can now be understood as the power expansion

$$\det\left(\frac{\delta f(\varphi)}{\delta \varphi}\right) = \det\left(1 + \frac{\delta \widetilde{f}}{\delta \varphi}\right) = \exp\left\{\operatorname{Tr}\ln\left(1 + \frac{\delta \widetilde{f}}{\delta \varphi}\right)\right\}$$
$$= \exp\left\{\int d^4x\, \left.\frac{\delta \widetilde{f}(x)}{\delta \varphi(y)}\right|_{x=y} + \frac{1}{2}\int d^4x\, d^4y\, \frac{\delta \widetilde{f}(x)}{\delta \varphi(y)}\frac{\delta \widetilde{f}(y)}{\delta \varphi(x)} + \cdots\right\}. \quad (3.1.195)$$

Here we have used the well-known formula for the determinant of a matrix $\mathbf{K}$:
$$\det \mathbf{K} = e^{\operatorname{Tr}\ln \mathbf{K}}.$$

Using definition (3.1.186) and integration by parts (cf problem 3.1.10, page 46) the left-hand side of (3.1.191) with the basic functional
$$F[\varphi] = \exp\left\{\frac{i}{2}\int d^4x\, d^4y\, \varphi(x)K(x-y)\varphi(y)\right\}$$

can be cast into the form
$$\int \mathcal{D}\lambda(x)\, \exp\left\{-\frac{i}{2}\int d^4x\, d^4y\, \lambda(x)K^{-1}(x-y)\lambda(y)\right\} B[g,\lambda] \quad (3.1.196)$$

where
$$B[g,\lambda] \stackrel{\text{def}}{\equiv} \det\left[1 + \frac{\delta \widetilde{f}}{\delta \varphi}\left(\frac{1}{i}\frac{\delta}{\delta \lambda}\right)\right] \overleftarrow{\exp}\left\{i\int dx\, \widetilde{f}\left(-\frac{1}{i}\frac{\delta}{\delta \lambda}\right)\lambda(x)\right\}$$
$$\times \exp\left\{-i\int dx\, [g(x) - c_0(x)]\lambda(x)\right\}. \quad (3.1.197)$$

Here the symbol $\overleftarrow{\exp}$ means that in the expansion of the exponential we should place all the operators $\delta/\delta\lambda$ to the left of $\lambda(x)$. It is readily seen that this functional satisfies the equation
$$\frac{\delta B}{\delta \lambda(x)} = i\left[c_0(x) - g(x) + \widetilde{f}\left(\frac{1}{i}\frac{\delta}{\delta \lambda}\right)\right]B \quad (3.1.198)$$

with the initial condition $B[g,0]$. It is natural to seek the solution of (3.1.198) in the form
$$B[g,\lambda] = B[g,0]\exp\left\{-i\int d^4x\, \lambda(x)\varphi(g)\right\}. \quad (3.1.199)$$

Substitution of (3.1.199) into (3.1.198) gives the condition
$$\varphi(x) = g(x) - c_0 - \widetilde{f}(\varphi). \quad (3.1.200)$$

Thus, integral (3.1.196) takes the form
$$B[g,0]\exp\left\{\frac{i}{2}\int d^4x\, d^4y\, \widetilde{\varphi}(x)K(x-y)\widetilde{\varphi}(y)\right\}$$

so that the required statement is proved if $B[g, 0] = 1$. To show the latter equality, we rewrite functional (3.1.197) as follows:

$$B[g, \lambda] = \overleftarrow{\exp}\left\{-i \int d^4x \frac{\delta}{\delta g(x)} \widetilde{f}(g - c_0)\right\}$$
$$\times \det\left(1 + \frac{\delta \widetilde{f}}{\delta \varphi}(g - c_0)\right) \exp\left\{-i \int d^4x \, \lambda(x)[g(x) - c_0(x)]\right\}$$

and therefore

$$B[g, 0] = \det\left[\overleftarrow{\exp}\left\{-i \int d^4x \frac{\delta}{\delta g(x)} \widetilde{f}(g - c_0)\right\}\left(1 + \frac{\delta \widetilde{f}}{\delta \varphi}(g - c_0)\right)\right] \cdot 1$$
$$= \det\left[\sum_n \frac{(-1)^n}{n!} \frac{\delta^n}{\delta g^n(x)} \widetilde{f}^n(g - c_0)\left(1 + \frac{\delta \widetilde{f}}{\delta g}(g - c_0)\right)\right] \cdot 1. \quad (3.1.201)$$

The last step of the proof is a verification that the second term of the nth binomial in the sum (3.1.201) cancels the first term of the $(n + 1)$th binomial.

Problem 3.1.13. With the help of the perturbative definition of path integral (3.1.186) prove the formula for changing the integration variables:

$$\varphi = f(\varphi') \qquad f(\varphi') = c_0(x) + \varphi'(x) + \widetilde{f}(\varphi')$$

($\widetilde{f}$ is defined as in (3.1.194); φ' here is a new field variable, not a derivative of the field φ). Namely, show the validity of the relation

$$\int \mathcal{D}\varphi(x) \exp\left\{\frac{i}{2} \int d^4x \, d^4y \, \varphi(x) K(x - y) \varphi(y) + i \int d^4x \, \varphi(x) J(x)\right\}$$
$$= \int \mathcal{D}\varphi'(x) \det\left(1 + \frac{\delta \widetilde{f}}{\delta \varphi'}\right)$$
$$\times \exp\left\{\frac{i}{2} \int d^4x \, d^4y \, f(\varphi(x)) K(x - y) f(\varphi(y)) + i \int d^4x \, f(\varphi(x)) J(x)\right\}. \quad (3.1.202)$$

Hint. A possible way to prove the required equality is to integrate both parts of (3.1.202) over $J(x)$ with the functional

$$\exp\left\{-i \int d^4x \, J(x) \sigma(x)\right\}$$

where $\sigma(x)$ is a new field from the same class of functions (this is the functional analog of the Fourier transform). Using the result of the preceding problem 3.1.12, it is easy to show that the results of such an integration of the left- and right-hand sides of (3.1.202) indeed coincide.

3.2 Path-integral quantization of gauge-field theories

So far, we have dealt only with *spacetime* (in particular, relativistic) transformations of physical systems. However, transformations which leave the spacetime coordinates unchanged but changing only the wavefunctions, $\Psi(x) \to \Psi'(x)$, and/or fields, $\varphi(x) \to \varphi'(x)$ exist. Such transformations, called *internal transformations*, are related to the internal properties of fields and elementary particles and are described by *internal symmetry groups*. If the group is given, the infinitesimal transformations of a collection of (for simplicity, scalar) fields φ_i forming a representation of the group read as

$$\varphi_i(x) \to \varphi'_i(x) = \varphi_i + \delta\varphi_i(x) \qquad (3.2.1)$$

$$\delta\varphi_i(x) = T^a_{ij}\varepsilon_a\varphi_j. \qquad (3.2.2)$$

Here T^a_{ij} ($a = 1, \ldots, N, i, j = 1, \ldots, r$) are N generators (more precisely, matrices of the representation of the generators), ε_a are the infinitesimal parameters of the group, N is the group dimension and r is the dimension of the representation (for basic notions on group theory, see supplement IV).

The condition $\delta S = 0$ of invariance of the action S of some field theoretical model under a group of internal transformations defined by generators T^a_{ij} has the form (see, e.g., Itzykson and Zuber (1980) and Chaichian and Nelipa (1984))

$$\frac{\partial\mathcal{L}}{\partial\varphi_i}T^a_{ij}\varphi_j + \frac{\partial\mathcal{L}}{\partial(\partial_\mu\varphi_i)}T^a_{ij}\partial_\mu\varphi_j = 0 \qquad a = 1, \ldots, N \qquad (3.2.3)$$

where $\mathcal{L}$ is the Lagrangian density of the system. These identities express the necessary and sufficient conditions for the Lagrangian and action to be invariant under the transformations of an arbitrary *global* group $\mathfrak{G}$ of internal symmetry.

A *globally* invariant Lagrangian can be non-invariant under a certain generalization of the notion of *Lie groups*, called *group of local transformations* or *gauge group*. To obtain a locally invariant Lagrangian, new fields have to be introduced. These are called *gauge fields*. In modern elementary particle theory, in high-energy (small-distance) physics, as well as in condensed matter physics, gauge invariance is the basic guiding principle for theoretical model building.

After a very short introduction to the structure of gauge-invariant Lagrangians and the geometry of gauge fields, we shall describe their quantization via path integrals which proves to be very convenient for this purpose since the standard operator canonical quantization meets serious combinatorial technical difficulties and is, in general, very cumbersome (in particular, it is rather difficult to control relativistic invariance at each step of the quantization in this case). Without exaggeration, we can say that the 'second birth' of path integrals in quantum mechanics and their recognition as a very powerful method for quantization of systems with complicated symmetries started in the 1970s with the construction of realistic gauge models.

As we shall discuss, gauge invariance leads to constraints in the theory. As a preliminary step, we shall describe the path-integral quantization of quantum-mechanical systems with constraints in the case of a finite number of degrees of freedom. Then we generalize this consideration to *quantum gauge-field theory*, i.e. to a quantized field theory invariant with respect to a gauge (local) group.

3.2.1 Gauge-invariant Lagrangians

In the case of a group of global transformations the parameters ε_a in (3.2.2) are independent of the coordinates. Suppose now that the parameters of the group are coordinate dependent. The functions of the field then transform according to

$$\delta\varphi_i(x) = T^a_{ij}\varepsilon_a(x)\varphi_j(x). \qquad (3.2.4)$$

The group of such transformations is called the *local* or *gauge group*. Even if a Lagrangian satisfies the condition of global invariance (3.2.3), it is not invariant under the local transformations (3.2.4), the variation being proportional to the derivatives of the parameters:

$$\delta\mathcal{L} = \frac{\partial\mathcal{L}}{\partial(\partial_\mu\varphi_i)}T^a_{ij}\varphi_j\partial_\mu\varepsilon_a \neq 0. \qquad (3.2.5)$$

◇ **Gauge fields and gauge-field tensors**

To achieve the invariance of the Lagrangian under the transformations (3.2.4), new vector fields A_μ^a must be introduced in addition to the initial fields φ_i, to compensate the right-hand side of (3.2.5). This results in a new Lagrangian, invariant under the transformations (3.2.4). The fields A_μ^a thus introduced are called *gauge fields*.

To construct a locally invariant Lagrangian from a globally invariant Lagrangian $\mathcal{L}_m(\varphi_i, \partial_\mu \varphi_i)$ we should substitute in the latter all the partial derivatives $\partial_\mu \varphi_i$ with the so-called *covariant derivatives* $D_\mu \varphi_i$:

$$\mathcal{L}_m(\varphi_i, \partial_\mu \varphi_i) \longrightarrow \mathcal{L}_m(\varphi_i, D_\mu \varphi_i) \tag{3.2.6}$$

where

$$D_\mu \varphi_i \stackrel{\text{def}}{\equiv} \partial_\mu - T_{ij}^a A_\mu^a \varphi_j. \tag{3.2.7}$$

We have added here the subscript 'm' to stress that this is a *matter-field Lagrangian*. Now if we postulate that under the local infinitesimal gauge transformations the gauge fields A_μ^a acquire the addition

$$\delta A_\mu^a = f_{ba}^c A_\mu^b \varepsilon_c(x) + \partial_\mu \varepsilon_a(x) \tag{3.2.8}$$

(f_{ba}^c are the structure constants of the global Lie group under consideration; we shall assume in this section that it is a *semisimple* Lie group; cf supplement IV), the Lagrangian on the right-hand side of (3.2.6) proves to be *gauge invariant*. The point is that (3.2.8) provides that

$$\delta(D_\mu \varphi_i) = T_{ij}^a \varepsilon_a(x) D_\mu \varphi_j. \tag{3.2.9}$$

The latter equality justifies the name *covariant* derivative ($D_\mu \varphi_i$ transforms under the gauge group in the same way as φ_i, i.e. $D_\mu \varphi_i$ is a covariant quantity). Now the matter Lagrangian $\mathcal{L}_m(\varphi_i, D_\mu \varphi_i)$ is not a free one since the covariant derivative introduces an interaction with the gauge fields A_μ^a. If we want to consider the latter as dynamical fields, we should add the kinematical part for them.

The gauge-invariant *Lagrangian for the gauge fields* is made of a specific combination $F_{\mu\nu}^a$, called the *gauge-field tensor*:

$$F_{\mu\nu}^a \stackrel{\text{def}}{\equiv} \partial_\mu A_\nu^a - \partial_\nu A_\mu^a - \tfrac{1}{2} f_{bc}^a (A_\mu^b A_\nu^c - A_\nu^b A_\mu^c). \tag{3.2.10}$$

This tensor is obviously antisymmetric in the indices μ and ν. In contrast to the field A_μ^a, the combination $F_{\mu\nu}^a$ is transformed *homogeneously* under the gauge group:

$$\delta F_{\mu\nu}^a = f_{ba}^c \varepsilon_c(x) F_{\mu\nu}^b. \tag{3.2.11}$$

There are many possibilities to construct a gauge-invariant quantity from $F_{\mu\nu}^a$. The additional condition of renormalizability of the complete gauge-field theory (see later) distinguishes the only appropriate Lagrangian. This is a Lagrangian quadratic in $F_{\mu\nu}^a$, which was proposed in the pioneering work by Yang and Mills (1954):

$$\mathcal{L}_{\text{YM}} = -\frac{1}{4g^2} F_{\mu\nu}^a F^{a\mu\nu}. \tag{3.2.12}$$

Recall that we adopt here and in what follows the standard convention for summation over group and relativistic indices:

$$F_{\mu\nu}^a F^{a\mu\nu} \stackrel{\text{def}}{\equiv} \sum_{a=1}^{N} \sum_{\mu,\nu,\rho,\sigma=0}^{3} F_{\mu\nu}^a g^{\mu\rho} g^{\nu\sigma} F_{\rho\sigma}^a.$$

Note that it is customary to name the gauge fields and the corresponding field theory *Yang–Mills fields* and *Yang–Mills theory*, respectively. From (3.2.10) and (3.2.12) it is seen that for non-zero structure constants f^a_{bc} (non-Abelian group) the Lagrangian $\mathcal{L}_{YM}$ contains the interaction (non-quadratic) terms for the gauge fields A^a_μ. Hence, even a pure Yang–Mills Lagrangian (without matter fields) for non-Abelian Lie groups corresponds to a non-trivial (non-free) field system. The constant g is the coupling constant for the gauge fields (it may occupy a more familiar place as a factor in higher-than-quadratic terms after the field rescaling: $A^a_\mu \to g A^a_\mu$).

The *full Lagrangian* $\mathcal{L}$ of the system of the matter fields φ_i and gauge fields A^a_μ is given by the sum of the Lagrangian $\mathcal{L}_{YM}(A^a_\mu)$ and $\mathcal{L}_m(\varphi_i, D_\mu \varphi_i)$ (the latter contains the Lagrangian of the matter fields as well as the interaction Lagrangian between the matter and gauge fields):

$$\mathcal{L}(A^a_\mu, \varphi_i) = \mathcal{L}_{YM}(A^a_\mu) + \mathcal{L}_m(\varphi_i, D_\mu \varphi_i). \tag{3.2.13}$$

◇ **Simplest examples of Yang–Mills theories**

Example 3.3 (Yang–Mills theory with Abelian group $U(1)$: quantum electrodynamics). Let us start from the free Lagrangian for a single spinor field $\psi(x)$ with mass m:

$$\mathcal{L}_{0,m}(\psi) = \frac{i}{2}(\bar\psi \gamma^\mu \partial_\mu \psi - \partial_\mu \bar\psi \gamma^\mu \psi) - m\bar\psi \psi. \tag{3.2.14}$$

This Lagrangian is invariant under the global one-parameter Abelian group $U(1)$ of phase transformations

$$\psi \to \psi' = e^{-ig\varepsilon}\psi \qquad \bar\psi \to \bar\psi' = e^{ig\varepsilon}\bar\psi \tag{3.2.15}$$

where ε is the (constant) parameter of the group and g is the coupling constant (as will be seen later). The infinitesimal version of (3.2.15) reads

$$\delta\psi = -i\varepsilon g \psi \qquad \delta\bar\psi = i\varepsilon g \bar\psi. \tag{3.2.16}$$

By comparing (3.2.16) with (3.2.4), we find that the matrix of the transformation generators is diagonal:

$$T_{11} = -ig \qquad T_{22} = ig \qquad T_{12} = T_{21} = 0. \tag{3.2.17}$$

Here the indices 1 and 2 refer to ψ and $\bar\psi$, respectively. Of course, the structure constants of this group vanish (as for any one-parameter group).

Consider the corresponding *local gauge* group with the coordinate-dependent parameter $\varepsilon(x)$. As we have discussed previously, the gauge-invariant Lagrangian for the matter fields is constructed via substitution of ordinary derivative by the covariant derivative:

$$\partial_\mu \psi \longrightarrow D_\mu \psi = \partial_\mu \psi + ig A_\mu \psi$$
$$\partial_\mu \bar\psi \longrightarrow D_\mu \bar\psi = \partial_\mu \bar\psi - ig A_\mu \bar\psi.$$

The Lagrangian part, $\mathcal{L}_e$, for the gauge field A_μ is expressed, according to (3.2.10) and (3.2.12), as follows:

$$\mathcal{L}_e = -\tfrac{1}{4} F_{\mu\nu} F^{\mu\nu} \tag{3.2.18}$$

where

$$F_{\mu\nu} = \partial_\mu A_\nu - \partial_\nu A_\mu. \tag{3.2.19}$$

Of course, the Lagrangian for the pure Yang–Mills field in the case of an Abelian group does not contain interaction terms. An infinitesimal transformation of the field A_μ is defined by (3.2.8):

$$\delta A_\mu = \partial_\mu \varepsilon(x). \tag{3.2.20}$$

Thus, the complete gauge-invariant Lagrangian takes the form

$$\mathcal{L}(A_\mu, \psi) = \mathcal{L}_e(A_\mu) + \mathcal{L}_m(\psi, \bar\psi, D_\mu\psi, D_\mu\bar\psi)$$
$$= -\frac{1}{4}F_{\mu\nu}F^{\mu\nu} + \frac{i}{2}(\bar\psi\gamma^\mu\partial_\mu\psi - \partial_\mu\bar\psi\gamma^\mu\psi) - m\bar\psi\psi - g\bar\psi\gamma^\mu\psi A_\mu \quad (3.2.21)$$

and coincides with the Lagrangian of *quantum electrodynamics* (QED) if we put $g = e$, where e is the electrical charge of an electron.

Example 3.4 (Yang–Mills theory for the non-Abelian SU(2) group). Let us consider the fundamental $SU(2)$ representation

$$\psi_i = \begin{pmatrix} \psi_1 \\ \psi_2 \end{pmatrix} \quad (3.2.22)$$

where ψ_1 and ψ_2 is a doublet of spinor fields. The free Lagrangian for them is given by

$$\mathcal{L}_{0,m}(\psi) = \frac{i}{2}(\bar\psi_i\gamma^\mu\partial_\mu\psi_i - \partial_\mu\bar\psi_i\gamma^\mu\psi_i) - m\bar\psi_i\psi_i. \quad (3.2.23)$$

This is invariant under the global non-Abelian group of $SU(2)$ transformations

$$\psi_i \longrightarrow \psi_i' = \left[\exp\left\{-\frac{i}{2}g\varepsilon_a\sigma_a\right\}\right]_{ij}\psi_j \quad (3.2.24)$$

$$\bar\psi_i \longrightarrow \bar\psi_i' = \bar\psi_j\left[\exp\left\{\frac{i}{2}g\varepsilon_a\sigma_a\right\}\right]_{ji} \quad (3.2.25)$$

where ε_a, $a = 1, 2, 3$, denote the (constant) parameters of the group and σ_a denote the Pauli matrices

$$\sigma_1 = \begin{pmatrix} 0 & 1 \\ 1 & 0 \end{pmatrix} \quad \sigma_2 = \begin{pmatrix} 0 & -i \\ i & 0 \end{pmatrix} \quad \sigma_3 = \begin{pmatrix} 1 & 0 \\ 0 & -1 \end{pmatrix}. \quad (3.2.26)$$

The matrices

$$T^a_{ij} = -\frac{ig}{2}(\sigma_a)_{ij} \quad (3.2.27)$$

are the generators of the group in the doublet (fundamental) representation and satisfy the commutation relations:

$$[T^a, T^b] = g\epsilon_{abc}T_c \quad (3.2.28)$$

with the group structure constants having the form $f^a_{bc} = g\epsilon_{abc}$. Here ϵ_{abc} is the totally antisymmetric tensor with $\epsilon_{123} = 1$.

Let us turn to the group of local gauge transformations. The Lagrangian (3.2.23) becomes invariant, provided that the substitution

$$\partial_\mu\psi_i \longrightarrow D_\mu\psi_i = \partial_\mu\psi_i + \frac{ig}{2}(\sigma_a)_{ij}\psi_j A^a_\mu \quad (3.2.29)$$

has been made, according to (3.2.6). As can be seen, in this case we have a triplet of vector gauge fields A^a_μ.

The Lagrangian for the gauge fields has, according to (3.2.12) and (3.2.10), the form

$$\mathcal{L}_{\text{YM}} = -\frac{1}{4g^2}F^a_{\mu\nu}F^{a\mu\nu} \quad (3.2.30)$$

where

$$F^a_{\mu\nu} \stackrel{\text{def}}{\equiv} \partial_\mu A^a_\nu - \partial_\nu A^a_\mu - \tfrac{1}{2}\epsilon_{abc}(A^b_\mu A^c_\nu - A^b_\nu A^c_\mu) \qquad (3.2.31)$$

is the tensor of the Yang–Mills fields. Equation (3.2.31) contains, besides the quadratic terms, cubic and quartic terms in the fields A^a_μ, i.e. the *self-interaction of the Yang–Mills fields*.

For the complete locally invariant Lagrangian we obtain

$$\begin{aligned}\mathcal{L}(A^a_\mu, \psi) &= \mathcal{L}_{\text{YM}}(A^a_\mu) + \mathcal{L}_m(\psi_i, \bar{\psi}_i, D_\mu\psi_i, D_\mu\bar{\psi}_i) \\ &= -\frac{1}{4g^2} F^a_{\mu\nu} F^{a\,\mu\nu} + \frac{i}{2}(\bar{\psi}_i \gamma^\mu \partial_\mu \psi_i - \partial_\mu \bar{\psi}_i \gamma^\mu \psi_i) - m\bar{\psi}_i\psi_i - \frac{1}{2}\bar{\psi}_i \gamma^\mu (\sigma_a)_{ij} \psi_j A^a_\mu.\end{aligned} \qquad (3.2.32)$$

The constant g plays the role of a coupling constant for the gauge field interacting with the spinor field and with itself. This is clearly seen after the field rescaling $A^a_\mu \to g A^a_\mu$, which casts the Lagrangian into the form:

$$\mathcal{L}(A^a_\mu, \psi) = -\frac{1}{4} F^a_{\mu\nu} F^{a\,\mu\nu} + \frac{i}{2}(\bar{\psi}_i \gamma^\mu \partial_\mu \psi_i - \partial_\mu \bar{\psi}_i \gamma^\mu \psi_i) - m\bar{\psi}_i\psi_i - \frac{g}{2}\bar{\psi}_i \gamma^\mu (\sigma_a)_{ij} \psi_j A^a_\mu \qquad (3.2.33)$$

with

$$F^a_{\mu\nu} = \partial_\mu A^a_\nu - \partial_\nu A^a_\mu - \frac{g}{2}\epsilon_{abc}(A^b_\mu A^c_\nu - A^b_\nu A^c_\mu).$$

─── ○ ───

◇ The Gauss law in electrodynamics as an example of constraints in Yang–Mills theories

The Lagrangian (3.2.18) for the purely electromagnetic field can be rewritten equivalently as follows

$$\mathcal{L}_e = -\tfrac{1}{2}[E_k \partial_0 A_k - \tfrac{1}{2}(E_k^2 + B_k^2) + A_0(\partial_k E_k)] \qquad (3.2.34)$$

where $E_k = F_{k0}$ is the electric field and $B_k = \tfrac{1}{2}\epsilon_{ijk} F_{ij}$ is the magnetic field (here $i, j, k = 1, 2, 3$ are the space part of the spacetime indices μ, ν of the tensor $F_{\mu\nu}$). It is seen that the time component of the electromagnetic field A_0 plays the role of a Lagrange multiplier. Variation of the corresponding action with respect to the latter gives the constraint

$$\partial_k E_k = 0 \qquad (3.2.35)$$

which expresses the Gauss law for a pure electromagnetic field (i.e. in the absence of any charged particles). This is a *constraint equation* but not an equation of motion because it does not contain any time derivatives and hence does not define a time evolution.

The existence of constraints is a general feature of any Yang–Mills gauge theory. We shall discuss this fact in somewhat more detail a bit later, but before, we shall consider systems with constraints and their path-integral quantization in the case of a finite number of degrees of freedom.

3.2.2 Constrained Hamiltonian systems and their path-integral quantization

Constraints are well known from classical mechanics where they are usually realized by surfaces which restrict the motion of some particles. In the simple case of *holonomic constraints*, i.e. constraints of the type

$$\phi_a(q_i) = 0. \qquad (3.2.36)$$

they can easily be used to reduce the number of coordinates to the number of *physical* degrees of freedom. Then the remaining coordinates are independent of each other and in the subsequent Lagrangian or Hamiltonian formulation, we no longer have to consider the constraints. In field theory, the concept of constraints proves to be more complicated. Non-physical degrees of freedom are introduced and kept in order to obtain a manifestly Lorentz and gauge-invariant formulation of a theory. The presence of these non-physical degrees of freedom leads to constraints. In most cases, these are not holonomic constraints of the type (3.2.36), but they have the more general form

$$\phi_a(q_i, p_i) = 0. \tag{3.2.37}$$

Even in this case it is possible to eliminate the constraints by reducing the number of variables to the number of physical degrees of freedom. However, in general, such a procedure is not desired because it leads to the loss of a manifestly Lorentz or gauge-invariant formulation of the theory. Instead, we consider the constraints within *generalized Hamiltonian dynamics* which was studied for the first time by Dirac (1950, 1958) (see also Dirac (1964), Faddeev and Slavnov (1980), Gitman and Tyutin (1990) and references therein).

◇ Constrained systems with a finite number of degrees of freedom

Consider a physical system given by the Lagrangian L which is a function of the coordinates q_i ($i = 1, \ldots, d$) and their first time derivatives. This Lagrangian is said to be *singular* if

$$\det\left(\frac{\partial^2 L}{\partial \dot{q}_i \partial \dot{q}_j}\right) = 0. \tag{3.2.38}$$

In this case, not all of the equations that define the momenta

$$p_i = \frac{\partial L}{\partial \dot{q}_i} \tag{3.2.39}$$

can be solved for the velocities $\dot{q}_i$. Instead, some of these relations yield the *primary constraints*

$$\phi_a^{(1)}(q_i, p_i) = 0. \tag{3.2.40}$$

Now we can construct the Hamiltonian

$$H = \dot{q}_i p_i - L \tag{3.2.41}$$

where the $\dot{q}_i$ have to be expressed in terms of the q_i and p_i by applying (3.2.39). Although (3.2.39) cannot be solved for all $\dot{q}_i$, we can show (see, e.g., Gitman and Tyutin (1990)) that, due to the presence of constraints (3.2.40), all the $\dot{q}_i$ can be eliminated from H, i.e. H only depends on q_i and p_i.

As in the unconstrained case, the Hamiltonian equations of motion follow from the least action principle

$$\delta \int [\dot{q}_i p_i - H] \, dt = 0 \tag{3.2.42}$$

however, the variations δq_i and δp_i are not independent of each other but they are restricted by the constraints (3.2.40). This case can be treated by the method of Lagrange multipliers, which yields the equations of motions for any observable, i.e. for a function $f(q_i, p_i)$ of q_i and p_i, of the form

$$\dot{f} = \{H^{(1)}, f\}|_{\phi_a^{(1)}=0} \tag{3.2.43}$$

with
$$H^{(1)} \equiv H + \lambda_a \phi_a^{(1)} \tag{3.2.44}$$

where the λ_a are the (*a priori* undetermined) Lagrange multipliers.

The primary constraints have to be consistent with the equations of motions, i.e. the time derivative of (3.2.40) also has to vanish:

$$\dot{\phi}_a^{(1)} = \{H^{(1)}, \phi_a^{(1)}\}|_{\phi_a^{(1)}=0} = \{H, \phi_a^{(1)}\}|_{\phi_a^{(1)}=0} + \lambda_b \{\phi_b^{(1)}, \phi_a^{(1)}\}|_{\phi_a^{(1)}=0} = 0. \tag{3.2.45}$$

If these relations are not fulfilled automatically, they can be written in the form

$$\phi_a^{(2)}(q_i, p_i) = 0. \tag{3.2.46}$$

In this case, constraints (3.2.46) also have to be satisfied in order to ensure consistency with the equations of motions and these are called *secondary constraints*.

This procedure has to be iterated, i.e. the demand that the time derivatives of the secondary constraints have to vanish may imply further constraints until finally a set of constraints which is consistent with the equations of motions is obtained. Although the various constraints are obtained at different stages of the procedure, there is no essential difference between them. In fact, they can be treated on the same level: we obtain an equivalent physical formulation of a constrained theory if we rewrite (3.2.43) and (3.2.44) as

$$\dot{f} = \{H_\text{T}, f\}|_{\phi_a=0} \tag{3.2.47}$$

with the total Hamiltonian

$$H_\text{T} \equiv H + \lambda_a \phi_a \tag{3.2.48}$$

where ϕ_a denotes *all* the constraints (Gitman and Tyutin 1990).

Another classification of the constraints is related to a determination of the Lagrange multipliers in (3.2.48). If the matrix

$$\{\phi_a, \phi_b\}|_{\phi_a=0} \tag{3.2.49}$$

is non-singular, the constraints are called *second class*. In this case, relations of the type (3.2.45), i.e.

$$\dot{\phi}_a = \{H_\text{T}, \phi_a\}|_{\phi_a=0} = \{H, \phi_a\}|_{\phi_a=0} + \lambda_b \{\phi_b, \phi_a\}|_{\phi_a=0} = 0 \tag{3.2.50}$$

can be solved for the λ_a:

$$\lambda_a = -\{\phi_a, \phi_b\}^{-1}\{H, \phi_b\}|_{\phi_a=0}. \tag{3.2.51}$$

Inserting this into (3.2.47) with (3.2.48), the equations of motion can be written in the simple form

$$\dot{f} = \{H, f\}_\text{DB} \tag{3.2.52}$$

where the *Dirac bracket* $\{\cdot, \cdot\}_\text{DB}$ is defined as (Dirac 1964)

$$\{f, g\}_\text{DB} \equiv \{f, g\}|_{\phi_a=0} - \{f, \phi_a\}\{\phi_a, \phi_b\}^{-1}\{\phi_b, g\}|_{\phi_a=0}. \tag{3.2.53}$$

Substitution of the Poisson brackets by the Dirac brackets allows us to formulate the dynamics of a second-class constrained system analogously to the dynamics of an unconstrained system.

A different situation arises if the matrix (3.2.49) is singular. Assuming that (3.2.49) has the rank r, we can order the constraints such that the upper left $r \times r$ submatrix of (3.2.49) has a non-vanishing determinant. Then only the first r constraints are second class and the remaining ones are *first-class* constraints. The Lagrange multipliers corresponding to the first-class constraints cannot be determined from (3.2.50). Thus, the equations of motion (3.2.47) with (3.2.48) contain undetermined Lagrange multipliers and therefore their solution (for given initial conditions) is not unique: two solutions f and f' of the equations of motion (3.2.47) with (3.2.48), with the same initial condition at $t = 0$ (but with distinct choices of the Lagrange multipliers corresponding to the first-class constraints), differ after an infinitesimal time interval dt by

$$\Delta f(dt) = dt(\lambda_a - \lambda'_a)\{\phi^a_{\text{f.c.}}, f\}. \qquad (3.2.54)$$

Here $\phi^a_{\text{f.c.}}$ denotes the first-class constraints. A transformation of the canonical variables, which relates different solutions of the equations of motion, is called a *gauge transformation*. This is the genuine and general definition of the gauge transformations, examples of which we encountered in the preceding section in the case of field theoretical models. Equation (3.2.54) shows that the first-class constraints are the generators of the (infinitesimal) gauge transformations. All solutions of the equations of motion with the same initial conditions describe the same physical process; in other words, all points in the phase space, which are related by gauge transformations, describe the same physical state of the system. Thus, a first-class constrained theory has a gauge freedom and is called *degenerate*.

The choice of a unique solution for the equations of motion for given initial conditions in a degenerate theory is achieved by imposing a *gauge* on the original theory, i.e. by introducing additional *gauge-fixing conditions*

$$\chi_a(q_i, p_i) = 0. \qquad (3.2.55)$$

The number of gauge-fixing conditions is equal to the number of first-class constraints and together with the latter, the gauge-fixing conditions must form a set of second-class constraints, that is

$$\det|\{\phi_a, \chi_b\}_P| \neq 0. \qquad (3.2.56)$$

Now the relations analogous to (3.2.50), but including both ϕ_a and χ_a, determine the Lagrange multipliers corresponding to the first-class constraints and the ambiguity in the solution of the equations of motions is removed.

The standard examples of a first- and a second-class constrained system are the massless and the massive vector fields, respectively. It turns out that a massive vector field is subject to two second-class constraints. This means that among the four field components in the four-dimensional spacetime and the four conjugate momenta, there are only six independent degrees of freedom (three fields and three generalized momenta). In the Hamiltonian treatment of a massless vector field, two first-class constraints arise and therefore two gauge-fixing conditions have to be introduced. Thus, there are only two physical field components and two physical momenta. We shall consider this system in the next subsection.

◇ **Transition to physical variables for systems with first-class constraints and for a specific choice of gauge conditions**

Let us consider in more detail the case of a constrained system with only first-class constraints ϕ_a ($a = 1, \ldots, r$). This case is important because many gauge-field models belong to this type. The first-class constraints satisfy the *condition of involution*:

$$\{\phi_a, \phi_b\} = \sum_d c_{abd}(p, q)\phi_d \qquad (3.2.57)$$

where the coefficients c_{abd} may depend, in general, on p_i, q_i. This involution condition follows from the fact that if all the constraints are first class, there are no submatrices of $\{\phi_a, \phi_b\}$ with non-zero determinant (on the surface defined by the constraints $\phi_a = 0$ ($a = 1, \ldots, r$)). We assume that the ϕ_a ($a = 1, \ldots, r$)

include all the constraint generations (i.e. primary, secondary, etc). Thus, they satisfy the consistency condition (cf (3.2.45))
$$\{H, \phi_a\} = c_{ab}(p, q)\phi_b \tag{3.2.58}$$
with some coefficients $c_{ab}(p, q)$. In order to separate out the physical variables, we have to impose r gauge conditions
$$\chi_a(p, q) = 0 \tag{3.2.59}$$
satisfying (3.2.56). Let us choose them in such a way that their mutual Poisson brackets vanish:
$$\{\chi_a, \chi_b\} = 0. \tag{3.2.60}$$
Then, the *physical* canonical variables $\widetilde{q}_i, \widetilde{p}_i$ parametrize the surface (subspace)
$$\chi_a(p, q) = 0 \qquad \phi_a(p, q) = 0 \qquad a = 1, \ldots, r \tag{3.2.61}$$
and the complete set of canonical (both physical and non-physical) variables can be chosen as follows:
$$q = (\chi_a, \widetilde{q}_i) \qquad p = (p_a, \widetilde{p}_i) \qquad a = 1, \ldots, r;\ i = 1, \ldots, n-r \tag{3.2.62}$$
where p_a are the momenta canonically conjugated to χ_a. Condition (3.2.56) now takes the form
$$\det\left|\frac{\partial \phi_a}{\partial p_b}\right| \neq 0 \tag{3.2.63}$$
so that the constraints $\phi_a = 0$ allow us to express the variables p_a ($a = 1, \ldots, r$) in terms of other variables. Thus the physical subspace $\Gamma^{2(n-r)}$ of the total canonical phase space $\mathbb{R}^{2n}$ is defined by the equations
$$q_a \equiv \chi_a = 0 \qquad p_a = p_a(\widetilde{p}_i, \widetilde{q}_i) \qquad a = 1, \ldots, r;\ i = 1, \ldots, n-r \tag{3.2.64}$$
and the variables $\widetilde{p}_i, \widetilde{q}_i$ are the physical canonical variables. The physical Hamiltonian H_{ph} reads
$$H_{\text{ph}}(\widetilde{p}_i, \widetilde{q}_i) = H(p, q)|_{\phi=\chi=0}. \tag{3.2.65}$$
The initially constrained system with Hamiltonian H and the reduced system with H_{ph} are totally equivalent (problem 3.2.1, page 98). Different choices of gauge conditions lead only to canonical transformations in the physical subspace $\Gamma^{2(n-r)}$ and do not affect the physical results.

◇ The Hamiltonian path integral for constrained systems

Now we are ready to quantize a constrained system with the help of path integrals. As just explained, a constrained system is equivalent to an unconstrained one with a reduced phase space consisting of the variables $\widetilde{q}_i$ and $\widetilde{p}_i$. The Hamiltonian path integral for this unconstrained system has the well-known simple form
$$\int \mathcal{D}\widetilde{q}_i\, \mathcal{D}\widetilde{p}_i\, \exp\left\{i \int dt\, [\dot{\widetilde{q}}_i \widetilde{p}_i - H_{\text{ph}}]\right\}. \tag{3.2.66}$$
However, as we have already mentioned, we do not normally use this choice of unconstrained parameters because in the primordial constrained parametrization it is easier to find a manifestly Lorentz or gauge-invariant formulation of the theory. Thus, it is practically important to rewrite path integral (3.2.66) in

terms of the original variables p_A, q_A ($A = 1, \ldots, n$). It is not difficult to verify that the integral

$$\int \prod_A \mathcal{D}p_A \mathcal{D}q_A \prod_a \mathcal{D}\lambda_a \prod_{\tau,a} \delta(\chi_a) \prod_\tau \det |\{\phi_a, \chi_b\}|$$

$$\times \exp\left\{i \int d\tau [p_A \dot{q}_A - H(p,q) - \lambda_a \phi_a(p,q)]\right\} \qquad (3.2.67)$$

$$A = 1, \ldots, n; \ a = 1, \ldots, r$$

fits the requirement. Indeed, integrating over the Lagrange multipliers λ_a, we obtain, from (3.2.67),

$$K(q, t; q_0, t_0) = \int \prod_A \mathcal{D}p_A \mathcal{D}q_A \left[\prod_{\tau,a} \delta(\chi_a)\delta(\phi_a)\right] \prod_\tau \det |\{\phi_a, \chi_b\}|$$

$$\times \exp\left\{i \int d\tau [p_A \dot{q}_A - H(p,q)]\right\}. \qquad (3.2.68)$$

In terms of the variables $\widetilde{p}_i, \widetilde{q}_i, p_a, q_a$ (cf (3.2.64)) the pre-exponential factor in the integrand can be cast into the form

$$\prod_{\tau,a} \delta(\chi_a)\delta(\phi_a) \prod_\tau \det |\{\phi_a, \chi_b\}| = \prod_{\tau,a} \delta(\chi_a)\delta(\phi_a) \prod_\tau \det \left|\frac{\partial \phi_a}{\partial p_b}\right|$$

$$= \prod_{\tau,a} \delta(q_a)\delta(p_a - p_a(\widetilde{p}, \widetilde{q})). \qquad (3.2.69)$$

Thus, after an integration over p_a, q_a, the path integral (3.2.67) is reduced to (3.2.66).

This is the general form of the Hamiltonian path integral for a first-class constrained system. This result can be extended to a system with both first- and second-class constraints. This generalization is not difficult because *after* introducing the gauge-fixing conditions, even a first-class constrained system formally becomes a second-class one. If the set of constraints $\{\phi_a\}$ consists of first-class constraints ϕ_{1st}^a and second-class constraints ϕ_{2nd}^a, we find (Gitman and Tyutin 1990) that the evolution operator or generating functional for an arbitrary system with constraints is given by the path integral

$$K(q, t; q_0, t_0) = \int \prod_A \mathcal{D}q_A \mathcal{D}p_A \left[\prod_{\tau,a} \delta(\phi_{2nd}^a)\delta(\phi_{1st}^a)\delta(\chi^a)\right]$$

$$\times \left[\prod_\tau \det^{\frac{1}{2}}(\{\phi_{2nd}^a, \phi_{2nd}^b\}) \det(\{\phi_{1st}^a, \chi^b\})\right] \exp\left\{i \int dt \, [\dot{q}_A p_A - H_T]\right\} \qquad (3.2.70)$$

(H_T is the total Hamiltonian, cf (3.2.48)). This is the Hamiltonian path integral for an arbitrary constrained system. It has the following properties:

- it is invariant under canonical transformations;
- it is invariant with respect to the choice of an equivalent set of constraints and
- it is independent of the choice of the gauge-fixing conditions.

We shall not discuss systems with second-class constraints and therefore we drop a detailed derivation of (3.2.70). The consideration in the subsequent subsections will be based on a field theoretical generalization of path integral (3.2.68).

3.2.3 Yang–Mills fields: constrained systems with an infinite number of degrees of freedom

Let us now turn to constrained systems with an infinite number of degrees of freedom. The examples we have in mind are the gauge fields: the Abelian gauge theory corresponding to the electromagnetic field and the non-Abelian Yang–Mills fields. The transition amplitudes for the quantum versions of these theories can be obtained as for unconstrained field systems: we start from a constrained system with a finite number of degrees of freedom considered in the preceding subsection and then generalize the result to the corresponding field models (with an infinite number of degrees of freedom). But first of all, we should clarify the structure of constraints in the gauge-field models.

◇ **Electrodynamics as a system with constraints**

We have already mentioned that electrodynamics is an example of a field system with constraints (cf (3.2.34) and (3.2.35)). Here, we shall consider the Hamiltonian structure of this important theory in more detail with the further aim of quantizing it and constructing transition amplitudes in terms of path integrals.

As initial variables we can choose the fields $A_\mu(x)$ with the Lagrangian

$$\mathcal{L}_e = -\tfrac{1}{4}(\partial_\mu A_\nu - \partial_\nu A_\mu)^2 \qquad (3.2.71)$$

or the fields $A_\mu(x)$ *together with* the tensor $F_{\mu\nu}(x)$ considered as independent variables and with the Lagrangian

$$\mathcal{L}_e = -\tfrac{1}{2}(\partial_\mu A_\nu - \partial_\nu A_\mu - \tfrac{1}{2}F_{\mu\nu})F^{\mu\nu}. \qquad (3.2.72)$$

These alternatives are said to be the *second-* and *first-order formalisms* in the case of Lagrangians (3.2.71) and (3.2.72), respectively. Of course, they both lead to the same physical result: the variation with respect to $F^{\mu\nu}$ in (3.2.72) gives the constraint which is nothing other than expression (3.2.19) for the electromagnetic tensor and its substitution into (3.2.72) gives Lagrangian (3.2.71). We shall use the first-order formalism, i.e. Lagrangian (3.2.72).

Let us rewrite (3.2.72) in three-dimensional notation, i.e. separate the four-dimensional indices μ, ν into time- ($\mu, \nu = 0$) and spacelike ($\mu, \nu = k, l = 1, 2, 3$) parts. Then, omitting the total divergence, (3.2.72) can be presented in the form

$$\mathcal{L}_e = E_k \dot{A}_k - H(E_k, A_k) + A_0 C(E_k) \qquad (3.2.73)$$

where

$$\dot{A}_k = \partial_0 A_k \qquad E_k = F_{k0} \qquad B_k = \tfrac{1}{2}\epsilon_{ijk}F_{ji} \qquad (3.2.74)$$

$$H(\boldsymbol{E}, \boldsymbol{A}) = \tfrac{1}{2}(E_k^2 + B_k^2) \qquad C(\boldsymbol{E}) = \partial_k E_k. \qquad (3.2.75)$$

The magnetic field F_{kl} is supposed to be expressed in terms of A_k:

$$F_{kl} = \partial_l A_k - \partial_k A_l. \qquad (3.2.76)$$

It is clear that $H(E_k, A_k)$ is the Hamiltonian of the system, while E_k and A_k ($k = 1, 2, 3$) are pairs of canonically conjugate momenta and generalized coordinates, respectively, and thus we can postulate Poisson brackets for them in the form

$$\{E_k(x), A_l(y)\} = \delta_{kl}\delta(x - y). \qquad (3.2.77)$$

As we have already pointed out, the timelike potential $A_0(x)$ plays the role of a Lagrange multiplier and its variation in the action corresponding to (3.2.73) produces the constraint (cf (3.2.35))

$$\partial_k E_k = 0. \qquad (3.2.78)$$

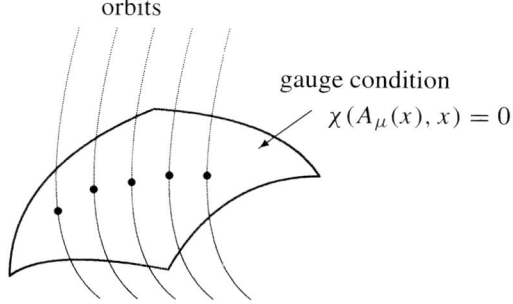

Figure 3.10. Graphical representation of the orbits generated by a gauge group and of the surface defined by a gauge condition.

This constraint satisfies the self-consistency conditions (cf (3.2.57) and (3.2.58)):

$$\{\partial_k E_k(t, \boldsymbol{x}), \partial_l E_l(t, \boldsymbol{y})\} = 0 \qquad (3.2.79)$$

$$\left\{\left(\int d^3x\, H(E_k(t, \boldsymbol{x}), A_k(t, \boldsymbol{x}))\right), \partial_k E_k(t, \boldsymbol{y})\right\} = 0 \qquad (3.2.80)$$

so that there are no additional higher-order (secondary, etc) constraints in this case.

The next step is to fix the subsidiary *gauge condition* or simply *gauge* $\chi = 0$ (cf (3.2.55)). For electrodynamics, the following gauges are most commonly used:

$$\partial_k A_k = 0 \qquad \text{the Coulomb gauge} \qquad (3.2.81)$$

$$\partial_\mu A^\mu = 0 \qquad \text{the Lorentz gauge} \qquad (3.2.82)$$

(recall that index k in (3.2.81) runs over 1, 2, 3, while the index μ in (3.2.82) runs over 0, 1, 2, 3). Both conditions ((3.2.81) and (3.2.82)) obviously satisfy the necessary condition $\det\{C(E_k), \chi\} \neq 0$, where χ stands for a gauge condition (cf (3.2.56)). For example, for the Coulomb gauge, we have

$$\{\partial_k A_k(t, \boldsymbol{x}), \partial_l A_l(t, \boldsymbol{y})\} = 0 \qquad (3.2.83)$$

$$\{\partial_k E_k(t, \boldsymbol{x}), \partial_l A_l(t, \boldsymbol{y})\} = \partial_i \partial_i \delta^{(3)}(\boldsymbol{x} - \boldsymbol{y}). \qquad (3.2.84)$$

The three-dimensional Laplacian in the right-hand side of (3.2.84) is reversible and has non-zero determinant.

Thus the gauge conditions lead to an equation which defines the parameter $\varepsilon(x)$ of the gauge transformations (cf (3.2.20))

$$A_\mu(x) \longrightarrow A_\mu(x) + \partial_\mu \varepsilon(x). \qquad (3.2.85)$$

A class of fields related by these transformations for all $\varepsilon(x)$ is called an *orbit* of the gauge group. Gauge invariance means that the fields A_μ and $A_\mu + \partial_\mu \varepsilon$ describe the same physical state for any $\alpha(x)$. A gauge condition chooses a representative from each class of physically (gauge) equivalent fields. An orbit can be depicted schematically as a line with points which are all physically equivalent and can be converted into each other by means of the gauge transformations. The gauge condition can be represented as a surface which crosses each orbit once (see figure 3.10).

Note that we have presented a simplified version of the analysis of electrodynamics as a system with constraints. This is quite sufficient for our purpose. A more rigorous and complete approach (Gitman and Tyutin 1990) considers the Lagrange multiplier A_0 on equal footing with A_k and inputs the *primary constraint*

$$E_0 = 0 \qquad (3.2.86)$$

as a consequence of the equality

$$\frac{\partial \mathcal{L}_e}{\partial (\partial_t A_0)} = 0 \qquad (3.2.87)$$

(E_0 is considered as a momentum which is canonically conjugate to A_0). Then we should consider the total Hamiltonian

$$H_T \stackrel{\text{def}}{\equiv} H(E_k, A_k) - A_0 C(E_k) \qquad (3.2.88)$$

and its Poisson bracket with the primary constraint (3.2.86) gives the *secondary constraint*

$$\left\{ \int d^3x \, H_C(E_k, E_0, A_k, A_0), E_0 \right\} = \partial_k E_k = 0 \qquad (3.2.89)$$

which coincides with (3.2.78). Thus, in this approach we have two constraints and, therefore, have to input two subsidiary gauge conditions: e.g., in addition to the Coulomb gauge (3.2.81) we can input the condition $\partial_k A_k - \partial_l \partial_l A_0 = 0$ (Gitman and Tyutin 1990). We shall not follow this way of analysing gauge systems. For the relatively simple gauge models which we shall encounter in this book, this simplified formalism outlined here is quite enough.

◇ Constrained Hamiltonian mechanics of non-Abelian Yang–Mills fields

First, we shall introduce some convenient notation which will simplify the formula-writing for Hamiltonians, constraints, gauge conditions, gauge transformations, etc. It is convenient to consider the gauge field for some compact Lie group $\mathfrak{G}$ as a field $\mathsf{A}_\mu(x)$ taking values in the corresponding Lie algebra $\mathfrak{g}$. Of course, $\mathsf{A}_\mu(x)$ is defined by its coefficients $A_\mu^a(x)$:

$$\mathsf{A}_\mu(x) = A_\mu^a(x) T^a \qquad (3.2.90)$$

where T^a, $a = 1, \ldots, \dim \mathfrak{g}$, form the basis of the Lie algebra $\mathfrak{g}$ (the basis of generators of the group $\mathfrak{G}$). We shall assume that T^a are *anti-Hermitian* operators. Then, if $g(x)$ is a gauge group element in the adjoint representation (see supplement IV), the gauge transformations can be written as follows:

- infinitesimal transformations (3.2.8) corresponding to elements $g(x)$ close to unity,

$$\mathsf{g}(x) = \mathbb{I} + \boldsymbol{\varepsilon}(x) \equiv \mathbb{I} + \varepsilon^a(x) T^a \qquad (3.2.91)$$

take the form

$$\delta \mathsf{A}_\mu = \partial_\mu \boldsymbol{\varepsilon} - [\mathsf{A}_\mu, \boldsymbol{\varepsilon}] = D_\mu \boldsymbol{\varepsilon} \qquad (3.2.92)$$

where $[\cdots, \cdots]$ is the Lie algebra commutator (recall also that D_μ is the covariant derivative, see (3.2.9)); and
- transformations with an arbitrary $g(x)$ read as

$$\mathsf{A}_\mu(x) \longrightarrow \mathsf{A}_\mu^g(x) = \mathsf{g}(x) \mathsf{A}_\mu(x) \mathsf{g}^{-1}(x) + (\partial_\mu \mathsf{g}(x)) \mathsf{g}^{-1}(x). \qquad (3.2.93)$$

The gauge-field tensor $\mathsf{F}_{\mu\nu}$ in this notation is expressed through the gauge field via the matrix relation

$$\mathsf{F}_{\mu\nu} = \partial_\nu \mathsf{A}_\mu - \partial_\mu \mathsf{A}_\nu + g[\mathsf{A}_\mu, \mathsf{A}_\nu]. \qquad (3.2.94)$$

This notation allows us to avoid writing a lot of indices and to present expressions for the Yang–Mills theory with an arbitrary group in compact form which is rather similar to the one for electrodynamics (Abelian $U(1)$ gauge theory).

In the first-order formalism (cf (3.2.72)), the Lagrangian for a non-Abelian theory reads as

$$\mathcal{L}_{YM} = \tfrac{1}{4}\mathrm{Tr}\{(\partial_\nu \mathsf{A}_\mu - \partial_\mu \mathsf{A}_\nu + g[\mathsf{A}_\mu, \mathsf{A}_\nu] - \tfrac{1}{2}\mathsf{F}_{\mu\nu})\mathsf{F}_{\mu\nu}\} \qquad (3.2.95)$$

where $\mathsf{F}_{\mu\nu}$ and A_μ are considered to be canonical variables. Passing to the three-dimensional notation as in the case of electrodynamics, we can write (cf (3.2.74)–(3.2.76))

$$\mathcal{L}_{YM} = -\tfrac{1}{2}\mathrm{Tr}[\mathsf{E}_k \dot{\mathsf{A}}_k - \tfrac{1}{2}H(\mathsf{E}_k, \mathsf{A}_k) + \mathsf{A}_0 \mathsf{C}] \qquad (3.2.96)$$

where

$$\mathsf{E}_k = \mathsf{F}_{k0} \qquad \mathsf{B}_k = \tfrac{1}{2}\epsilon_{ijk}\mathsf{F}_{ji} \qquad H(\mathsf{E}_k, \mathsf{A}_k) = \tfrac{1}{2}(\mathsf{E}_k^2 + \mathsf{B}_k^2) \qquad (3.2.97)$$
$$\mathsf{C}(x) = \partial_k \mathsf{E}_k(x) - g[\mathsf{A}_k(x), \mathsf{E}_k(x)] \qquad (3.2.98)$$
$$\mathsf{F}_{kl} = \partial_l \mathsf{A}_k - \partial_k \mathsf{A}_l + g[\mathsf{A}_k, \mathsf{A}_l]. \qquad (3.2.99)$$

The same Lagrangian can be presented in component form:

$$\mathcal{L}_{YM} = -\tfrac{1}{2}[E_k^a \dot{A}_k^a - \tfrac{1}{2}H(E_k^b, A_k^c) + A_0^a C^a]. \qquad (3.2.100)$$

Introducing the Poisson brackets

$$\{E_k^a(x), A_l^b(y)\} = \delta_{kl}\delta^{ab}\delta(x-y) \qquad (3.2.101)$$

we can easily verify that

$$\{C^a(x), C^b(y)\} = gf^{abc}C^c(x)\delta(x-y) \qquad (3.2.102)$$
$$\left\{\int d^3x\, H(E_k^b, A_k^c), C^a(y)\right\} = 0 \qquad (3.2.103)$$

where f^{abc} are the structure constants of the gauge Lie algebra $\mathfrak{g}$. Thus C^a is the set of the first-class constraints and they produce no new higher-order (secondary, etc) constraints.

According to the general method for quantizing systems with non-holonomic first-class constraints, we have to add a subsidiary gauge condition. For the Yang–Mills fields, quite a number of gauges have been invented; the most common are:

$$\partial_k A_k^a = 0 \qquad \text{the Coulomb gauge} \qquad (3.2.104)$$
$$\partial^\mu A_\mu^a = 0 \qquad \text{the Lorentz gauge} \qquad (3.2.105)$$
$$n^\mu A_\mu^a = 0 \qquad \text{the axial gauge} \qquad (3.2.106)$$

where n_μ is a unit four-vector.

Let us consider Coulomb gauge condition (3.2.104). Conditions (3.2.60) and (3.2.56) for this concrete case read as follows:

$$\{\partial_k A_k^a(x), \partial_j A_j^a(y)\} = 0 \qquad (3.2.107)$$
$$\det\{C^a(x), \partial_k A_k^b(y)\} = \det[\partial_k(\partial_k - gf^{abc}A_k^c(x))\delta(x-y)] \neq 0. \qquad (3.2.108)$$

The correctness of (3.2.107) is obvious. For (3.2.108), in the Coulomb gauge the operator

$$\widehat{M}_{\text{C}} \stackrel{\text{def}}{\equiv} \partial_k\partial_k\delta^{ab} - gf^{abc}A_k^c(x)\partial_k \qquad (3.2.109)$$

(C standing for Coulomb) is invertible in the framework of perturbation theory in the coupling constant g. Indeed, the inverse operator can be determined from the integral equation for its kernel:

$$(M_{\text{C}}^{-1})^{ab}(x,y) = \frac{1}{4\pi}\frac{\delta^{ab}}{|x-y|} + \frac{g}{4\pi}\int dx'\, f^{acd}\frac{A_k^c(x')}{|x-x'|}\partial_k(M_{\text{C}}^{-1})^{db}(x',y) \qquad (3.2.110)$$

which can be solved by iterations in g. Note that for large fields A_k^a (so that a perturbative solution is not valid) the operator $\widehat{M}_{\text{C}}$ acquires zero eigenvalues so that condition (3.2.108) proves to be violated. In fact, this is a general situation for any gauge condition and it is called the *Gribov multi-valuedness* or the *Gribov ambiguity* (Gribov 1978). Since this multi-valuedness appears at large values of fields, it does not influence the perturbation theory for gauge models. We shall not discuss this problem any further in this book (for more details about this problem see, e.g., in Halpern and Koplik (1978) and Dell' Antonio and Zwanziger (1989)).

3.2.4 Path-integral quantization of Yang–Mills theories

We continue the study of Yang–Mills theory in the Coulomb gauge. The very form of this gauge (cf (3.2.104)) prompts the orthogonal separation into longitudinal and transversal parts:

$$\mathsf{A}_k = \mathsf{A}_k^{\text{L}} + \mathsf{A}_k^{\text{T}} \qquad (3.2.111)$$

where

$$\mathsf{A}_k^{\text{L}}(x) \stackrel{\text{def}}{\equiv} \partial_k \mathsf{a}(x) \qquad \text{for an appropriate } \mathsf{a}(x) \qquad (3.2.112)$$

$$\partial_k \mathsf{A}_k^{\text{T}} = 0. \qquad (3.2.113)$$

It is seen that the transversal part A_k^{T} plays the role of the unconstrained physical coordinates $\widetilde{q}$ (see section 3.2.2). The momenta conjugate to them are the transversal components of the gauge tensor E_k^{T} and the constraint condition is imposed on the longitudinal part $\mathsf{E}^{\text{L}}(x)$: if we introduce $\mathsf{L}(x)$ by the relation

$$\mathsf{E}_k^{\text{L}}(x) = \partial_k \mathsf{L}(x)$$

constraint (3.2.98) proves to be

$$\partial_k\partial_k\mathsf{L} - g[\mathsf{A}_k, \partial_k\mathsf{L}] - g[\mathsf{A}_k, \mathsf{E}_k^{\text{T}}] = \widehat{M}_{\text{C}}\mathsf{L} - g[\mathsf{A}_k, \mathsf{E}_k^{\text{T}}] = 0, \qquad (3.2.114)$$

where we meet again the operator $\widehat{M}_{\text{C}}$ defined in (3.2.109). This equation allows us to express the longitudinal component E_k^{L} through E_k^{T} and A_k^{T}. After substituting this solution into the Hamiltonian $H(\mathsf{A},\mathsf{E})$, we obtain the Hamiltonian $H_{\text{ph}}(\mathsf{A}_k^{\text{T}},\mathsf{E}_k^{\text{T}})$ in terms of the unconstrained variables A_k^{T} and E_k^{T}. Thus the true physical variables for a Yang–Mills field are the components A_k^{T} of the three-vector A_k subjected to the constraint (3.2.113). This means that the Yang–Mills field has only two possible states of polarization.

Now we are ready to write down the path integral for the corresponding S-matrix. In terms of the unconstrained physical variables, it is a direct generalization of that for a scalar field (cf (3.1.85))

$$K_S(a^*(\boldsymbol{k}), a(\boldsymbol{k}); t, t_0) = \lim_{\substack{t\to\infty \\ t_0\to -\infty}}\int \prod_{j,b}\mathcal{D}a_j^{*b}(\boldsymbol{k},\tau)\mathcal{D}a_j^b(\boldsymbol{k},\tau)$$

$$\times \exp\left\{i\int d^3k\, \frac{1}{2}\left[\sum_{b=1}^{\dim\mathfrak{G}}\sum_{j=1}^{2} a_j^{*b}(\boldsymbol{k},t)a_j^b(\boldsymbol{k},t) + a_j^{*b}(\boldsymbol{k},t_0)a_j^b(\boldsymbol{k},t_0)\right]\right\}$$

$$\times \exp\left\{i\int_{t_0}^{t} d\tau \int d^3x \left[\frac{1}{4}\mathrm{Tr}(\dot{\mathsf{E}}_l^{\mathrm{T}}(\boldsymbol{x},\tau)\mathsf{A}_l^{\mathrm{T}}(\boldsymbol{x},\tau) - \mathsf{E}_l^{\mathrm{T}}(\boldsymbol{x},\tau)\dot{\mathsf{A}}_l^{\mathrm{T}}(\boldsymbol{x},\tau))\right.\right.$$

$$\left.\left. - H_{\mathrm{ph}}(\mathsf{E}_l^{\mathrm{T}}, \mathsf{A}_l^{\mathrm{T}})\right]\right\} \qquad (3.2.115)$$

where (cf (3.1.77))

$$(A^{\mathrm{T}})_l^b(\boldsymbol{r},\tau) = \frac{1}{(2\pi)^{3/2}}\sum_{j=1}^{2}\int d^3k\, \frac{1}{\sqrt{2\omega}}(a_j^{*b}(\boldsymbol{k},\tau)u_l^j(-\boldsymbol{k})e^{-i\boldsymbol{k}\cdot\boldsymbol{r}} + a_j^b(\boldsymbol{k},\tau)u_l^j(\boldsymbol{k})e^{i\boldsymbol{k}\cdot\boldsymbol{r}})$$
(3.2.116)

$$(E^{\mathrm{T}})_l^b(\boldsymbol{r},\tau) = \frac{i}{(2\pi)^{3/2}}\sum_{j=1}^{2}\int d^3k\, \sqrt{\frac{\omega}{2}}(a_j^{*b}(\boldsymbol{k},\tau)u_l^j(-\boldsymbol{k})e^{-i\boldsymbol{k}\cdot\boldsymbol{r}} - a_j^b(\boldsymbol{k},\tau)u_l^j(\boldsymbol{k})e^{i\boldsymbol{k}\cdot\boldsymbol{r}})$$

$$\omega = \sqrt{\boldsymbol{k}^2 + m^2} \qquad (3.2.117)$$

with the two polarization vectors u_l^j, $j = 1, 2$ (any two orthonormal vectors orthogonal to $\boldsymbol{k}$). The asymptotic conditions have the form (cf (3.1.86))

$$a_j^{*b}(\boldsymbol{k},t) \xrightarrow[t\to\infty]{} a_j^{*b}(\boldsymbol{k})\exp\{i\omega_k t\} \qquad a_j^b(\boldsymbol{k},t_0) \xrightarrow[t\to-\infty]{} a_j^b(\boldsymbol{k})\exp\{-i\omega_k t_0\}. \qquad (3.2.118)$$

Generally speaking, this formula solves the problem of constructing the S-matrix for the Yang–Mills theories. However, it is not practically convenient. The main obstruction to its direct application is the necessity of knowing the Hamiltonian H_{ph}. An explicit derivation of this Hamiltonian requires the solution of equation (3.2.114), in other words, the inversion of operator $\widehat{M}_{\mathrm{C}}$. In fact, we can only do this perturbatively, i.e. we can present $\widehat{M}_{\mathrm{C}}^{-1}$ as an infinite series in the coupling constant g. Although this is just a technical difficulty, it prevents actual practical calculations in gauge theories as well as troubling their general analysis (e.g., proof of their renormalizability). To overcome this difficulty, we can use formulae (3.2.67)–(3.2.69) (more precisely, their generalization to an infinite number of degrees of freedom) and present the S-matrix as the path integral over *all* fields $\mathsf{A}_k, \mathsf{E}_k$ ($k = 1, 2, 3$):

$$K_{\mathrm{S}}(a^*(\boldsymbol{k}), a(\boldsymbol{k}); t, t_0) = \lim_{\substack{t\to\infty \\ t_0\to-\infty}} \int \prod_{a=1}^{\dim\mathfrak{G}} \left[\mathcal{D}A_0^a(x) \prod_{k=1}^{3}(\mathcal{D}A_k^a(x)\mathcal{D}E_k^a(x))\right]$$

$$\times \exp\left\{i\int d^3k\, \frac{1}{2}\left[\sum_{b=1}^{\dim\mathfrak{G}}\sum_{j=1}^{2} a_j^{*b}(\boldsymbol{k},t)a_j^b(\boldsymbol{k},t) + a_j^{*b}(\boldsymbol{k},t_0)a_j^b(\boldsymbol{k},t_0)\right]\right\}$$

$$\times \exp\left\{\frac{i}{4}\int_{t_0}^{t} d\tau \int d^3x\, \mathrm{Tr}[\dot{\mathsf{E}}_l(\boldsymbol{x},\tau)\mathsf{A}_l(\boldsymbol{x},\tau) - \mathsf{E}_l(\boldsymbol{x},\tau)\dot{\mathsf{A}}_l(\boldsymbol{x},\tau)\right.$$

$$\left. + \mathsf{E}_l^2(\boldsymbol{x},\tau) + \mathsf{B}_l^2(\boldsymbol{x},\tau) - 2\mathsf{A}_0(\boldsymbol{x},\tau)(\partial_l\mathsf{E}_l(\boldsymbol{x},\tau) - g[\mathsf{A}_l(\boldsymbol{x},\tau), \mathsf{E}_l(\boldsymbol{x},\tau)])]\right\}$$

$$\times \prod_{\boldsymbol{r},\tau}\delta(\partial_l\mathsf{A}_l)\det\widehat{M}_{\mathrm{C}}[\mathsf{A}]. \qquad (3.2.119)$$

Here the boundary terms $a_j^{*b}(\boldsymbol{k},t)$, $a_j^b(\boldsymbol{k},t_0)$ are defined by the preceding formula (3.2.118) (i.e. they are constructed only via the transversal components $\mathsf{A}_l^{\mathrm{T}}$ of the Yang–Mills field).

66   *Quantum field theory: the path-integral approach*

The momenta E_l enter the exponent of (3.2.119) quadratically and we can integrate over them. For the *normal symbol* of the S-matrix, this yields:

$$S = \mathfrak{N}^{-1} \int \prod_{a=1}^{\dim \mathfrak{G}} \prod_{\mu=0}^{3} \mathcal{D}A_\mu^a(x) \prod_x \delta(\partial_l A_l) \det \widehat{M}_C[A] \exp\left\{i \int dx \, \frac{1}{8} \text{Tr}(F_{\mu\nu} F^{\mu\nu})\right\} \qquad (3.2.120)$$

where the integration goes over all the fields A_μ, with a fixed asymptotic behaviour for their transversal (in the three-dimensional sense) components:

$$A_l^T(x) \xrightarrow[t\to\pm\infty]{} A_l^T(x, \,^{\text{in}}_{\text{out}}) \qquad (3.2.121)$$

$$(A^T)_l^b(x; \,^{\text{in}}_{\text{out}}) = \frac{1}{(2\pi)^{3/2}} \sum_{j=1}^{2} \int d^3k \, \frac{1}{\sqrt{2\omega}} (a_j^{*b}(\mathbf{k}; \,^{\text{in}}_{\text{out}}) u_l^j(\mathbf{k}) e^{-i\mathbf{k}\cdot\mathbf{r}+i\omega t} + a_j^b(\mathbf{k}; \,^{\text{in}}_{\text{out}}) u_l^j(\mathbf{k}) e^{i\mathbf{k}\cdot\mathbf{r}-i\omega t})$$

$$\qquad (3.2.122)$$

$$a_j^b(\mathbf{k}; \text{in}) = a_j^b(\mathbf{k}) \qquad a_j^{*b}(\mathbf{k}; \text{out}) = a_j^{*b}(\mathbf{k}) \qquad j = 1, 2.$$

The normalization factor $\mathfrak{N}^{-1}$ has appeared due to the integration over the momenta E_l.

In path integral (3.2.120), the δ-functional together with the determinant select one representative from each class (orbit) of gauge-equivalent fields. Note that the asymptotic conditions (3.2.121) are also adjusted to the Coulomb gauge condition.

◇ **Diagram technique for the Yang–Mills theory in Coulomb gauge**

Separating out terms higher than second order in the exponent of the integrand in (3.2.120) and expanding the exponential in the perturbation series generate the Feynman diagram techniques for Yang–Mills theory. The propagator is defined by the Gaussian integral (i.e. the integral (3.2.120) with all higher-order terms being dropped out)

$$\mathcal{Z}_0[J] \equiv \mathcal{Z}[J^a]|_{g=0} = \mathfrak{N}^{-1} \int \prod_{a=1}^{\dim \mathfrak{G}} \prod_{\mu=0}^{3} \mathcal{D}A_\mu^a(x) \prod_x \delta(\partial_l A_l) \det \widehat{M}_C[A]$$
$$\times \exp\left\{i \int dx \, \text{Tr}\left[\frac{1}{8}(\partial_\nu A_\mu - \partial_\mu A_\nu)^2 - \frac{1}{2} J^\mu A_\mu\right]\right\} \qquad (3.2.123)$$

where the class of functions to be integrated over is defined by the boundary conditions (3.2.121) and (3.2.122) imposed on A_k^T.

The new feature of this free generating functional in comparison with that for a field theory without constraints (cf (3.1.100)) is the presence of the δ-function in the integrand of (3.2.123). This integral reminds us of the problem of a Brownian particle with inertia (see section 1.2.4, equation (1.2.70)). As we learned there, in order to solve integral (3.2.123), we have to find the extremum of the exponent under the condition defined by the δ-functional with the help of the Lagrange multiplier method or, equivalently, we just use the path-integral representation (3.1.193) for the δ-functional. The latter method gives

$$\mathcal{Z}_0[J^b] = \mathfrak{N}^{-1} \int \prod_{a=1}^{\dim \mathfrak{G}} \prod_{\mu=0}^{3} \mathcal{D}A_\mu^a(x) \, \mathcal{D}\lambda(x) \exp\left\{i \int dx \left[-\frac{1}{4}(\partial_\nu A_\mu^a - \partial_\mu A_\nu^a)^2 + J^{a\mu} A_\mu^a + \lambda^a \partial_k A_k^a\right]\right\}.$$
$$\qquad (3.2.124)$$

Solution of the extremality equation yields (problem 3.2.2, page 98)

$$\mathcal{Z}_0[J^b] = \exp\left\{\frac{i}{2}\int dx\, dy\, J^{a\mu}(x) D_C^{\mu\nu}(x-y) J^{a\nu}(y)\right\} \qquad (3.2.125)$$

where $D_C^{\mu\nu}(x-y)$ is the propagator of the Yang–Mills field in the Coulomb gauge:

$$\begin{aligned}
D_C^{ml}(x) &= -\frac{1}{(2\pi)^4}\int dk\, e^{-ikx}\frac{1}{k^2+i\varepsilon}\left(\delta^{ml}-\frac{k_m k_l}{|\boldsymbol{k}|^2}\right) \quad (m,l=1,2,3) \\
D_C^{00}(x) &= -\frac{1}{(2\pi)^4}\int dk\, e^{-ikx}\frac{1}{|\boldsymbol{k}|^2} \\
D_C^{m0}(x) &= D_C^{0m}(x) = 0.
\end{aligned} \qquad (3.2.126)$$

This expression clearly shows that only the transversal components $\mathsf{A}_k^{\mathrm{T}}$ really propagate in time.

Using the usual methods of the path-integral formalism, i.e. representing the higher-order (interaction) terms via functional derivatives, we could now develop the complete perturbation theory technique, including the diagram representation. However, the essential shortcoming of the Coulomb gauge and the corresponding perturbation theory expansion is the absence of an *explicit* relativistic invariance. Therefore, instead of dealing with the physically transparent but technically inconvenient (because of the absence of an *explicit* Lorentz covariance) Coulomb gauge, we shall learn, in the next subsection, a method for a transition to any gauge condition in the path integral and then develop the perturbation theory rules for the relativistically invariant Lorentz gauge (3.2.82) and its generalizations.

3.2.5 Covariant generating functional in the Yang–Mills theory

In order to construct an explicitly relativistic invariant S-matrix in each order of the perturbation theory, all the ingredients of the perturbation expansion should have simple transformation properties with respect to the Poincaré group. Such an expansion is called *covariant perturbation theory*. In particular, for Yang–Mills theories, such a technique must be based on some relativistically invariant gauge condition. In the case of pure Yang–Mills theory (i.e. without matter fields) the simplest condition is the Lorentz gauge (3.2.82). In this subsection we shall show that, using the known path-integral representation for the generating functional (for the S-matrix or Green functions) in the Coulomb gauge, we can pass to the Lorentz (or any other suitable) gauge condition (Faddeev and Popov 1967, De Witt 1967). From a geometrical point of view, we have to transfer the path-integral measure defined on the surface specified by the Coulomb gauge $\partial_k \mathsf{A}_k = 0$ to the surface specified by the Lorentz gauge $\partial_\mu \mathsf{A}_\mu = 0$.

◇ **Faddeev–Popov trick**

Let us introduce the functional $\Delta_{\mathrm{L}}[\mathsf{A}]$ defined by the equality

$$\Delta_{\mathrm{L}}[\mathsf{A}]\int \mathcal{D}u(x)\,\delta[\partial^\mu \mathsf{A}_\mu^u] = 1 \qquad (3.2.127)$$

where A_μ^u denotes the gauge transformed field: $\mathsf{A}_\mu^u(x) \stackrel{\mathrm{def}}{\equiv} u(x)\mathsf{A}_\mu u^{-1}(x) + (\partial_\mu u(x))u^{-1}(x)$ and the integration is carried out with the measure

$$\mathcal{D}u(x) = \prod_x d_{\mathrm{H}} u(x) \qquad (3.2.128)$$

where $u(x)$ are elements of the gauge group $\mathfrak{G}$ and $d_H u(x)$ at each x is the left- and right-invariant measure (the so-called *invariant Haar measure*, see supplement IV) on the group $\mathfrak{G}$, i.e.

$$d_H(u_0 u) = d_H(u u_0) = d_H u \qquad u, u_0 \in \mathfrak{G}. \tag{3.2.129}$$

Due to the latter property, the functional $\Delta_L[A]$ is gauge invariant:

$$\Delta_L[A^u] = \Delta_L[A]. \tag{3.2.130}$$

Since the left-hand side of (3.2.127) is equal to unity, we can harmlessly insert it into the integrand of (3.2.120), so that the S-matrix symbol now takes the form

$$S = N^{-1} \int \prod_{\mu=0}^{3} \mathcal{D}A_\mu^a(x)\, \mathcal{D}u(x)\, \delta[\partial_l A_l]\, \det \widehat{M}_C[A]$$
$$\times \Delta_L[A] \delta[\partial^\mu A_\mu^u] \exp\left\{ i \int dx\, \tfrac{1}{8} \mathrm{Tr}(\mathsf{F}_{\mu\nu} \mathsf{F}^{\mu\nu}) \right\}. \tag{3.2.131}$$

The next step is to introduce, in analogy with (3.2.127), one more gauge-invariant functional $\Delta_C[A]$, defined by the equality

$$\Delta_C[A] \int \mathcal{D}u(x)\, \delta[\partial_k A_k^u] = 1. \tag{3.2.132}$$

It is readily seen that on the gauge surface $\partial_k A_k = 0$, the functional $\Delta_C[A]$ coincides with $\det \widehat{M}_C[A]$. Indeed, if A_k satisfies $\partial_k \mathsf{A}_k = 0$, the only contribution to the integral in (3.2.132) (at least in the framework of the perturbation theory) comes from the infinitesimal vicinity of $u(x) = 1$. Hence, we can put $u(x) \approx 1 + \alpha(x)$ and, therefore,

$$\partial_k \mathsf{A}_k^u = \partial_k \partial_k \alpha(x) - g[\mathsf{A}_k(x), \partial_k \alpha(x)] \equiv \widehat{M}_C \alpha(x). \tag{3.2.133}$$

Taking into account that the substitution $u(x) \to \alpha(x) = u(x) - 1$ has unit Jacobian, $\mathcal{D}u(x) = \mathcal{D}\alpha(x)$, we can calculate integral (3.2.132) explicitly (problem 3.2.3, page 99)

$$\Delta_C[A]\big|_{\partial_k \mathsf{A}_k = 0} = \prod_\tau \det \widehat{M}_C(A). \tag{3.2.134}$$

After changing the integration variables

$$\mathsf{A}_\mu \longrightarrow \mathsf{A}_\mu^{u^{-1}} \tag{3.2.135}$$

(with unit Jacobian) and using the equality (3.2.134), integral (3.2.131) can be rewritten as follows:

$$S = \mathfrak{N}^{-1} \int \prod_{a=1}^{\dim \mathfrak{G}} \prod_{\mu=0}^{3} \mathcal{D}A_\mu^a(x)\, \mathcal{D}u(x)\, \delta[\partial_\mu A_\mu] \Delta_L[A] \delta[\partial_l A_l^{u^{-1}}] \Delta_C[A]$$
$$\times \exp\left\{ i \int dx\, \tfrac{1}{8} \mathrm{Tr}(\mathsf{F}_{\mu\nu} \mathsf{F}^{\mu\nu}) \right\}. \tag{3.2.136}$$

The definition (3.2.132) of the functional $\Delta_C[A]$ together with the substitution (the change of the variables $u^{-1} \to u$ in the integral over u),

$$\mathsf{A}^{u^{-1}} \longrightarrow \mathsf{A}^u \tag{3.2.137}$$

show that (3.2.131) is, in fact, the expression for the S-matrix in the Lorentz gauge:

$$S = \mathfrak{N}^{-1} \int \prod_{\mu=0}^{3} \mathcal{D}A_\mu^a(x)\, \delta[\partial^\mu A_\mu] \Delta_L[A] \exp\left\{i \int dx\, \tfrac{1}{8} \text{Tr}(F_{\mu\nu}F^{\mu\nu})\right\}. \quad (3.2.138)$$

Quite similarly to the case of the functional $\Delta_C[A]$ (cf (3.2.134)), we can show that

$$\Delta_L[A]\big|_{\partial^\mu A_\mu = 0} = \det \widehat{M}_L(A) \quad (3.2.139)$$

where

$$\widehat{M}_L \alpha(x) \stackrel{\text{def}}{\equiv} \Box \alpha - g\partial^\mu [A_\mu, \alpha]. \quad (3.2.140)$$

This method of transition from the Coulomb gauge condition to any other one (in particular, to the covariant Lorentz gauge condition) is called the *Faddeev–Popov trick* (Faddeev and Popov 1967).

◇ Asymptotic boundary conditions in the Lorentz gauge: justification of the Faddeev–Popov trick

When performing the manipulations which have led us to the covariant gauge condition in the path integral (3.2.138) for the S-matrix, we did not pay any attention to the asymptotic conditions, so that our consideration looks a little formal. In fact, we need two types of asymptotic condition:

(i) First, we did not clarify the way to calculate $\det \widehat{M}_L(A)$: the complete definition of operator $\widehat{M}_L(A)$ requires the determination of asymptotic conditions for $\tau \to \pm\infty$. Indeed, to define the determinant explicitly, it is convenient to use the formula

$$\det \widehat{M}_L(A) = \exp\{\text{Tr} \ln \widehat{M}_L(A)\}$$
$$= \exp\{\text{Tr} \ln \Box + \text{Tr} \ln(1 + \Box^{-1} K(A))\} \quad (3.2.141)$$

where the operator K is the second term in the operator $\widehat{M}_L$ in (3.2.140): $K(A)\mathfrak{f} \stackrel{\text{def}}{\equiv} -g\partial^\mu [A_\mu, \mathfrak{f}]$. The trace operation in (3.2.141) also implies integration over the coordinates. The first term in the exponential (3.2.141) gives a non-essential contribution to the normalization constant (since it is independent of the gauge fields). The second term gives a contribution to the action which can be written as the series

$$\text{Tr} \ln(1 + \Box^{-1} K(A)) = \sum_n \frac{(-1)^{n+1}}{n} \text{Tr}(\Box^{-1} K)^n$$
$$= \sum_{n=1}^{\infty} (-1)^{n+1} \frac{g^n}{n} \int d^4x_1 \cdots d^4x_n\, \text{Tr}(A^{\mu_1}(x_1) \cdots A^{\mu_n}(x_n))$$
$$\times \partial_{\mu_1} D(x_1 - x_2) \cdots \partial_{\mu_n} D(x_n - x_1). \quad (3.2.142)$$

Here $D(x)$ is the Green function of the d'Alembert operator $\Box$, which, of course, is not uniquely defined unless appropriate boundary conditions are imposed. Any Green function of the d'Alembert operator can be presented via the Fourier transform

$$D = -\frac{1}{(2\pi)^4} \int d^4k\, e^{-ikx} \frac{1}{k^2} \quad (3.2.143)$$

but without a rule for pole bypassing in the integrand, this expression is formal. Boundary conditions just determine a way of bypassing the singularity and select one of the possible Green functions.

(ii) Essentially the same problem appears for the Green function $D^L_{\mu\nu}(x-y)$ corresponding to the quadratic form in the Yang–Mills action with a Lorentz gauge (the superscript 'L' is related to 'Lorentz', not to 'longitudinal'). The formal Fourier transform for $D^L_{\mu\nu}(x-y)$ reads as

$$D^L_{\mu\nu}(x) = -\frac{1}{(2\pi)^4}\int d^4k\, e^{-ikx}\left(g_{\mu\nu} - \frac{k_\mu k_\nu}{k^2}\right)\frac{1}{k^2}. \qquad (3.2.144)$$

Again, we have to clarify the rule for bypassing the singularities.

In order to derive the asymptotic conditions correctly, we have to transform a path integral with the Coulomb gauge condition into a path integral with the Lorentz gauge condition *before* the transition to the limit $t \to \infty$, $t_0 \to -\infty$ in the expression for the S-matrix. Note that the change of variables (3.2.135)

$$A_\mu \longrightarrow A_\mu^{u^{-1}} = u^{-1}A_\mu u + (\partial_\mu u^{-1})u \qquad (3.2.145)$$

used in the transformation of the path integral to the Lorentz gauge should not violate the Coulomb condition $\partial_k A_k = 0$, as well as the boundary conditions (3.2.118) for the transversal components A^T_k, at the boundary times t and t_0. This implies a restriction for the group elements:

$$u(\mathbf{r},t) = u(\mathbf{r},t_0) = 1 \qquad (3.2.146)$$

or, equivalently, a restriction for the corresponding Lie algebra elements

$$\alpha(\mathbf{r},t) = \alpha(\mathbf{r},t_0) = 0 \qquad u(x) = e^{\alpha(x)}. \qquad (3.2.147)$$

Thus, the operator $\widehat{M}_L(A)$ acts in the space of the functions $\alpha(x)$ (with values in the Lie algebra) which satisfy conditions (3.2.147). Hence, we have to look for the Green function entering the expansion (3.2.142) subjected to the same conditions (3.2.147). Such a Green function has the form

$$D_1(x,y) = \frac{1}{(2\pi)^3}\int d^3k\, e^{ik(x-y)}\frac{\sin(|k|(x^0-t_0))\sin(|k|(y^0-t))}{|k|\sin(|k|(t-t_0))} \quad \text{for } x^0 < y^0 \qquad (3.2.148)$$

$$D_1(x,y) = D_1(y,x) \qquad \text{for } x^0 \geq y^0.$$

With this definition, the operator $\widehat{M}_L(A)$ proves to be positively defined and this fact justifies the absence of the absolute value sign on the right-hand side of (3.2.139) (otherwise, we should write $|\det \widehat{M}_L(A)|$).

We shall return soon (see (3.2.156)) to the question about pole bypassing in (3.2.143), but before that, let us treat the Green function $D^L_{\mu\nu}(x)$ (3.2.144) in the Lorentz gauge in a similar way. First, we have to solve the equations

$$\Box A_\mu = J_\mu \qquad (3.2.149)$$
$$\partial^\mu A_\mu = 0 \qquad (3.2.150)$$

where the source J_μ satisfies the consistency condition

$$\partial^\mu J_\mu = 0 \qquad (3.2.151)$$

with the boundary conditions at *finite* times t, t_0 (cf (3.2.116) and (3.2.118)):

$$\begin{aligned}&\mathsf{a}^*_j(\mathbf{k},t) = \mathsf{a}^*_j(\mathbf{k})\exp\{i\omega_k t\}\\&\mathsf{a}_j(\mathbf{k},t_0) = \mathsf{a}_j(\mathbf{k})\exp\{-i\omega_k t_0\} \qquad j=1,2\\&\partial_k A_k(\mathbf{r},t) = \partial_k A_k(\mathbf{r},t_0) = 0\\&\partial_0 A_0(\mathbf{r},t) = \partial_0 A_0(\mathbf{r},t_0) = 0\end{aligned} \qquad (3.2.152)$$

(the last condition follows from the actual system (3.2.149)). The solution of this system has the form

$$A_l^T(x) = (A^{(0)})_l^T(x) + \int d^4y\, D_c(x,y) J_l^T(y) \qquad t_0 \le x^0,\ y^0 \le t$$

$$(A^{(0)})_l^T(x) \stackrel{\text{def}}{\equiv} \frac{1}{(2\pi)^{3/2}} \sum_{j=1}^{2} \int d^3k\, \frac{1}{\sqrt{2\omega}} (a_j^*(k) u_l^j(-k) e^{-ik\cdot r + i\omega x^0} + a_j(k) u_l^j(k) e^{ik\cdot r - i\omega x^0}) \quad (3.2.153)$$

$$A_0(x) = \int d^4y\, D_2(x,y) J_0(y)$$

$$A_l^L(x) = \int d^4y\, D_2(x,y) J_l^L(y) \qquad (3.2.154)$$

where $D_2(x,y)$ is another Green function of the d'Alembertian, which this time acts in the space of functions $f(x)$ with the boundary conditions

$$\partial_0 f(x)|_{x^0=t} = \partial_0 f(x)|_{x^0=t_0} = 0.$$

This Green function reads as

$$D_2(x,y) = \frac{1}{(2\pi)^3} \int d^3k\, e^{ik(x-y)} \frac{\cos(|\mathbf{k}|(x^0-t))\cos(|\mathbf{k}|(y^0-t_0))}{|\mathbf{k}|\sin(|\mathbf{k}|(t-t_0))} \qquad \text{for } x^0 < y^0 \qquad (3.2.155)$$

$$D_2(x,y) = D_2(y,x) \qquad \text{for } x^0 \ge y^0.$$

The Green functions D_1 and D_2 look somewhat disturbing because they obviously do not have well-defined limits at $t \to \infty$, $t_0 \to -\infty$, while such limits exist for the transversal components of the Yang–Mills field (see (3.2.153); this relates to the fact that A_l^T corresponds to the *physical* polarizations of the Yang–Mills field). On the other hand, the path integral for the S-matrix in the Lorentz gauge does exist, since it is, by construction, equal to that in the Coulomb gauge for which the infinite time limit is well defined. Hence, the combined contribution of the functions D_1 and D_2 to the perturbation expansion leads to a well-defined limit expression. A straightforward proof of this fact in all orders of the perturbation theory is not easy. Formally, this limit can be found by the simultaneous identical regularization of the Green functions D_1 and D_2 (Faddeev and Slavnov 1980). The most convenient way is to add an infinitesimal imaginary quantity to the momentum variable:

$$k^2 \longrightarrow k^2 - i\varepsilon. \qquad (3.2.156)$$

After this substitution, the oscillating function in (3.2.148) and (3.2.155) will either increase or decrease and have infinite time limits. Moreover, in this case, the infinite time limits of both D_1 and D_2 coincide with the standard causal Green function $D_c(x)$.

This defines the rule of pole bypassing for the function D in (3.2.143) (cf (3.1.93)) and from (3.2.153) and (3.2.154) we can read off the covariant Green function (3.2.144), together with the bypassing rule:

$$D_{\mu\nu}^L(x) = -\frac{1}{(2\pi)^4} \int d^4k\, e^{-ikx} \left(g_{\mu\nu} - \frac{k_\mu k_\nu}{k^2 + i\varepsilon}\right) \frac{1}{k^2 + i\varepsilon}. \qquad (3.2.157)$$

In the Lorentz gauge, it is this Green function that appears, instead of the Coulomb function $D_{\mu\nu}^C$, in an expression analogous to (3.2.125) for the generating functional $\mathcal{Z}_0[J]$ in the zero coupling constant approximation.

Thus, we conclude that all singularity bypasses are fixed by the infinitesimal shift of the momentum variables in the complex plane: $(k^2 + i\varepsilon)^{-1}$, and the path-integral representation for the S-matrix of the

72   *Quantum field theory: the path-integral approach*

Yang–Mills theories in the Lorentz gauge has the form (3.2.138), where the integration goes over fields with the following asymptotic behaviour:

$$\mathsf{A}_\mu(x) \xrightarrow[x^0 \to -\infty]{} \mathsf{A}_\mu(x; \text{in}) \qquad \mathsf{A}_\mu(x) \xrightarrow[x^0 \to \infty]{} \mathsf{A}_\mu(x; \text{out}) \qquad (3.2.158)$$

with $\mathsf{A}_\mu(x; \text{in})$, $\mathsf{A}_\mu(x; \text{out})$ being the solutions of the equations

$$\Box \mathsf{A}_\mu(x) = 0, \qquad \partial^\mu \mathsf{A}_\mu(x) = 0 \qquad (3.2.159)$$

parametrized by the amplitudes $\mathsf{a}^*_\mu(k)$, $\mathsf{a}_\mu(k)$, such that

$$\mathsf{a}_0 = 0 \qquad k_l \mathsf{a}_l = 0$$
$$\mathsf{a}^*_0 = 0 \qquad k_l \mathsf{a}^*_l = 0.$$

Note that in $\mathsf{A}_\mu(x; \text{in})$ the amplitude a_l is fixed, while in $\mathsf{A}_\mu(x; \text{out})$ it is the amplitude a^*_l that is fixed.

◇ **S-matrix for Yang–Mills theory in the α-gauge**

Formula (3.2.138) is not the only possible relativistically invariant expression for the S-matrix in Yang–Mills theory. The point is that we can integrate over gauge-equivalent classes of fields choosing not a single representative from each class, but some *compact* subset of representatives. The only requirement is that the resulting path integral be convergent, in which case this generalized approach may only change the normalization factor. An explicit transition to the corresponding expression can be produced in much the same way as that used when passing from the Coulomb to the Lorentz gauge.

Let us insert in the integrand of (3.2.120) unity represented in the form

$$1 = \Delta_B[A] \int \mathcal{D}u(x)\, B[A^u_\mu] \qquad (3.2.160)$$

where $B[A^u_\mu]$ is some gauge non-invariant functional, such that the integral on the right-hand side of (3.2.160) is convergent. Acting as in the case of the transition from Coulomb to Lorentz gauges, we obtain the path-integral representation for the S-matrix or generating functional of the type (3.2.138) where $\delta(\partial^\mu A_\mu) \det \widetilde{M}_L[A]$ is substituted by $\Delta_B[A] \int \mathcal{D}u(x)\, B[A^u_\mu]$. The choice of the functional $B[A]$ in the form

$$B[A^u_\mu] = \exp\left\{-\frac{i}{4\alpha} \int d^4x\, \text{Tr}(\partial^\mu A_\mu)^2\right\} \qquad (3.2.161)$$

($\alpha \in \mathbb{R}$ is a parameter) leads to the perturbation theory with the following free Green function

$$D^\alpha_{\mu\nu}(x) = -\frac{1}{(2\pi)^4} \int d^4k\, e^{-ikx} \left(g_{\mu\nu} - \frac{k_\mu k_\nu(1-\alpha)}{k^2 + i\varepsilon}\right) \frac{1}{k^2 + i\varepsilon}. \qquad (3.2.162)$$

Varying the parameter α, we obtain the important particular cases:

(i) at $\alpha = 0$, we return to the Lorentz gauge (note that in the limit $\alpha \to 0$, the functional $\alpha B[A]$ in (3.2.161) is converted to the δ-functional);
(ii) at $\alpha = 1$, we obtain the diagonal (in spacetime indices) Green function which is very convenient for practical calculations.

The simplest way to derive the corresponding path-integral representation for the S-matrix and to prove formula (3.2.162) goes as follows. Let us, first, transform the path integral (3.2.138) in the usual Lorentz gauge into that in the *generalized Lorentz gauge*

$$\partial^\mu A_\mu(x) = \mathsf{a}(x) \qquad (3.2.163)$$

where $\mathbf{a}(x)$ is an arbitrary matrix function. To do this, we repeat the Faddeev–Popov trick using the corresponding functional $\Delta_a[A]$ defined by the equality

$$\Delta_a[A] \int \mathcal{D}u(x)\, \delta[\partial^\mu \mathbf{A}_\mu^u - \mathbf{a}] = 1. \tag{3.2.164}$$

Note that, on the surface

$$\partial^\mu \mathbf{A}_\mu - \mathbf{a} = 0$$

the functional $\Delta_a[A]$ coincides with $\det \widehat{M}_L$. Thus the generating functional for the S-matrix can be written as

$$S = \mathfrak{N}^{-1} \int \prod_{\mu=0}^{3} \mathcal{D}A_\mu^a(x)\, \delta[\partial^\mu \mathbf{A}_\mu - \mathbf{a}]\Delta_L[A] \exp\left\{i \int dx\, \tfrac{1}{8}\,\mathrm{Tr}(F_{\mu\nu}F^{\mu\nu})\right\}. \tag{3.2.165}$$

Since the initial S-matrix functional does not depend on the function $\mathbf{a}(x)$, we can integrate it over $\mathbf{a}(x)$ with the weight

$$\exp\left\{-\frac{i}{4\alpha}\int d^4 x\, \mathrm{Tr}\, \mathbf{a}^2(x)\right\} \tag{3.2.166}$$

which leads to a simple change in the normalization factor. This integration yields the S-matrix in the form

$$S = \mathfrak{N}^{-1} \int \prod_{\mu=0}^{3} \mathcal{D}A_\mu^a(x)\, \det \widehat{M}_L[A] \exp\left\{i \int dx\, \mathrm{Tr}\left[\frac{1}{8}F_{\mu\nu}F^{\mu\nu} - \frac{1}{4\alpha}(\partial^\mu \mathbf{A}_\mu)^2\right]\right\} \tag{3.2.167}$$

and, hence, produces the free Green function (3.2.162).

Extending the concept of gauge conditions, the functional (3.2.167) is called the *S-matrix in the α-gauge*.

3.2.6 Covariant perturbation theory for Yang–Mills models

Having at our disposal the covariant generating functionals, obtained in the preceding subsection, we are almost ready to develop a covariant perturbation expansion and the corresponding diagram techniques. The only non-standard peculiarity of the functionals (3.2.167) or (3.2.138) is the presence of the non-local functional $\det \widehat{M}_L$, so that they do not have the customary form of the Feynman functional $\exp\{iS\}$ under the sign of the path integral, where S is the action of a system.

- Note, however, that in the case of quantum electrodynamics, the operator $\widehat{M}_L$ reduces to the ordinary d'Alembertian (cf (3.2.140); for the Abelian case, the commutator in this formula identically vanishes), so that $\det \widehat{M}_L^{(\mathrm{QED})}$ does not depend on the gauge (electromagnetic) fields. Therefore, we can just remove the determinant from the integral sign and readily develop the standard perturbation theory expansion for quantum electrodynamics (see, e.g., Itzykson and Zuber (1980); for the standard operator approach, see any textbook on quantum field theory, e.g., Bogoliubov and Shirkov (1959)).

In the general case of a non-Abelian Yang–Mills theory, the determinant can be expressed as a path integral over anticommuting scalar fields which are commonly referred to as *Faddeev–Popov ghosts*.

◇ Faddeev–Popov ghosts

Using the results of sections 2.6 and 3.1.3, the determinant $\det \widehat{M}_L$ can be expressed via the integral representation

$$\det \widehat{M}_L = \int \prod_{d=1}^{\dim \mathfrak{G}} \mathcal{D}\bar{c}^d \, \mathcal{D}c^d \, \exp\left\{ i \int d^4 x \, \bar{c}^a(x) M_{ab} c^b(x) \right\} \quad (3.2.168)$$

where $\bar{c}^a$, c^b ($a, b = 1, \ldots, \dim \mathfrak{G}$) are anticommuting scalar functions (generators of the Grassmann algebra). The boundary conditions for the fields $\bar{c}^a$, c^b have essentially the same form as for the Yang–Mills fields:

$$\begin{aligned} c(x) &\xrightarrow[t \to \pm\infty]{} c(x; \substack{\text{in} \\ \text{out}}) \\ \bar{c}(x) &\xrightarrow[t \to \pm\infty]{} \bar{c}(x; \substack{\text{in} \\ \text{out}}) \end{aligned} \quad (3.2.169)$$

$$\begin{aligned} c_j^a(x; \substack{\text{in} \\ \text{out}}) &= \frac{1}{(2\pi)^{3/2}} \int d^3 k \, \frac{1}{\sqrt{2\omega}} (\lambda_j^{*a}(\boldsymbol{k}; \substack{\text{in} \\ \text{out}}) e^{-i\boldsymbol{k} \cdot \boldsymbol{r} + i\omega t} + \beta_j^a(\boldsymbol{k}; \substack{\text{in} \\ \text{out}}) e^{i\boldsymbol{k} \cdot \boldsymbol{r} - i\omega t}) \\ \bar{c}_l^a(x; \substack{\text{in} \\ \text{out}}) &= \frac{1}{(2\pi)^{3/2}} \int d^3 k \, \frac{1}{\sqrt{2\omega}} (\beta_j^{*a}(\boldsymbol{k}; \substack{\text{in} \\ \text{out}}) e^{-i\boldsymbol{k} \cdot \boldsymbol{r} + i\omega t} + \lambda_j^a(\boldsymbol{k}; \substack{\text{in} \\ \text{out}}) e^{i\boldsymbol{k} \cdot \boldsymbol{r} - i\omega t}). \end{aligned} \quad (3.2.170)$$

To obtain the chosen determinant, we have to input the zero boundary conditions for the anticommuting amplitudes:

$$\begin{aligned} \beta^{*a}(\boldsymbol{k}; \text{out}) &= 0 & \lambda^{*a}(\boldsymbol{k}; \text{out}) &= 0 \\ \beta^a(\boldsymbol{k}; \text{in}) &= 0 & \lambda^a(\boldsymbol{k}; \text{in}) &= 0. \end{aligned}$$

Using this integral representation, the S-matrix generating functional (3.2.167) can be cast into the form

$$S = \mathfrak{N}^{-1} \int \prod_{\mu=0}^{3} \mathcal{D}\mathsf{A}_\mu(x) \, \mathcal{D}\mathsf{c}(x) \, \mathcal{D}\bar{\mathsf{c}}(x) \, \exp\left\{ i \int d^4 x \, \mathcal{L}_\alpha^{\text{YM}} \right\} \quad (3.2.171)$$

here

$$\mathcal{L}_\alpha^{\text{YM}} = \text{Tr}\left[\frac{1}{8} \mathsf{F}_{\mu\nu} \mathsf{F}^{\mu\nu} - \frac{1}{4\alpha} (\partial^\mu \mathsf{A}_\mu)^2 - \frac{1}{2} \bar{\mathsf{c}}(\Box \mathsf{c} - g \partial^\mu [\mathsf{A}_\mu, \mathsf{c}]) \right] \quad (3.2.172)$$

$$\begin{aligned} \mathsf{A}_\mu(x) &\xrightarrow[t \to \pm\infty]{} \mathsf{A}_\mu(x; \substack{\text{in} \\ \text{out}}) \\ \mathsf{c}(x) &\xrightarrow[t \to \pm\infty]{} \mathsf{c}(x; \substack{\text{in} \\ \text{out}}) \\ \bar{\mathsf{c}}(x) &\xrightarrow[t \to \pm\infty]{} \bar{\mathsf{c}}(x; \substack{\text{in} \\ \text{out}}). \end{aligned} \quad (3.2.173)$$

Due to the introduction of the fictitious ('ghost') fields $\bar{\mathsf{c}}$, c, called the *Faddeev–Popov ghosts* (Faddeev and Popov 1967), we have managed to present the generating functional for the non-Abelian Yang–Mills theory in the standard form of the path integration of $\exp\{iS_{\text{eff}}\}$, though with the *effective quantum action* S_{eff} including the non-physical anticommuting fields. This result allows us to develop the perturbation theory for an arbitrary Yang–Mills model in much the same way as for any field theory.

◇ Perturbation theory expansion and diagram techniques for the Yang–Mills theory

Let us pass in the usual way (cf section 3.1.1) to the generating functional $\mathcal{Z}[J, \bar{\eta}, \eta]$ for Green functions:

$$\mathcal{Z}[J, \bar{\eta}, \eta] = \mathfrak{N}^{-1} \int \prod_{\mu=0}^{3} \mathcal{D}\mathsf{A}_\mu(x)\,\mathcal{D}\mathsf{c}(x)\,\mathcal{D}\bar{\mathsf{c}}(x)$$
$$\times \exp\left\{ i \int d^4x\, \mathcal{L}_\alpha^{\text{YM}} + J^{a\mu} A_\mu^a + \bar{\eta}^a c^a + \bar{c}^a \eta^a \right\}. \tag{3.2.174}$$

Here the sources $\bar{\eta}^a$ and η^a anticommute with each other and with the fields $\bar{c}^a$ and c^a. The representation of the higher-order (i.e. higher than quadratic) terms of the exponential in (3.2.174), which read as

$$S_{\text{int}}[\mathsf{A}_\mu, \bar{c}, c] \equiv \tfrac{1}{8}\,\text{Tr} \int d^4x\, (2g(\partial_\nu \mathsf{A}_\mu - \partial_\mu \mathsf{A}_\nu)[\mathsf{A}^\mu, \mathsf{A}^\nu] + g^2[\mathsf{A}_\mu, \mathsf{A}_\nu][\mathsf{A}^\mu, \mathsf{A}^\nu] + g\bar{c}\partial^\mu[\mathsf{A}_\mu, c]) \tag{3.2.175}$$

by the variational derivatives and followed by integration of the remaining Gaussian integrand yields

$$\mathcal{Z}[J, \bar{\eta}, \eta] = \exp\left\{ i S_{\text{int}}\left[\frac{1}{i}\frac{\delta}{\delta J_\mu}, \frac{1}{i}\frac{\delta}{\delta \bar{\eta}}, \frac{1}{i}\frac{\delta}{\delta \eta}\right] \right\}$$
$$\times \exp\left\{ -\frac{i}{2} \int d^4x\, d^4y [J_\mu^a(x) D_{\mu\nu}^{\alpha ab}(x-y) J_\nu^b + 2\bar{\eta}^a(x) D^{ab}(x-y)\eta^b] \right\}. \tag{3.2.176}$$

Note that the derivatives with respect to $\bar{\eta}$ act from the *left*, and those with respect to η act from the *right*.

The expansion of the first exponential in (3.2.176) in the Taylor series produces the Feynman diagram technique listed in table 3.3.

Each Feynman diagram constructed of the elements from table 3.3 gives a contribution to the corresponding Green function. The contribution of any diagram is accompanied by the combinatorial factor (which can be straightforwardly derived from (3.2.176))

$$\frac{(-1)^s}{r}\left(\frac{i}{(2\pi)^4}\right)^{l-V-1} \tag{3.2.177}$$

where V is the number of vertices in the diagram, l is the number of internal lines, s is the number of ghost loops and r is the diagram symmetry factor, counting the number of possible transpositions of the internal lines of a diagram at fixed vertices (see also supplement III).

◇ Yang–Mills fields interacting with matter fields

The Yang–Mills fields and the corresponding particles in realistic physical models (in particular, photons, gluons, $W^\pm$, Z^0-bosons; see below section 3.2.8 for a short discussion of the physical applications of gauge theories) serve as interaction mediators between other fields, called *matter fields*. The latter correspond to the particles (e.g., electrons, muons, quarks) which are the building blocks of any kind of matter in nature. In fact, this terminology is rather conditional: for example, we can truly state that such a basic 'matter building block' as the proton (as well as other *hadrons*) consists both of quarks and gluons. However, nowadays this terminology is customary and we shall follow it.

The addition of matter fields (i.e. spinor or scalar fields) to a Yang–Mills model does not bring up any new problems. The gauge group still acts on the gauge fields in the same way as in the absence of matter fields and the classical initial Lagrangian still remains gauge invariant (see section 3.2.1). Therefore, path-integral quantization requires a gauge condition which fixes the choice of representatives in the classes of

Table 3.3. Correspondence rules for the Yang–Mills theory.

Physical quantity	Mathematical expression	Diagram element
Propagator of gauge fields	$D_{\mu\nu}^{\alpha\,ab}(k) = -\dfrac{\delta^{ab}}{k^2 + i\varepsilon}\left(g_{\mu\nu} - \dfrac{(1-\alpha)k_\mu k_\nu}{k^2 + i\varepsilon}\right)$	$\mu, a \;\; k \;\; \nu, b$ (gluon line)
Propagator of ghost fields	$D_g^{ab} = -\dfrac{\delta^{ab}}{k^2 + i\varepsilon}$	$a \;\; k \;\; b$ (ghost line)
Three-interaction vertex	$V_{A^3} = -ig f_{abc}[(p-k)_\rho g_{\mu\nu} + (k-q)_\mu g_{\nu\rho} + (q-p)_\nu g_{\mu\rho}]$	three-gluon vertex
Four-interaction vertex	$V_{A^3} = g^2[f_{abe}f_{cde}(g_{\mu\rho}g_{\nu\sigma} - g_{\mu\sigma}g_{\nu\rho}) + f_{ace}f_{bde}(g_{\mu\nu}g_{\rho\sigma} - g_{\mu\sigma}g_{\nu\rho}) + f_{ade}f_{cbe}(g_{\mu\rho}g_{\sigma\nu} - g_{\mu\nu}g_{\sigma\rho})]$	four-gluon vertex
Ghost $\leftrightarrow$ Yang–Mills field vertex	$V_{\bar c c A} = -i\dfrac{g}{2} f_{abc}(k-q)_\mu$	ghost-gluon vertex

gauge-equivalent Yang–Mills fields and provides the convergence of the path integral for the generating functionals. Path integration over matter fields (spinor $\bar\psi$, ψ or scalar φ) occurs without any peculiarities. Of course, a rigorous derivation, which we drop here (the gap is partially filled in problem 3.2.4, page 99), must be based on the Hamiltonian formalism; essentially, it repeats the consideration previously outlined for the pure Yang–Mills theory (see Faddeev and Slavnov (1980) and Gitman and Tyutin (1990)).

The matter fields are transformed according to the representations $T(u(x))$ of the gauge group $\mathfrak{G}$:

$$\begin{aligned}\psi_i(x) &\to T_{ij}^{(1)}(u(x))\psi_j(x) & i,j=1,\ldots,d_1\\ \varphi_i(x) &\to T_{ij}^{(2)}(u(x))\varphi_j(x) & i,j=1,\ldots,d_2\end{aligned} \qquad (3.2.178)$$

where $T_{ij}^{(1)}(u(x))$ and $T_{ij}^{(2)}(u(x))$ are some matrix representations of dimensions d_1 and d_2, respectively, for the elements $u(x) \in \mathfrak{G}$ of the gauge group $\mathfrak{G}$.

Consider a typical model with non-Abelian fields which contains a multiplet of spinor fields $\psi^i(x)$, a multiplet of charged scalar fields $\varphi^j(x)$ and the gauge fields A_μ corresponding to a gauge group $\mathfrak{G}$. The

Lagrangian of the model has the form

$$\mathcal{L} = -\frac{1}{4g^2} F^a_{\mu\nu} F^{a\mu\nu} + \bar{\psi}^i \gamma^\mu (\partial_\mu \delta_{ij} + \mathrm{i} t^{(1)a}_{ij} A^a_\mu) \psi^j - m_\psi \bar{\psi}^i \psi^i$$
$$+ \sum_i |(\partial_\mu \delta_{ij} + \mathrm{i} t^{(2)a}_{ij} A^a_\mu) \varphi^j|^2 - m_\varphi^2 \varphi^{*i} \varphi^i$$
$$- h_{ijk} \bar{\psi}^i \psi^j \varphi^k - h^*_{ijk} \bar{\psi}^i \psi^j \varphi^{*k} - \tfrac{1}{4} \lambda_{ijkl} \varphi^{*i} \varphi^{*j} \varphi^k \varphi^l \qquad (3.2.179)$$

where $t^{(1)a}_{ij}$ and $t^{(2)a}_{ij}$ are the generators of the group $\mathfrak{G}$ in the representations $T^{(1)}_{ij}$ and $T^{(2)}_{ij}$ (cf (3.2.178)), respectively. The first term in (3.2.179) is the pure Yang–Mills action for the gauge fields, while the rest of the terms in the first line describe the spinor fields and their interactions with the gauge fields; the terms in the second line are responsible for the scalar matter fields and their interaction with the gauge bosons; finally, the terms in the third line represent the Yukawa interaction between the spinor and scalar fields, as well as the scalar self-interaction (the very last term).

The starting point for deriving the perturbation theory expansion for the generating functional $\mathcal{Z}[J,\ldots]$ (or for the S-matrix symbol) is the path integral (in the Coulomb gauge)

$$\mathcal{Z}[J, \rho, \bar{\eta}, \eta] = \int \mathcal{D}A_\mu \, \mathcal{D}\bar{\psi} \, \mathcal{D}\psi \, \mathcal{D}\varphi \, \delta(\partial_k A_k) \det \widehat{M}_{\mathrm{C}}$$
$$\times \exp\left\{ \mathrm{i} \int d^4 x \, [\mathcal{L} + J^{a\mu} A^a_\mu + \bar{\eta}\psi + \eta\bar{\psi} + \rho\varphi] \right\}. \qquad (3.2.180)$$

The repetition of the steps discussed earlier for the pure Yang–Mills theory leads to the following expression for the generating functional in the α-gauge:

$$\mathcal{Z}[J, \rho, \bar{\eta}, \eta, \bar{\chi}, \chi] = \exp\left\{ \mathrm{i} \int d^4 x \, \mathcal{L}_{\mathrm{int}}\left(\frac{1}{\mathrm{i}} \frac{\delta}{\delta J^a_\mu}, \ldots, \frac{1}{\mathrm{i}} \frac{\delta}{\delta \chi^a} \right) \right\}$$
$$\times \exp\left\{ \int d^4 x \, d^4 y \, [J^a_\mu(x) D^{\alpha ab}_{\mu\nu}(x-y) J^b_\nu(y) + \bar{\eta}^i(x) S^{ij}_{\mathrm{c}}(x-y) \eta^j(y) \right.$$
$$\left. + \rho^i(x) D^{ij}_{\mathrm{c}}(x-y) \rho^j(y) + \bar{\chi}^a(x) D^{ab}_g \chi^b(y)] \right\}. \qquad (3.2.181)$$

Here we have extended the generating functional via inclusion of the sources $\bar{\chi}$, χ for the Faddeev–Popov ghost fields (of course, the ghost fields do not appear in the physical amplitudes as in- and out-particles); $\mathcal{L}_{\mathrm{int}}$ is the interaction part of the Lagrangian (3.2.179) (i.e. all the terms higher than second order in the fields); $D^{\alpha ab}_{\mu\nu}$, S^{ij}_{c}, D^{ij}_{c}, D^{ab}_g are the propagators of the corresponding fields; see (3.2.162), (3.1.106), (3.1.93) and (3.2.176), respectively. As a result, we have the new diagram elements depicted in table 3.4, in addition to those presented in table 3.3.

◇ **Lagrangian, path-integral representation for the generating functional and the Feynman diagrams for quantum electrodynamics**

If we consider electromagnetic interactions between *leptons* only (i.e. electrons, positrons, muons etc, see section 3.2.8), the corresponding Lagrangian has the form (3.2.21), in general, with several spinor fields, corresponding to the different sorts of charged lepton. (In the case of *hadrons*, the Lagrangian becomes more involved; in particular it contains the so-called form-factors (see, e.g., Feynman (1972b).) Let us consider, for simplicity, the interaction of photons with only one sort of lepton field, e.g., with fields

Table 3.4. Correspondence rules for the Yang–Mills theory with matter fields, in addition to table 3.3.

Physical quantity	Mathematical expression	Diagram element
Propagator of spinor fields	$S_c^{ij}(k) = -\dfrac{\delta^{ij}}{\not{k} - m_\psi + i\varepsilon}$	$\mu, a \quad k \quad \nu, b$
Propagator of scalar fields	$D_c^{ij} = -\dfrac{\delta^{ij}}{k^2 - m_\varphi^2 + i\varepsilon}$	$a \quad k \quad b$
Spinor $\leftrightarrow$ gauge-field interaction vertex	$V_{\bar\psi A\psi} = g\gamma_\mu t_{ij}^{(1)a}$	
Scalar $\leftrightarrow$ gauge-field four-interaction vertex	$V_{A^2\varphi^2} = ig^2 g_{\mu\nu}(t_{ik}^{(2)a}t_{kj}^{(2)b} + t_{ik}^{(2)b}t_{kj}^{(2)a})$	
Scalar $\leftrightarrow$ gauge-field three-interaction vertex	$V_{\varphi^2 A} = g(t_{ij}^{(2)a}(p_\mu - q_\mu))$	
Scalar $\leftrightarrow$ scalar four-interaction vertex	$V_{\varphi^4} = -\dfrac{i}{4}(\lambda_{ijkl} + \lambda_{jikl} + \lambda_{ijlk} + \lambda_{jilk})$	
Yukawa coupling (spinor $\leftrightarrow$ scalar field vertex)	$V_{\bar\psi\varphi\psi} = h_{ijk}$	

Path-integral quantization of gauge-field theories

Table 3.5. Correspondence rules for quantum electrodynamics.

Physical quantity	Mathematical expression	Diagram element
Propagator of photon fields	$D^\alpha_{\mu\nu}(k) = -\dfrac{1}{k^2 + i\varepsilon}\left(g_{\mu\nu} - \dfrac{(1-\alpha)k_\mu k_\nu}{k^2 + i\varepsilon}\right)$	$\mu \;\;\; k \;\;\; \nu$ (wavy line)
Spinor ↔ photon interaction vertex	$V_{\bar\psi A\psi} = g\gamma_\mu$	(vertex diagram)

describing electrons and positrons. The standard calculations (in the α-gauge) discussed earlier cast the Green function generating functional for QED into the form

$$\mathcal{Z}[J_\mu, \bar\eta, \eta] = \exp\left\{-i\int d^4y\, \mathcal{L}_{\text{int}}\left(\frac{1}{i}\frac{\delta}{\delta J_\mu(y)}, \frac{1}{i}\frac{\delta}{\delta\bar\eta(y)}, \frac{1}{i}\frac{\delta}{\delta\eta(y)}\right)\right\}$$

$$\times \int \mathcal{D}A_\mu(x)\, \mathcal{D}\bar\eta(x)\, \mathcal{D}\eta(x) \exp\left\{iS_{\text{QED}} + \int d^4x\left[J^\mu A_\mu + \bar\eta\psi + \bar\psi\eta\right]\right\}$$

$$= \exp\left\{-i\int d^4y \left(\frac{1}{i}\frac{\delta}{\delta J_\mu(y)}\right)\left(\frac{1}{i}\frac{\delta}{\delta\bar\eta(y)}\right)\gamma_\mu\left(\frac{1}{i}\frac{\delta}{\delta\eta(y)}\right)\right\}$$

$$\times \exp\left\{-i\int d^4x\, d^4x'\, [\tfrac12 J^\mu D^\alpha_{\mu\nu}(x-x')J^\nu + \bar\eta S_c(x-x')\eta]\right\} \quad (3.2.182)$$

$$S_{\text{QED}} = \int d^4x\, [\tfrac12 A_\mu(g^{\mu\nu}\Box - (1-1/\alpha)\partial^\mu\partial^\nu)A_\nu + \bar\psi(i\gamma^\mu\partial_\mu - m)\psi]. \quad (3.2.183)$$

We recall that quantizing electrodynamics is easier than that of a non-Abelian Yang–Mills theory because the Faddeev–Popov determinant $\det\widehat{M}$ does not depend on the integration variables and, hence, can be absorbed into the normalization constant (see the remark on page 73). The standard graphical representation for the photon propagator and the electromagnetic interaction vertex is depicted in table 3.5. Note that due to the Abelian nature of the electrodynamics gauge group, i.e. the group $U(1)$, there is no self-interaction of photons.

◇ **Lowest-order ('tree') approximation in QED: an example**

As an illustration of a *practical* application of the covariant perturbation techniques in QED, let us consider, briefly, the calculations for the so-called *Compton scattering* of photons on electrons:

$$\gamma(k) + e^-(p) \longrightarrow \gamma(k') + e^-(p') \quad (3.2.184)$$

i.e. for the process of (elastic) scattering in which both the initial and final states correspond to a free electron and photon with different momenta. The Feynman diagrams for this process in the lowest

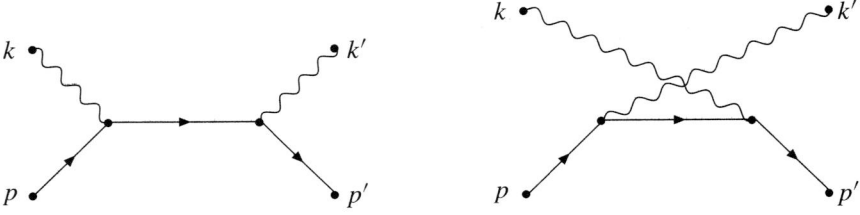

Figure 3.11. Diagrams for the Compton scattering on the electrons.

(second-order) approximation are depicted in figure 3.11. Using the Feynman rules, we can immediately write the expression for the amplitude of the Compton scattering:

$$\langle p', k'; \text{out}|p, k; \text{in}\rangle = (-\mathrm{i})^2 e^2 \Bigg[\bar{v}_{r'}^{(+)}(\boldsymbol{p}') \not{\varepsilon}^{\lambda'} \frac{\not{p} + \not{k} + m}{(p-k)^2 - m^2} \not{\varepsilon}^{\lambda} v_r^{(-)}(\boldsymbol{p}) $$
$$+ \bar{v}_{r'}^{(+)}(\boldsymbol{p}') \not{\varepsilon}^{\lambda} \frac{\not{p} - \not{k}' + m}{(p-k')^2 - m^2} \not{\varepsilon}^{\lambda'} v_r^{(-)}(\boldsymbol{p}) \Bigg]. \qquad (3.2.185)$$

Here $v_r^{(-)}(\boldsymbol{p})$ and $\bar{v}_r^{(+)}(\boldsymbol{p})$ are the initial and final electron states (Dirac spinors).

The probability of a transition upon which the momenta of the final particles (photon and electron in the Compton scattering case) fall within the intervals $(\boldsymbol{k}', \boldsymbol{k}' + d\boldsymbol{k}')$ and $(\boldsymbol{p}', \boldsymbol{p}' + d\boldsymbol{p}')$ is given by the expression (see, e.g., Bogoliubov and Shirkov (1959))

$$w = (2\pi)^4 N_c^2 |\langle p', k'|p, k\rangle|^2 \delta(p + k - p' - k') \frac{d\boldsymbol{k}'}{(2\pi)^3} \frac{d\boldsymbol{p}'}{(2\pi)^3} \qquad (3.2.186)$$

where N_c is a normalization constant. The practically measurable quantity is the so-called *cross section*, which is equal to the probability (3.2.186) divided by the flux of the initial particles.

3.2.7 Higher-order perturbation theory and a sketch of the renormalization procedure for Yang–Mills theories

The perturbation theory described in the preceding sections allows us to calculate Green functions, amplitudes and probabilities with an arbitrary precision in an expansion parameter (usually, a coupling constant g). However, the direct application of the Feynman rules in higher-order perturbation theory, which corresponds to Feynman diagrams with loops, leads to meaningless infinite expressions (divergent integrals). To recover the physical meaning of higher-order terms in the perturbation expansion and eliminate the divergences, we must apply the so-called *renormalization* procedure.

We shall not go into all the details of this involved technique which deals mainly with Feynman diagrams and the corresponding amplitudes rather than directly with path integrals (see, e.g., Bogoliubov and Shirkov (1959) and Itzykson and Zuber (1980)). However, in gauge theories, the independence of the renormalized amplitude of a specific choice of gauge condition should be proven. The basis of such a proof (Faddeev and Slavnov 1980) is provided by the generalized Ward–Takahashi identities (cf section 3.1.5). The derivation of these identities (called in the case of non-Abelian theories *Slavnov–Taylor–Ward–Takahashi identities*), as well as the revelation of some hidden and very helpful symmetry of the effective gauge action (*Becchi–Rouet–Stora–Tyutin (BRST) symmetry*), prove to be simplest in the path-integral formalism. Therefore, we shall only briefly describe the renormalization of gauge theories stressing mainly the role of the path-integral technique.

◇ Divergences of matrix elements

Suppose we want to calculate the fermion Green function in QED. Then, in second-order perturbation theory, we shall encounter the diagram:

$$\begin{array}{c} k \\ p \longrightarrow \underset{p-k}{} \longrightarrow p. \end{array}$$

This diagram corresponds to the following mathematical expression

$$S_c^{(2)} = S_c(p)\Sigma^{(2)}(p)S_c(p) \qquad (3.2.187)$$

where the *electron self-energy* $\Sigma^{(2)}(p)$ in second-order perturbation theory is expressed via the integral

$$\Sigma^{(2)}(p) = \frac{e^2}{(2\pi)^3} \int d^4k \, \gamma_\nu \frac{\not{p} - \not{k} + m}{(p-k)^2 - m^2} \gamma_\nu \frac{1}{k^2} \qquad (3.2.188)$$

(for simplicity, we have used the α-gauge with $\alpha = 1$ for the photon propagator). Counting the powers immediately shows that the integral over k in (3.2.188) is divergent. Indeed, an integral of the type

$$\int_{-\infty}^{\infty} d^n k \, \frac{Ak^m + Bk^{m-1} + \cdots}{Ck^l + Dk^{l-1} + \cdots} \qquad m, l \geq 0$$

converges only if $l > n + m + 1$. In contrast to this, the integral in (3.2.188) contains the fifth order of k (including the integration measure d^4k) in the numerator and only fourth order in the denominator. Thus, we have a linear divergence in this case. Other higher-order Feynman diagrams also contain divergences of different powers.

Two primary problems arise in connection with the occurrence of divergences in higher-order contributions to the matrix elements:

(i) to find all possible types of divergent Feynman diagram and
(ii) to elucidate whether the number of these types of divergence depends on the order of the perturbation theory.

The solution of these problems depends exclusively on the type of interaction term in the corresponding Lagrangian. A characteristic which conveniently discriminates between the different situations and different types of quantum field theory is the so-called *superficial divergence index* ω. This is ascribed to any diagram according to a definite rule, so that negative values of ω correspond to convergent diagrams while positive or zero values correspond to divergent expressions. The adjective 'superficial' is used because a Feynman diagram may contain divergent *subgraphs*, although the overall index ω is negative. Thus, to be sure that a diagram is convergent, we must check the values of ω for all the subgraphs of the given Feynman diagram.

In electrodynamics, the superficial divergence index for a Feynman diagram of an *arbitrary* order has the form

$$\omega = 4 - \tfrac{3}{2}N_e - N_{\text{ph}} \qquad (3.2.189)$$

where N_e is the number of *external* spinor (electron–positron) lines in the diagrams and N_{ph} is the number of *external* photon lines. It is immediately seen that only a limited class of diagrams with a restricted number of external lines (less than five photon lines and four spinor lines) have a non-negative index ω. This is a characteristic property of *renormalizable quantum field theories*. In contrast, *non-renormalizable quantum field theories* have divergent diagrams with a differing number of external lines depending on the order of the perturbation expansion. Non-Abelian Yang–Mills theories also belong to the class of renormalizable field models: the superficial divergence index ω for them reads as follows:

$$\omega = 4 - \tfrac{3}{2} N_\psi - N_{\text{YM}} - N_{\text{gh}} \qquad (3.2.190)$$

where N_ψ is number of *external* spinor lines in the diagrams and N_{YM}, N_{gh} are the numbers of *external* lines corresponding to the gauge and ghost fields, respectively.

◇ **Renormalization procedure**

The divergences of renormalizable theories can be eliminated. More precisely, they can be absorbed into a finite number of constants which can be associated with physically measurable quantities, such as the masses and charges of particles. After substituting *finite* values of the quantities taken from experimental data instead of a combination of the initial parameters introduced into the field Lagrangian and the divergent parts of the Feynman diagrams, all other calculations in the theory become meaningful and we can calculate any other quantities (elements of S-matrix, energy levels etc) and, hence, we can predict any other measurements *theoretically*. A special procedure, referred to as *renormalization*, has been devised for this purpose.

- The first step of this procedure is the *regularization* of divergent diagrams. It is awkward to perform calculations with divergent integrals and therefore it is necessary to temporarily modify the theory so as to make all the integrals finite. At the final stage, the regularization is removed.
- After regularization, we can proceed to eliminate the divergences. A special technique developed for this is called the *R-operation*. This operation enables one to obtain physically meaningful expressions which remain finite after the regularization is removed. In fact, the R-operation consists of substituting the divergent parts and the initial parameters with experimentally measured quantities.
- Additional problems arise for gauge theories. As a matter of fact, the renormalization is equivalent to a redefinition of the initial Lagrangian. Therefore, in the case of gauge fields:
  - it is desirable to choose an intermediate regularization which does not violate the invariance under the gauge transformations;
  - the renormalization procedure (R-operation) should not violate the gauge invariance and
  - the independence of the renormalized amplitudes on the specific choice of the gauge conditions should be proven.

  In proving the renormalizability of gauge-field models, we use Lagrangians containing non-physical fields (ghost fields and longitudinal components of gauge fields). This leads to a loss of an explicit unitarity of the S-matrix, while the unitarity is a necessary condition for the self-consistency of any field theory. Fortunately, there are also gauges leading to Lagrangians which do not contain non-physical fields and have explicitly unitary amplitudes (but which are highly inconvenient for the proof of renormalizability). Thus, the invariance of a renormalized amplitude with respect to the choice of the gauge means that the theory is both *unitary* and *renormalizable*.

◇ **Remarks on regularization methods**

There are several regularization methods: the Pauli–Villars procedure, the method of higher covariant derivatives, the dimensional regularization etc. Perhaps the most natural regularization is the *lattice one*, i.e. the discretization of space and time and transition to the case of a large but finite number of degrees of freedom. In fact, in this chapter we introduced field theory itself using this regularization. Unfortunately, this regularization has the essential disadvantage of losing an explicit relativistic invariance. Nevertheless, it plays an essential role in field theoretical calculations and we shall consider this in chapter 4.

In the present section, we shall illustrate the idea of regularization with the example of *dimensional regularization*. This is based on the fact that the superficial divergence index of a diagram significantly depends on the dimension d of the space, e.g., this index for QED (cf (3.2.189)) can be presented in the form

$$\omega = \sum_{l=1}^{L}(r_l + n - 2) - d(m - 1) \tag{3.2.191}$$

where the summation over all internal lines of the diagram is carried out; L is the number of internal lines, r_l is the order of the polynomial corresponding to the internal line and m is the number of vertices. Therefore, the integrals which are divergent in a four-dimensional space may prove to be convergent in a space of smaller dimension. The number d can be thought of as being a not necessarily positive integer, but also of being a non-integer and even complex number.

Before starting actual calculations, it is necessary to formulate the rules for treating tensor quantities and the γ-matrices in a d-dimensional space with an arbitrary d. This is done by the continuation of the usual rules for the summation of tensor indices and γ-matrix commutation relations, e.g.,

$$g^{\mu\nu} p_\nu = p^\mu$$
$$\gamma_\mu \gamma_\nu + \gamma_\nu \gamma_\mu = 2 g_{\mu\nu} \mathbb{1} \tag{3.2.192}$$

where μ, ν are now *formal* indices corresponding to non-integer or complex-dimensional spaces. The technical details of the dimensional regularization may be found, e.g., in Itzykson and Zuber (1980) and Faddeev and Slavnov (1980). We note only that the dimensional regularization, at least in its simple form, is not applicable to theories in which the matrix γ_5 is involved. Indeed, the genuine definition of the γ_5 matrix,

$$\gamma_5 \stackrel{\text{def}}{\equiv} i \gamma_0 \gamma_1 \gamma_2 \gamma_3 \tag{3.2.193}$$

is heavily based on the concrete value (four) of the spacetime dimension. It can easily be generalized to any even integer dimension, but there is no way for a consistent generalization to non-integer dimensions. This fact may cause the so-called *quantum anomalies*, i.e. the violation of classical symmetries at quantum level; we shall consider this phenomenon in the path-integral formalism in section 3.3.4.

To obtain an idea about dimensional regularization, let us consider the integral

$$I(p^2) = \int d^d k \frac{1}{(k^2 - m^2)[(k-p)^2 - m^2]} \tag{3.2.194}$$

which comes from the purely scalar field theory (cf diagram (b) in figure 3.3, page 27). This integral is divergent for $d = 4$ and convergent for $d < 4$. Let us calculate it assuming that $d < 4$.

Using the formula (*Feynman parametrization*)

$$\frac{1}{ab} = \int_0^1 dx \frac{1}{[ax + b(1-x)]^2} \tag{3.2.195}$$

we can rewrite (3.2.194) in the form (after the change of variables $k - p(1 - x) \to k$):

$$I(p^2) = \int_0^1 dx \int d^d k \frac{1}{[k^2 + p^2 x(1-x) - m^2]^2}. \tag{3.2.196}$$

Rotating the integration contour by 90° (the so-called Wick rotation) and again changing the variable $k_0 \to i k_0$, we obtain the integral over the d-dimensional Euclidean space:

$$I(p^2) = i \int_0^1 dx \int d^d k \frac{1}{[k^2 + p^2 x(1-x) + m^2]^2}. \tag{3.2.197}$$

If d is integer, we can use, for the calculation, the known result for the standard integral

$$\int d^d k \, \frac{1}{(k^2 + M^2)^\alpha} = \pi^{d/2} M^{2(d/2-\alpha)} \frac{\Gamma(\alpha - d/2)}{\Gamma(\alpha)}. \qquad (3.2.198)$$

where $\Gamma(\cdot)$ is the gamma-function. For an arbitrary d, we just postulate formula (3.2.198) and, using it, we obtain the expression for the *regularized integral*

$$I(p^2) = i\pi^{d/2} \Gamma\left(2 - \frac{d}{2}\right) \int_0^1 dx \, [m^2 - p^2 x(1-x)]^{d/2-2}. \qquad (3.2.199)$$

The integral over the x-variable is obviously convergent but the Γ-factor leads to a divergence at $d = 4$, since the function $\Gamma(2 - d/2)$ has a pole at this point. More complicated divergent integrals can be regularized and calculated along the same line.

An important property of dimensional regularization is that it does not violate gauge invariance and since all properties of Lorentz-covariant tensors (except the γ_5-matrix) are straightforwardly generalized to an arbitrary dimension d, any expression in the dimensional regularization has formally a relativistically covariant form.

◇ **Renormalization (R-operation)**

Removing the regularization, i.e. setting $d = 4$ in (3.2.199), gives rise to a pole. This corresponds to the divergence of the initial integral over the four-dimensional space. The expansion of the regularized integral (3.2.199) in the Laurent series with respect to d in the vicinity of the point $d/2 = 2$ yields

$$I(p^2) = \frac{-i\pi^2}{d/2 - 2} - i\pi^2 \int_0^1 dx \, \ln[m^2 - p^2 x(1-x)] + C + \mathcal{O}(d/2 - 2) \qquad (3.2.200)$$

where C is a finite constant. Then let us expand (3.2.200) in the Taylor series with respect to p^2 at some point $p^2 = \lambda^2$ (referred to as a *renormalization point*). Subtracting $I(p^2)|_{p^2=\lambda^2}$ from (3.2.200), we arrive at an expression which does not contain divergences:

$$I_R(p^2) = -i\pi^2 \int_0^1 dx \, \ln \frac{m^2 - p^2 x(1-x)}{m^2 - \lambda^2 x(1-x)}. \qquad (3.2.201)$$

This is the renormalized expression for the integral (3.2.194). The choice of the renormalization point is arbitrary. By taking another renormalization point, we obtain an expression which differs from (3.2.201) by a finite polynomial in p^2. In realistic physical models, this arbitrariness is fixed by the requirement that the particles in the theory possess experimentally known charges and masses.

In a similar way, the renormalization of other integrals can be performed. But for more complicated integrals, including integrations over many momenta, the simple renormalization method discussed so far is insufficient. In such cases, the *R-operation* developed by Bogoliubov, Parasiuk, Hepp and Zimmermann should be applied (see, e.g., Bogoliubov and Shirkov (1959) and Hepp (1969)).

The renormalization procedure can also be formulated in a different language. The point is that the replacement of divergent integrals by the renormalized ones is equivalent to the inclusion of some additional terms in the initial Lagrangian. These are called *counter-terms*. Therefore, the renormalization can be carried out by introducing counter-terms into Lagrangians (see, e.g., Bogoliubov and Shirkov (1959) and Itzykson and Zuber (1980)).

◇ Generalized Ward–Takahashi identities

As already mentioned, the renormalization of gauge theories brings about additional problems associated with the requirement that the gauge invariance of the theory should not be violated. In this case, the renormalization is also carried out by means of the R-operation. To prove that the gauge invariance is not violated by that procedure, it is convenient to utilize the *generalized Ward–Takahashi identities* also called the *Slavnov–Taylor–Ward–Takahashi identities*. The derivation of these identities in the case of non-Abelian Yang–Mills theory is heavily based on the path-integral representation for the Green function generating functional and we shall consider this derivation in some more detail.

Let us discuss as an example the pure Yang–Mills theory (i.e. without matter fields). We start from the generating functional $\mathcal{Z}[J^a]$ in the α-gauge written in the form

$$\mathcal{Z}[J^a] = \int \prod_{a=1}^{\dim \mathfrak{G}} \mathcal{D}A_\mu^c(x)\,\mathcal{D}a^c(x)\,\delta(\partial^\mu A_\mu^c - a^c)\Delta_\alpha(A_\mu^c)$$
$$\times \exp\left\{ iS_{\text{YM}} - i\int d^4x \left[\frac{1}{2\alpha}(A^b)^2 + J^{b\mu}A_\mu^b \right] \right\} \qquad (3.2.202)$$

where S_{YM} is the gauge-invariant Yang–Mills action (3.2.12).

The general idea of deriving the Ward–Takahashi-type identities was discussed in section 3.1.5. Practically, we can proceed as follows. Let us perform the gauge transformation $\mathbf{A}_\mu \to \mathbf{A}_\mu^u$, so that the Yang–Mills field now satisfies the condition

$$\partial^\mu (A^u)_\mu^c(x) - a^c(x) - b^c(x) = 0. \qquad (3.2.203)$$

The transition to this new gauge is fulfilled via the standard Faddeev–Popov trick: we introduce the gauge-invariant functional $\widetilde{\Delta}_\alpha(\mathbf{a}_\mu)$, defined by

$$\widetilde{\Delta}_\alpha[\mathbf{A}] \int \mathcal{D}u(x)\,\delta[\partial^\mu \mathbf{A}_\mu^u - \mathbf{a} - \mathbf{b}] = 1. \qquad (3.2.204)$$

The substitution of (3.2.204) into (3.2.202) and the change of variables $\mathbf{A} \to \mathbf{A}^u$, $u \to u^{-1}$ yield:

$$\mathcal{Z}[J^a] = \int \mathcal{D}\mathbf{A}_\mu(x)\,\mathcal{D}\mathbf{a}(x)\,\mathcal{D}u(x)\,\Delta_\alpha(A_\mu^a)\widetilde{\Delta}_\alpha(A_\mu^a)\delta(\partial^\mu \mathbf{A}_\mu - \mathbf{a})\delta(\partial^\mu \mathbf{A}_\mu - \mathbf{a} - \mathbf{b})$$
$$\times \exp\left\{ iS_{\text{YM}} - i\int d^4x \frac{1}{2}\operatorname{Tr}\left[\frac{1}{2\alpha}\mathbf{a}^2 + \mathbf{J}^\mu \mathbf{A}_\mu^u \right] \right\}. \qquad (3.2.205)$$

At this point, we have essentially used the gauge invariance of the classical Yang–Mills action S_{YM}. To perform the integration over $u(x)$ and $\mathbf{a}$, we make use of the fact that the term $\frac{1}{2}\operatorname{Tr}\mathbf{J}^\mu \mathbf{A}_\mu^u$ in (3.2.205) can be represented, due to the presence of the two δ-functionals, as $\frac{1}{2}\operatorname{Tr}\mathbf{J}^\mu \mathbf{A}_\mu^{u_0}$, where $u_0(x; \mathbf{A}_\mu, \mathbf{b})$ is the solution of the following set of equations:

$$\partial^\mu (A^{u_0})_\mu^c(x) - a^c(x) = 0$$
$$\partial^\mu A_\mu^c(x) - a^c(x) - b^c(x) = 0. \qquad (3.2.206)$$

For infinitesimally small functions $b^a(x)$, the first equation in (3.2.206) can be written as follows:

$$\partial^\mu A_\mu^c(x) - a^c(x) + \widehat{M}^{cb}\varepsilon_b(x) = 0 \qquad (3.2.207)$$

where $\varepsilon_b(x)$ denotes the parameters of the infinitesimal gauge transformation u_0 and

$$\widehat{M}^{ab} = \partial^\mu D_\mu^{ab} = \partial^\mu(\delta_{ab}\partial_\mu + gf_{abc}A_\mu^c). \qquad (3.2.208)$$

The obvious solution of (3.2.206) in this case reads as

$$\varepsilon^a(x) = (\widehat{M}^{-1})^{ac}b^c. \qquad (3.2.209)$$

Thus, the term $\frac{1}{2}\operatorname{Tr} \mathsf{J}^\mu \mathsf{A}_\mu^u$ takes the form

$$J^{a\mu}(x)(A^{u_0})_\mu^a(x) = J^{a\mu}(x)A_\mu^a(x) + J^{a\mu}(x)D_\mu^{ab}\varepsilon_b(x)$$
$$= J^{a\mu}(x)A_\mu^a(x) + J^{a\mu}(x)D_\mu^{ab}\int d^4y\,(M^{-1})^{bc}(x,y)b^c(y) \qquad (3.2.210)$$

and the whole generating functional (3.2.205) now becomes

$$\mathcal{Z}[J^a] = \int \prod_\mu \mathcal{D}\mathsf{A}_\mu(x)\,\det\widehat{M}\,\exp\left\{\mathrm{i}S_{\mathrm{YM}} - \mathrm{i}\int d^4x\left[\frac{1}{2\alpha}(\partial^\mu A_\mu^a - b^a)^2 + J^{a\mu}(x)A_\mu^a(x)\right.\right.$$
$$\left.\left. + \mathrm{i}\int d^4x\,d^4y\,J^{a\mu}(x)D_\mu^{ab}(M^{-1})^{bc}(x,y)b^c(y)\right]\right\} \qquad (3.2.211)$$

(recall that on the surface defined by the gauge condition, there exists the equality $\widetilde{\Delta}_\alpha(\mathsf{A}) = \det\widehat{M}$).

Let us now differentiate both sides of (3.2.211) with respect to $b^a(x)$. Since the initial functional (3.2.202), coinciding with (3.2.211), does not depend on $b^a(x)$, we have

$$\left.\frac{\delta\mathcal{Z}[J]}{\delta b^a(y)}\right|_{b^a=0} = \int \prod_\mu \mathcal{D}\mathsf{A}_\mu(x)\,\det\widehat{M}\left[\frac{1}{\alpha}\partial^\mu A_\mu^a(y) + \int dy'\,J^{b\mu}(y')D_\mu^{bc}(M^{-1})^{ca}(y',y)\right]$$
$$\times \exp\left\{\mathrm{i}S_{\mathrm{YM}} - \mathrm{i}\int d^4x\left[\frac{1}{2\alpha}(\partial^\mu A_\mu^a)^2 + J^{a\mu}(x)A_\mu^a(x)\right]\right\} = 0. \qquad (3.2.212)$$

This system of identities can be rewritten in terms of functional derivatives of the generating functional $\mathcal{Z}[J]$ with respect to the currents

$$\left\{\frac{1}{\alpha}\partial^\mu\left[\frac{1}{\mathrm{i}}\frac{\delta}{\delta J^{a\mu}(x)}\right] + \int d^4y\,J_\mu^b(y)\left[D_\mu^{bc}\left(y,\frac{1}{\mathrm{i}}\frac{\delta}{\delta J_\mu^d(y)}\right)(M^{-1})^{ca}\left(y,x;\frac{1}{\mathrm{i}}\frac{\delta}{\delta J_\mu^d}\right)\right]\right\}\mathcal{Z}[J] = 0. \qquad (3.2.213)$$

In a similar way, the generalized Ward–Takahashi identities can be found for other cases, e.g., for gauge fields interacting with (spinor or scalar) fields of matter.

The generalized Ward–Takahashi identities stem from the physical equivalence of various gauges. As can be seen from (3.2.213), they lead to certain relations between the Green functions.

◇ BRST symmetry of the Yang–Mills effective action and another way of deriving the generalized Ward–Takahashi identity

We still consider the case of pure Yang–Mills fields (without matter fields). This time we shall use the generating functional with the Faddeev–Popov ghost fields:

$$\mathcal{Z}[J] = \int \mathcal{D}A_\mu^a(x)\,\mathcal{D}\bar{c}^a(x)\,\mathcal{D}c^a(x)\,\exp\left\{\mathrm{i}\int dx\,[\mathcal{L}_{\mathrm{eff}} + J^{a\mu}A_\mu^a + \bar{c}^a\eta + \bar{\eta}^a c^a]\right\} \qquad (3.2.214)$$

with
$$\mathcal{L}_{\text{eff}} = -\frac{1}{4}(F^a_{\mu\nu})^2 - \frac{1}{2\alpha}(\partial^\mu A^a_\mu)^2 - \partial^\mu \bar{c}^a \partial_\mu c^a - g f_{abd} \partial^\mu \bar{c}^a A^b_\mu c^d. \tag{3.2.215}$$

It can be directly proven that the *effective* Lagrangian $\mathcal{L}_{\text{eff}}$ is invariant under the following combined transformations of the gauge fields $A^a_\mu(x)$ and the ghost fields $c^a(x)$ which are called *Becchi–Rouet–Stora–Tyutin (BRST) transformations* (Becchi *et al* 1975, 1976, Tyutin 1975):

$$\begin{aligned} A^a_\mu(x) &\longrightarrow A^a_\mu(x) + (D_\mu c^a(x))\zeta \\ c^a(x) &\longrightarrow c^a(x) - \tfrac{1}{2} f_{abd} c^b(x) c^d(x) \zeta \\ \bar{c}^a(x) &\longrightarrow \bar{c}^a(x) - \frac{1}{\alpha}(\partial^\mu A^a_\mu(x))\zeta. \end{aligned} \tag{3.2.216}$$

The parameter ζ does not depend on the coordinates, i.e. these are *global* transformations. In addition, the parameter ζ cannot be an ordinary number because the Yang–Mills fields A^a_μ obey commutation relations (the variables A^a_μ in the path integral commute), while the ghost fields $\bar{c}^a$, c^a are anticommuting variables. Thus, for consistency, the parameter ζ must be Grassmannian and satisfy the relations:

$$\zeta^2 = 0 \qquad [\zeta, A^a_\mu] = 0 \qquad \{\zeta, c^a\} = 0 \qquad \{\zeta, \bar{c}^a\} = 0.$$

Making the change of variables (3.2.216) in the path integral (3.2.214), we arrive at an expression for the generating functional which contains the parameter ζ. Since the initial path integral does not contain ζ and since an integral does not depend on the choice of integration variables (note that the Jacobian of the substitution (3.2.216) is equal to unity), the differentiation of the resulting path integral over ζ gives the identity

$$\int \mathcal{D}A^a_\mu(x)\,\mathcal{D}\bar{c}^a(x)\,\mathcal{D}c^a(x) \left[\int d^4y\, J^{a\mu}(y) D_\mu c^a(y) - \frac{1}{\alpha}\partial^\mu A^a_\mu(y)\eta^a(y) - \frac{1}{2}\bar{\eta}^a f_{abd} c^b(y) c^d(y) \right]$$
$$\times \exp\left\{ i \int dx\, [\mathcal{L}_{\text{eff}} + J^a_\mu A^a_\mu + \bar{c}^a \eta + \bar{\eta}^a c^a] \right\} = 0. \tag{3.2.217}$$

This is just another form of the generalized Ward–Takahashi identity (3.2.213) (problem 3.2.6, page 100).

Likewise, Ward–Takahashi identities of the type (3.2.217) for more general models than pure gauge fields can be obtained.

Two remarks are in order:

- The role of the Ward–Takahashi identities in the renormalization of Yang–Mills theories can be briefly described as follows. The general renormalization procedure (R-operation) prescribes counter-terms (subtractions) which potentially may violate the gauge invariance of the theory. However, if the gauge model under consideration allows some gauge-invariant regularization, the generalized Ward–Takahashi identities establish certain relations between the counter-terms leaving only their gauge-invariant combinations. After this, we can remove the regularization and due to the explicit gauge invariance at each step of the renormalization procedure, the resulting quantum theory without divergences also proves to be gauge invariant.
- If a model does not allow gauge-invariant regularizations (e.g., theories containing γ_5-matrices), there is no guarantee that the 'naive' Ward–Takahashi identity (without taking into account field theoretical divergences) is still valid for the regularized theory. Instead, we obtain what is called the *anomalous Ward–Takahashi identity*. Correspondingly, the renormalized theory may lose the gauge invariance of its classical counterpart. In this case the theory is said to have *quantum anomalies*. We shall consider this situation in somewhat more detail in section 3.3.4.

Further applications and detailed discussion of the BRST symmetry may be found in Nakanishi and Ojima (1990).

3.2.8 Spontaneous symmetry-breaking of gauge invariance and a brief look at the standard model of particle interactions

In order to illustrate the actual physical application of the quantum gauge-field theory which, in turn, is heavily based on the path-integral formalism, we shall briefly discuss the so-called *standard model of electroweak and strong interactions*.

Recall that, in spite of the apparent diversity of interactions between physical objects in nature (depending on the interactions of the elementary particles forming these objects), only *four* types of *fundamental interaction* of elementary particles (the interactions are enumerated in the order of their increasing strength) exist:

(i)   *gravitational* interactions;
(ii) *weak* interactions (responsible for most decays and many transformations of elementary particles);
(iii) *electromagnetic* interactions;
(iv) *strong* interactions (providing, in particular, the bounds of particles in atomic nuclei, so that sometimes they are also called *nuclear* interactions).

Up to the beginning of the 1970s, quantum electrodynamics was the only successful example of a physical application of gauge-field theories. However, by that time experimental study of the weak interactions had revealed a considerable similarity between weak and electromagnetic interactions. Among other things, there was strong evidence that weak interactions are mediated by vector particles (similar to the photon) and are characterized by a single coupling constant (the so-called universality of the weak interactions). All these features acquire natural explanations if we assume that both electromagnetic and weak interactions are described by a gauge-invariant theory, the Yang–Mills field being the mediator of the interactions (Schwinger 1957, Glashow 1961). However, along with the similarity between the two types of fundamental interaction, there are essential differences. The most obvious one is that, in contrast to the *long-range* electromagnetic interaction (which corresponds to *massless* intermediate particles, i.e. photons), the weak interaction has a very short interaction radius and, hence, must be based on *massive* intermediate particles. For a long time, this and some other differences prevented the construction of a unified field theoretical model of weak and electromagnetic interactions. Fortunately, this problem can be overcome with the help of the so-called *Higgs mechanism* based, in turn, on the very important phenomenon of *spontaneous symmetry-breaking* (Weinberg 1967, Salam 1968). The unified model of weak and electromagnetic interactions based on the non-Abelian Yang–Mills theory with the gauge group $SU(2) \times U(1)$, together with the spontaneous symmetry-breaking and the Higgs mechanism, is called the *Glashow–Salam–Weinberg model* or *standard model*. A nice property of this model is that it proved to be a *renormalizable* quantum field theory ('t Hooft 1971).

At first sight, the dynamics of *strong* interactions looks too complicated to be described by some Lagrangian field theory. The first attempts to construct such a theory were not even in qualitative agreement with the experimental facts. However, later (in the early 1970s), experiments on *deep inelasting scattering* have shown that at small distances, hadrons (strongly interacting particles) behave as if they were made of non-interacting pointlike constituents (*partons* or *quarks*). This fact has led to the conjecture that hadrons are composite objects, the constituents being weakly interacting particles at small distances and strongly interacting at large distances. This phenomenon has acquired the name *asymptotic freedom*. It has been shown (Gross and Wilczek 1973, Politzer 1973) that only the non-Abelian Yang–Mills theory possesses such a property and provides the desired behaviour of quarks. The resulting physical theory, based on the gauge group $SU(3)$, is called *quantum chromodynamics* (QCD).

It is necessary to note that nowadays it is customary to ascribe the name '*standard model*' to a combination of the Glashow–Salam–Weinberg electroweak theory *and* chromodynamics, i.e. to the theory based on the gauge group $SU(3) \times SU(2) \times U(1)$ (with the appropriate spontaneous symmetry-breaking).

Another important remark is that the standard model cannot be considered as a truly *unified model* for the three fundamental interactions. The point is that it is based on a non-simple gauge group and therefore contains a few independent coupling constants (actually, it contains several dozen free parameters such as the masses of particles etc). Many advanced attempts to improve this model, based either on simple gauge groups like, e.g., $SU(5)$ (*grand unified theory*), or on supersymmetric generalizations of the main idea, have been made.

In this subsection, we shall consider only the basic idea of the standard model and other unified theories, namely, spontaneous symmetry-breaking and its implication for the path-integral representation of generating functionals in quantum field theories. The reader may find further details about the standard model, QCD and other unified models in Itzykson and Zuber (1980), Okun (1982), Chaichian and Nelipa (1984), Cheng and Li (1984), West (1986), Bailin and Love (1993), Peskin and Schroeder (1995) and Weinberg (1996, 2000).

◇ The concept of spontaneous symmetry-breaking

First, we shall explain what spontaneous symmetry-breaking is and then discuss a concrete model with local symmetry-breaking.

Consider a quantum-mechanical system with a Hamiltonian $\widehat{H}$. The system can be in various energy states E_n, determined by the stationary Schrödinger equation

$$\widehat{H}\psi_n = E_n\psi_n.$$

If there is a single vacuum state corresponding to the minimum eigenvalue E_0, this is called a non-degenerate vacuum state, otherwise it is called *degenerate*.

Let a certain transformation group $\mathfrak{G}$ be given. The *vacuum* state is *invariant* under the group $\mathfrak{G}$ if it transforms into itself and *non-invariant* otherwise. In the framework of the local relativistic quantum field theory, there exists a connection between the invariance of the vacuum state under a group of transformations and the invariance of the Lagrangian under the same group. This is given in the *Coleman theorem*, that states:

(i) If the vacuum state is invariant, the Lagrangian must necessarily be invariant, too (the case of *exact symmetry*).
(ii) If the vacuum state is non-invariant, the Lagrangian may be either non-invariant or invariant; in both these cases, the symmetry as a whole is broken:
   - in the case of non-invariance of both the vacuum state and the Lagrangian we speak of *explicit* symmetry-breaking;
   - if the vacuum state is non-invariant, whereas the Lagrangian is invariant, the symmetry-breaking is called *spontaneous*.

It can be shown that the case of spontaneous symmetry-breaking necessarily leads to the occurrence of zero-mass particles. This statement is known as the *Goldstone theorem*. Accordingly, the massless particles are called *goldstones*.

◇ A simple model with spontaneous symmetry-breaking of a global symmetry

Consider a model described by the Lagrangian

$$\mathcal{L} = (\partial^\mu \varphi^*)(\partial_\mu \varphi) - m^2 \varphi^* \varphi - \tfrac{1}{4}\lambda(\varphi^*\varphi)^2 \qquad (3.2.218)$$

where $\varphi(x)$ is a complex scalar field, λ is the coupling constant ($\lambda > 0$) and m is the mass of the scalar particle ($m^2 > 0$).

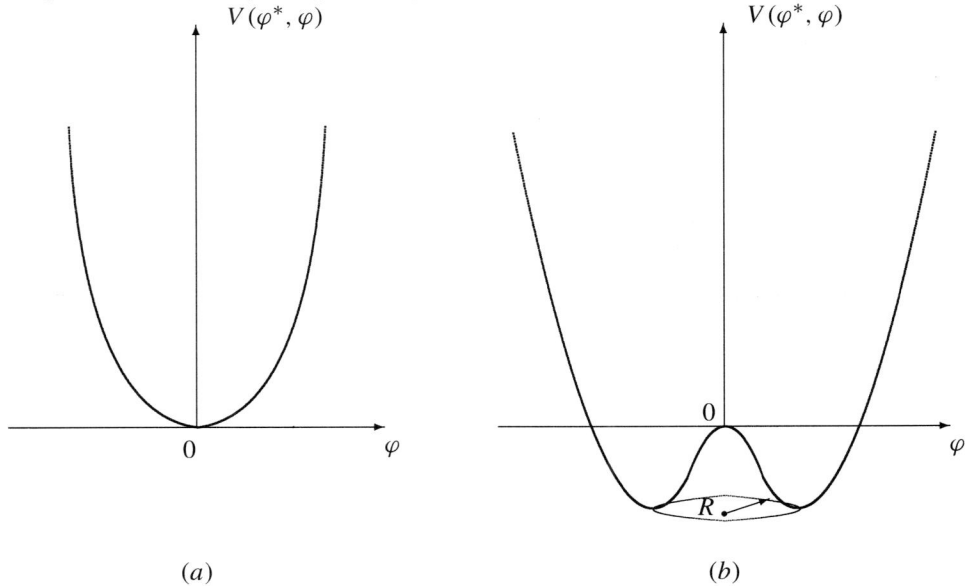

Figure 3.12. Examples of potentials with non-degenerate (*a*) and degenerate (*b*) vacuum states.

This Lagrangian is invariant under the global group $U(1)$ of the phase transformations

$$\varphi(x) \longrightarrow \varphi'(x) = e^{-i\lambda\varepsilon}\varphi(x) \qquad \varphi^*(x) \longrightarrow \varphi'^*(x) = e^{i\lambda\varepsilon}\varphi^*(x). \tag{3.2.219}$$

The conserved energy of the field system is given by the expression

$$E = \int d^3r \, [\partial_0 \varphi^*(t, r)\partial_0 \varphi(t, r) + \partial_i \varphi^*(t, r)\partial_i \varphi(t, r) + m^2\varphi^*(t, r)\varphi(t, r) + \tfrac{1}{4}\lambda(\varphi^*(t, r)\varphi(t, r))^2]. \tag{3.2.220}$$

In the class of static and translationally invariant fields (i.e. $\varphi(x) = $ constant; we shall consider more general solutions later, see section 3.3.3) the energy minimum coincides with the minimum of the function

$$V(\varphi^*, \varphi) = m^2 \varphi^*\varphi + \tfrac{1}{4}\lambda(\varphi^*\varphi)^2. \tag{3.2.221}$$

This minimum is obviously located at the origin of the field space, $\varphi^* = \varphi = 0$, see figure 3.12(*a*). Hence, the vacuum state of the model is non-degenerate and invariant under the transformations (3.2.219). The Lagrangian (3.2.218) is also invariant under the transformations of the group $U(1)$. The model thus has exact $U(1)$-symmetry. In the quantum theory, the *vacuum expectation value* of the field $\varphi(x)$ is zero:

$$\langle 0|\hat{\varphi}(x)|0\rangle = \langle 0|\hat{\varphi}^*(x)|0\rangle = 0. \tag{3.2.222}$$

Now consider a model described by almost the same Lagrangian but with the opposite sign for the quadratic term:

$$\mathcal{L}_{\text{SSB}} = (\partial^\mu \varphi^*)(\partial_\mu \varphi) + m^2\varphi^*\varphi - \tfrac{1}{4}\lambda(\varphi^*\varphi)^2. \tag{3.2.223}$$

The energy of the system in this case takes the form

$$E = \int d^3r \, [\partial_0 \varphi^*(t, r)\partial_0 \varphi(t, r) + \partial_i \varphi^*(t, r)\partial_i \varphi(t, r) - m^2\varphi^*(t, r)\varphi(t, r) + \tfrac{1}{4}\lambda(\varphi^*(t, r)\varphi(t, r))^2] \tag{3.2.224}$$

which, for static and translationally invariant fields, reduces to the function (see figure 3.12(b))

$$V(\varphi^*, \varphi) = -m^2\varphi^*\varphi + \tfrac{1}{4}\lambda(\varphi^*\varphi)^2. \tag{3.2.225}$$

Thus, the energy has a minimum at $\varphi^*\varphi = 2m^2/\lambda$, that is

$$|\varphi_{\min}| = \sqrt{\frac{2}{\lambda}} m. \tag{3.2.226}$$

The system has an infinite set of vacuum states, each of them corresponding to a point on the circle of radius $R = \sqrt{2}m/\sqrt{\lambda}$ on the complex plane φ (see figure 3.12(b)). Thus, the vacuum states are infinitely degenerate. Let us make a few important remarks:

(i) the transformations (3.2.219) convert a certain vacuum state (a point of the circle) into any other state; this means that an arbitrarily chosen vacuum state is *not invariant* under the transformations (3.2.219);
(ii) the Lagrangian $\mathcal{L}_{\text{SSB}}$ in (3.2.223) is *invariant* under the transformations (3.2.219) and
(iii) in order to construct a quantum theory, a definite vacuum state, i.e. a definite point on the circle (3.2.226) has to be chosen; we should bear in mind that different degenerate vacuum states are not related to each other and no superposition can be formed from them (there is no such physical state). This fact is sometimes expressed by the words: 'to different vacuum states there correspond different worlds'.

Thus, the system described by the Lagrangian $\mathcal{L}_{\text{SSB}}$ has a *spontaneously broken $U(1)$-symmetry*.

For the quantization of a theory with spontaneously broken symmetry it is convenient to shift the field variables. Choosing the vacuum state corresponding to the intercept of the circumference with the real axis in the plane φ, we introduce the new field variables φ_1, φ_2 via the relation

$$\varphi(x) = \frac{1}{\sqrt{2}}\left(\frac{2m}{\sqrt{\lambda}} + \varphi_1(x) + i\varphi_2(x)\right) \tag{3.2.227}$$

so that the $\varphi_{1,2}$ describe fluctuations *around the chosen vacuum state*. It is clear that the vacuum expectations of the fields $\hat{\varphi}_{1,2}(x)$ are zero, while that of the field $\operatorname{Re}\varphi(x)$ is non-zero: $\langle 0|\operatorname{Re}\varphi(x)|0\rangle = \sqrt{2}m/\sqrt{\lambda} \neq 0$. Substituting (3.2.227) into (3.2.223) we find

$$\mathcal{L}_{\text{SSB}} = \tfrac{1}{2}(\partial_\mu\varphi_1)^2 - m_1^2\varphi_1^2 + \tfrac{1}{2}(\partial_\mu\varphi_2)^2 - \frac{\lambda}{16}(\varphi_1^4 + 2\varphi_1^2\varphi_2^2 + \varphi_2^4) - \frac{m\sqrt{\lambda}}{2}(\varphi_1^2 + \varphi_2^2)\varphi_1 \tag{3.2.228}$$

where $m_1 = \sqrt{2}m$ is the mass of the particle $\varphi_1(x)$. This Lagrangian does not contain a term proportional to $\varphi_2^2(x)$, i.e. the scalar particle described by the quantum field $\hat{\varphi}_2(x)$ is massless; it emerged as a result of the spontaneous symmetry-breaking and is a *goldstone*.

◇ **Spontaneous breaking of local symmetry**

Let us now modify the Lagrangian $\mathcal{L}_{\text{SSB}}$ by introducing the Abelian gauge fields $A_\mu(x)$, so that it is invariant with respect to the *local $U(1)$-transformations*:

$$\mathcal{L}_{\text{SSB}}^{YM} = -\tfrac{1}{4}F_{\mu\nu}^2 + (\partial^\mu\varphi^* - igA^\mu\varphi^*)(\partial_\mu\varphi + igA_\mu\varphi) + m^2\varphi^*\varphi - \tfrac{1}{4}\lambda(\varphi^*\varphi)^2. \tag{3.2.229}$$

This Lagrangian is invariant under the transformations

$$\begin{aligned}\varphi(x) &\longrightarrow \varphi'(x) = e^{-i\lambda\varepsilon(x)}\varphi(x) \\ \varphi^*(x) &\longrightarrow \varphi'^*(x) = e^{i\lambda\varepsilon(x)}\varphi^*(x) \\ A_\mu(x) &\longrightarrow A'_\mu(x) = A_\mu(x) + \partial_\mu\varepsilon(x).\end{aligned} \tag{3.2.230}$$

The circle of radius R, as in figure 3.12(b), again corresponds to the vacuum states (in the vacuum states, $A_\mu = 0$). Making the shift (3.2.227), we now obtain

$$\mathcal{L}_{\text{SSB}}^{\text{YM}} = -\frac{1}{4}F_{\mu\nu}^2 + \frac{2g^2 m^2}{\lambda}A_\mu^2 + \frac{1}{2}(\partial_\mu \varphi_1)^2 - m^2 \varphi_1^2 + \frac{1}{2}(\partial_\mu \varphi_2)^2 + \frac{2mg}{\sqrt{\lambda}}A^\mu \partial_\mu \varphi_2 + \mathcal{L}_{\text{int}} \qquad (3.2.231)$$

where $\mathcal{L}_{\text{int}}$ is the interaction Lagrangian for the fields $A_\mu, \varphi_1, \varphi_2$:

$$\mathcal{L}_{\text{int}} = g A^\mu (\varphi_1 \partial_\mu \varphi_2 - \varphi_2 \partial_\mu \varphi_1) + \frac{2g^2 m}{\sqrt{\lambda}} A_\mu^2 \varphi_1 + \frac{g^2}{2} A_\mu^2 (\varphi_1^2 + \varphi^2) + \frac{m^4}{\lambda}$$
$$- \tfrac{1}{16}\lambda(\varphi_1^4 + \varphi_2^4 + 2\varphi_1^2 \varphi_2^2) - \tfrac{1}{2} m\sqrt{\lambda}(\varphi_1^2 + \varphi_2^2)\varphi_1. \qquad (3.2.232)$$

The free Lagrangian in (3.2.231) is non-diagonal because of the term $(2mg/\sqrt{\lambda})A^\mu \partial_\mu \varphi_2$. To determine the mass term, we should diagonalize the free Lagrangian in (3.2.231). This is not difficult to do, but there is an even easier way: we may use the freedom in choosing the gauge condition to fix the gauge transformations (3.2.230) and setting

$$\varphi_2(x) = 0. \qquad (3.2.233)$$

This gauge condition is called the *unitary gauge*. In this gauge, the spectrum of free particles in the model is clearly exhibited: the model contains one vector massive field $A_\mu(x)$ and one scalar massive field $\varphi_1(x)$. Thus, a remarkable feature of Lagrangian (3.2.231) is that it contains a *massive* vector particle with the mass $2gm/\sqrt{\lambda}$. A direct introduction of a term proportional to $A^\mu A_\mu$ in the Lagrangian (3.2.229) is not allowed because of its non-invariance under the local gauge transformations (3.2.230). In contrast to this, the mass term in (3.2.231) emerges due to the *spontaneous* breaking of the invariance, while the Lagrangian remains invariant under the transformations (3.2.230). The massless scalar field (goldstone), which certainly appears in models with spontaneously broken *global* symmetry, disappears here. The corresponding degree of freedom is 'eaten' by the massive vector field (recall that a massless vector field has *two* possible polarizations, while a massive vector field has *three* possible polarizations). This phenomenon of the transition of a degree of freedom initially attributed to the scalar field into that of a gauge vector field is called the *Higgs mechanism* and the surviving physical scalar field $\varphi_1(x)$ is called the *Higgs boson*.

◇ A non-Abelian gauge model with spontaneous symmetry-breaking and its path-integral quantization

The Lagrangians of models with spontaneous breaking of non-Abelian gauge symmetry are constructed in essentially the same way as in the case of Abelian groups. As an example, let us consider the $SU(2)$-invariant gauge theory with the doublet of scalar fields

$$\varphi = \begin{pmatrix} \varphi_1 \\ \varphi_2 \end{pmatrix} \qquad \varphi^\dagger = (\varphi_1^*, \varphi_2^*) \qquad (3.2.234)$$

which are transformed according to the fundamental representation of the group $SU(2)$:

$$\varphi(x) \longrightarrow \varphi'(x) = e^{-i\sigma^a u^a(x)/2} \varphi(x) \qquad (3.2.235)$$

(σ^a are the Pauli matrices). The gauge-invariant Lagrangian reads as

$$\mathcal{L} = \mathcal{L}_{\text{YM}} + (D^\mu \varphi)^\dagger D_\mu \varphi - \lambda^2 (\varphi^\dagger \varphi - \mu^2)^2 \qquad (3.2.236)$$

where
$$D_\mu \varphi = \partial_\mu \varphi + \frac{i}{2} g \sigma^a A_\mu^a \varphi. \tag{3.2.237}$$

In the same way as in the preceding (Abelian) case, we find that the stable minimum of the potential energy corresponds to the constant field φ satisfying the condition:
$$\varphi^\dagger \varphi = \mu^2. \tag{3.2.238}$$

It is easy to see that the variety of the vacuum states forms a three-dimensional sphere S^3. The choice of one vacuum state, e.g.,
$$\varphi = \begin{pmatrix} 0 \\ \mu \end{pmatrix} \tag{3.2.239}$$

removes the degeneration and corresponds to the admissible gauge condition
$$\varphi_1(x) = 0 \qquad \text{Im } \varphi_2(x) = 0. \tag{3.2.240}$$

In this gauge, there exists only one scalar field, Re φ_2. It is convenient to introduce the following shifted field
$$\sigma(x) \stackrel{\text{def}}{\equiv} \sqrt{2}(\text{Re } \varphi_2 - \mu). \tag{3.2.241}$$

In terms of the field $\sigma(x)$, the Lagrangian (3.2.236) reads as
$$\mathcal{L} = -\frac{1}{4} F^{a\mu\nu} F_{\mu\nu}^a + \frac{1}{2} m_1^2 A^{a\mu} A_\mu^a + \frac{1}{2} \partial^\mu \sigma \partial_\mu \sigma - \frac{1}{2} m_2^2 \sigma^2$$
$$+ \frac{1}{2} m_1 g \sigma A^{a\mu} A_\mu^a + \frac{1}{8} g^2 \sigma^2 A^{a\mu} A_\mu^a - \frac{g m_2^2}{4 m_1} \sigma^3 - \frac{g^2 m_2^2}{32 m_1^2} \sigma^4 \tag{3.2.242}$$

where
$$m_1 = \frac{\mu g}{\sqrt{2}} \qquad m_2 = 2\lambda\mu. \tag{3.2.243}$$

Thus, this model describes three massive vector fields (with the equal masses m_1) interacting with one scalar field σ of mass m_2.

Since the Lagrangian (3.2.236) (or (3.2.242)) is locally invariant, the corresponding equations of motion contain constraints. To derive the latter explicitly, let us rewrite (3.2.236) in the first-order form:
$$\mathcal{L} = F_{0k}^a \partial_0 A_k^a + \varphi_0^\dagger \partial_0 \varphi + (\partial_0 \varphi^\dagger) \varphi_0 - H(F_{0k}^a, A_k^a, \varphi_0, \varphi)$$
$$+ A_0^a \left[\partial_k F_{0k}^a - g \varepsilon^{abc} A_k^b F_{0k}^c + i \frac{g}{2} (\varphi_0^\dagger \sigma^a \varphi - \varphi^\dagger \sigma^a \varphi_0) \right] \tag{3.2.244}$$

where $H(F_{0k}^a, A_k^a, \varphi_0, \varphi)$ is the Hamiltonian of the system the explicit form of which is not important at the moment (cf problem 3.2.7, page 100) and $\varphi_0(x) \stackrel{\text{def}}{\equiv} D_0 \varphi(x)$. It is seen that the pairs (F_{0k}^a, A_k^a) and (φ_0, φ) are canonically conjugate momenta and coordinates, A_0^a are Lagrange multipliers and
$$C^a \stackrel{\text{def}}{\equiv} -\partial_k F_{0k}^a - g \varepsilon^{abc} A_k^b F_{0k}^c + i \frac{g}{2} (\varphi_0^\dagger \sigma^a \varphi - \varphi^\dagger \sigma^a \varphi_0) \tag{3.2.245}$$

are the sought constraints. The reader is invited to verify that conditions (3.2.57) and (3.2.59) are fulfilled for this model (problem 3.2.7) and that gauge condition (3.2.240) satisfies (3.2.56). Thus we can use, for this model, the standard Faddeev–Popov method for path-integral quantization which, for the kernel of the S-matrix, gives

$$K_S(a^*(\boldsymbol{k}), a(\boldsymbol{k}); t, t_0) = \lim_{\substack{t\to\infty \\ t_0\to-\infty}} \int \mathcal{D}F^a_{0k}(x)\,\mathcal{D}A^a_k(x)\,\mathcal{D}\sigma_0(x)\,\mathcal{D}\sigma(x)\,\mathcal{D}\upsilon_0(x)\,\mathcal{D}\upsilon(x)$$

$$\times \prod_x \delta(\upsilon^a(x)) \left(m_1 + \tfrac{1}{2}g\sigma\right)^3$$

$$\times \exp\left\{\frac{i}{2}\int d^3k \left[\sum_{j=1}^{3}(a^{*b}_j(\boldsymbol{k},t)a^b_j(\boldsymbol{k},t) + a^{*b}_j(\boldsymbol{k},t_0)a^b_j(\boldsymbol{k},t_0))\right.\right.$$

$$\left.\left. + (a^*_\sigma(\boldsymbol{k},t)a_\sigma(\boldsymbol{k},t) + a^*_\sigma(\boldsymbol{k},t_0)a_\sigma(\boldsymbol{k},t_0))\right]\right\}$$

$$\times \exp\left\{i\int_{t_0}^{t} d\tau \int d^3x \left[\tfrac{1}{2}(F^a_{0k}\dot{A}^a_k - \dot{F}^a_{0k}A^a_k + \sigma_0\dot\sigma - \dot\sigma_0\sigma)\right.\right.$$

$$\left.\left. - H(F^a_{0k}, A^a_k, \upsilon_0, \upsilon, \sigma_0, \sigma)\right]\right\}. \quad (3.2.246)$$

Here we have used the more convenient field variables $v^a(x)$, related to $\varphi^a(x)$ through

$$\varphi^1 = \frac{iv^1 + v^2}{\sqrt{2}} \qquad \varphi^2 = \mu + \frac{\sigma - iv^3}{\sqrt{2}}.$$

The asymptotic conditions have the form (cf (3.1.86) and (3.2.118)):

$$a^{*b}_j(\boldsymbol{k},t) \xrightarrow[t\to\infty]{} a^{*b}_j(\boldsymbol{k})\exp\{i\omega_1 t\} \qquad a^b_j(\boldsymbol{k},t_0) \xrightarrow[t\to-\infty]{} a^b_j(\boldsymbol{k})\exp\{-i\omega_1 t_0\}$$

$$a^*_\sigma(\boldsymbol{k},t) \xrightarrow[t\to\infty]{} a^*_\sigma(\boldsymbol{k})\exp\{i\omega_2 t\} \qquad a_\sigma(\boldsymbol{k},t_0) \xrightarrow[t\to-\infty]{} a_\sigma(\boldsymbol{k})\exp\{-i\omega_2 t_0\} \quad (3.2.247)$$

where $\omega_1 = \sqrt{\boldsymbol{k} + m_1^2}$ and $\omega_2 = \sqrt{\boldsymbol{k} + m_2^2}$. The holomorphic variables are introduced in the usual way (cf (3.2.116)):

$$A^b_l(\boldsymbol{r},\tau) = \frac{1}{(2\pi)^{3/2}} \sum_{j=1}^{3} \int d^3k \frac{1}{\sqrt{2\omega_1}} (a^{*b}_j(\boldsymbol{k},\tau) u^j_l(-\boldsymbol{k}) e^{-i\boldsymbol{k}\cdot\boldsymbol{r}} + a^b_j(\boldsymbol{k},\tau) u^j_l(\boldsymbol{k}) e^{i\boldsymbol{k}\cdot\boldsymbol{r}})$$

$$F^b_{0l}(\boldsymbol{r},\tau) = \frac{1}{(2\pi)^{3/2}} \sum_{j=1}^{3} \int d^3k \sqrt{\frac{\omega_1}{2}} (a^{*b}_j(\boldsymbol{k},\tau) \tilde{u}^j_l(-\boldsymbol{k}) e^{-i\boldsymbol{k}\cdot\boldsymbol{r}} - a^b_j(\boldsymbol{k},\tau) \tilde{u}^j_l(\boldsymbol{k}) e^{i\boldsymbol{k}\cdot\boldsymbol{r}})$$

$$(3.2.248)$$

where the polarization vectors are defined as follows: $u^1_l = \tilde{u}^1_l$ and $u^2_l = \tilde{u}^2_l$ are two arbitrary orthonormal vectors, also orthogonal to $\boldsymbol{k}$, while

$$u^3_l = \frac{k_l \omega_1}{|\boldsymbol{k}| m_1} \qquad \tilde{u}^3_l = \frac{k_l m_1}{|\boldsymbol{k}| \omega_1}. \quad (3.2.249)$$

The momenta variables enter the Lagrangian (3.2.244) quadratically and can be integrated out (see problem 3.2.8, page 100), resulting in the explicitly relativistically invariant expression for the normal symbol of the S-matrix, for the model with spontaneously broken local symmetry $SU(2)$:

$$S(A^{(0)}_\mu, \sigma^{(0)}) = \mathfrak{N}^{-1} \int_{\substack{A\to A_{\text{in}}, A_{\text{out}} \\ \sigma\to\sigma_{\text{in}},\sigma_{\text{out}}}} \mathcal{D}A_\mu(x)\,\mathcal{D}\sigma(x) \prod_x (m_1 + \tfrac{1}{2}g\sigma)^3 \exp\left\{i\int d^4x\,\mathcal{L}(x)\right\} \quad (3.2.250)$$

where

$$\mathcal{L} = -\frac{1}{4}(\partial_\nu A_\mu^a - \partial_\mu A_\nu^a + g\varepsilon^{abc}A_\mu^b A_\nu^c)^2 + \frac{1}{2}m_1^2 A_\mu^2 + \frac{1}{2}(\partial_\mu \sigma)^2 - \frac{1}{2}m_2^2\sigma^2$$
$$+ \frac{1}{2}m_1 g\sigma A_\mu^2 + \frac{1}{8}g^2\sigma^2 A_\mu^2 - \frac{gm_2}{4m_1}\sigma^3 - \frac{g^2 m_2^2}{32 m_1^2}\sigma^4. \qquad (3.2.251)$$

The asymptotic conditions are analogous to those in (3.2.122):

$$A_\mu^b(x; \text{in/out}) = \frac{1}{(2\pi)^{3/2}} \sum_{j=1}^{2} \int d^3k \frac{1}{\sqrt{2\omega}} (a_j^{*b}(\boldsymbol{k}; \text{in/out}) u_\mu^j(\boldsymbol{k}) e^{-i\boldsymbol{k}\cdot\boldsymbol{r}+i\omega t} + a_j^b(\boldsymbol{k}; \text{in/out}) u_\mu^j(\boldsymbol{k}) e^{i\boldsymbol{k}\cdot\boldsymbol{r}-i\omega t})$$
$$(3.2.252)$$

$$a_j^b(\boldsymbol{k}; \text{in}) = a_j^b(\boldsymbol{k}) \qquad a_j^{*b}(\boldsymbol{k}; \text{out}) = a_j^{*b}(\boldsymbol{k}) \qquad u_\mu^j \stackrel{\text{def}}{=} (0, u_k^j) \qquad j = 1, 2 \qquad u_\mu^3 \stackrel{\text{def}}{=} \left(i\frac{|\boldsymbol{k}|}{m}, u_k^3\right)$$

$$\sigma(x; \text{in/out}) = \frac{1}{(2\pi)^{3/2}} \int d^3k \frac{1}{\sqrt{2\omega_2}} (a_\sigma^*(\boldsymbol{k}; \text{in/out}) e^{-i\boldsymbol{k}\cdot\boldsymbol{r}+i\omega t} + a_\sigma(\boldsymbol{k}; \text{in/out}) e^{i\boldsymbol{k}\cdot\boldsymbol{r}-i\omega t}) \qquad (3.2.253)$$

$$a_\sigma(\boldsymbol{k}; \text{in}) = a_\sigma(\boldsymbol{k}) \qquad a_\sigma^*(\boldsymbol{k}; \text{out}) = a_\sigma^*(\boldsymbol{k}).$$

As usual, the functions $a_j^b(\boldsymbol{k}, \text{out})$, $a_j^{*b}(\boldsymbol{k}, \text{in})$, $a_\sigma(\boldsymbol{k}, \text{out})$, $a_\sigma^*(\boldsymbol{k}, \text{in})$ are not fixed by boundary conditions. The quadratic form in (3.2.251) is defined as follows:

$$\tfrac{1}{2}\int d^4x \, (A_\mu^a - A_\mu^{(0)a})(g^{\mu\nu}\Box - \partial^\mu\partial^\nu + g^{\mu\nu}m_1^2)(A_\nu^a - A_\nu^{(0)a}). \qquad (3.2.254)$$

The Green function of the operator

$$(g_{\mu\nu}\Box - \partial_\mu\partial_\nu + g_{\mu\nu}m_1^2) \qquad (3.2.255)$$

and, hence, the perturbation expansion (Feynman diagram techniques) are defined by the asymptotic conditions (3.2.252).

However, the perturbation theory in this *unitary gauge* (3.2.240) is rather cumbersome for the following reasons:

- the Green function of the operator (3.2.255) is more singular than the functions we have met so far (e.g., $D_\alpha(x)$) and
- the integral (3.2.250) contains in its measure the factor $\prod_x (m_1 + \tfrac{1}{2}g\sigma)^3$ which, when exponentiated, produces other singularities.

Therefore, for practical calculations, it is better to pass to another gauge, e.g. the α-gauge, with the help of the standard Faddeev–Popov trick. The result, for the normal symbol of the S-matrix, reads as

$$S = \mathfrak{N}^{-1} \int_{\substack{A\to A_{\text{in}}, A_{\text{out}} \\ \sigma \to \sigma_{\text{in}}, \sigma_{\text{out}}}} \mathcal{D}A_\mu(x)\, \mathcal{D}\sigma(x)\, \mathcal{D}v^a(x)\, \det M_\alpha \, \exp\left\{i\int d^4x \left(\mathcal{L}(x) + \frac{1}{2\alpha}(\partial_\mu A_\mu)^2\right)\right\} \qquad (3.2.256)$$

where

$$\mathcal{L} = -\frac{1}{4}(\partial_\nu A_\mu^a - \partial_\mu A_\nu^a + g\varepsilon^{abc}A_\mu^b A_\nu^c)^2 + \frac{1}{2}m_1^2 A_\mu^2 + \frac{1}{2}\partial^\mu\sigma\partial_\mu\sigma - \frac{1}{2}m_2^2\sigma^2$$
$$+ \frac{1}{2}\partial^\mu v^a \partial_\mu v^a + m_1 A_\mu^a \partial^\mu v^a$$

$$+ \frac{1}{2}gA^a_\mu(\sigma\partial^\mu v^a - v^a\partial^\mu\sigma - \varepsilon^{abc}v^b\partial^\mu v^c) + \frac{1}{2}m_1 g\sigma A^2_\mu + \frac{1}{8}g^2(\sigma^2 A^2_\mu + v^2)$$
$$- \frac{gm_2}{4m_1}\sigma(\sigma^2 + v^2) - \frac{g^2 m_2^2}{32 m_1^2}(\sigma^2 + v^2)^2 \qquad (3.2.257)$$

and

$$\widehat{M}_\alpha \mathsf{f}(x) = \Box \mathsf{f}(x) - g\partial^\mu[\mathsf{A}_\mu(x), \mathsf{f}(x)]. \qquad (3.2.258)$$

In this gauge, the spectrum of the system is not explicit (and, hence, the unitarity of the theory is not obvious), but realization of the renormalization procedure is much easier than in the explicitly unitary gauge (3.2.240). Thus, due to the equivalence of the S-matrix in all gauges, the model possesses both of the necessary properties of a physically meaningful theory: unitarity and renormalizability.

◇ Unified theories based on gauge theories: the standard model of electroweak and strong interactions

As we mentioned at the very beginning of this section, the gauge model with spontaneous symmetry-breaking describes in a unified way the strong, weak and electromagnetic interactions of elementary particles. As we shall see later (section 3.4), the gravitational interaction is also described by a gauge theory, though one which is more involved.

To construct the Lagrangian for a unified model, the following steps are required:

(i) Choose the gauge group which determines the interaction-mediating fields; the number of gauge fields is equal to the dimension of the adjoint representation of this group.
(ii) Choose primary fermions to underlie the model and the representations of the gauge group in which the fermions are placed; the lowest representations are usually chosen.
(iii) Introduce an appropriate number of multiplets of scalar fields as well as the interaction terms of these multiplets with fermions (the Yukawa coupling) to obtain massive particles via spontaneous symmetry-breaking and the Higgs mechanism.
(iv) Write the corresponding locally gauge-invariant Lagrangian.
(v) Quantize the model in path-integral formalism with the help of the Faddeev–Popov approach; the spectrum of free particles is exhibited in the unitary gauge after diagonalization of the free (quadratic) part of the Lagrangian, while renormalization is carried out in a suitable α-gauge (and after the introduction of the ghost fields).

The standard model is characterized by the following selections:

(i) The total gauge group is $SU(3) \times SU(2) \times U(1)$. The $SU(3)$ factor and the corresponding gauge fields are responsible for the strong interactions of particles, while the $SU(2) \times U(1)$ part provides the weak and electromagnetic interactions.
(ii) Matter spinor fields are divided into two parts:
  - There are fields describing *leptons*, i.e. particles which only participate in electroweak interactions. These are the electrons e^-, the μ^--leptons, the τ^--leptons, the neutrinos ν_e, ν_μ, ν_τ and their antiparticles. In fact, to fit experimental data, the Lagrangian of the standard model is constructed out of left-handed and right-handed projections of the fields, the projector operators L and R being made by means of the Dirac γ_5-matrix: $L = (1 + \gamma_5)/2$, $R = (1 - \gamma_5)/2$. Left- (right-) handed fields are obtained from an initial spinor field ψ as follows: $\psi_L = \frac{1}{2}(1 + \gamma_5)\psi$, $\psi_R = \frac{1}{2}(1 - \gamma_5)\psi$. The standard model contains three left-handed lepton $SU(2)$-doublets

$$\begin{pmatrix} \nu_e \\ e^- \end{pmatrix}_L \qquad \begin{pmatrix} \nu_\mu \\ \mu^- \end{pmatrix}_L \qquad \begin{pmatrix} \nu_\tau \\ \tau^- \end{pmatrix}_L \qquad (3.2.259)$$

and three right-handed lepton $SU(2)$-singlets

$$e_R^- \quad \mu_R^- \quad \tau_R^-. \tag{3.2.260}$$

All the lepton fields are singlets with respect to the $SU(3)$ gauge group; this reflects the fact that leptons do not participate in strong interactions.

- The quark fields constitute *hadrons*, i.e. strongly interacting particles. Correspondingly, all quarks have non-trivial transformation properties with respect to $SU(3)$, namely they are transformed according to the fundamental three-dimensional representation of $SU(3)$. The index labelling the fields in the $SU(3)$-triplets has acquired the name '*color*'. This explains the name of the $SU(3)$ gauge theory of the quark interactions: *quantum chromodynamics* (QCD). (In the currently accepted terminology, QCD is part of the standard model.) Besides the strong interactions, the quarks also participate in electroweak interactions. Therefore, they have non-trivial transformation properties with respect to the $SU(2) \times U(1)$ part of the standard-model gauge group. Again, as for the leptons, there are three left-handed $SU(2)$-doublets and *all* quark fields have right-handed parts, giving a total of six right-handed singlets.

(iii) The total gauge group $SU(3) \times SU(2) \times U(1)$ of the standard model is spontaneously broken down to $SU(3) \times U(1)$. This is achieved via the introduction of a $SU(2)$-doublet of scalar complex fields (Higgs fields), similarly to the model we previously considered in this section.

(iv) The gauge-invariant Lagrangian is constructed by the usual rules, with the help of the covariant derivatives D_μ; the potential energy of the Higgs fields has degenerate minima (as in the previous model), providing the spontaneous symmetry-breaking. This, in turn, gives masses to three gauge bosons and to fermions (excluding neutrinos). The fermions acquire masses due to the Yukawa couplings with the scalar fields.

(v) The path-integral quantization of the standard model and the development of the perturbation theory are carried out as we have described in this section for general Yang–Mills theories. Note that the massless photon (electromagnetic) field appears as a linear combination of the gauge boson corresponding to the factor $U(1)$ in $SU(3) \times SU(2) \times U(1)$ and one of the gauge bosons corresponding to $SU(2)$. Another (linearly independent) combination of these bosons becomes massive and responsible (together with two other massive $SU(2)$ gauge bosons) for the weak interactions. One more peculiarity of the standard model is that its Lagrangian essentially contains γ_5-matrices, because the spinor fields enter the Lagrangian via their left- and right-handed projection. This fact might potentially cause quantum anomalies (see section 3.3.4) and the theory may prove to be non-renormalizable. However, the whole structure of the spinor multiplets (both leptons and quarks) is such that the anomalies in different sectors of the model cancel each other out and the complete theory is well defined, non-anomalous and renormalizable.

At present, there is no single experimental result which contradicts the standard model. Moreover, almost all the ingredients of the standard model have experimental confirmation. In particular, all the particles, except the Higgs particle, have been successfully detected. The Higgs particle is expected to be detected in the near future.

Of course, our description of the standard model is very far from being complete. We shall consider some more aspects of this model (or the non-perturbative properties of the Yang–Mills theory, in general) in the next section. The reader may find more details about the standard model in, e.g., Itzykson and Zuber (1980), Okun (1982), Chaichian and Nelipa (1984), Cheng and Li (1984), West (1986), Peskin and Schroeder (1995) and Weinberg (1996).

3.2.9 Problems

Problem 3.2.1. Show the equivalence of a classical system with some Hamiltonian $H(p, q)$ plus first-class constraints (cf (3.2.57) and (3.2.59)), after imposing gauge conditions (cf (3.2.59)) which satisfy (3.2.60), and the reduced physical subsystem with the Hamiltonian defined as in (3.2.65).

Hint. The equations of motion for the initial system are

$$\dot{p}_A + \frac{\partial H}{\partial q_A} + \lambda^a \frac{\partial \phi_a}{\partial q_A} = 0,$$

$$\dot{q}_A - \frac{\partial H}{\partial p_A} - \lambda^a \frac{\partial \phi_a}{\partial p_A} = 0 \quad (3.2.261)$$

$$\phi_a = 0 \quad A = 1, \ldots, n, \ a = 1, \ldots, r.$$

A solution of these equations contains the arbitrary functions $\lambda^a(t)$ (the Lagrange multipliers). The gauge conditions $\chi_a = 0$ allow us to express $\lambda_a(t)$ through the canonical variables. Let us choose the canonical variables according to (3.2.62) and (3.2.64). Then, equations $\dot{q}_a = 0$ together with (3.2.261) give

$$\frac{\partial H}{\partial p_a} + \lambda^b \frac{\partial \phi_b}{\partial p_a} = 0 \quad a, b = 1, \ldots, r. \quad (3.2.262)$$

The equations of motion for the physical coordinates in (3.2.261) have the form

$$\dot{\tilde{q}}_i = \frac{\partial H}{\partial \tilde{p}_i} + \lambda^a \frac{\partial \phi_a}{\partial \tilde{p}_i}. \quad (3.2.263)$$

On the other hand, if we start from the physical Hamiltonian defined by (3.2.65), the equation of motion for the same coordinates is

$$\dot{\tilde{q}}_i = \frac{\partial H_{\text{ph}}}{\partial \tilde{p}_i} = \frac{\partial H}{\partial \tilde{p}_i} + \frac{\partial H}{\partial p_a} \frac{\partial p_a}{\partial \tilde{p}_i}. \quad (3.2.264)$$

The right-hand sides of (3.2.263) and (3.2.264) are equal to each other if

$$\lambda_a \frac{\partial \phi_a}{\partial \tilde{p}_i} = \frac{\partial H}{\partial p_a} \frac{\partial p_a}{\partial \tilde{p}_i}.$$

Using (3.2.262), this condition can be rewritten as

$$\lambda_a \left(\frac{\partial \phi_a}{\partial \tilde{p}_i} + \frac{\partial \phi_a}{\partial p_b} \frac{\partial p_b}{\partial \tilde{p}_i} \right) = \lambda_a \frac{d}{d\tilde{p}_i} \phi_a = 0.$$

The latter form shows that this condition holds automatically due to the constraints $\phi_a = 0$. Thus the physical coordinates have the same equations of motion both in the initial and in the reduced system. The momenta are considered quite similarly and with the same result which proves the required statement.

Problem 3.2.2. Calculate path integral (3.2.124) which defines the free propagator of the Yang–Mills field in the Coulomb gauge and prove that the latter has the form (3.2.125).

Hint. The extremality equations for the Gaussian integral (3.2.124) read as

$$\partial^\nu (\partial_\nu A_k^a - \partial_k A_\nu^a) + J_k^a + \partial_k \lambda^a = 0$$
$$\partial^\nu (\partial_\nu A_0^a - \partial_0 A_\nu^a) + J_0^a = 0$$
$$\partial_k A_k^a = 0$$

or, equivalently,

$$\Box A_k^a - (\partial_k \lambda^a + \partial_0 \partial_k A_0^a) + J_k^a = 0$$
$$\partial_k \partial_k A_0^a - J_0^a = 0$$
$$\partial_k A_k^a = 0.$$

The latter equation shows that the longitudinal component (in the three-dimensional sense) of the Yang–Mills field vanishes at the extremal field. Thus we can choose the source field also to be transversal:

$$\partial_k J_k^a = 0.$$

As a result, this system has a unique solution for fields satisfying the boundary conditions (3.2.121) and (3.2.122) imposed on A_k^T:

$$A^{a\mu}(x) = \int dy \, D_C^{\mu\nu}(x-y) J_\nu^a(y),$$

where $D_C^{\mu\nu}(x-y)$ is the Coulomb propagator (3.2.126).

Problem 3.2.3. Calculate the integral (3.2.132) in the vicinity of $g(x) = 1$ (i.e. verify formula (3.2.134)).

Hint. Use the basis of eigenfunctions $\alpha_n(x)$ of the operator $\widehat{M}_C(A)$

$$\widehat{M}_C(A)\alpha_n(x) = \lambda_n \alpha_n(x)$$

and the δ-function property: $\delta(\lambda\alpha) = |\lambda^{-1}|\delta(\alpha)$.

Problem 3.2.4. Derive the constraints and Hamiltonian equations of motion for the gauge fields interacting with a matter spinor field according to the Lagrangian

$$\mathcal{L} = \frac{1}{8g^2} \text{Tr}(\mathsf{F}_{\mu\nu}\mathsf{F}^{\mu\nu}) + i\bar{\psi}\gamma^\mu D_\mu \psi - m\bar{\psi}\psi \tag{3.2.265}$$

$$D_\mu \equiv \partial_\mu + it^a A_\mu^a$$

where t_{ij}^a are the generators of the gauge group in the appropriate representation. Show that the constraints are first class and that they generate the gauge transformations of the fields.

Hint. The constraints have the form

$$C^a(x) = \partial_k F_{k0}^a - f^{abc} A_k^b F_{0k}^c + i\bar{\psi}\gamma_0 t^a \psi. \tag{3.2.266}$$

A straightforward calculation of the Poisson brackets gives

$$\{C^a(x), C^b(y)\} = f^{abc}\delta(x-y)C^b(x)$$

so that these are indeed first-class constraints and the Poisson bracket relations

$$\{C^a(x), A_k^b(y)\} = \delta^{ab}\partial_k\delta(x-y) - f^{abc}A_k^c\delta(x-y)$$
$$\{C^a(x), \psi(y)\} = t^a\psi(x)\delta(x-y)$$
$$\{C^a(x), \bar{\psi}(y)\} = -t^a\bar{\psi}(x)\delta(x-y)$$

exactly correspond to the infinitesimal gauge transformations of the fields entering the Lagrangian.

Problem 3.2.5. Draw the Feynman diagrams and write the corresponding amplitude for the electron–electron scattering process: $e^-(p_1) + e^-(p_2) \longrightarrow e^-(p_1) + e^-(p_2)$.

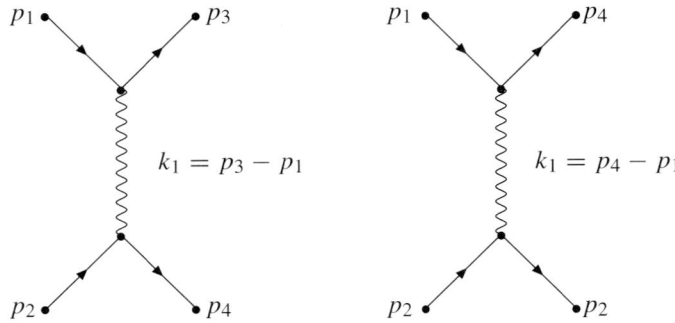

Figure 3.13. Feynman diagrams for the electron–electron scattering process.

Hint. The diagrams are depicted in figure 3.13. The corresponding amplitude has the form

$$\langle p_3, p_4; \text{out}|p_1, p_2; \text{in}\rangle = (-\mathrm{i})^2 e^2 \Bigg[\bar{v}_{r'}^{(+)}(\boldsymbol{p}_3)\gamma_\mu v_r^{(-)}(\boldsymbol{p}_1)\frac{1}{(p_3-p_1)^2}\bar{v}_{r'}^{(+)}(\boldsymbol{p}_4)\gamma_\mu v_r^{(-)}(\boldsymbol{p}_2)$$
$$- \bar{v}_{r'}^{(+)}(\boldsymbol{p}_4)\gamma_\mu v_r^{(-)}(\boldsymbol{p}_1)\frac{1}{(p_4-p_1)^2}\bar{v}_{r'}^{(+)}(\boldsymbol{p}_3)\gamma_\mu v_r^{(-)}(\boldsymbol{p}_2) \Bigg]. \quad (3.2.267)$$

Here the Dirac spinors $v_r^{(-)}(\boldsymbol{p})$ and $\bar{v}_r^{(+)}(\boldsymbol{p})$ represent the initial and final electron states, respectively.

Problem 3.2.6. Show that the generalized Ward–Takahashi identity (3.2.217) with ghost fields and ghost sources can be reduced to the Ward–Takahashi identity in the form (3.2.213), derived without the use of the BRST transformations.

Hint. Differentiate (3.2.217) with respect to $\bar{\eta}^a$ and η^a, set $\bar{\eta}^a = \eta^a = 0$ and subsequently integrate over c^a and $\bar{c}^a$. The result exactly coincides with (3.2.213).

Problem 3.2.7. Verify that constraints (3.2.245) satisfy conditions (3.2.57) and (3.2.59) so that they are first-class constraints. Show also that gauge (3.2.240) is admissible.

Hint. Derive the explicit form of the Hamiltonian in (3.2.244) and, using the canonical Poisson brackets for conjugate variables (F_{0k}^a, A_k^a) and (φ_0, φ), verify the required equalities.

To prove the admissibility of the gauge condition (3.2.240), show that the matrix of the Poisson brackets is non-degenerate (in the framework of perturbation theory). More precisely, show that the brackets have the form

$$\{C^a(x), v^b(y)\} = (m_1 + \tfrac{1}{2}g\sigma(x))\delta^{ab}\delta(x-y) + \cdots \quad (3.2.268)$$

where

$$v^1 = \operatorname{Im}\varphi_1/\sqrt{2} \qquad v^2 = \operatorname{Re}\varphi_1/\sqrt{2} \qquad v^3 = -\operatorname{Im}\varphi_2/\sqrt{2}$$

and the dots denote terms vanishing when $v^a = 0$. The right-hand side of (3.2.268) is obviously invertible if $|g\sigma| \leq m_1$.

Problem 3.2.8. With the help of integration over momentum variables convert the phase-space path integral (3.2.246) for the S-matrix of the model with spontaneous symmetry-breaking into the configuration path integral (3.2.250), exhibiting explicit relativistic invariance.

Hint. For the integration over v_0 we should make the shift of variables:

$$v_0 \longrightarrow v_0 - m_1 A_0^a$$

while the integration over F_{0k}^a is performed after the shift:

$$F_{0k}^a \longrightarrow F_{0k}^a + \partial_0 A_k^a - \partial_k A_0^a.$$

3.3 Non-perturbative methods for the analysis of quantum field models in the path-integral approach

Many phenomena in quantum systems, including field theoretical systems, cannot be described in the framework of the perturbation theory which we have discussed in the preceding sections of this chapter. This is especially important in the case of the strong coupling regime of the theory when the perturbative expansion in powers of the coupling constant is absolutely unreliable. For example, this is true for quantum chromodynamics at relatively large distances. That is why the quantitative description of quark dynamics, in particular the explanation of *quark confinement* (absence of free quarks) is an extremely complicated problem. But even in the weak coupling regime (small coupling constants) there are many phenomena, e.g. the existence of *solitons* and *instantons*, which cannot be described by perturbation theory (because the appropriate quantities describing these phenomena are non-analytic functions of the coupling constants). Sometimes, straightforward perturbation theory calculations lead to meaningless results and some non-trivial rearrangement and partial summation of the perturbation expansion is required (for instance, in the case of the so-called *infrared catastrophe*).

Thus, the problem of developing non-perturbative methods of analysis and calculation in quantum field theory is extremely important. The path-integral formalism has proved to be very useful for this aim. In this section, we shall consider the most powerful and well-developed non-perturbative path-integral methods for quantum field theories in continuous spacetime. The discussion of such methods in discretized space and time (i.e. for quantum field theories on lattices) pertains to chapter 4.

The physical problem which we shall discuss in the last part of this section (section 3.3.5) is somehow outside the main topic of this chapter. While the main subject of the current chapter is relativistically invariant quantum field theory (with applications to the theory of elementary particles and fundamental interactions), section 3.3.5 is devoted to non-perturbative path-integral methods in a *non-relativistic* field theory. Physically, this field theory describes an electron moving in a crystal and interacting with the vibrations of the crystal lattice (the so-called *polaron problem*). We shall see that, though the physical situation in this case is quite different from that for elementary particle models, the essence of the field theoretical and path-integral methods remains the same.

It is worth mentioning that, in fact, we have already dealt with non-perturbative applications of path-integral methods in quantum field theory: as we learned in section 3.1.5, the path-integral formalism is very convenient for deriving the Schwinger–Dyson equations, which contain complete information about the field model under consideration. Then we may try to solve these equations by some non-perturbative (though, of course, in most cases still approximate) method.

3.3.1 Rearrangements and partial summations of perturbation expansions: the $1/N$-expansion and separate integration over high and low frequency modes

If standard perturbation theory cannot be applied to the calculation of some physical characteristics in a field theory (for example, because the corresponding coupling constant is not small), we may look for new and unorthodox parameters which could serve to define a new perturbation expansion. An example of such a modification, which consists in employing a series expansion with respect to the parameter $1/N$ (N is the number of field components entering the Lagrangian of a model; it is supposed that $N \geq 1$), instead of the usual coupling constant, is referred to as the $1/N$-*expansion*.

◇ The $1/N$-expansion

We shall illustrate the basics of the $1/N$-expansion technique by considering, as an example, the four-dimensional φ^4-interaction model which is invariant under the (global) $O(N)$ group. Let the set of scalar fields $\varphi^a(x)$ be transformed according to the *fundamental* representation of the $O(N)$ group, so that the scalar field forms a multiplet φ^a with N components: $a = 1, 2, \ldots, N$.

The Lagrangian of such a model can be written as

$$\mathcal{L}' = \frac{1}{2}\partial^\mu \varphi^a \partial_\mu \varphi^a - \frac{1}{2}m^2 \varphi^a \varphi^a - \frac{1}{8}\frac{\lambda}{N}(\varphi^a \varphi^a)^2. \tag{3.3.1}$$

Here we have explicitly introduced the parameter $1/N$ redefining the usual φ^4-coupling constant λ as λ/N. Physically, this is an non-essential redefinition (especially as the coupling constant is subjected to renormalization). But this trick allows us to separate diagrams of different orders in $1/N$ in an easier way. For example, to separate the diagrams of zero order in this parameter among all the diagrams of the usual perturbation theory, we may just put $N \to \infty$ (i.e. $1/N = 0$). If we did not modify the coupling constant, parts of the diagrams would be proportional to *positive* powers of N and the $N \to \infty$ limit would be meaningless.

As will be seen, it is convenient to pass to another Lagrangian for the *same* system:

$$\mathcal{L} = \mathcal{L}' + \frac{1}{2}\frac{N}{\lambda}\left(\sigma - \frac{1}{2}\frac{\lambda}{N}\varphi^a \varphi^a\right)^2 \tag{3.3.2}$$

$$= \frac{1}{2}\varphi^a K(\sigma)\varphi^a + \frac{1}{2}\frac{N}{\lambda}\sigma^2 \tag{3.3.3}$$

where $\sigma(x)$ is an auxiliary *one-component* field and

$$K(\sigma) = -(\partial^\mu \partial_\mu + m^2 + \sigma).$$

The additional term in (3.3.2) does not change the dynamics of the system. In fact, the change of variables $\sigma \to \sigma(\lambda/N)\varphi^a \varphi^a$, $\varphi^a \to \varphi^a$, in the path integral corresponding to the Lagrangian (3.3.3), yields

$$\int \mathcal{D}\sigma(x)\,\mathcal{D}\varphi^a(x)\exp\left\{i\int d^4x\,\mathcal{L}\right\} = \int \mathcal{D}\sigma(x)\exp\left\{i\int d^4x\,\sigma^2(x)\right\}\mathcal{D}\varphi^a(x)\exp\left\{i\int d^4x\,\mathcal{L}'\right\}. \tag{3.3.4}$$

The integration over $\sigma(x)$ only changes the normalization constant of the generating functional $\mathcal{Z}[J]$ for the Green functions and, consequently, the Lagrangians $\mathcal{L}$ and $\mathcal{L}'$ describe systems with the same dynamics.

With the use of (3.3.3), the expression for the generating functional $\mathcal{Z}[J]$ for the Green functions of the fields φ^a takes the form

$$\mathcal{Z}[J] = \mathfrak{N}^{-1}\int \mathcal{D}\sigma(x)\,\mathcal{D}\varphi^a(x)\exp\left\{i\int d^4x\,\left[\frac{1}{2}\varphi^a K(\sigma)\varphi^a + \frac{1}{2}\frac{N}{\lambda}\sigma^2 + J^a(x)\varphi^a(x)\right]\right\} \tag{3.3.5}$$

where J^a are the auxiliary external currents associated with the fields φ^a. The Gaussian integration over the fields φ^a gives

$$\mathcal{Z}[J] = \int \mathcal{D}\sigma(x)\exp\left\{iN\int d^4x\,\mathcal{L}_{\text{eff}}(\sigma)\right\} \tag{3.3.6}$$

where

$$\mathcal{L}_{\text{eff}}(\sigma) = \frac{1}{2}\left[i\operatorname{Tr}\ln K(\sigma) + \frac{1}{\lambda}\int d^4x\,\sigma^2(x) - \frac{1}{N}\int d^4x\,d^4y\,J^a(x)K^{-1}(x,y)J^a(y)\right]. \quad (3.3.7)$$

It is seen that the integral over the multiplet φ^a is reduced to an integral over a single scalar function $\sigma(x)$, which was the actual goal when introducing the field $\sigma(x)$. The main achievement of such a representation of the generating functional is that the parameter N proves to be explicitly extracted as the overall factor in front of the exponent in the path integral (note that the last term in the effective Lagrangian (3.3.7) contains, together with the factor $1/N$, the summation over N terms and, hence, it is also of zero order in $1/N$). Since we assume that N is a large quantity ($N \geq 1$), we can use for the calculation of (3.3.6) the *stationary-phase method* (see section 2.2.3).

First, we have to find the stationary point, i.e. the solution σ_0 of the equation

$$\frac{\delta}{\delta\sigma}\int d^4x\,\mathcal{L}_{\text{eff}}(\sigma) = 0 \quad (3.3.8)$$

and then we expand the action around this stationary value. Note that the stationary point σ_0 depends on the currents J^a: $\sigma_0 = \sigma_0(J)$. The quadratic approximation gives

$$\mathcal{Z}^{(\text{quadr})}[J, j] = \exp\{iNS^{\sigma}_{\text{eff}}[\sigma_0(J)]\}\int \mathcal{D}\sigma(x)\exp\left\{iN\int d^4x\,\frac{1}{2}\mathcal{L}''_{\text{eff}}(\sigma_0)\sigma^2(x) + \frac{1}{N}j(x)\sigma(x)\right\}$$

$$= [\det\mathcal{L}''_{\text{eff}}(\sigma_0)]^{-1/2}\exp\{iNS^{\sigma}_{\text{eff}}[\sigma_0(J)]\}\exp\left\{-\frac{i}{2}Nj(x)[\mathcal{L}''_{\text{eff}}(\sigma_0)]^{-1}j(x)\right\} \quad (3.3.9)$$

where $j(x)$ is the auxiliary current corresponding to the field $\sigma(x)$ and

$$\mathcal{L}''_{\text{eff}}(\sigma_0) = \left.\frac{\partial^2 \mathcal{L}_{\text{eff}}(\sigma)}{\partial\sigma^2}\right|_{\sigma=\sigma_0}.$$

Expression (3.3.9) gives the leading contribution in $1/N$ to the generating functional. In the standard way, we can take into account all higher orders in $1/N$ (i.e. higher orders of the expansion around the stationary point σ_0):

$$\mathcal{Z}[J, j] = [\det\mathcal{L}''_{\text{eff}}(\sigma_0)]^{-1/2}\exp\{iNS^{\sigma}_{\text{eff}}[\sigma_0(J)]\}$$

$$\times \exp\left\{iN\sum_{n=3}^{\infty}\frac{1}{n!}\left[\frac{\partial^n\mathcal{L}_{\text{eff}}}{\partial\sigma^n}\right]_{\sigma=\sigma_0}\frac{1}{i^n}\frac{\delta^n}{\delta j^n(x)}\right\}\exp\left\{-\frac{i}{2}Nj(x)[\mathcal{L}''_{\text{eff}}(\sigma_0)]^{-1}j(x)\right\}. \quad (3.3.10)$$

Functional differentiation of $\mathcal{Z}[J, j]$ with respect to the currents J and j gives the corresponding Green functions. In particular, according to (3.3.10), for the propagator $D(x, y)$ of the field $\sigma(x)$ in the first order in $1/N$, we have

$$D^{(\sigma)}_{1/N}(x, y) = \left.\frac{\delta^2\mathcal{Z}[J, j]}{\delta j(x)\delta j(y)}\right|_{J=j=0} = -iN^{-1}[\mathcal{L}''_{\text{eff}}(\sigma_0)]^{-1}. \quad (3.3.11)$$

However, the effective Lagrangian $\mathcal{L}_{\text{eff}}$ contains the term $i\operatorname{Tr}\ln K(\sigma)$ with an implicit dependence on the field $\sigma(x)$. In order to make this dependence explicit, let us present this term in the following way:

$$\operatorname{Tr}\ln K(\sigma) = \operatorname{Tr}\ln[-(\partial^\mu\partial_\mu + m^2) - \sigma]$$

$$= \operatorname{Tr}\ln[-(\partial^\mu\partial_\mu + m^2)] + \operatorname{Tr}\ln[\mathbb{1} + D_c(x, y)\sigma(y)]. \quad (3.3.12)$$

Here $D_c(x, y)$ is the usual Green function (3.1.93) for a scalar field (i.e. the kernel of the operator $(\partial_\mu \partial_\mu + m^2)^{-1}$: $(\partial_\mu \partial_\mu + m^2) D_c(x, y) = \delta(x - y)$). Note that the field $\sigma(x)$ entering the operator $K(\sigma)$ is considered to be a diagonal operator $\sigma(x)\delta(x - y)$ and in the form of an integral kernel it acts on an arbitrary function $f(x)$ as follows

$$\hat{\sigma} f(x) = \int dy\, \sigma(x)\delta(x - y) f(y) = \sigma(x) f(x).$$

The operator $D_c(x, y)\sigma(y)$ is also understood as an integral kernel (infinite-dimensional matrix with its elements labeled by x and y), so that its trace has the form: $\int dx\, D_c(x, x)\sigma(x)$. The first term in (3.3.12) gives an inessential constant. The Taylor expansion of the second term gives the explicit dependence of $\mathcal{L}_{\text{eff}}$ on σ:

$$\mathrm{Tr}\ln(\mathbb{I} + D_c(x, y)\sigma(y)) = \int dx\, D_c(x, x)\sigma(x)$$
$$- \tfrac{1}{2} \int dx\, dy\, D_c(x, y)\sigma(y) D_c(y, x)\sigma(x)$$
$$+ \tfrac{1}{3} \int dx\, dy\, dz\, D_c(x, y)\sigma(y) D_c(y, z)\sigma(z) D_c(z, x)\sigma(x) + \cdots. \quad (3.3.13)$$

In general, the stationary point σ_0 is not zero (and depends on the external source $J(x)$) and the explicit form of the σ-propagator, even in the leading $1/N$-approximation, is rather complicated, being represented by an infinite number of Feynman diagrams of the ordinary perturbation theory. To illustrate this, let us assume that $\sigma_0 = 0$, so that (3.3.13) gives

$$\mathcal{L}''_{\text{eff}}|_{\sigma_0 = 0} = \lambda^{-1}\delta(x - y) - \frac{i}{2} D_c(x, y) D_c(y, x)$$

and the σ-field propagator in the leading $1/N$-approximation becomes

$$D^{(\sigma)}_{1/N}(x, y) = -\frac{\lambda}{N} \left[1 + \frac{1}{2}\lambda \Sigma(x, y)\right]^{-1} \quad (3.3.14)$$

where

$$\Sigma(x, y) = D_c(x, y) D_c(y, x)$$
$$= \frac{1}{(2\pi)^8} \int dp\, e^{ip(x-y)} \left[\int dk\, \frac{1}{k^2 - m^2} \frac{1}{(p+k)^2 - m^2}\right].$$

The expansion of (3.3.14) as a power series in the coupling constant λ is depicted graphically in figure 3.14.

Let us consider an arbitrary connected diagram which contains E external lines, I internal lines and V vertices corresponding to the field σ. According to (3.3.6), each vertex of the diagram involves the parameter N. Since the propagator is a reciprocal quantity with respect to the quadratic part of the Lagrangian, to each internal or external line there corresponds a factor $1/N$. The diagram is thus characterized by the quantity N^{V-I-E}. The number of internal lines is equal to the number of momenta over which the integration is to be carried out. These momenta are, however, not independent, because the momenta meeting at each of the vertices V are interrelated through a conservation law; besides, one of the conservation laws (pertaining to the process as a whole) involves the external momenta so that the number of independent internal momenta is $L = I - (V - 1)$.

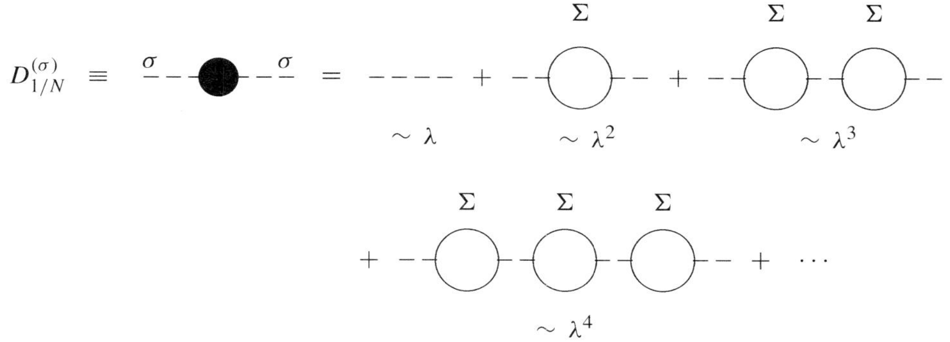

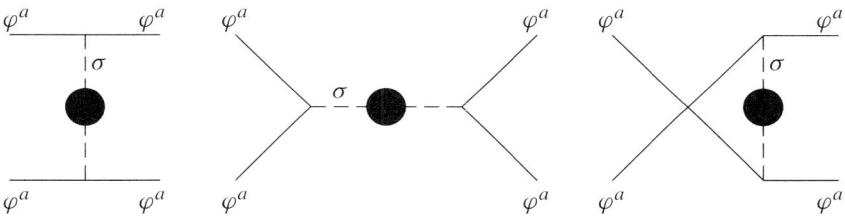

Figure 3.14. Expansion of the diagram for the propagator of the σ-field in the first order in $1/N$ into an infinite series corresponding to diagrams of different orders of the usual perturbation theory in the coupling constant λ.

Figure 3.15. Diagrams for the process $\varphi^a \varphi^a \to \varphi^a \varphi^a$ in the first order in $1/N$.

Taking this into account we find for the power of N:

$$N^{V-I-E} = N^{-E-L+1}. \tag{3.3.15}$$

In particular, the diagrams containing two external lines ($E = 2$) and no loops ($L = 0$, the tree approximation) contribute to the leading order in $1/N$.

As an example, the typical diagrams for $\varphi^a \varphi^a \to \varphi^a \varphi^a$ scattering in the leading order in the parameter $1/N$ are depicted in figure 3.15. The filled circle denotes the two-point Green function $D_{1/N}^{(\sigma)}$ of the σ-field and is presented in figure 3.14. It can be seen that an infinite number of the usual Feynman diagrams with increasing powers of the coupling constant λ^n, $n = 1, 2, 3, \ldots$, contribute to the leading-order contribution in the parameter $1/N$ (we assume that the Lagrangian is presented in the normal form so that tadpole diagrams do not appear).

It should be emphasized that the $1/N$-expansion technique is based on the possibility of reducing the integration over the field $\varphi^a(x)$ in the generating functional to integration over the field $\sigma(x)$. This allows us to express the generating functional in the form (3.3.6), which contains N in front of the action as a common factor. Unfortunately, this possibility can only be realized if the Lagrangian comprises fields which transform according to the fundamental representation of a symmetry group. The method of $1/N$-expansion, as outlined here, cannot be applied directly to the gauge fields of the $O(N)$ groups, because these transform according to the adjoint rather than the fundamental representation of the $O(N)$ group (the number of gauge fields of the group $O(N)$ being equal to $N(N-1)/2$ and not to N). Other (topological) methods have been developed for the classification of diagrams in gauge theories according to the powers of $1/N$. However, at least in the four-dimensional case, these methods have not produced impressive results on the non-perturbative structure of the gauge theories. The reader may find further details on $1/N$-expansion (together with analysis of other models containing fields in the fundamental representations of the symmetry groups) in Zakrzewski (1989).

◇ Separate integration over lower and higher modes and infrared asymptotics of Green functions

Infrared divergences in quantum field theories appear in the perturbation theory calculations of transition amplitudes, due to the integration over a region of small energies of particles (virtual or real) entering the process (see Bogoliubov and Shirkov (1959) and Itzykson and Zuber (1980)). An important example of a theory where such problems exist is quantum electrodynamics (QED). The physical reason for the infrared divergences in QED is the existence of photons of arbitrarily small energies (since photons are massless). Some diagrams of the standard perturbation theory, in this case, have (along with the ultraviolet divergences which we have discussed earlier and which require a renormalization procedure for their removal), the specific infrared singularity. This means that the expression corresponding to a given diagram is singular at zero external momenta (i.e. momenta attributed to the external lines of the diagram). The origin of infrared singularities is quite different from that of ultraviolet ones. The former are removed not via renormalization but with the help of an appropriate rearrangement of the perturbation theory. One possible method of such a rearrangement is based on the path-integral approach and consists in successive integration, at first over the higher and then over lower momentum modes of the quantum fields.

We shall consider the Green function $S_c^{(tot)}(p)$, that is the total two-point electron Green function in QED taking into account the interactions (as distinct from the free-spinor Green function S_c; cf (3.1.106)). In this case, the infrared problem reveals itself as an appearance of a *power singularity* in $S_c^{(tot)}(p)$ at the mass shell $p^2 = m^2$, instead of a simple pole. As we have already mentioned, this is explained by the fact that a physical electron is surrounded by a 'cloud' of photons of arbitrarily small energy (arbitrarily large wavelength). The form of this singularity was found for the first time by Landau *et al* (1954) by the summation of a special infinite subset of the usual Feynman diagrams. The path-integral approach allows us to do this in a more natural and technically easier way (see Popov (1983) and references therein).

The words 'higher' and 'lower' modes mean that we should separate the Fourier modes $\widetilde{\psi}(k)$ of the photon field

$$A_\mu(x) = \int d^4k\, e^{ikx} \widetilde{A}_\mu(k) \qquad (3.3.16)$$

into two parts, corresponding to the small and large values of the momentum k. To do this, it is convenient to pass, first, to the Euclidean version of quantum field theory via the analytic continuation to imaginary time. In this case, we may define the lower modes as $\{\widetilde{A}_\mu(k),\ k^2 \equiv \sum_{\mu=1}^{4} k_\mu^2 \leq k_0^2\}$ for some appropriate k_0 and the higher modes as $\{\widetilde{A}_\mu(k),\ k^2 \equiv \sum_{\mu=1}^{4} k_\mu^2 > k_0^2\}$. The transition to the physical case of pseudo-Euclidean metrics is carried out by backward analytic continuation to real time in the final expression for the Green functions. Note that we now use the *Euclidean* γ-matrices, with the defining relations

$$\gamma_\mu \gamma_\nu + \gamma_\nu \gamma_\mu = 2\delta_{\mu\nu} \qquad \mu, \nu = 1, 2, 3, 4. \qquad (3.3.17)$$

The electron Green function (Euclidean version) is defined in the usual way

$$S_c^{tot}(x, y) = \frac{\delta}{\delta\bar{\eta}(x)} \frac{\delta}{\delta\eta(y)} \mathcal{Z}_{\text{QED}}[J_\mu, \eta, \bar{\eta}]\bigg|_{J=\bar{\eta}=\eta=0} = \langle 0|\bar{\psi}(x)\psi(y)|0\rangle$$
$$= \mathfrak{N}^{-1} \int \mathcal{D}A_\mu(x')\, \mathcal{D}\bar{\psi}(x')\, \mathcal{D}\psi(x')\, \delta[\partial^\mu A_\mu(x')]$$
$$\times \exp\{-S_{\text{QED}}[A(x'), \bar{\psi}(x'), \psi(x')]\}\bar{\psi}(x)\psi(y) \qquad (3.3.18)$$

where S_{QED} is given in (3.2.183). The calculation of this Green function can be carried out along the following steps:

(i) Integrate over the spinor electron–positron fields ψ, $\bar\psi$ using the fact that they enter the exponent in the path integral (3.3.18) quadratically.
(ii) Calculate the integral over the higher modes of the electromagnetic field using standard perturbation theory (as $k^2 > k_0^2$ for the higher modes, the infrared problems do not appear at this stage).
(iii) Calculate the remaining path integral over the lower modes $k^2 \leq k_0^2$ using the specific approximation outlined later.

The usual calculation of the Gaussian-like Grassmann integral over the fields ψ and $\bar\psi$ gives the following general structure for the (causal) Green function (3.3.18)

$$S_c^{\text{tot}}(x, y) = \mathfrak{N}^{-1} \int \mathcal{D}A_\mu(x') \, \delta[\partial_\mu A_\mu(x')] e^{-S_e[A]} \det(\gamma_\mu(\partial_\mu + ieA_\mu) - m) S_c(x, y; A) \quad (3.3.19)$$

where $S_e[A]$ is the action (cf (3.2.71)) of the free electromagnetic field, $S_c(x, y; A)$ is the electron Green function in the presence of an *external* electromagnetic field A_μ, i.e. $S_c(x, y; A)$ is the causal solution of the equation

$$[\gamma_\mu(\partial_\mu + ieA_\mu) - m] S_c(x, y; A) = \delta(x - y)$$

and

$$\det(\gamma_\mu(\partial_\mu + ieA_\mu) - m) \quad (3.3.20)$$

is the determinant of the Dirac operator in the *external* field. As usual the functional determinant must be regularized. The most natural way is to divide it by the free Dirac determinant (cf the regularization of the corresponding determinant for harmonic oscillator in the external field, section 2.2.2). Thus we use the ratio

$$\det(\gamma_\mu(\partial_\mu + ieA_\mu) - m) \rightarrow \frac{\det(\gamma_\mu(\partial_\mu + ieA_\mu) - m)}{\det(\gamma_\mu(\partial_\mu - m))} = \exp\left\{ \text{Tr} \ln\left(\mathbb{1} + \frac{1}{\slashed\partial - m} ie\slashed A \right) \right\}$$

$$= \exp\left\{ -\sum_{n=1}^\infty \frac{(-1)^n e^{2n}}{2n} \int dx_1 \cdots dx_n \, \text{Tr}[S_c^{(0)}(x_1 - x_2) \slashed A(x_2) \right.$$

$$\left. \cdots S_c^{(0)}(x_n - x_1) \slashed A(x_1)] \right\}. \quad (3.3.21)$$

The non-trivial terms in the expansion of this exponential (i.e. all terms except unity) represent the so-called *vacuum polarization* effect. Note that the sum in (3.3.21) goes only over even powers. This is a consequence of the *Furry theorem* (see problem 3.3.1, page 144).

Explicit and exact integration over the electromagnetic fields in (3.3.19) is impossible and we need some approximation methods. The first non-trivial approximation for the electron Green function can be obtained under the following assumptions:

(i) Let us set the Dirac determinant divided by the normalization constant (the free Dirac determinant) equal to unity, i.e. let us neglect all the vacuum polarization terms in (3.3.21).
(ii) After separating the electromagnetic field into the high-frequency part $A_\mu^{(\text{hf})}$

$$A_\mu^{(\text{hf})}(x) \stackrel{\text{def}}{\equiv} \int_{k^2 > k_0^2} d^4k \, e^{ikx} \tilde A_\mu(k)$$

and the low-frequency part $A_\mu^{(\text{lf})}$

$$A_\mu^{(\text{lf})}(x) \stackrel{\text{def}}{\equiv} \int_{k^2 \leq k_0^2} d^4k\, e^{ikx}\, \widetilde{A}_\mu(k)$$

(for some constant k_0), let us use for the Green function of an electron in the external field the approximate formula

$$S_c(x, y; A) \approx S_c(x, y; A^{(\text{hf})}) \exp\left\{ ie \int_x^y dx'_\mu\, A_\mu^{(\text{lf})}(x') \right\} \qquad (3.3.22)$$

where the line integral in the exponent goes along the straight contour connecting the points x and y. It can be shown that, for fixed x, y, this expression asymptotically converges to the exact one in the limit $k_0 \to 0$.

These assumptions allow us to present the path integral for the electron Green function as a product of two factors:

$$S_c^{(\text{tot})}(x, y) = \mathfrak{N}^{-1} \int \mathcal{D}A_\mu^{(\text{hf})}(x')\, \delta[\partial_\mu A_\mu^{(\text{hf})}(x')] S_c(x, y; A^{(\text{hf})}) \exp\{-S_e[A^{(\text{hf})}]\}$$
$$\times \int \mathcal{D}A_\mu^{(\text{lf})}(x')\, \delta[\partial_\mu A_\mu^{(\text{lf})}(x')] \exp\left\{ -S_{\text{el}}[A^{(\text{lf})}] + ie \int_x^y dx'_\mu\, A_\mu^{(\text{lf})}(x') \right\}. \qquad (3.3.23)$$

The first factor, containing only the high-frequency electromagnetic field, can be calculated by the usual perturbation theory and presented in the form

$$\frac{1}{(2\pi)^4} \int d^4p\, e^{ipx}\, \frac{-i\slashed{p} + m - \Sigma(p)}{p^2 + (m - \Sigma(p))^2} \qquad (3.3.24)$$

where the electron self-energy $\Sigma(p)$ is calculated in the lowest (second-order) approximation corresponding to the Feynman diagram (cf (3.2.187))

The momentum of the internal photon propagator is bounded from below by the constant k_0 which separates, according to our agreement, the higher and lower frequencies, so that this part of the self-energy is given by the integral

$$-\frac{1}{(2\pi)^4} \int_{k > k_0} d^4k\, \frac{k^2 \delta_{\mu\nu} - k_\mu k_\nu}{k^4} \gamma_\mu\, \frac{m - i(\slashed{p} - \slashed{k})}{(p-k)^2 + m^2} \gamma_\nu. \qquad (3.3.25)$$

Its calculation is performed in the vicinity of the mass-shell, $p^2 \approx -m^2$, and a subsequent Fourier transformation yields the infrared asymptotics of the first factor $S_c^{k > k_0}$ in (3.3.23):

$$S_c^{k > k_0} \approx \frac{1}{2} C_{k_0} \left(\frac{m_{k_0}}{2\pi \sqrt{x^2}} \right)^{3/2} (1 + \slashed{n}) \exp\left\{ -m_{k_0} \sqrt{x^2} \right\} \qquad (3.3.26)$$

where $\not{n} \equiv n_\mu \gamma_\mu$, $n_\mu \overset{\text{def}}{\equiv} x_\mu/\sqrt{x^2}$, and

$$C_{k_0} \overset{\text{def}}{\equiv} 1 - \frac{7e^2}{16\pi^2} - \frac{3e^2}{8\pi^2} \ln \frac{k_0}{m} \tag{3.3.27}$$

$$m_{k_0} \overset{\text{def}}{\equiv} m\left[1 + \frac{3e^2}{16\pi^2}\left(\ln \frac{\Lambda^2}{m^2} + 1\right)\right] - \frac{e^2}{4\pi^2} k_0. \tag{3.3.28}$$

The quantity m_{k_0} is the electron mass calculated in the second order of the usual perturbation theory, taking into account only the high-frequency electromagnetic field. The parameter Λ in (3.3.28) defines the ultraviolet regularization (e.g., ultraviolet cutoff). Of course, the usual ultraviolet renormalization is necessary.

Let us now turn to the second factor in (3.3.23), containing the low-frequency fields, as they are particularly important in the infrared region. The usual perturbation, as we have already mentioned, is inapplicable in this situation. Fortunately, with our assumptions, the second path integral in (3.3.23) becomes *Gaussian* and can be calculated straightforwardly. The answer is given by the following contour (line) integral:

$$\exp\left\{-\frac{e^2}{2}\int_x^y \int_x^y dx'_\mu dx''_\nu D^{(\text{lf})\perp}_{\mu\nu}(x'-x'')\right\} \tag{3.3.29}$$

where $D^{(\text{lf})\perp}_{\mu\nu}(x)$ is the (free) transversal low-frequency photon propagator:

$$D^{(\text{lf})\perp}_{\mu\nu}(x) = \frac{1}{(2\pi)^4} \int_{k<k_0} \delta^4 k\, e^{ikx} \left(\frac{k^2\delta_{\mu\nu} - k_\mu k_\nu}{k^4}\right). \tag{3.3.30}$$

An approximate calculation (at $k_0\sqrt{x^2} \gg 1$) of the ordinary contour integrals (3.3.29) yields

$$\exp\left\{-\frac{e^2}{4\pi^2}k_0\sqrt{x^2} + \frac{3e^2}{8\pi^2}\ln k_0\sqrt{x^2} + \frac{3e^2}{16\pi^2}(1 + 2\mathfrak{C} - 2\ln 2)\right\} \tag{3.3.31}$$

where $\mathfrak{C} = 0.5772\ldots$ is the Euler constant. The combination of this 'infrared' exponential with the first factor (3.3.28) finally results in the asymptotic expression for the Green function of an electron in the infrared region:

$$S_c^{(\text{tot})}(x)|_{p^2 \approx -m^2} \approx \frac{1}{2} a \frac{m^3}{(2\pi)^{3/2}} (m\sqrt{x^2})^{-\frac{3}{2}+\frac{3e^2}{8\pi^2}} (1 + \not{n}) e^{-m\sqrt{x^2}}. \tag{3.3.32}$$

Here a is the normalization factor

$$a = 1 + \frac{3e^2}{8\pi^2}(\mathfrak{C} - 2 - \ln 2)$$

and m is the physical (renormalized) electron mass. In terms of the regularization parameter Λ, i.e. before renormalization, the mass m can be expressed as follows

$$m = m_0\left[1 - \frac{3e^2}{16\pi^2}\left(\ln \frac{\Lambda^2}{m_0} + 1\right)\right].$$

The transition to the momentum representation via the Fourier transform gives, for the electron Green function near the mass-shell ($p^2 \approx -m^2$), the expression

$$\widetilde{S}_c^{(\text{tot})}(p)|_{p^2 \approx -m^2} \approx \left(1 - \frac{3e^2}{4\pi^2}\right) \frac{m - i\not{p}}{m^2\left(1 + \frac{p^2}{m^2}\right)^{1+3e^2/(8\pi^2)}}. \tag{3.3.33}$$

Recall that, together with the non-perturbative calculation, we have partially used ordinary perturbation theory, so that the infrared asymptotics (3.3.33) of the electron Green function are correct only up to the higher-order corrections (in the coupling constant e) of ordinary perturbation theory. We stress, however, that the expression contains all the powers of the coupling constant e (cf the denominator in (3.3.33)), so that this expression corresponds to the summation of an infinite subset of Feynman diagrams.

As we expected, the electron Green function has a *power singularity* on the mass-shell (instead of the ordinary pole, as it would in any finite order of the usual perturbation theory).

3.3.2 Semiclassical approximation in quantum field theory and extended objects (solitons)

So far, we have considered field theories quantized in the vicinity of the trivial field configuration $\varphi(x) = 0$ or (in spontaneous symmetry-breaking) in the vicinity of the non-zero *constant* field $\varphi(x) = a = $ constant. However, the field theoretical equations of motion, in many models with appropriate interaction terms, admit coordinate- and even time-dependent classical solutions with finite energy localized in a restricted area of the space. Such solutions are called '*solitons*'. The existence of solitons explains many important phenomena in particle physics, cosmology, solid state physics etc, which cannot be described within ordinary perturbation theory. In particular, their existence in non-Abelian gauge theories provides a natural way of introducing *magnetic monopoles* into field theory. These are hypothetical particles which possess a quantized magnetic charge (perhaps along with the ordinary electric charge). The modern theory of magnetic charges was first formulated by Dirac many years ago (Dirac 1931) in the framework of quantum electrodynamics. Their existence has been under active investigation ever since. The Dirac monopole is described by a field configuration with a singularity and has to be introduced into the standard QED 'by hand'. An intriguing aspect of non-Abelian gauge theories is that they have intrinsically existing solitons with the properties of magnetic monopoles, the so-called *'t Hooft–Polyakov monopoles* (Polyakov 1974, 't Hooft 1974).

We shall not discuss here the special topic of monopoles (the reader may find an elementary introduction in Cheng and Li (1984) and a more profound consideration in Rajaraman (1982)). Instead, we shall discuss, using some simpler examples, the general problem of field theoretical quantization in the vicinity of solitons.

A different direction in the study of extended objects in quantum field theory was advocated by Polyakov (1974) and Belavin *et al* (1975), who pointed out the importance of the classical solutions of finite action in the Euclidean space, obtained after the continuation to purely imaginary time. Such solutions, called *instantons*, exist only in theories with degenerate vacua and they signal tunneling between these different vacua. We shall consider this type of extended object in the next subsection.

In both *solitons* and *instantons*, the usual perturbation theory fails to describe the phenomena adequately and we have to use the semiclassical WKB approximation (Dashen *et al* 1974, Polyakov 1974, Belavin *et al* 1975). As we have already discussed in chapters 2 and 3, the path integral is the most natural and practically convenient tool for developing such an approximation. In this case, we will be simply making a change of variables in the path integral and the general idea is to consider the contribution of the fluctuations around the non-trivial minima of the action.

◇ Solitons in the two-dimensional scalar field theory

To provide an elementary introduction to the theory of classical localized solutions with finite energy for the field equations of motion (i.e. *solitons*), we shall briefly consider an example of $\lambda\varphi^4$-theory in one space and one time dimension. The Lagrangian is given by

$$\mathcal{L} = \int dx \, [\tfrac{1}{2}(\partial_t\varphi)^2 - \tfrac{1}{2}(\partial_x\varphi)^2 - V(\varphi)] \tag{3.3.34}$$

where
$$V(\varphi) = \frac{\lambda}{2}(\varphi^2 - a^2)^2 \qquad (3.3.35)$$
and
$$a^2 = m^2/\lambda.$$

The Hamiltonian is given by
$$H = \int dx \, [\tfrac{1}{2}(\partial_t \varphi)^2 + \tfrac{1}{2}(\partial_x \varphi)^2 + V(\varphi)]. \qquad (3.3.36)$$

As we have discussed in section 3.2.8, the classical ground-state configuration for the case $m^2 > 0$ is
$$\varphi = \pm a = \pm \sqrt{\frac{m^2}{\lambda}} \qquad (3.3.37)$$

and the ground-state energy is $E = 0$. An interesting feature of this model is that, in addition, there exists a static (time-independent) *finite-energy* solution of the equation of motion, that is, the *soliton*. The time-independent solution can be obtained from the Lagrangian $\mathcal{L}$ through the variational principle:
$$-\delta \mathcal{L} = \delta \int dx \, [\tfrac{1}{2}(\partial_x \varphi(x))^2 + V(\varphi(x))] = 0. \qquad (3.3.38)$$

Mathematically, this is equivalent to the problem of the motion of a particle of unit mass in a potential $-V(q)$, where the equation of motion is derived from
$$-\delta \mathcal{L}' = \delta \int dx \left[\frac{1}{2}\left(\frac{dq}{dt}\right)^2 + V(q) \right] = 0 \qquad (3.3.39)$$

(q is the coordinate of this fictitious particle which should not be confused with the space coordinate x; the field φ depends on the latter: $\varphi = \varphi(x)$). Any classical motion of the particle in the potential $-V(x)$ corresponds to a time-independent solution of the field equation. However, not all of these solutions are of finite energy. To get a finite-energy solution, we must require φ to go to a zero of $V(\varphi)$ as $x \to \pm\infty$, so that the energy integral in (3.3.36) is finite. In the fictitious particle problem, this corresponds to the condition that the particle must go to the zeros of the potential as $t \to \pm\infty$. Of course, the ground states where the particle sits at $x = a$ or $-a$ for all times satisfy this requirement, but there are also non-trivial motions which satisfy this requirement. The finiteness of energy requires the solution to take on the vacuum value ($\pm a$) at $t = \pm\infty$, but since we have a system of degenerate vacua, the solution may take on different minima ($+a$ or $-a$) at different infinity points ($+\infty$ or $-\infty$). Thus, for example, there are motions where the particle starts at the top of one hill and moves to the top of the other and has zero energy, see figure 3.16. We use this property of zero-energy motion to find the explicit form of the finite-energy solution in the field theory case. From energy conservation in the motion of the fictitious particle with zero total energy, we have
$$\frac{1}{2}\left(\frac{dq}{dt}\right)^2 + [-V(q)] = 0$$

which corresponds to the static field equation
$$\frac{1}{2}\left(\frac{d\varphi}{dx}\right)^2 = V(\varphi). \qquad (3.3.40)$$

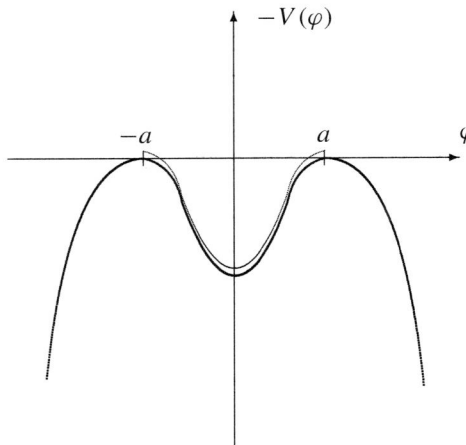

Figure 3.16. The potential $-V(\varphi)$ for the fictitious particle problem which is equivalent to the soliton problem; the thin line shows the motion of the 'particle' which starts at the top of one hill and moves to the top of another one.

Equation (3.3.40) can be solved easily by integration and the result is

$$x = \pm \int_{\varphi_0}^{\varphi} d\varphi' \, [2V(\varphi')]^{-1/2} \qquad (3.3.41)$$

where φ_0 is the value of φ at $x = 0$ and can be any number between a and $-a$. The presence of the arbitrary parameter φ_0 is due to the translational invariance of equation (3.3.40), i.e. if $\varphi = f(x)$ is a solution, then $\varphi = f(x - c)$ is also a solution where c is an arbitrary constant. In $\lambda\varphi^4$-theory, the potential is given by (3.3.35) and the finite-energy solutions in (3.3.41) can be written as

$$\varphi_+(x) = a \tanh(mx) \qquad \varphi_-(x) = -a \tanh(mx). \qquad (3.3.42)$$

The solution φ_+ is usually called the *kink* and φ_- the *anti-kink*. The energy of the kink (or anti-kink) can be calculated from (3.3.42) and (3.3.36) to give

$$E = 4m^3/3\lambda \qquad (3.3.43)$$

which is indeed finite. It is clear that, as $x \to \pm\infty$, φ_+ (or φ_-) approaches the zeros of $V(\varphi)$, i.e.

$$\varphi_+(x) \to \pm a \qquad \text{as } x \to \pm\infty. \qquad (3.3.44)$$

This behaviour is illustrated in figure 3.17. These solutions can be shown to be stable with respect to small perturbations even though they are not the absolute minima of the potential energy $V(\varphi)$ (i.e. $\varphi \neq \pm a$ for all x and t). The physical interest in these finite-energy solutions of the equation of motion comes from the fact that they resemble a particle with structure for the following reasons.

- Its energy is concentrated in a finite region of space. This is because these solutions $\varphi_\pm$ deviate from the ground-state configuration, $\varphi = \pm a$ (zero energy), only in a small region around the origin.
- It can be made to move with any speed less than unity (i.e. less than the speed of light). This is due to the fact that the equation of motion is Lorentz covariant and we can apply a Lorentz boost to obtain a solution with non-zero speed.

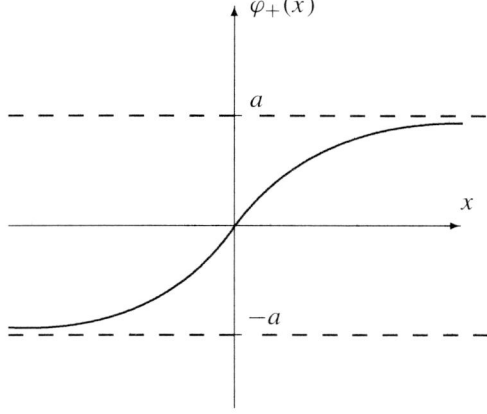

Figure 3.17. The form of the kink solution in two-dimensional $\lambda\varphi^4$-field theory.

The kink and anti-kink solutions in $\lambda\varphi^4$-theory in two-dimensional spacetime can be characterized by integers because, from the finite-energy requirement, we have at the spatial infinities:

$$\varphi(\infty) - \varphi(-\infty) = 2an \qquad n \in \mathbb{Z}$$

where $n = 0$ corresponds to the ground state ($\varphi = \pm a$), $n = 1$ to the kink solution, and $n = -1$ to the anti-kink solution. The value of the number n cannot be changed by smooth deformations of the solutions and, hence, it is a conserved quantum number, called the *topological charge*. The adjective 'topological' just reflects the fact that the charge depends on the global properties of the solution and cannot be varied by a smooth deformation of the field. The existence of this number provides the stability of the solitons. This consideration can be generalized to more complicated theories in higher dimensions (see, e.g., Rajaraman (1982)).

◇ **Quantization in the vicinity of solitons**

As we have just seen, there are four sectors of field configurations separated by the topological quantum number n. Thus, in the corresponding path integral we have to integrate separately over fields obeying the boundary conditions appropriate for a given sector.

As we have explained previously, the kink solution has an arbitrary parameter due to the translational invariance. This is a general property of all soliton solutions. Hence, the change of variables $\varphi(x) \to \varphi_{cl} + \Phi(x)$ which we have to perform to carry out the semiclassical calculation is not well defined. In other words, a variation of the translational parameters in φ_{cl} does not lead to a variation of the action and φ_{cl} belongs to a continuous family of stationary solutions, whereas the standard stationary-phase approximation works well only if the stationary points are sufficiently widely separated. Practically, the existence of such a degeneracy leads to a zero determinant when we calculate the fluctuation factor (and since the determinant appears in the denominator, this produces a meaningless infinite factor). The solution of this problem (also called the *zero-mode problem*) reduces to treating these free parameters of φ_{cl} as *dynamical variables*, which therefore should be determined from φ itself (see, e.g., Gervais (1977), Rajaraman (1982) and references therein). We shall consider this method in more detail later.

◇ **Quantization of one soliton in two dimensions**

We shall discuss the general two-dimensional Lagrangian for a scalar field

$$\mathcal{L} = -\tfrac{1}{2}(\partial_\mu \varphi)^2 - V(\varphi) \qquad (3.3.45)$$

with only the following condition for the potential:

$$V(\varphi) = \frac{1}{\lambda^2} V(\lambda \varphi) \qquad (3.3.46)$$

(restricting ourselves to the specific example of the kink, which we have studied earlier, does not essentially simplify the consideration). The meaning of condition (3.3.46) is as follows. Let us define the field $\varphi' = \lambda \varphi$. Then the Lagrangian and action take the form

$$\mathcal{L}(\varphi) = \frac{1}{\lambda^2} \mathcal{L}'(\varphi') \qquad S[\varphi] = \frac{1}{\lambda^2} S'[\varphi']$$

where $\mathcal{L}'(\varphi')$, $S'[\varphi']$ do not contain any coupling constant. This implies that all the dependence of the integrand in the corresponding path integral $\sim \int \mathcal{D}\varphi' \exp\{iS'[\varphi']/(\lambda^2 \hbar)\}$ on the coupling constant comes in the combination $(\lambda^2 \hbar)^{-1}$ in the exponent. The validity of the stationary-phase approximation requires that this combination $(\lambda^2 \hbar)$ be small compared with the action $S'[\varphi'_{cl}]$ of the classical solutions. This is the correct criterion, which is certainly satisfied when $\hbar$ and the coupling λ are both small. Note that this property of the action (factorization of the coupling constant) is shared by most of the models we have considered (e.g., by the Yang–Mills action).

The equation of motion corresponding to the Lagrangian (3.3.45) has a classical solitary wave solution with finite energy:

$$\varphi_{cl}(x,t) = \varphi_0\left(\frac{x - vt - x_0}{\sqrt{1-v^2}}\right)$$

where $\varphi_0(x)$ satisfies the equation

$$-\frac{\partial^2}{\partial x^2}\varphi_0(x) + \frac{\delta V}{\delta \varphi_0(x)} = 0. \qquad (3.3.47)$$

The transition amplitude between the initial and final states described by the wavefunctionals $\Psi_i[\varphi]$ and $\Psi_f[\varphi]$ is given by the path integral

$$\begin{aligned}
S_{fi} &= \langle \Psi_f | e^{-i\widehat{H}(t_f - t_i)} | \Psi_i \rangle \\
&= \int \prod_{x'} d\varphi'(x', t_f) \prod_x d\varphi(x, t_f) \, \langle \Psi_f | \varphi'(x', t_f) \rangle \langle \varphi'(x', t_f) | e^{-i\widehat{H}(t_f - t_i)} | \varphi(x, t_i) \rangle \langle \varphi(x, t_i) | \Psi_i \rangle \\
&= \int \mathcal{D}\varphi(x, \tau) \mathcal{D}\pi(x, \tau) \, \bar{\Psi}_f[\varphi(x, t_f)] \Psi_f[\varphi(x, t_i)] \\
&\quad \times \exp\left\{ i \int_{t_i}^{t_f} d\tau \int dx \, [\pi \dot\varphi - H(\pi, \varphi)] \right\}
\end{aligned} \qquad (3.3.48)$$

with a Hamiltonian of the form

$$H = \int dx \, [\tfrac{1}{2}\pi^2 + \tfrac{1}{2}(\partial_x \varphi)^2 + V(\varphi)].$$

If, in order to develop a perturbation expansion for the one-soliton sector, we simply expand around the classical solution φ_0, as in the case of spontaneous symmetry-breaking, we encounter divergences originating from the translation invariance of the theory. Namely, the propagator of this perturbation expansion would be the inverse of the following differential operator:

$$-\frac{\partial^2}{\partial t^2} + \widehat{\Omega}^2 \equiv -\frac{\partial^2}{\partial t^2} + \frac{\partial^2}{\partial x^2} + V''(\varphi_0) \qquad (3.3.49)$$

where

$$\widehat{\Omega}^2 \stackrel{\text{def}}{\equiv} \frac{\partial^2}{\partial x^2} + V''(\varphi_0) \qquad V''(\varphi_0) = \left.\frac{\partial^2 V}{\partial \varphi^2}\right|_{\varphi=\varphi_0}.$$

Taking the space derivative of field equation (3.3.47) satisfied by φ_0, we immediately see that $\partial_x \varphi_0$ is an eigenstate of Ω^2 with zero eigenvalue. Thus the propagator is ill defined, since the differential operator (3.3.49) has a zero eigenvalue (the *zero-mode problem*).

◇ Separation of the centre-of-mass coordinate (the zero-mode) in the one-soliton sector

To solve this difficulty and to develop a consistent perturbation expansion for the one-soliton sector, we should, first, separate the centre-of-mass coordinate (such an approach is called the *method of collective coordinates*). This can be achieved by a variant of the Faddeev–Popov trick. To this aim, we insert the following identities into the path-integral expression for the S-matrix element (3.3.48):

$$\int \mathcal{D}p(\tau) \prod_\tau \delta(p(\tau) + P) = 1 \qquad P \equiv \int dx\, \pi \partial_x \varphi \qquad (3.3.50)$$

$$\int \mathcal{D}X(\tau) \prod_\tau \delta(Q[\varphi(\tau, x + X(\tau)), \pi(\tau, x + X(\tau))]) \frac{\partial Q}{\partial X} = 1.$$

The first identity is the constraint which serves to identify the variable $p(\tau)$ with the total momentum of the system (recall that the Noetherian momentum for a field system with the Lagrangian (3.3.45) reads as follows: $\int dx\,(-\dot\varphi \partial_x \varphi) = \int dx\,(-\pi \partial_x \varphi)$, see, e.g., Bogoliubov and Shirkov (1959)), while the second identity is the gauge condition associated with this constraint. The τ-dependent functional Q (i.e. it is a functional only with respect to the dependence of φ and π on the space coordinate x) can be arbitrary, but here we shall consider a concrete example. Note that $\partial Q/\partial X$ is given by the Poisson bracket:

$$\frac{\partial Q}{\partial X} = \{Q, P\} \equiv \int dx \left[\frac{\partial Q}{\partial \varphi}\frac{\partial P}{\partial \pi} - \frac{\partial Q}{\partial \pi}\frac{\partial P}{\partial \varphi}\right]. \qquad (3.3.51)$$

Next, let us make a change of variables $\varphi \to \widetilde\varphi$, $\pi \to \widetilde\pi$:

$$\varphi(\tau, x) = \widetilde\varphi(\tau, x - X(\tau)) \equiv \widetilde\varphi(\tau, \rho)$$
$$\pi(\tau, x) = \widetilde\pi(\tau, x - X(\tau)) \equiv \widetilde\pi(\tau, \rho) \qquad (3.3.52)$$
$$\rho = x - X(\tau)$$

so that, using $\dot\varphi = \dot{\widetilde\varphi} - \dot X \partial_x \widetilde\varphi$ and the constraint, we obtain

$$\int dx\,[\pi\dot\varphi - H(\pi, \varphi)] = -p(\tau)\dot X(\tau) + \int d\rho\,[\widetilde\pi \dot{\widetilde\varphi} - H(\widetilde\pi, \widetilde\varphi)] \qquad (3.3.53)$$

$$\Psi_{i,f}[\varphi] = \exp\{-i p_{i,f} X(t_{i,f})\} \Psi_{i,f}[\widetilde\varphi].$$

From the first expression, we see that X is the variable conjugate to p, i.e. the centre-of-mass position and (3.3.52) is a transition to the moving frame attached to this centre of mass. The latter equality in (3.3.53) reflects the fact that the soliton as a whole (i.e. its centre of mass) moves as a free particle. Thus, we have explicitly exhibited the total momentum and centre-of-mass position associated with a given field configuration. If the latter corresponds to a quantum fluctuation around the one-soliton classical solution, X and p automatically become the position and momentum of this soliton.

Since X appears only in the term $p\dot{X}$, we can immediately integrate over X and p which leads to

$$S_{fi} = \delta(p_f - p_i) \int \mathcal{D}\tilde{\pi}\,\mathcal{D}\tilde{\varphi}\,\bar{\Psi}_f[\tilde{\varphi}]\Psi_i[\tilde{\varphi}]\delta(p+P)\delta(Q)$$

$$\times \{Q, P\} \exp\left\{i \int d\rho\, d\tau\,[\tilde{\pi}\dot{\tilde{\varphi}} - H(\tilde{\pi}, \tilde{\varphi})]\right\} \quad (3.3.54)$$

where p is not fixed: $p = p_i + p_f$. The stationary point of the action with constraints is given by the following variational equation:

$$\delta \int d\tau \left[\int d\rho\,(\tilde{\pi}\dot{\tilde{\varphi}} - \mathcal{H}) + \alpha(\tau)(p+P) \right] = 0$$

where α is a Lagrange multiplier. We obtain, for the lowest energy stationary point ($\dot{\varphi}_{\text{cl}} = 0$), exactly the soliton solution

$$\varphi_{\text{cl}} = \varphi_0\left(\sqrt{1 + \frac{p^2}{M_0^2}}(\rho - a)\right) \qquad \pi_{\text{cl}} = \frac{-p}{\sqrt{p^2 + M_0^2}} \partial_x \varphi_{\text{cl}} \quad (3.3.55)$$

where φ_0 is a solution of the equation

$$-\frac{\partial^2 \varphi_0}{\partial x^2} + \frac{\delta V}{\delta \varphi_0} = 0 \quad (3.3.56)$$

$M_0 = \int dx\,(\partial\varphi_0/\partial x)^2$ and the constant a is fixed by the gauge condition. The corresponding classical energy is found to be

$$E_{\text{cl}} = \sqrt{p^2 + M_0^2}. \quad (3.3.57)$$

Next, we have to choose an explicit form for the gauge condition. Although an arbitrary choice leads to a consistent perturbation expansion, free of infrared divergences, we choose a linear gauge condition

$$Q[\varphi(\tau, x + X(\tau))] \equiv \int dx\, f(x)\varphi(\tau, x + X(\tau))$$

$$\frac{\partial Q}{\partial X} = \int dx\, f(x)\partial_x \varphi(\tau, x + X) \quad (3.3.58)$$

in order to eliminate the zero-energy mode in the simplest possible way. Here f is still an arbitrary function but, identifying it later with the zero-frequency eigenfunction, we can completely eliminate the zero-frequency mode from the path integral. Now, before making the shift $\tilde{\varphi} = \varphi_0 + \chi$, it is convenient to linearize constraint (3.3.50), which is quadratic in fields, by making the following change of variables:

$$\tilde{\pi}(\tau, \rho) = -f(\rho)\frac{p + \int d\rho\,\varpi(\tau, \rho)[\partial_x\tilde{\varphi} - fc]}{\int d\rho\, f\partial_x\tilde{\varphi}} + \varpi(\tau, \rho) \quad (3.3.59)$$

(c is some constant to be defined later). Then the constraint becomes

$$\delta\left(p + \int d\rho\,\tilde{\pi}\partial_x\tilde{\varphi}\right) = \delta\left(c\int d\rho\, f(\rho)\varpi(\tau, \rho)\right).$$

Computing the Jacobian of this transformation, we get (see problem 3.3.2, page 145)

$$\det\left(\frac{\delta\widetilde{\pi}}{\delta\varpi}\right) = \prod_\tau \left(\int d\rho\, f\partial_x\widetilde{\varphi}\right)^{-1}$$

which exactly cancels out the $\partial Q/\partial X$ given by (3.3.58). Now the Hamiltonian becomes more complicated:

$$\int d\rho\, H = \frac{(p + \int d\rho\, \varpi(\tau,\rho)[\partial_x\widetilde{\varphi} - fc])^2}{2(\int d\rho\, f\partial_x\widetilde{\varphi})^2} + \int d\rho \left[\frac{1}{2}\varpi^2 + \frac{1}{2}(\partial_x\widetilde{\varphi})^2 + V(\widetilde{\varphi})\right]. \quad (3.3.60)$$

Here we have to input the normalization $\int d\rho\, f^2(\rho) = 1$. The transition amplitude now takes the form

$$S_{fi} = \delta(p_f - p_i) \int \mathcal{D}\varpi\, \mathcal{D}\widetilde{\varphi}\, \bar{\Psi}_f[\widetilde{\varphi}]\Psi_i[\widetilde{\varphi}]\delta\left(\int d\rho\, f\widetilde{\varphi}\right)\delta\left(\int d\rho\, f\varpi\right)$$

$$\times \exp\left\{i\int d\tau \int d\rho\, [\widetilde{\pi}\dot{\widetilde{\varphi}} - H]\right\} \quad (3.3.61)$$

and since both the gauge condition and the constraint are linear in the fields, we can easily develop a perturbation expansion.

◇ Perturbation expansion in the one-soliton sector

At this point, we observe that, due to the property (3.3.46) of our potential, φ_{cl} is of the order of $1/\lambda$; accordingly, M_0 is of the order of $1/\lambda^2$. We can develop the perturbation expansion in λ around the classical solution

$$\widetilde{\varphi} = \varphi_{\text{cl}} + \chi(\tau,\rho). \quad (3.3.62)$$

Here χ represents small quantum fluctuations around the classical solution. In general, we can also consider a shift of the momentum variable $\widetilde{\pi} = \pi_{\text{cl}}(\rho) + \vartheta(\tau,\rho)$. This leads to a relativistic form for the soliton energy and for the perturbation theory. However, we shall restrict ourselves only to the shift (3.3.62), because in this case the corresponding perturbation expansion is much simpler.

With the choice of f and c as

$$f = \frac{1}{\sqrt{M_0}}\partial_x\varphi_0 \equiv \Psi_0 \qquad c = \sqrt{M_0} \quad (3.3.63)$$

the Hamiltonian reads as

$$\int d\rho\, H = M_0 + \frac{p + \int d\rho\, \varpi\partial_x\chi}{2M_0(1 + \xi/M_0)^2} + \int d\rho \left[\frac{1}{2}\varpi^2 + \frac{1}{2}\partial_x\chi^2 + V - V(\varphi_0) - \left.\frac{\delta V}{\delta\varphi}\right|_{\varphi_0}\chi\right] + \Delta V \quad (3.3.64)$$

with

$$\xi = \int d\rho\, \partial_x\varphi_0(\rho)\partial_x\chi(\tau,\rho) \quad (3.3.65)$$

$$\Delta V = \frac{1}{8}\left[-3\frac{\langle\partial_x\Psi_0|\partial_x\Psi_0\rangle}{\langle\partial_x\widetilde{\varphi}|\Psi_0\rangle^2} + 2\frac{\langle\partial_x\Psi_0|\partial_x^2\widetilde{\varphi}\rangle}{\langle\Psi_0|\partial_x\widetilde{\varphi}\rangle^3} + \frac{\langle\partial_x\Psi_0|\partial_x\varphi\rangle^2}{\langle\Psi_0|\partial_x\varphi\rangle^4} + \sum_{n,m\neq 0}\frac{|\langle\Psi_n|\partial_x\Psi_m\rangle|^2}{\langle\Psi_0|\partial_x\varphi\rangle}\right] \quad (3.3.66)$$

where Ψ_m are the eigenfunctions of the operator $\widehat{\Omega}^2$:

$$\widehat{\Omega}^2 \Psi_m = \omega_m \Psi_m$$

and the scalar product for any two functions $h_1(x), h_2(x)$ is defined as usual:

$$\langle h_1|h_2\rangle \stackrel{\text{def}}{\equiv} \int dx\, h_1(x)h_2(x).$$

The additional potential ΔV starts contributing at the two-loop level (because it is proportional to $\hbar^2$ after explicit recovery of the Planck constant). As we learned in chapter 2 (see section 2.5.2), the additional term ΔV always arises if a change of variables, in particular (3.3.52), (3.3.62), is carried out with enough care.

The Feynman rules for the perturbation expansion can now be obtained. The propagator is determined from the quadratic part of the Hamiltonian by expanding it in terms of the eigenfunctions of the following differential equations:

$$\Omega^2 \Psi_m \equiv \left(-\frac{\partial^2}{\partial \rho^2} + V''(\varphi_0)\right)\Psi_m = \omega_m^2 \Psi_m. \tag{3.3.67}$$

The zero-energy eigenfunction is given by Ψ_0. We have chosen f to be given precisely by Ψ_0, so that the $\omega_0 = 0$ mode disappears from the eigenfunction expansion of ϖ and χ because of the δ-condition in (3.3.61).

Since we have used the first-order formalism, this perturbation expansion involves three different propagators:

$$\langle 0|\mathbf{T}(\chi(t_1, x_1)\chi(t_2, x_2))|0\rangle \qquad \langle 0|\mathbf{T}(\varpi(t_1, x_1)\varpi(t_2, x_2))|0\rangle \qquad \text{and} \qquad \langle 0|\mathbf{T}(\chi(t_1, x_1)\varpi(t_2, x_2))|0\rangle.$$

The Hamiltonian (3.3.64) contains products of χ and ϖ at the same point and therefore there are ordering problems if we want to write H as an operator. In the path-integral formalism, this ordering problem also appears in practice because the perturbation expansion contains the mixed propagator $\langle 0|\mathbf{T}(\chi(t, x_1)\varpi(t, x_2))|0\rangle$ with zero-time separation, which is ambiguous. Using the discrete-time approximation for the path integral, we find that the expression (3.3.66) corresponds to the mid-point definition. Namely, we have to choose the field variables $\chi(\tau_{2l}, x)$, $\varpi(\tau_{2l+1}, x)$ and write

$$\int d\tau\, \varpi\, \dot\chi \equiv \sum_l \int dx\, \varpi(\tau_{2l+1}, x)[\chi(\tau_{2l+2}, x) - \chi(\tau_{2l}, x)].$$

As we know (see section 2.5), this implies that in the operator formalism, ΔV is the term associated with the Weyl ordering for expression (3.3.64) of H. For the perturbation theory this means, in turn, that the mixed propagator $\langle 0|\mathbf{T}(\chi(t, x_1)\varpi(t, x_2))|0\rangle$ for zero-time separation is taken to be zero, i.e. all closed loops of the mixed propagator are to be dropped.

Note that in this perturbation expansion, Lorentz invariance is not manifest, but we can show that higher-order corrections in the coupling constant sum up to restore Lorentz invariance at least at the tree level (Gervais *et al* 1975).

The renormalization of the one-soliton sector can be carried out in a straightforward manner, by adding counter terms (see section 3.2.7). With this systematic perturbation expansion, we can perform perturbation calculations of any desirable quantities, e.g., energy or field matrix elements in the one-soliton sector. We refer the reader for these results and their discussion to Gervais *et al* (1975),

Gervais (1977) and Rajaraman (1982). Note that expression (3.3.61) *a priori* looks like a highly non-renormalizable Hamiltonian since it involves vertices with an arbitrary number of legs. It is remarkable that finite results are in fact obtained to any order just by using the same counter terms as in the usual sector. Already at the two-loop level, this involves a remarkable cancelation among highly divergent integrals.

◇ Remarks on the quantization of several solitons

For the case of several solitons, we obviously need to extract more collective coordinates. A general method for doing this (see, e.g., Gervais *et al* (1976)) which can be applied to any problem in which collective coordinates are relevant exists. For the particular case considered earlier, the generalization can be achieved quite straightforwardly by introducing into the corresponding path integral the unity in the form

$$\int \prod_{\alpha,\beta=1}^{n} dX_\alpha \, dP_\beta \, J \delta(p_\alpha + P_\alpha[\pi,\varphi])\delta(Q_\beta[\pi,\varphi]) = 1$$

$$J = \prod_\tau \det\{P_\alpha, Q_\beta\}_p = \prod_\tau \det\left(\frac{\delta Q_\beta}{\delta X_\gamma}\right)$$

where n is the number of solitons, $\alpha, \beta = 1, \ldots, n$. It is worth noting that for small λ, the soliton position moves much more slowly than the other degrees of freedom. This is the standard criterion for introducing collective coordinates (the so-called adiabatic approximation). In this case we can determine an effective potential by first solving the dynamics of the other degrees of freedom with fixed X_α, P_β. In the functional formalism, this is formally done by assuming that $\Psi_{i,f}$ are eigenstates of P_α with eigenvalues $p_{\alpha,i}, p_{\alpha,f}$, and computing for fixed X_α, p_β the effective Hamiltonian defined by the relation

$$\exp\left\{-i\int d\tau \, H_{\text{eff}}(X,P)\right\} \equiv \int \mathcal{D}\tilde{\pi} \, \mathcal{D}\tilde{\varphi} \, \delta(p_\alpha + P_\alpha)\delta(Q_\beta) \det\{P,Q\} \Psi_f^*[\tilde{\varphi}]\Psi_i[\tilde{\varphi}]$$

$$\times \exp\left\{i \int d\tau \, dx [\tilde{\pi}\dot{\tilde{\varphi}} - H[\tilde{\pi}, \tilde{\varphi}, X]]\right\} \quad (3.3.68)$$

In the adiabatic approximation, the transition probability is given by

$$\int \mathcal{D}X_\alpha \, \mathcal{D}P_\beta \, e^{-ip_{\alpha,f}X_\alpha(t_f)} e^{ip_{\beta,i}X_\alpha(t_i)} \exp\left\{i\int_{t_i}^{t_f} dt \, (P_\alpha \dot{X}_\alpha - H_{\text{eff}}\right\}.$$

If $\{P_\alpha, H\} = 0$, we have $H = H[\tilde{\pi}, \tilde{\varphi}]$, which is independent of X and the dynamics of X_α, P_β is trivial. The eigenstates are plane waves. In this case, it is better to reverse the method used for the one-soliton case. Namely, again choosing $\Psi_{i,f}$ to be eigenstates of $\widehat{P}_\alpha$, we first integrate over X_α and P_β, immediately obtaining

$$\prod_\alpha \delta(p_{\alpha,i} - p_{\alpha,f}) \int \mathcal{D}\tilde{\pi} \, \mathcal{D}\tilde{\varphi} \, \delta(p_{\alpha,i} + P_\alpha)\delta(Q_|b) \det\{P,Q\}$$

$$\times \bar{\Psi}_f[\tilde{\varphi}]\Psi_i[\tilde{\varphi}] \exp\left\{i \int d\tau \, dx \, [\tilde{\pi}\dot{\tilde{\varphi}} - H[\tilde{\pi}, \tilde{\varphi}]]\right\}. \quad (3.3.69)$$

In order to apply the semiclassical method, we look for the minimum of the action taking into account the constraints. Technically convenient modifications of the method of collective coordinates have been developed by using the BRST invariance (cf section 3.2.7) of the appropriately constructed effective action (see, e.g., Alfaro and Damgaard (1990)) which are more suitable for the study of solitons in more complicated field theoretical models.

It is necessary to add that semiclassical calculations of the sort considered in this subsection reveal a hidden but physically very important symmetry, known as the *duality*, of some models with topological solitons. Such a symmetry was first conjectured by Montonen and Olive (1977) for a model in which a simple gauge group is spontaneously broken into a $U(1)$ electromagnetic gauge group. They noted that the semiclassical approximation gives the mass of particles with mass

$$m = \sqrt{2} \left| v \left(ne + \frac{4\pi i \ell}{e} \right) \right|$$

(v is the vacuum expectation value of the Higgs field, e is the gauge coupling constant; n and ℓ are integers of any sign), which is invariant under the transformations:

$$\ell \to n \qquad n \to -\ell \qquad e \to 4\pi/e.$$

On this basis, they suggested that the theory with a weak gauge coupling e is fully equivalent to one with a strong coupling $4\pi/e$. Unfortunately, the purely bosonic theory does not really have this property (Osborn 1979). However, there are strong indications that theories with the so-called *extended* (more precisely, $N = 4$) *supersymmetry* (see, e.g., Weinberg (2000)) are indeed invariant under the interchange of electric and magnetic quantum numbers and of e and $4\pi/e$. Moreover, even theories with smaller ($N = 2$) supersymmetry proved to be invariant with respect to duality transformations of more subtle sort (Seiberg and Witten 1994). Using this property, Seiberg and Witten were able to carry out prominent non-perturbative calculations in this type of field theory (for a review, see Intriligator and Seiberg (1996)).

3.3.3 Semiclassical approximation and quantum tunneling (instantons)

Instantons are a special type of vacuum fluctuation in non-Abelian gauge theories and classical solutions of the Euclidean equations of motion. The instanton, being a solution in Euclidean field theory, is a minimum of the action in which all kinetic terms are positive. Hence, an *instanton* solution in a d-dimensional spacetime is also a time-independent *soliton* solution in $d + 1$ dimensions and possible solutions can be classified simultaneously for both cases. This is based on homotopy theory which we will not present here (see Rajaraman (1982)). The instantons are characterized by a topological quantum number (similar to the topological solitons) and correspond to tunneling events between degenerate classical vacua in the Minkowski space. The existence of the non-zero topological number means that it is impossible to deform the instanton field configuration into a zero field, keeping the value of the field action finite. A distinguishing property of the instanton solution is their finite size in the space and time directions, so that they are localized configurations and remind us of particle-like behaviour in four dimensions. In fact, they can indeed be treated formally as a kind of particle in four-dimensional quantum statistical mechanics. However, we should remember that the genuine physical meaning of the solution is *tunnel transitions* taking part in the complicated vacuum (ground state) of a non-Abelian quantum field theory. The transition processes take a finite period of time and therefore the instantons have a finite 'longitude' in time. That is why these solutions were called 'instan*ton*', the first part of the word reflecting their time behaviour and the end of the word reflecting their particle-like nature in four dimensions.

The simplest situation relating to instantons happens in non-relativistic quantum mechanics and corresponds to the tunneling of a particle in the potential $V(x) \sim (x^2 - a^2)^2$ with degenerate vacuum states and we shall use this case as our basic example. Then we shall briefly consider the generalization of the quantum tunneling phenomenon to the case of Yang–Mills theory (quantum chromodynamics). It is necessary to stress that instanton calculations are formally very close to the hopping path approximation which we considered in the first chapter (section 1.2.6). The reader may find more details on instanton calculations in Rajaraman (1982) and Schäfer and Shuryak (1998).

◇ Tunneling in quantum mechanics and instantons: double-well potential

For an introduction to instanton methods we start with a relatively simple quantum-mechanical problem, which does not suffer from any of the divergences that occur in field theory. Tunneling is a quantum-mechanical phenomenon, a particle penetrating a classically forbidden region. Nevertheless, after continuing the transition amplitude to imaginary time, the tunneling process can be described by classical equations of motion.

Let us consider an anharmonic oscillator with a Euclidean action

$$S_E = \int d\tau \left[\frac{\dot{x}^2}{2} + V(x) \right]$$

where $V(x)$ is a double-well potential (cf problem 3.1.9, page 44).

Continuing $\tau = it$, the classical equation of motion is given by

$$\frac{d^2x}{d\tau^2} = +\frac{dV}{dx} \qquad (3.3.70)$$

where the sign of the potential energy term has changed in comparison with the real-time case. This means that the classically forbidden regions (for the real time) are now classically allowed. The special role of the classical tunneling path becomes clear if we consider the Feynman path integral. Although any path is allowed in quantum mechanics, the path integral is dominated by paths that maximize the weight factor $\exp(-S[x_{cl}(\tau)])$, or minimize the Euclidean action.

Now we choose the concrete form of the potential as follows:

$$V = \lambda(x^2 - \eta^2)^2 \qquad (3.3.71)$$

with minima at $\pm\eta$, the two 'classical vacua' of the system. Quantizing around the two minima, we would find two degenerate states localized at $x = \pm\eta$. Of course, we know that this is not the correct result. Tunneling mixes the two states, the true ground state being (approximately) the symmetric combination, while the first excited state is the antisymmetric combination of the two states.

Formally, equation (3.3.70) is the same as in two-dimensional $\lambda\varphi^4$-theory (cf the preceding section, equation (3.3.39)) which contains the soliton (kink) solutions. Therefore, we can immediately write the solution:

$$x_{cl}(\tau) = \eta \tanh\left[\frac{\omega}{2}(\tau - \tau_0)\right] \qquad (3.3.72)$$

which goes from $x(-\infty) = -\eta$ to $x(\infty) = \eta$. Here, τ_0 is a free parameter (the instanton centre) and $\omega^2 = 8\lambda\eta^2$. The action of the solution is $S_0 = \omega^3/(12\lambda)$. We will refer to path (3.3.72) as the *instanton*, since (unlike the soliton) the solution is localized in time. An *anti*-instanton solution is given by $x_{cl}^A(\tau) = -x_{cl}(\tau)$.

The semiclassical approximation for the path integral is obtained by systematically expanding the action around the classical solution, similarly to the case of solitons

$$\langle -\eta|e^{-H\tau}|\eta\rangle = e^{-S_0} \int \mathcal{D}X(\tau) \exp\left\{ -\frac{1}{2} \int d\tau\, X(\tau) \left.\frac{\delta^2 S}{\delta x^2}\right|_{x_{cl}} X(\tau) + \cdots \right\}. \qquad (3.3.73)$$

Note that we implicitly assumed τ to be large, but smaller than the typical lifetime for tunneling. If τ is larger than the lifetime, we have to take into account multi-instanton configurations (i.e. multiple movements of the particle from one extremum to another). It is seen that the hopping path method which we briefly discussed in chapter 1 (section 1.2.6), in fact, coincides with the instanton calculations in quantum mechanics. Clearly, the tunneling amplitude is proportional to $\exp\{-S_0\}$. The pre-exponent requires the calculation of fluctuations around the classical instanton solution.

◇ Tunneling amplitude at one-loop order

In order to take into account the fluctuations around the classical path, we have to calculate the path integral

$$\int \mathcal{D}X(\tau) \exp\left\{-\frac{1}{2}\int d\tau\, X(\tau)\widehat{O}X(\tau)\right\} \tag{3.3.74}$$

where $\widehat{O}$ is the differential operator

$$\widehat{O} = -\frac{1}{2}\frac{d^2}{d\tau^2} + \left.\frac{d^2V}{dx^2}\right|_{x=x_{\text{cl}}}. \tag{3.3.75}$$

This calculation is carried out in the standard way (see, in particular, section 1.2.7 and the case of solitons in the preceding section) and it provides a very good illustration of the steps that are required to solve the more difficult field theory problem (Polyakov 1977).

Expanding the differential operator $\widehat{O}$ in some basis $\{x_i(\tau)\}$, we have

$$\int \left(\prod_n dx_n\right) \exp\left(-\frac{1}{2}\sum_{ij} x_i O_{ij} x_j\right) = \prod_n (2\pi)^{n/2} (\det \widehat{O})^{-1/2}. \tag{3.3.76}$$

The determinant can be calculated by diagonalizing $\widehat{O}$, $\widehat{O}x_n(\tau) = \epsilon_n x_n(\tau)$. This eigenvalue equation reads as

$$\left(-\frac{d^2}{d\tau^2} + \omega^2\left[1 - \frac{3}{2\cosh^2(\omega\tau/2)}\right]\right) x_n(\tau) = \epsilon_n x_n(\tau). \tag{3.3.77}$$

Formally, this equation coincides with the one-dimensional Schrödinger equation (where τ plays the role of a *space* coordinate) for the so-called modified *Pöshl–Teller* potential. In the standard quantum-mechanical context, this Schrödinger equation is discussed in, e.g., Flügge (1971) (vol I, problem 39) and in Landau and Lifshitz (1981). There are two bound states plus a continuum of scattering states. The lowest eigenvalue is $\epsilon_0 = 0$, and the other bound state is at $\epsilon_1 = \frac{3}{4}\omega^2$. The appearance of a zero mode is related, similarly to the case of solitons, to translational invariance (the fact that the action does not depend on the location τ_0 of the instanton). The normalized eigenfunction ($\int d\tau\, x_n^2 = 1$) of the zero energy state is

$$x_0(\tau) = \sqrt{\frac{3\omega}{8}}\frac{1}{\cosh^2(\omega\tau/2)} \tag{3.3.78}$$

which is just the derivative of the instanton solution over τ_0 (see the explanation after equation (3.3.49)):

$$x_0(\tau) = -S_0^{-1/2}\frac{d}{d\tau_0}x_{\text{cl}}(\tau - \tau_0). \tag{3.3.79}$$

Recall that the presence of a zero mode also indicates that there is one direction in the functional space in which fluctuations are large, so the integral is not Gaussian. This means that the integral in that direction should not be performed directly in the Gaussian approximation, but has to be treated with care using the collective coordinate method (see the preceding subsection).

In a quantum-mechanical (not a field theoretical) system, the transition to collective coordinates is not complicated and can be achieved by replacing the integral over the expansion parameter c_0 associated with the zero-mode direction (we use the mode expansion: $x(\tau) = \sum_n c_n x_n(\tau)$) with an integral over the collective coordinate τ_0. Using

$$dx = \frac{dx_{\text{cl}}}{d\tau_0} d\tau_0 = -\sqrt{S_0} x_0(\tau) d\tau_0 \qquad (3.3.80)$$

and $dx = x_0 dc_0$, we have $dc_0 = \sqrt{S_0} d\tau_0$. The functional integral over the quantum fluctuation is now given by

$$\int \mathcal{D}X(\tau)] \exp\{-S\} = \left[\prod_{n>0}\left(\frac{2\pi}{\epsilon_n}\right)\right]^{1/2} \sqrt{S_0} \int d\tau_0 \qquad (3.3.81)$$

where the first factor is the determinant with the zero mode excluded. The result shows that the tunneling amplitude grows linearly with time, i.e. there is a finite transition probability *per unit of time*.

The next step is to calculate the non-zero-mode determinant. For this purpose we make the spectrum discrete by considering a finite-time interval $[-T/2, T/2]$ and imposing boundary conditions at $\pm T/2$: $x_n(\pm T/2) = 0$. The product of all eigenvalues is divergent, but the divergence is related to large eigenvalues, independent of the detailed shape of the potential. The determinant can be renormalized by taking the ratio over the determinant of the harmonic oscillator (similarly to what we did in section 2.2.2, where we used the ratio of the determinants for a harmonic oscillator and a free particle). The result is

$$\left(\frac{\det\left[-\frac{d^2}{d\tau^2} + V''(x_{\text{cl}})\right]}{\det\left[-\frac{d^2}{d\tau^2} + \omega^2\right]}\right)^{-1/2} = \sqrt{\frac{S_0}{2\pi}} \omega \int d\tau_0 \left(\frac{\det'\left[-\frac{d^2}{d\tau^2} + V''(x_{\text{cl}})\right]}{\omega^{-2}\det\left[-\frac{d^2}{d\tau^2} + \omega^2\right]}\right)^{-1/2} \qquad (3.3.82)$$

where we have eliminated the zero mode from the instanton determinant, denoting this by a prime: $\det \to \det'$, and replaced it by the integration over τ_0. We also have to extract the lowest mode from the harmonic oscillator determinant, which is given by ω^2. The next eigenvalue of the fluctuation operator (3.3.75) is $3\omega^2/4$, while the corresponding oscillator mode is ω^2 (up to corrections of order $1/T^2$, that are not important as $T \to \infty$). The rest of the spectrum is continuous as $T \to \infty$. The contribution from these states can be calculated as follows.

The potential $V''(x_{\text{cl}})$ is localized, so for $\tau \to \pm\infty$ the eigenfunctions are just plane waves. This means that we can take one of the two linearly independent solutions to be $x_p(\tau) \sim \exp(ip\tau)$ as $\tau \to \infty$. The effect of the potential is to give a phase shift

$$x_p(\tau) = \exp(ip\tau + i\delta_p) \qquad \tau \to -\infty \qquad (3.3.83)$$

where, for this particular potential, there is no reflected wave. The phase shift is given (Landau and Lifshitz 1981) by

$$\exp(i\delta_p) = \frac{1 + ip/\omega}{1 - ip/\omega} \frac{1 + 2ip/\omega}{1 - 2ip/\omega}. \qquad (3.3.84)$$

The second independent solution is obtained by $\tau \to -\tau$. The spectrum is determined by the quantization condition $x(\pm T/2) = 0$, which gives

$$p_n T - \delta_{p_n} = \pi n \qquad (3.3.85)$$

while the harmonic oscillator modes are determined by $p_n T = \pi n$. If we denote the solutions of (3.3.85) by $\tilde{p}_n$, the ratio of the determinants is given by

$$\prod_n \left[\frac{\omega^2 + \tilde{p}_n^2}{\omega^2 + p_n^2}\right] = \exp\left\{\sum_n \log\left[\frac{\omega^2 + \tilde{p}_n^2}{\omega^2 + p_n^2}\right]\right\} = \exp\left\{\frac{1}{\pi}\int_0^\infty \frac{2p\,dp\,\delta_p}{p^2 + \omega^2}\right\} = \frac{1}{9} \qquad (3.3.86)$$

where we have expanded the integrand in the small difference $\tilde{p}_n - p_n = \delta_{p_n}/T$ and changed from the summation over n to an integral over p. In order to perform the integral, it is convenient to integrate by parts and use the result for $(d\delta_p)/(dp)$. Collecting everything, we finally get

$$\langle -\eta | e^{-\widehat{H}T} | \eta \rangle = \left[\sqrt{\frac{\omega}{\pi}}\exp\left\{-\frac{\omega T}{2}\right\}\right]\left[\sqrt{\frac{6S_0}{\pi}}\exp\{-S_0\}\right](\omega T) \qquad (3.3.87)$$

where the first factor comes from the harmonic oscillator amplitude (cf the calculations in section 2.2.2) and the second is the ratio of the two determinants.

Recall that in terms of stationary states the ground-state wavefunction is the symmetric combination $\Psi_0(x) = (\phi_{-\eta}(x) + \phi_\eta(x))/\sqrt{2}$, while the first excited state $E_1 = E_0 + \Delta E$ is antisymmetric, $\Psi_1(x) = (\phi_{-\eta}(x) - \phi_\eta(x))/\sqrt{2}$ (see, e.g., Landau and Lifshitz (1981)). Here, $\phi_{\pm\eta}$ are the harmonic oscillator wavefunctions around the two classical minima. For times satisfying $T \ll 1/\Delta E$, the tunneling amplitude is given by

$$\langle -\eta | e^{-HT} | \eta \rangle = \Psi_0^*(-\eta)\Psi_0(\eta)e^{-E_0 T} + \Psi_1^*(-\eta)\Psi_1(\eta)e^{-E_1 T} + \cdots$$

$$= \frac{1}{2}\phi_{-\eta}^*(-\eta)\phi_\eta(\eta)(\Delta E T)e^{-\omega T/2} + \cdots. \qquad (3.3.88)$$

For large times $T > 1/\Delta E$, we have to take into account the multi-instanton paths, that is, the classical solution corresponding to multiple movements of the fictitious 'particle' from one extremum to another and backward, as depicted in figure 1.13, page 93, volume I (recall that in the case of classical stochastic processes, the analogs of the instanton solutions are called hopping paths). If we ignore the 'interaction' between instantons, multi-instanton contributions can easily be summed:

$$\langle -\eta | e^{-HT} | \eta \rangle = \sqrt{\frac{\omega}{\pi}}e^{-\omega T/2}\sum_{n \text{ odd}}\int_{-T/2 < \tau_1 < \cdots < T/2}\left[\prod_{i=1}^n \omega\,d\tau_i\right]\left(\sqrt{\frac{6S_0}{\pi}}\exp\{-S_0\}\right)^n \qquad (3.3.89)$$

$$= \sqrt{\frac{\omega}{\pi}}e^{-\omega T/2}\sum_{n \text{ odd}}\frac{(\omega T d)^n}{n!} = \sqrt{\frac{\omega}{\pi}}e^{-\frac{\omega T}{2}}\sinh(\omega T d)$$

where $d = (6S_0/\pi)^{1/2}\exp\{-S_0\}$. Under the 'interaction' between instantons (which are, of course, fictitious particles) we understand an account of the difference between the true classical solution of equation (3.3.70) with multiple movements (from one extremum to another) from just a 'sewing' together of a number of one-instanton solutions (3.3.78). Summing over all instantons simply leads to the exponentiation of the tunneling rate. Now we can directly read off the level splitting from (3.3.87) and (3.3.88)

$$\Delta E = \sqrt{\frac{6S_0}{\pi}}\omega\exp(-S_0). \qquad (3.3.90)$$

If the tunneling rate increases, $1/\Delta E \simeq 1/\omega$, the interactions between instantons become important.

It is worth mentioning that in the gauge theory of strong interactions (i.e. in quantum chromodynamics) the multi-instanton configurations play an important role in attempts to explain the phenomenon of quark confinement (Callan *et al* (1979), see also Schäfer and Shuryak (1998) and references therein).

◇ **Fermions coupled to the double-well potential**

Let us now add one fermionic degree of freedom ψ_α ($\alpha = 1, 2$) coupled to the double-well potential (still in the framework of non-relativistic quantum mechanics). This model provides additional insight into the vacuum structure, not only of quantum mechanics, but also of gauge theories: we will see that fermions are intimately related to tunneling, and that the fermion-induced interaction between instantons leads to strong instanton–anti-instanton correlations. The names instanton and *anti*-instanton are attributed to movements from one extremum to another in opposite directions.

The model is defined by the action

$$S = \tfrac{1}{2} \int dt \, (\dot{x}^2 + W'^2 + \psi\dot{\psi} + cW''\psi\sigma_2\psi) \qquad (3.3.91)$$

where ψ_α ($\alpha = 1, 2$) is a two-component spinor, the dots denote time derivatives and primes the spatial derivatives, and $W' = x(1 - \lambda x)$. We will see that the vacuum structure depends crucially on the Yukawa coupling constant c. For $c = 0$, fermions decouple and we recover the double-well potential studied in the previous sections, while for $c = 1$, the classical action is supersymmetric (see, e.g., Weinberg (2000) and supplement VI). The supersymmetry transformation is given by

$$\delta x = \zeta \sigma_2 \psi \qquad \delta \psi = \sigma_2 \zeta \dot{x} - W' \zeta \qquad (3.3.92)$$

where ζ is a Grassmann variable. For this reason, W is usually referred to as the superpotential.

As before, the potential $V = \tfrac{1}{2} W'^2$ has degenerate minima connected by the instanton solution. The tunneling amplitude is given by

$$\int d\tau \, J \, \frac{\sqrt{\det \widehat{O}_F}}{\sqrt{\det' \widehat{O}_B}} e^{-S_{cl}} \qquad (3.3.93)$$

where S_{cl} is the classical action, $\widehat{O}_B$ is the bosonic operator (3.3.75) and $\widehat{O}_F$ is the fermionic operator

$$\widehat{O}_F = \frac{d}{dt} + c\sigma_2 W''(x_{cl}). \qquad (3.3.94)$$

As explained earlier, $\widehat{O}_B$ has a zero mode (related to the translational invariance) which has to be treated separately by introducing the corresponding collective coordinate. The fermion determinant also has a zero mode, given by

$$\chi(t) = N \exp\left\{\mp \int_{-\infty}^{t} dt' \, cW''(x_{cl})\right\} \frac{1}{\sqrt{2}} \begin{pmatrix} 1 \\ \mp i \end{pmatrix}. \qquad (3.3.95)$$

Since the fermion determinant appears in the numerator of the tunneling probability, the presence of a zero mode implies that the tunneling rate vanishes. This can be explained by the fact that the two vacua have different eigenvalues of the fermionic number operator $\hat{\psi}_+\hat{\psi}_-$, where $\hat{\psi}_\pm = (\hat{\psi}_1 \pm i\hat{\psi}_2)/\sqrt{2}$. Thus, the corresponding two ground states $|0, \pm\rangle$ (where $\hat{\psi}_+\hat{\psi}_-|0, \pm\rangle = \pm|0, \pm\rangle$) cannot be connected by a bosonic operator. The tunneling amplitude is non-zero only if a fermion is created during the process, $\langle 0, +|\hat{\psi}_+|0, -\rangle$. Formally, we get a finite result because the fermion creation operator absorbs the zero mode in the fermion determinant. For $c = 1$, the tunneling rate is given by

$$\langle 0, +|\hat{\psi}_+|0, -\rangle = \frac{1}{\sqrt{\pi \lambda^2}} e^{-1/(6\lambda^2)}. \qquad (3.3.96)$$

This result can be checked by performing a direct calculation using the Schrödinger equation.

◇ Tunneling and instantons in Yang–Mills theory: classical vacua in non-Abelian gauge theory and topology

Before discussing tunneling phenomena in the Yang–Mills theory, we have to become more familiar with the classical vacuum (i.e. the extremum of the classical action functional) of the theory. In the Hamiltonian formulation, it is convenient to use the temporal gauge $A_0 = 0$ ($A_i = A_i^a \lambda^a/2$, where the $SU(N)$ generators satisfy $[\lambda^a, \lambda^b] = 2if^{abc}\lambda^c$ and are normalized according to $\text{Tr}(\lambda^a \lambda^b) = 2\delta^{ab}$). In this case, the momentum conjugate to the field variables $A_i(x)$ is just the electric field $E_i = \partial_0 A_i$. The Hamiltonian is given by

$$H = \frac{1}{2g^2} \int d^3x \, (E_i^2 + B_i^2) \qquad (3.3.97)$$

where E_i^2 is the kinetic and B_i^2 the potential energy term. The classical vacuum corresponds to configurations with zero field strength $F_{\mu\nu}$. For non-Abelian gauge fields this does not imply that the potential has to be constant, but limits the gauge fields to be 'pure gauge'

$$\mathsf{A}_i = i\mathsf{U}(x)\partial_i \mathsf{U}(x)^\dagger. \qquad (3.3.98)$$

In order to enumerate the classical vacua we have to classify all possible gauge transformations $U(x)$. This means that we have to study equivalence classes of maps from the space $\mathbb{R}^3$ (spacelike part of the four-dimensional spacetime) into the gauge group $SU(N)$. In practice, we can restrict ourselves to matrices satisfying $U(x) \to 1$ as $x \to \infty$. Such mappings can be classified using an integer called the *winding* (or *Pontryagin*) *number*, which counts how many times the group manifold is covered:

$$n_{\mathrm{W}} = \frac{1}{24\pi^2} \int d^3x \, \epsilon^{ijk} \, \text{Tr}[(\mathsf{U}^\dagger \partial_i \mathsf{U})(\mathsf{U}^\dagger \partial_j \mathsf{U})(\mathsf{U}^\dagger \partial_k \mathsf{U})]. \qquad (3.3.99)$$

In terms of the corresponding gauge fields, this number is the *Chern–Simons characteristic* (problem 3.3.3, page 146)

$$n_{\mathrm{CS}} = \frac{1}{16\pi^2} \int d^3x \, \epsilon^{ijk} (A_i^a \partial_j A_k^a + \tfrac{1}{3} f^{abc} A_i^a A_j^b A_k^c). \qquad (3.3.100)$$

Because of its topological meaning, continuous deformations of the gauge fields do not change n_{CS}. In the case of $SU(2)$, an example of a mapping with winding number n can be found from the 'hedgehog' ansatz:

$$\mathsf{U}(x) = \exp(if(r)\tau^a \hat{x}^a) \qquad (3.3.101)$$

where $r = |x|$, τ^a are the generators of $SU(2)$ in the adjoint representation and $\hat{x}^a = x^a/r$. For this mapping, we find

$$n_{\mathrm{W}} = \frac{2}{\pi} \int dr \, \sin^2(f) \frac{df}{dr} = \frac{1}{\pi} \left[f(r) - \frac{\sin(2f(r))}{2} \right]_0^\infty. \qquad (3.3.102)$$

In order for $U(x)$ to be uniquely defined, $f(r)$ has to be a multiple of π at both zero and infinity, so that n_{W} is indeed an integer. Any smooth function with $f(r \to \infty) = 0$ and $f(0) = n\pi$ provides an example of a function with winding number n.

We conclude that there is an infinite set of classical vacua enumerated by an integer n. Since they are topologically different, we cannot go from one vacuum to another by means of a continuous gauge transformation. Therefore, there is no path from one vacuum to another, such that the energy remains *zero* all the way. In other words, the vacuum (extremal) field configurations are separated by non-extremal configurations.

◇ Belavin–Polyakov–Schwartz–Tyupkin instantons in the Yang–Mills theory

Having found the infinite set of vacua, we may look for a tunneling path in the gauge theory, which connects topologically different classical vacua. From the quantum-mechanical example, we know that we have to look for classical solutions of the Euclidean equations of motion. The best tunneling path is the solution with the minimal Euclidean action connecting vacua with different Chern–Simons numbers. To find these solutions, it is convenient to exploit the following identity:

$$S_{\text{YM}} = \frac{1}{4g^2} \int d^4x \, F^a_{\mu\nu} F^{a\mu\nu} = \frac{1}{4g^2} \int d^4x \left[\pm F^a_{\mu\nu} \tilde{F}^{a\mu\nu} + \frac{1}{2} (F^a_{\mu\nu} \mp \tilde{F}^a_{\mu\nu})^2 \right] \quad (3.3.103)$$

where $\tilde{F}^{\mu\nu} = 1/2 \epsilon^{\mu\nu\rho\sigma} F_{\rho\sigma}$ is the dual field strength tensor (the field tensor in which the roles of electric and magnetic fields are interchanged). The crucial fact is that the first term in the last expression for S_{YM} is the *topological charge* (cf explanation after (3.3.44)) called the four-dimensional *Pontryagin index* Q:

$$Q = \frac{1}{32\pi^2} \int d^4x \, F^a_{\mu\nu} \tilde{F}^{a\mu\nu}. \quad (3.3.104)$$

For finite-action field configurations, Q has to be an integer. This can be seen from the fact that the integrand is a total derivative:

$$Q = \frac{1}{32\pi^2} \int d^4x \, F^a_{\mu\nu} \tilde{F}^{a\mu\nu} = \int d^4x \, \partial_\mu K^\mu = \int d\sigma_\mu \, K^\mu \quad (3.3.105)$$

$$K^\mu = \frac{1}{16\pi^2} \epsilon^{\mu\alpha\beta\gamma} \left(A^a_\alpha \partial_\beta A^a_\gamma + \frac{1}{3} f^{abc} A^a_\alpha A^b_\beta A^c_\gamma \right). \quad (3.3.106)$$

For finite-action configurations, the gauge potential has to be a pure gauge at infinity $A_\mu \to i U \partial_\mu U^\dagger$. Similar to the arguments given after equation (3.3.98), all maps from the three sphere $S^{(3)}$ (corresponding to $|x| \to \infty$) into the gauge group can be classified by a winding number n. Inserting $A_\mu = i U \partial_\mu U^\dagger$ into (3.3.105) we find that $Q = n$.

Since the first term in (3.3.103) is a topological invariant and the last term is always positive, it is clear that the action is minimal if the field is *(anti-) self-dual*:

$$F^a_{\mu\nu} = \pm \tilde{F}^a_{\mu\nu}. \quad (3.3.107)$$

This is a useful observation, because in contrast to the equation of motion, the self-duality equation (3.3.107) is a first-order differential equation. In addition to this, we can show that the energy–momentum tensor vanishes for self-dual fields. In particular, self-dual fields have zero (Minkowski) energy density.

From (3.3.103) we can see that the action of a self-dual field configuration is determined by the topological charge: $S = (8\pi^2 |Q|)/g^2$. Furthermore, if the gauge potential falls off sufficiently rapidly at spatial infinity, the Pontryagin index and the Chern–Simons characteristics are related as follows:

$$Q = \int dt \, \frac{d}{dt} \int d^3x \, K_0 = n_{\text{CS}}(t = \infty) - n_{\text{CS}}(t = -\infty) \quad (3.3.108)$$

which shows that field configurations with $Q \neq 0$ connect different topological vacua. In order to find an explicit solution with $Q = 1$, it is useful to start from the simplest configuration with the winding number $n = 1$. Similarly to (3.3.101), we can take $A_\mu = i U \partial_\mu U^\dagger$, with $U = i\hat{x}_\mu \tau^+_\mu$, where $\tau^\pm_\mu = (\boldsymbol{\tau}, \mp i\mathbb{1})$. Then $A^a_\mu = 2\eta_{a\mu\nu} x_\nu / x^2$, where we have introduced the 't Hooft symbol $\eta_{a\mu\nu}$, defined by

$$\eta_{a\mu\nu} = \begin{cases} \epsilon_{a\mu\nu} & \mu, \nu = 1, 2, 3 \\ \delta_{a\mu} & \nu = 4 \\ -\delta_{a\nu} & \mu = 4. \end{cases} \quad (3.3.109)$$

We also define $\bar{\eta}_{a\mu\nu}$ by changing the sign of the last two equations. We can now look for a solution of the self-duality equation (3.3.107) using the ansatz $A^a_\mu = 2\eta_{a\mu\nu} x_\nu f(x^2)/x^2$, where f has to satisfy the boundary condition $f \to 1$ as $x^2 \to \infty$. Inserting the ansatz in (3.3.107), we get

$$f(1-f) - x^2 f' = 0. \qquad (3.3.110)$$

This equation is solved by $f = x^2/(x^2 + \rho^2)$, which gives the Belavin–Polyakov–Schwartz–Tyupkin (BPST) instanton solution (Belavin *et al* 1975)

$$A^a_\mu(x) = \frac{2\eta_{a\mu\nu} x^\nu}{x^2 + \rho^2}. \qquad (3.3.111)$$

Here ρ is an arbitrary parameter characterizing the size of the instanton. A solution with topological charge $Q = -1$ can be obtained by replacing $\eta_{a\mu\nu}$ with $\bar{\eta}_{a\mu\nu}$. The corresponding field strength is

$$(F^a_{\mu\nu})^2 = \frac{192\rho^4}{(x^2 + \rho^2)^4}. \qquad (3.3.112)$$

The classical instanton solution has a number of degrees of freedom, which should be treated as collective coordinates (see the preceding subsection). In the case of $SU(2)$, the solution is characterized by the instanton size ρ, the instanton position z_μ and three parameters which determine the 'color' orientation (i.e. in the space of the $su(N)$ Lie algebra) of the instanton.

Thus, we have described the tunneling path that connects different topological vacua and from the value of the classical action for it, $S = (8\pi^2 |Q|)/g^2$, it is clear that the tunneling probability is

$$P_{\text{tunneling}} \sim \exp\{-8\pi^2/g^2\}. \qquad (3.3.113)$$

As in the quantum-mechanical example, the coefficient in front of the exponent is determined by a one-loop calculation (i.e. by the quadratic approximation).

◇ **The theta vacua**

We have seen that non-Abelian gauge theory has a periodic potential and that the instantons connect the different vacua. This means that the ground state of a non-Abelian Yang–Mills theory (in particular, of the $SU(3)$-theory, physically corresponding to quantum chromodynamics (QCD)) cannot be described by any of the topological vacuum states, but has to be a superposition of all vacua. This problem is similar to the motion of an electron in the periodic potential of a crystal (see, e.g., Davydov (1976) and Ashcroft and Mermin (1976)). It is well known that the solutions form a band ψ_θ, characterized by a phase $\theta \in [0, 2\pi]$ (sometimes referred to as quasi-momentum). The wavefunctions are Bloch waves, satisfying the periodicity condition $\psi_\theta(x + n) = e^{i\theta n} \psi_\theta(x)$.

Let us see how this band arises from tunneling events. If instantons are sufficiently dilute, then the amplitude to go from one topological vacuum $|i\rangle$ to another one $|j\rangle$, is given by

$$\langle j | \exp(-\widehat{H}\tau)|i\rangle = \sum_{M_+} \sum_{M_-} \frac{\delta_{(M_+ - M_- - j + i), 0}}{M_+! M_-!} (K\tau e^{-S})^{M_+ + M_-} \qquad (3.3.114)$$

where K is the pre-exponential factor in the tunneling amplitude and $M_\pm$ are the numbers of instantons and anti-instantons. Using the identity

$$\delta_{ab} = \frac{1}{2\pi} \int_0^{2\pi} d\theta \, e^{i\theta(a-b)} \qquad (3.3.115)$$

the sum over instantons and anti-instantons can be rewritten as

$$\langle j| \exp(-\widehat{H}\tau)|i\rangle = \frac{1}{2\pi} \int_0^{2\pi} d\theta \, e^{i\theta(i-j)} \exp[2K\tau \cos(\theta) \exp(-S)]. \tag{3.3.116}$$

This result shows that the true eigenstates are the theta vacua, $|\theta\rangle = \sum_n e^{in\theta}|n\rangle$. Their energy is

$$E(\theta) = -2K \cos(\theta) \exp(-S). \tag{3.3.117}$$

The width of the zone is of the order of the tunneling rate. The lowest state corresponds to $\theta = 0$ and has negative energy. This is as it should be: tunneling lowers the ground-state energy.

Note, however, that although we can construct stationary states for any value of θ, they are not excitations of the $\theta = 0$ vacuum, because in QCD the value of θ cannot be changed. As far as the strong interaction is concerned, different values of θ correspond to different worlds. Physical arguments show that the parameter θ must be very small. Current experiments imply that

$$\theta < 10^{-9}. \tag{3.3.118}$$

The question of why θ is so small is known as the *strong CP problem*, because the existence of a non-vanishing θ-parameter leads to the violation of the charge (denoted by C) and parity (P) symmetries of the world. The status of this problem is unclear. As long as we do not understand the source of CP violation in nature, it is not clear whether the strong CP problem should be expected to have a solution within the standard model or whether there is some mechanism outside the standard model that adjusts θ to be small.

◇ The tunneling amplitude: pre-exponential factor

The next natural step in the study of instanton contributions is the one-loop calculation of the pre-exponential in the tunneling amplitude. In gauge theory, this is a rather tedious calculation which was done by 't Hooft in a classical paper ('t Hooft 1976). Basically, the procedure is completely analogous to what we did in the context of quantum mechanics. The field is expanded around the classical solution, $A_\mu = A_\mu^{(cl)} + \delta A_\mu$. In QCD, we have to make a gauge choice. In this case, it is most convenient to work in the background field gauge: $D_\mu(A^{(cl)})\delta A^\mu = 0$, where $D_\mu^{ab}(A^{(cl)}) = (\delta^{ab}\partial_\mu + if_c^{ab} A_\mu^{(cl)c})$.

We have to calculate the one-loop determinants for gauge fields, ghosts and possible matter fields. The determinants are divergent both in the ultraviolet, like any other one-loop graph, and in the infrared region, due to the presence of zero modes.

We already know how to deal with the zero modes of the system: the integral over the zero mode must be converted into an integral over the corresponding collective variable. After an appropriate regularization and renormalization of the ultraviolet divergences, the differential one-instanton tunneling rate dn_I for the gauge group $SU(N)$ proves to be ('t Hooft 1976):

$$dn_I = \frac{0.466 \exp(-1.679N)}{(N-1)!(N-2)!} \left(\frac{8\pi^2}{g^2}\right)^{2N} \exp\left\{-\frac{8\pi^2}{g^2(\rho)}\right\} \frac{d^4z \, d\rho}{\rho^5}. \tag{3.3.119}$$

This tunneling rate corresponds to the contribution of the instantons with the parameters of their size and position being in the $d^4z \, d\rho$ vicinity of some chosen values of ρ, z^μ. Note that the collective coordinates corresponding to the 'colour' orientation form a compact manifold (as well as the whole group $SU(N)$) and can be simply integrated over (result (3.3.119) was obtained after such an integration).

3.3.4 Path-integral calculation of quantum anomalies

An anomaly expresses the breakdown of a classical symmetry by quantum effects. This view of an anomaly arises quite naturally in the path-integral formalism. As we have discussed in the preceding sections, the generating functional for Green functions of some field theoretical model is expressed in terms of the path integral, the integrand being the exponential of the classical action of the model or the corresponding effective action, i.e. modifications due to gauge fixing and ghost-field terms in gauge theories, for example. If the exponent (classical or effective action) is invariant under some symmetry transformations, the symmetry under consideration can still be violated by *the path-integral measure*, the quantity which reflects the quantum nature of the theory: in general, this does not remain invariant under the symmetry transformations. The associated Jacobian (after an appropriate regularization) of the transformations produces precisely the anomaly (Fujikawa 1979, 1980) (see also, e.g., Bertlmann (1996) and references therein). Although we introduced anomalies (see section 3.2.7) by considering the regularization and renormalization of Feynman diagrams, this path-integral treatment of the anomaly is independent of perturbation theory and, for this reason, is called the non-perturbative approach.

In this subsection, we shall perform a chiral transformation of the path integral and find the anomalous Ward–Takahashi identities. Then, we shall regularize the transformation Jacobian (following Fujikawa's work) and, in this way, derive the so-called *singlet anomaly* (related to one-parameter (Abelian) gauge transformations). We shall discuss the independence of the anomaly from the choice of the regularization and the alternative between gauge and chiral symmetry. We shall also present a generalization of the path-integral method to non-Abelian gauge transformations leading to the non-Abelian anomaly.

◇ Fermionic path-integral measure and chiral transformations

Historically, the first example of quantum anomalies was the well-known Adler–Bell–Jackiw chiral anomaly (Adler 1969, Bell and Jackiw 1969). This is connected with the so-called chiral $U(1)$ transformation

$$\delta \psi = i\alpha \gamma_5 \psi \qquad \delta \bar{\psi} = i\alpha \bar{\psi} \gamma_5 \qquad (3.3.120)$$

where α is an infinitesimal real parameter. The corresponding Noether current

$$j_\mu^5(x) = \bar{\psi} \gamma_\mu \gamma_5 \psi \qquad (3.3.121)$$

obeys classically the following equation

$$\partial^\mu j_\mu^5(x) = 2im\bar{\psi}\gamma_5\psi \qquad (3.3.122)$$

and it is conserved in the *chiral limit* $m \to 0$. In the quantum case, however, the anomalous term $\frac{1}{8\pi^2}\widetilde{F}^{\mu\nu}F_{\mu\nu}$ (where $\widetilde{F}_{\mu\nu}$ is the dual tensor to $F_{\mu\nu}$, see the definition after equation (3.3.103)) appears on the right-hand side of the previous equation, spoiling the chiral invariance of the theory. Although this term could be removed by a suitable counterterm added to the chiral current $j_\mu^5(x)$, this counterterm spoils the gauge invariance and is therefore not admissible in any reasonable gauge theory. This situation is typical: the anomalous symmetries appear in pairs and saving one of them (in the case under consideration, the gauge symmetry) necessarily spoils the other one (chiral symmetry (3.3.120)). A further example of an anomalous pair is the conflict between the scale and translational symmetries, leading to the so-called trace anomaly. In the gravitational background, there is also the so-called Lorentz anomaly, the consequence of which is the anomalous antisymmetric part of the energy–momentum tensor. For a comprehensive discussion of this and related topics see the books by Treiman *et al* (1985) and Bertlmann (1996), where an extensive list of references can also be found.

◇ **Singlet anomaly**

We start with the *singlet anomaly* related to the one-parameter transformations (3.3.120). Let us consider quantized Dirac fermions interacting with the non-Abelian field A_μ^a which may be Abelian (in this case, a is a redundant index) or non-Abelian. The Lagrangian is

$$\mathcal{L} = \bar{\psi}(i\slashed{D} - m)\psi \qquad (3.3.123)$$

with the Dirac operator

$$\slashed{D} = \gamma^\mu D_\mu = \gamma^\mu(\partial_\mu + \mathsf{A}_\mu) \qquad (3.3.124)$$

and the gauge potential $\mathsf{A}_\mu = A_\mu^a \mathsf{T}^a$ (T^a are the group generators). For the actual calculations, we perform a Wick rotation to Euclidean spacetime: $x^0 \to x^4 = ix^0$, together with the modification of the γ^0-matrix (see (3.3.17)) and zero-component gauge field:

$$\gamma^0 \to \gamma^4 = i\gamma^0 \qquad \mathsf{A}_0 \to \mathsf{A}_4 = -i\mathsf{A}_0. \qquad (3.3.125)$$

Note that now all γ-matrices become anti-Hermitian:

$$(\gamma^\mu)^\dagger = -\gamma^\mu \qquad \mu = 1, 2, 3, 4 \qquad (3.3.126)$$

while the γ_5-matrix, on the other hand, remains Hermitian:

$$\gamma_5 \stackrel{\text{def}}{\equiv} i\gamma^0\gamma^1\gamma^2\gamma^3 = \gamma^4\gamma^1\gamma^2\gamma^3 \qquad (\gamma_5)^\dagger = \gamma_5. \qquad (3.3.127)$$

Then, the Dirac operator turns out to be Hermitian in the Euclidean spacetime

$$\slashed{D}^\dagger = \slashed{D}. \qquad (3.3.128)$$

The metric becomes the following:

$$g_{\mu\nu} = -\delta_{\mu\nu}. \qquad (3.3.129)$$

Now, we perform a *local chiral transformation*:

$$\begin{aligned}\psi(x) &\longrightarrow \psi'(x) = e^{i\beta(x)\gamma_5}\psi(x) \\ \bar{\psi}(x) &\longrightarrow \bar{\psi}'(x) = e^{i\beta(x)\gamma_5}\bar{\psi}(x)\end{aligned} \qquad (3.3.130)$$

where $\beta(x)$ denotes some gauge function. For an infinitesimal β, the Lagrangian (3.3.123) changes to

$$\mathcal{L} \longrightarrow \mathcal{L}' = \mathcal{L} - (\partial_\mu \beta)j_\mu^5 - 2im\beta\rho_p \qquad (3.3.131)$$

where we have used the axial current

$$j_\mu^5 = \bar{\psi}\gamma_\mu\gamma_5\psi \qquad (3.3.132)$$

and the pseudoscalar density

$$\rho_p = \bar{\psi}\gamma_5\psi. \qquad (3.3.133)$$

Thus, the classical action $S = \int d^4x\, \mathcal{L}$ transforms as follows:

$$S \to S' = S + \int d^4x\, \beta(x)[\partial^\mu j_\mu^5(x) - 2im\rho_p(x)] \qquad (3.3.134)$$

and remains invariant if the axial current and density satisfy the relation

$$\partial^\mu j_\mu^5 = 2im\rho_p \qquad (3.3.135)$$

in particular, if in the limit $m = 0$, the classical conservation law for the axial current is fulfilled.

To approach the phenomena of quantum anomalies, let us consider the path-integral representation of the Dirac determinant:

$$\det(i\slashed{D} - m) = \int \mathcal{D}\psi(x)\,\mathcal{D}\bar{\psi}(x)\,\exp\left\{\int d^4x\,\bar{\psi}(i\slashed{D} - m)\psi\right\}. \qquad (3.3.136)$$

Performing here the chiral transformation (3.3.130), Fujikawa discovered that the path-integral measure transforms with a Jacobian containing the anomaly (Fujikawa 1979, 1980). More precisely, he established the following result:

- The path-integral measure transforms chirally as

$$\mathcal{D}\bar{\psi}'\,\mathcal{D}\psi' = \mathcal{D}\bar{\psi}\,\mathcal{D}\psi\,J[\beta, A_\mu] \qquad (3.3.137)$$

where the transformation Jacobian $J[\beta, A_\mu]$ reads as

$$J[\beta, A_\mu] = \exp\left\{-\int d^4x\,\beta(x)\mathfrak{A}(A_\mu(x))\right\} \qquad (3.3.138)$$

and contains precisely the *singlet anomaly* in the Euclidean space

$$\mathfrak{A}(A_\mu(x)) \stackrel{\text{def}}{\equiv} -\frac{i}{16\pi^2}\varepsilon^{\mu\nu\rho\sigma}\,\text{Tr}\,F_{\mu\nu}F_{\rho\sigma}. \qquad (3.3.139)$$

(Note that the imaginary unity i disappears in the Minkowski space.)

◇ The Jacobian of the chiral transformations

In order to derive expressions (3.3.138) and (3.3.139) for the singlet anomaly (i.e. to determine the Jacobian of the chiral transformations), it is convenient to write the path integral (3.3.136) in terms of the mode expansion (cf sections 1.2.3, 2.2.2 and 3.2.1):

$$\det(i\slashed{D} - m) = \int \prod_{n=0}^{\infty} da_n\,d\bar{b}_n\,\exp\left\{\sum_{n=0}^{\infty}(i\lambda_n - m)\bar{b}_n a_n\right\}$$

$$= \prod_{n=0}^{\infty}(i\lambda_n - m) \qquad (3.3.140)$$

where $\bar{b}_n$ and a_n are coefficients of the mode expansion

$$\psi(x) = \sum_{n=0}^{\infty} a_n \varphi_n(x)$$
$$\bar{\psi}(x) = \sum_{n=0}^{\infty} \varphi_n^\dagger(x)\bar{b}_n \qquad (3.3.141)$$

over the orthogonal and complete set of eigenfunctions $\varphi_n(x)$ of the Dirac operator

$$\slashed{D}\varphi_n(x) = \lambda_n \varphi_n(x). \tag{3.3.142}$$

Note that the coefficients $\bar{b}_n$ and a_n are *Grassmann elements*. The substitution of the mode expansion (3.3.141) into the transformation rule (3.3.130) shows that an *infinitesimal* chiral transformation of the expansion coefficients reads as

$$\begin{aligned} a'_n &= \sum_m O_{nm} a_m \\ \bar{b}'_n &= \sum_m O_{mn} \bar{b}_m \end{aligned} \tag{3.3.143}$$

with the transformation matrix

$$O_{nm} = \delta_{nm} + \mathrm{i}\int d^4x\, \beta(x) \varphi_n^\dagger(x) \gamma_5 \varphi_m(x). \tag{3.3.144}$$

Recall that the Grassmann measure transforms with the inverse determinant (see section 2.6, equation (2.6.203)), so that

$$\begin{aligned} \prod_n da'_n &= (\det \mathsf{O})^{-1} \prod_n da_n \\ \prod_n d\bar{b}'_n &= (\det \mathsf{O})^{-1} \prod_n d\bar{b}_n. \end{aligned} \tag{3.3.145}$$

As a result, the Jacobian of the chiral transformation is cast into the form

$$J[\beta] = (\det \mathsf{O})^{-2} = \exp\{-2\,\mathrm{Tr}\ln \mathsf{O}\} \tag{3.3.146}$$

and for an infinitesimal group parameter β, it can be rewritten as

$$\begin{aligned} J[\beta] &\approx \exp\left\{-2\,\mathrm{Tr}\ln\left(\delta_{nm} + \mathrm{i}\int d^4x\, \beta(x)\varphi_n^\dagger(x)\gamma_5\varphi_m(x)\right)\right\} \\ &\approx \exp\left\{-2\mathrm{i}\int d^4x\, \beta(x) \sum_n \varphi_n^\dagger(x)\gamma_5\varphi_n(x)\right\}. \end{aligned} \tag{3.3.147}$$

Applying the completeness relation $\sum_n \varphi_n(x)\varphi_n^\dagger(y) = \delta(x-y)$ for the Dirac operator eigenfunctions, we see that the sum in the exponential of (3.3.147) is ill defined:

$$\sum_n \varphi_n^\dagger(x)\gamma_5\varphi_n(x) = \mathrm{Tr}\,\gamma_5 \cdot \delta(0) \tag{3.3.148}$$

and requires a *regularization*.

◇ Fujikawa's regularization of the Dirac determinant: derivation of the singlet (Abelian) anomaly

Fujikawa suggested regularizing the determinant by a Gaussian cutoff:

$$\begin{aligned} \sum_n \varphi_n^\dagger(x)\gamma_5\varphi_n(x) &= \lim_{M\to\infty} \sum_n \varphi_n^\dagger(x)\gamma_5 \exp\left\{-\frac{\slashed{D}^2}{M^2}\right\} \varphi_n(x) \\ &= \lim_{M\to\infty} \sum_n \varphi_n^\dagger(x)\gamma_5 \exp\left\{-\frac{\lambda_n^2}{M^2}\right\} \varphi_n(x) \end{aligned} \tag{3.3.149}$$

where M is the (dimensional) parameter of the regularization. The Gaussian factors damp the contributions from the large eigenvalues, providing the convergence of the sum. It is convenient to evaluate of the regularized sum using the Fourier transform $\widetilde{\varphi}_n$ of the eigenfunctions

$$\varphi_n(x) = \frac{1}{(2\pi)^2} \int d^4k \, e^{ikx} \widetilde{\varphi}_n(k). \tag{3.3.150}$$

Then, using again the completeness of the eigenfunctions, the sum in (3.3.149) can be rewritten as follows:

$$\sum_n \varphi_n^\dagger(x) \gamma_5 \varphi_n(x) = \lim_{M \to \infty} \frac{1}{(2\pi)^4} \int d^4k \, d^4l \sum_n e^{-ilx} \widetilde{\varphi}_n^\dagger(l) \gamma_5 \exp\left\{-\frac{\slashed{D}^2}{M^2}\right\} e^{ikx} \widetilde{\varphi}_n(k)$$

$$= \lim_{M \to \infty} \frac{1}{(2\pi)^4} \int d^4k \, \mathrm{Tr} \, e^{-ikx} \gamma_5 \exp\left\{-\frac{\slashed{D}^2}{M^2}\right\} e^{ikx} \tag{3.3.151}$$

where the trace in the last expression is understood to be both over the γ-matrices and over the group generators. The calculation of this integral is carried out with the help of the following decomposition of the Dirac operator:

$$\slashed{D}^2 = \gamma^\mu \gamma^\nu D_\mu D_\nu = (\tfrac{1}{2}\{\gamma^\mu, \gamma^\nu\} + \tfrac{1}{2}[\gamma^\mu, \gamma^\nu]) D_\mu D_\nu$$

$$= D_\mu D^\mu + \tfrac{1}{4}[\gamma^\mu, \gamma^\nu] F_{\mu\nu} \tag{3.3.152}$$

where we have used the relation $[D_\mu, D_\nu] = F_{\mu\nu}$. The use of this decomposition in (3.3.151) yields the result (see problem 3.3.4)

$$\sum_n \varphi_n^\dagger(x) \gamma_5 \varphi_n(x) = -\frac{1}{32\pi^2} \varepsilon^{\mu\nu\rho\sigma} \, \mathrm{Tr} \, F_{\mu\nu} F_{\rho\sigma}. \tag{3.3.153}$$

This, together with (3.3.147), proves formulae (3.3.138) and (3.3.139) for the singlet anomaly.

◇ **Anomalous Ward–Takahashi identity**

The immediate consequence of the quantum anomaly in chiral theories is the appearance of anomalous relations between Green functions, i.e. the so-called *anomalous Ward–Takahashi identities*. The derivation follows the usual steps for deriving the Ward–Takahashi identities in any theory (cf sections 3.1.5 and 3.2.5). We start from the generating functional $\mathcal{Z}[\eta, \bar{\eta}]$ for Green functions. For brevity and simplicity we shall use the generating functional for spinor fields in an *external* Yang–Mills field, dropping the pure Yang–Mills Lagrangian and the corresponding functional integration over the gauge fields. The restoration of the dynamical quantum nature of the Yang–Mills fields adds nothing essential to the discussion of the anomalous terms in the Ward–Takahashi identity.

Thus, we consider the generating functional

$$\mathcal{Z}[\eta, \bar{\eta}, A_\mu] = \mathfrak{N}^{-1} \int \mathcal{D}\bar{\psi} \, \mathcal{D}\psi \, \exp\left\{\int d^4x \, [\mathcal{L} + \bar{\eta}\psi + \bar{\psi}\eta]\right\} \tag{3.3.154}$$

and perform the change of integration variable defined by the chiral transformation (3.3.130) with the parameter $\beta(x)$. Since the value of the integral does not depend on the choice of variables and hence on the value of the parameter $\beta(x)$, the differentiation over the latter gives the identity

$$\frac{\delta}{\delta \beta(x)} \mathcal{Z}[\eta, \bar{\eta}, A_\mu] = \mathfrak{N}^{-1} \int \mathcal{D}\bar{\psi} \, \mathcal{D}\psi \, \exp\left\{\int d^4x \, [\mathcal{L} + \bar{\eta}\psi + \bar{\psi}\eta]\right\}$$

$$\times [\partial^\mu j_\mu^5 - 2im\rho_p - \mathfrak{A}(A_\mu(x)) + i\bar{\eta}\gamma_5\psi + i\bar{\psi}\gamma_5\eta] = 0. \tag{3.3.155}$$

Condition (3.3.155) determines all (anomalous) Ward–Takahashi identities of the theory, by differentiation with respect to the sources. For example, differentiating (3.3.155) with respect to $\bar{\eta}$, η and setting the sources equal to zero, we obtain the following relation between the Green functions:

$$\partial_x^\mu \langle 0|\mathbf{T}j_\mu^5(x)\psi(x_1)\bar{\psi}(x_2)|0\rangle = 2im\langle 0|\mathbf{T}\rho_p(x)\psi(x_1)\bar{\psi}(x_2)|0\rangle + \langle 0|\mathbf{T}\mathfrak{a}(A_\mu(x))\psi(x_1)\bar{\psi}(x_2)|0\rangle$$
$$- \langle 0|\mathbf{T}\gamma_5\psi(x_1)\bar{\psi}(x_2)|0\rangle\delta(x-x_1) - \langle 0|\mathbf{T}\psi(x_1)\bar{\psi}(x_2)\gamma_5|0\rangle\delta(x-x_2). \quad (3.3.156)$$

We stress that path integral (3.3.154) is independent of parameter $\beta(x)$ of the chiral transformations only taking into account the non-trivial transformation of the functional integration measure (non-trivial Jacobian). This results in the appearance of the anomalous term $\mathfrak{A}(A_\mu(x))$ in Ward–Takahashi identity (3.3.155).

Fujikawa (1980) emphasized that the anomaly is independent of the chosen regularization for the large eigenvalue contributions in the sum (3.3.149). Instead of exponential damping we could also choose some other function $f(x)$ which is smooth and decreasing sufficiently rapidly at infinity:

$$\exp\left\{-\frac{\lambda_n^2}{M^2}\right\} \longrightarrow f\left(\frac{\lambda_n^2}{M^2}\right) \quad (3.3.157)$$

with

$$f(\infty) = f'(\infty) = f''(\infty) = \cdots = 0$$
$$f(0) = 1. \quad (3.3.158)$$

We suggest the reader checks this claim in problem 3.3.6, page 147. For instance, the choice of

$$f\left(\frac{\lambda_n^2}{M^2}\right) = \frac{1}{1+\lambda_n/M^2} \quad (3.3.159)$$

as the cutoff function makes the Fujikawa regularization similar to the well-known Pauli–Villars regularization (see, e.g., Bogoliubov and Shirkov (1959)).

◇ **Competition between gauge and chiral symmetries and the anomaly**

Roughly speaking, the calculation of the Jacobian is reduced to the following summation:

$$\sum_n \varphi_n^\dagger(x)\gamma_5\varphi_n(x) = \sum_{n=0}^\infty \sum_{\alpha,\beta=1}^4 (\gamma_5)_{\alpha\beta}\varphi_{n\beta}\varphi_{n\alpha} \simeq \mathrm{Tr}\,\gamma_5 \cdot \delta(x-x')$$
$$\simeq +1+1-1-1+1+1-1-1+1+1-1-1+\cdots. \quad (3.3.160)$$

The sign $\simeq$ is used here to stress the mathematical ambiguity of this chain of expressions: its clarification requires a regularization which we have just carried out. Here, we want to point out that series (3.3.160) is *conditionally* convergent so that it has a definite value depending on the way of summation. Fujikawa's Gaussian cutoff (3.3.149) or any regularizer with the properties (3.3.157) and (3.3.158) corresponds to a summation which preserves the *gauge invariance*. This leads to a chiral symmetry-breaking and the chiral trace becomes anomalous: $\mathrm{Tr}\,\gamma_5 \neq 0$, in contrast to the naive summation of (3.3.160) in the following way: $(+1+1-1-1)+(+1+1-1-1)+\cdots = 0$. Another option is to choose a regularization that preserves the exact *chiral* symmetry. This would lead to an anomaly in the *gauge symmetry*. This conflict of the two symmetries is the consequence of the following general fact:

- The gauge-covariant operator $\not{D}$ and the chiral matrix γ_5 do not commute, their commutator expectation value giving rise precisely to the anomaly

$$\langle 0|\bar{\psi}(x)[\not{D}, \gamma_5]\psi(x)|0\rangle = \mathfrak{A}[A_\mu](x). \tag{3.3.161}$$

Hence, $\not{D}$ and γ_5 cannot be diagonalized simultaneously.

This, in turn, implies:

- It is impossible to impose both symmetries (gauge symmetry and the chiral symmetry) simultaneously.

(Calculating the vacuum expectation (3.3.161) pertains to problem 3.3.7, page 147.) Thus the operators $\not{D}$ and γ_5 satisfy a type of 'uncertainty principle': they cannot be simultaneously diagonal and the corresponding symmetries, gauge and chiral, cannot be simultaneously exact.

◇ **Non-Abelian anomaly**

So far, we have treated the singlet (or Abelian) anomaly case. But the non-Abelian anomaly is also determined by the path-integral measure when performing a non-Abelian gauge transformation.

We consider the following non-Abelian Lagrangian (restricting ourselves to the massless case for simplicity):

$$\mathcal{L} = \bar{\psi}i\not{D}\psi \tag{3.3.162}$$

where the Dirac operator

$$\not{D} = \not{\partial} + \not{B} + \not{A}\gamma_5 \tag{3.3.163}$$

now contains a vector $B_\mu = B_\mu^a T^a$ and an axial $A_\mu = A_\mu^a T^a$ gauge potentials. This Dirac operator, however, is not Hermitian in the Euclidean space

$$\not{D}^\dagger(B, A) = \not{D}(B, -A). \tag{3.3.164}$$

As a result, the Dirac operator has no well-defined eigenvalue problem and we cannot use it for the regularization procedure. One way to overcome the problem is to work with the Laplacian operators $\not{D}^\dagger \not{D}$ or $\not{D}\not{D}^\dagger$, which have different sets of eigenfunctions:

$$\not{D}^\dagger \not{D}\varphi_n = \lambda^2 \varphi_n \qquad \not{D}\not{D}^\dagger \Phi_n = \lambda^2 \Phi_n. \tag{3.3.165}$$

Note that the Dirac operator and its conjugate transform one set of eigenfunctions into another:

$$\not{D}\varphi_n = \lambda_n \Phi_n \qquad \not{D}^\dagger \Phi_n = \lambda_n \varphi_n. \tag{3.3.166}$$

They are Hermitian and have well-defined eigenstates. The regularization is performed in a gauge-covariant way and thus the regularized Jacobian produces the covariant anomaly (Fujikawa 1984, 1985) (see also Bertlmann (1996) and references therein).

After the expansion of the spinor fields over the eigenfunctions as follows

$$\psi(x) = \sum_n a_n \varphi(x) \qquad \bar{\psi}(x) = \sum_n \varphi^\dagger(x)\bar{b}_m \tag{3.3.167}$$

the calculation of the Jacobian of the chiral transformations

$$\psi'(x) = \exp\{-\beta(x)\gamma_5\}\psi(x)$$
$$\bar{\psi}'(x) = \bar{\psi}(x)\exp\{-\beta(x)\gamma_5\} \qquad \beta(x) = \beta^a(x)T^a \tag{3.3.168}$$

goes in a similar way to the case of the Abelian (singlet) anomaly and yields the result

$$J[\beta] = \exp\left\{\int d^4x \sum_n (\varphi_n^\dagger(x)\beta\gamma_5\varphi_n(x) + \Phi_n^\dagger(x)\beta\gamma_5\Phi_n(x))\right\}. \quad (3.3.169)$$

The next step is the same as in the Abelian case: we have to regularize the sum in the exponential

$$\sum_n (\varphi_n^\dagger(x)\beta\gamma_5\varphi_n(x) + \Phi_n^\dagger(x)\beta\gamma_5\Phi_n(x))$$

$$= \lim_{M\to\infty} \exp\left\{-\frac{\lambda_n^2}{M^2}\right\} \sum_n (\varphi_n^\dagger(x)\beta\gamma_5\varphi_n(x) + \Phi_n^\dagger(x)\beta\gamma_5\Phi_n(x))$$

$$= \lim_{M\to\infty} \int \frac{d^4k}{(2\pi)^4} \operatorname{Tr}\left[\beta e^{-ikx}\gamma_5\left(\exp\left\{-\frac{\slashed{D}^\dagger\slashed{D}}{M^2}\right\} + \exp\left\{-\frac{\slashed{D}\slashed{D}^\dagger}{M^2}\right\}\right)e^{ikx}\right]. \quad (3.3.170)$$

The calculation of the right-hand side gives the so-called *covariant anomaly*:

$$\sum_n (\varphi_n^\dagger(x)\beta\gamma_5\varphi_n(x) + \Phi_n^\dagger(x)\beta\gamma_5\Phi_n(x)) = \frac{1}{32\pi^2}\varepsilon^{\mu\nu\alpha\beta} \operatorname{Tr}\beta(\mathsf{F}_{\mu\nu}^+\mathsf{F}_{\alpha\beta}^+ + \mathsf{F}_{\mu\nu}^-\mathsf{F}_{\alpha\beta}^-) \quad (3.3.171)$$

where $\mathsf{F}_{\mu\nu}^\pm$ denotes the field strength corresponding to the chiral gauge field $\mathsf{A}_\mu^\pm = \mathsf{B}_\mu \pm \mathsf{A}_\mu$.

When working with the fermionic path integral the problem of regularization always occurs. Fujikawa's Gaussian cutoff procedure is one possibility, but other techniques also exist, e.g. the *heat kernel* and *zeta function regularization* which are elegant and based on mathematically solid grounds. For details of these methods, we refer the reader to Bertlmann (1996) and references therein.

3.3.5 Path-integral solution of the polaron problem

An electron moving in a polar crystal polarizes the crystal lattice in its vicinity. Obviously, the perturbation of the crystal is not static but follows the electron. More precisely, this interaction of an electron with its surrounding ionic lattice induces *vibration* of the crystal lattice. In fact, we have already considered the simplest variant of lattice vibrations at the very beginning of chapter 3 (see section 3.1.1), as a prototype for a scalar field theory. However, in solid state physics, the quantum theory of lattice vibrations (in more complicated variants) is of great interest, both from the theoretical and practical points of view, in its own right. As we learned in section 3.1.1, the transition to normal modes allows us to describe the excitations of a lattice within the second-quantization formalism, in terms of quasi-particles. In the case of the crystal lattice, the corresponding quanta of excitations (quasi-particles) are called *phonons*. The interaction of an electron with these lattice excitations leads to the 'dressing' of the 'bare' (free, non-interacting) electron by 'clouds' of phonons. This both lowers the energy of the electron and increases its *effective mass* in comparison with the case when the electron interacts with a *rigid*, non-vibrating lattice, i.e. when it moves in a fixed external periodic potential (for a definition of the effective mass in the latter case and a general introduction to solid state physics see, e.g., Ashcroft and Mermin (1976)).

- An electron moving in a crystal together with the accompanying lattice distortion or, in other words, the physical state of an electron surrounded ('dressed') by a cloud of phonons is called *a polaron* (see, e.g., Kittel (1987) and references therein).

The polaron problem can also be considered as an interesting field-theory model of non-relativistic particles interacting with a scalar boson field, and it was widely studied in two contexts: the practical study of crystal properties and the abstract non-relativistic quantum field theory.

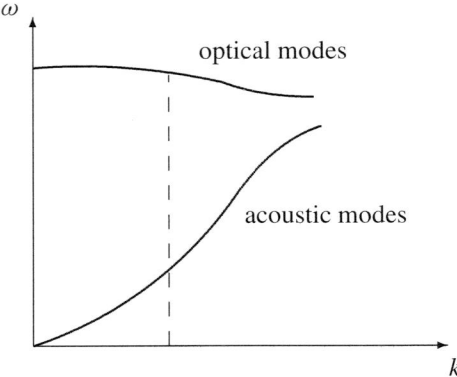

Figure 3.18. Qualitative behaviour of the dispersion curves for optical and acoustical phonons in crystals. The broken line shows the domain of interest: it is seen that in this domain, the frequencies ω of the optical phonons are almost independent of the wavevector k.

◇ Vibrations of a crystal lattice and the polaron Hamiltonian

It is worth mentioning that a real crystal lattice has an essentially more complicated structure than that depicted in figure 3.1 or its *straightforward* three-dimensional generalization. We shall not go into a detailed description, referring the reader to the previously cited books, but just note that a crystal lattice is formed by periodically arranged *cells*, each consisting of *a few* ions. Correspondingly, the structure of the possible vibrations of such lattices is richer than that of a simple cubic lattice with one ion at each site (as depicted in figure 3.1 for the one-dimensional case). In particular, there are the so-called *acoustic phonons* when the ions of a cell vibrate in the same phase (these phonons correspond to propagation of a sound in the crystal) and there also exist *optical phonons*. In the latter case, the ions of a crystal cell vibrate with opposite phases, so that the centrum of mass of the cell remains at rest. The formation of a polaron is caused mainly by the optical phonons. The distinctive property of the latter is that their frequency is almost independent of the wavevector, while acoustic phonons have an almost linear dispersion law, see figure 3.18.

Let us agree that the potential of the non-perturbed fixed lattice is taken into account by the change in the electron mass (i.e. the substitution of the mass of a free electron by the effective mass in the periodic potential of the fixed lattice). If the lattice is distorted because of the presence of an electron in the crystal, the potential $V(x)$ which acts on the electron due to this deformation is defined by the Laplace equation

$$\nabla^2 V(x) = e\rho(x) = -e\nabla \cdot P(x) \qquad (3.3.172)$$

where $P(x)$ is the polarization vector, e is the electron charge and $\rho(x)$ is the charge density caused by the polarization. The polarization vector (which is proportional to shifts of ions) can be written via the normal modes (via Fourier transform, cf section 3.1.1) as follows:

$$P(x) = C \int \frac{d^3k}{(2\pi)^3} \sum_{i=1}^{3} [a_i(k)e^{ikx}e_i + a_i^*(k)e^{-ikx}e_i] \qquad (3.3.173)$$

(e_i are three orthogonal vectors). After quantization, the modes $a_i(k)$, $a_i^\dagger(k)$ become the creation and annihilation operators of (optical) phonons. Equation (3.3.172) shows that only the longitudinal mode (along the wavevector $\boldsymbol{k}$) contributes to the polarization charge density. Therefore, we can neglect the two transversal modes.

Calculations of the charge density according to (3.3.172) yields the change of the potential energy of an electron caused by lattice vibrations (in other words, due to the interaction with phonons):

$$V(\boldsymbol{x}) = -\mathrm{i}(\sqrt{2}\pi\alpha)^{1/2} \left(\frac{\hbar^5\omega^3}{m}\right)^{1/4} \int \frac{d^3k}{(2\pi)^3} \frac{1}{|\boldsymbol{k}|} (a_{\boldsymbol{k}}^* \mathrm{e}^{-\mathrm{i}\boldsymbol{k}\boldsymbol{x}} - a_{\boldsymbol{k}} \mathrm{e}^{\mathrm{i}\boldsymbol{k}\boldsymbol{x}}). \qquad (3.3.174)$$

It can be shown (see, e.g., Kittel (1987)) that the dimensionless electron–phonon coupling constant is expressed via crystal and electron characteristics as follows:

$$\alpha = \frac{1}{2}\left(\frac{1}{\varepsilon_\infty} - \frac{1}{\varepsilon_0}\right) \frac{e^2}{\hbar\omega} \left(\frac{2m\omega}{\hbar}\right)^{1/2} \qquad (3.3.175)$$

where ε_0 and ε_∞ are the static and high-frequency dielectric constants, respectively; ω is the (constant) frequency of the optical phonons, m and e are the electron mass and charge. The potential (3.3.174) implies the following Hamiltonian for the electron–phonon system, suggested by Fröhlich (1937, 1954)

$$H = \frac{\boldsymbol{p}^2}{2} + \sum_{\boldsymbol{k}} a_{\boldsymbol{k}}^* a_{\boldsymbol{k}} + \frac{\mathrm{i}(\sqrt{2}\pi\alpha)^{1/2}}{L^3} \sum_{\boldsymbol{k}} \frac{1}{|\boldsymbol{k}|}[a_{\boldsymbol{k}}^* \mathrm{e}^{-\mathrm{i}\boldsymbol{k}\cdot\boldsymbol{x}} - a_{\boldsymbol{k}} \mathrm{e}^{\mathrm{i}\boldsymbol{k}\cdot\boldsymbol{x}}] \qquad (3.3.176)$$

where $\boldsymbol{p} = -\mathrm{i}\nabla$ is the electron momentum operator, $\boldsymbol{x}$ is its coordinate, and L^3 is the volume of the crystal, which tends to infinity. The $\boldsymbol{k}$s are the usual normal modes (e.g., $\boldsymbol{k} = 2\pi L^{-1}(n_1, n_2, n_3)$ for a cubic box) and, as usual, $L^{-3}\sum_{\boldsymbol{k}} \to (2\pi)^{-3}\int d^3k$. Also the $\boldsymbol{k} = 0$ mode is omitted (it describes the rigid lattice and we have agreed to include its effect in the effective mass). In expression (3.3.176), for simplicity, we have used such units that $\hbar = m = \omega = 1$.

As usual, we cannot find the *exact* eigenvectors and eigenvalues of the Fröhlich Hamiltonian (3.3.176) and have to develop some approximate method. The choice of the latter depends on the value of the coupling constant:

- if the coupling constant is small, $\alpha \ll 1$, we can use the perturbation theory;
- for real crystals, the coupling constant α takes values in the range 1–20 (e.g., for crystals of common salt, $\alpha \approx 5$); in this case, we have to use some non-perturbative variational methods.

The first case has no direct practical applications. From the technical point of view, we may use the ordinary *stationary* perturbation theory of non-relativistic quantum mechanics, so that the shift of the electron energy inside a crystal is given by the standard expression (see, e.g., Landau and Lifshitz (1981) and Davydov (1976)):

$$\Delta E_0 = \langle 0|\widehat{H}_{\mathrm{int}}|0\rangle + \sum_n \frac{\langle 0|\widehat{H}_{\mathrm{int}}|n\rangle \langle n|\widehat{H}_{\mathrm{int}}|0\rangle}{E_0^0 - E_n^0} + \cdots \qquad (3.3.177)$$

where we have separated the Fröhlich Hamiltonian (3.3.176) into two parts: the free Hamiltonian

$$H_0 = \frac{\boldsymbol{p}^2}{2} + \sum_{\boldsymbol{k}} a_{\boldsymbol{k}}^* a_{\boldsymbol{k}} \qquad (3.3.178)$$

and the interaction Hamiltonian

$$H_{\mathrm{int}} = \frac{\mathrm{i}(\sqrt{2}\pi\alpha)^{1/2}}{L^3} \sum_{\boldsymbol{k}} \frac{1}{|\boldsymbol{k}|}[a_{\boldsymbol{k}}^* \mathrm{e}^{-\mathrm{i}\boldsymbol{k}\cdot\boldsymbol{x}} - a_{\boldsymbol{k}} \mathrm{e}^{\mathrm{i}\boldsymbol{k}\cdot\boldsymbol{x}}]. \qquad (3.3.179)$$

The states $|n\rangle$, $n = 0, 1, 2, \ldots$, in (3.3.177), are the eigenstates of the free Hamiltonian $\widehat{H}_0$, while E_n^0 are the eigenvalues of H_0 corresponding to $|n\rangle$. In fact, the corresponding calculations are quite similar to those in the relativistic field theory. In particular, the contribution of the lowest order of the perturbation theory can be represented by a Feynman-like diagram similar to that on page 81

where the wavy line now represents the phonon propagator $\sim 1/k^2$ and the full line with an arrow corresponds to the non-relativistic electron propagator. The calculation in the first order in the coupling constant yields

$$\Delta E_0 = -\alpha \frac{\sqrt{2}}{p} \arcsin \frac{p}{\sqrt{2}} \xrightarrow[p \to 0]{} -\alpha. \qquad (3.3.180)$$

For small (but non-zero) momenta, expression (3.3.180) can be expanded in a series in p^2:

$$E \equiv E_0 + \Delta E_0 = \frac{p^2}{2} - \alpha - \frac{p^2}{12}\alpha + \cdots$$

$$= \frac{p^2}{2(1 + \alpha/6)} - \alpha + \cdots. \qquad (3.3.181)$$

This expression shows that the interaction with phonons increases the effective mass of an electron by a factor $(1 + \alpha/6)$.

◇ **Feynman variational method for the large coupling constant**

As we have already mentioned, in real crystals the electron–phonon coupling constant takes large values and the perturbation theory cannot be used. Thus, to estimate the polaron ground-state energy E_0, we have to use another approximation scheme. To this aim, let us first pass to the imaginary time $T = -it$, so that

$$e^{i\widehat{H}t} \to e^{-\widehat{H}T}$$

and then use the obvious formula

$$E_0 = \lim_{T \to \infty} \left[-\frac{1}{T} \ln(\operatorname{Tr} e^{-\widehat{H}T}) \right]. \qquad (3.3.182)$$

Now, the trace of the exponential can be represented by the path integral and we can use some non-perturbative approximation method for the evaluation of the path integral. In the present case, the *Feynman variational method* (see later) proves to be most suitable. In order to apply it, let us pass in the phonon Hamiltonian from the annihilation and creation operators to the corresponding coordinates and momenta. After the Gaussian integration over the momenta, we arrive at the configuration path integral:

$$\operatorname{Tr} e^{-\widehat{H}T} = \int_{\substack{\mathcal{C}\{x(0)=x(T)\} \\ \mathcal{C}\{q(0,k)=q(T,k)\}}} \mathcal{D}x(\tau)\, \mathcal{D}q(\tau, k)\, e^{-S} \qquad (3.3.183)$$

where the action S has the form

$$S = \int d\tau \left[\frac{1}{2}\dot{x}^2(\tau) + \int \frac{d^3k}{(2\pi)^3} \left(\frac{1}{2}(\dot{q}^2(\tau, k) + q^2(\tau, k)) + \sqrt{2}(\sqrt{2\pi}\alpha)^{1/2} \frac{1}{|k|} q(\tau, k) e^{ikx(\tau)} \right) \right] \qquad (3.3.184)$$

(x are the electron and q are the phonon coordinates). The path integral over the phonon variables is Gaussian and we can explicitly calculate it. In fact, we can use the available result of the path integral for a driven oscillator (see (1.2.262)), which in the present case yields

$$\text{Tr}\,e^{-\hat{H}T} \approx \int_{x(0)=x(T)} \mathcal{D}x(\tau)\,\exp\left\{-\frac{1}{2}\int_0^T d\tau\,\dot{x}^2(\tau)\right\}$$
$$\times \exp\left\{2\sqrt{2}\pi\alpha \int_0^T d\tau\,ds \int \frac{d^3k}{(2\pi)^2}\frac{1}{k^2}e^{ik(x(\tau)-x(s))}e^{-|\tau-s|}\right\} \quad (3.3.185)$$

where we have taken into account that we are interested in the expression at large values of the imaginary time T, so that we could drop terms which give relatively small contributions in this limit. The only difference between the integral (3.3.184) and (1.2.262) is that both the phonon coordinates and the 'external (time-depending) source' are complex. But the generalization of (1.2.262) to this case is quite straightforward and can be easily carried out by the reader (note that the results for real and complex quantities look identical). The integral over k in the exponent of the integrand in (3.3.185) can be performed (as it is the Fourier transform of $1/x$), yielding

$$\text{Tr}\,e^{-\hat{H}T} \approx \int_{x(0)=x(T)} \mathcal{D}x(\tau)\,\exp\left\{-\frac{1}{2}\int_0^T d\tau\,\dot{x}^2(\tau)\right\}\exp\left\{\frac{\alpha}{\sqrt{8}}\int_0^T d\tau\,ds \int \frac{e^{-|\tau-s|}}{|x(\tau)-x(s)|}\right\}. \quad (3.3.186)$$

This integral cannot be calculated exactly. To estimate the ground-state energy from above, we shall use the Feynman variational method (Feynman 1955). To this aim, we first write

$$\int \mathcal{D}x(\tau)\,e^{-S} = \frac{\int \mathcal{D}x(\tau)\,e^{-(S-S_0)}e^{-S_0}}{\int \mathcal{D}x(\tau)\,e^{-S_0}}\int \mathcal{D}x(\tau)\,e^{-S_0} \equiv \langle e^{-(S-S_0)}\rangle_{S_0}\int \mathcal{D}x(\tau)\,e^{-S_0} \quad (3.3.187)$$

where S is the exponent in (3.3.186):

$$S = \frac{1}{2}\int_0^T d\tau\,\dot{x}^2(\tau) - \frac{\alpha}{\sqrt{8}}\int_0^T d\tau\,ds \int \frac{e^{-|\tau-s|}}{|x(\tau)-x(s)|} \quad (3.3.188)$$

and S_0 is a test action to be chosen. Then using the *Jensen inequality* (see (C.2) and (C.5) in appendix C, volume I), we obtain

$$\int \mathcal{D}x(\tau)\,e^{-S} \geq \exp\{-\langle S-S_0\rangle_{S_0}\}\int \mathcal{D}x(\tau)\,e^{-S_0}$$

where

$$\langle S-S_0\rangle \stackrel{\text{def}}{\equiv} \frac{\int \mathcal{D}x(\tau)\,(S-S_0)e^{-S_0}}{\int \mathcal{D}x(\tau)\,e^{-S_0}}$$

(mathematically minded readers may easily generalize the proof of the Jensen inequality in chapter 1 to the functional case). On the other hand, at large values of T, the trace can be estimated by the ground state: $\text{Tr}\,e^{-T\hat{H}}|_{T\to\infty} \approx e^{-ET}$ (here E denotes the ground-state eigenvalue of the Hamiltonian $\hat{H}$). Thus we arrive at the estimation

$$E \leq E_0 + \frac{1}{T}\langle S-S_0\rangle_{S_0} \quad (3.3.189)$$

where E_0 is the energy corresponding to the test action S_0:

$$E_0 = -\lim_{T\to\infty}\frac{1}{T}\ln\int \mathcal{D}x(\tau)\,e^{-S_0}. \quad (3.3.190)$$

It is worth noting that the second term in the action S given by (3.3.188) corresponds to a *retarded* potential. The physical reason for this is that a perturbation of a crystal lattice, caused by the motion of an electron, propagates with a finite speed.

◇ Choice of the test action

The next step consists in the choice of an appropriate test action S_0 which gives the best (lowest) estimation (3.3.189) for the polaron ground energy. It is clear that an optimal choice depends on the value of the coupling constant α:

(i) For a weak electron–phonon coupling, $\alpha \ll 1$, it is reasonable to choose as S_0 the action without any potential term (i.e. to drop in (3.3.188) the second term completely): $S_0 = S|_{\alpha=0}$. In this case, the result coincides with that of perturbation theory:

$$E - E_0 \leq -\alpha. \qquad (3.3.191)$$

Note that the inequality (3.3.191) gives the *upper bound* for the energy shift and that it is not easy to obtain this estimation by the usual operator methods.

(ii) For the strong electron–phonon coupling, $\alpha \gg 1$, a crystal reacts to the electron movement very quickly and we can use the test action S_0 with the potential term $V(x)$ for the electron in a fixed external potential, e.g., in the harmonic potential. In this case, the Feynman variational method proves to be equivalent to the ordinary Ritz variational method in quantum mechanics (with the test function of the form $\sim e^{-Cx^2}$).

(iii) It can be shown that, for intermediate values of the coupling constant, $1 \leq \alpha < 6$, instant test functions do not provide a good energy estimation: no potential term proves to be better than just the zero potential, $V = 0$. For this practically important domain of values we have to use some kind of *retarded* potential. A natural choice is

$$S_0 = \frac{1}{2}\int_0^T d\tau\, \dot{x}(\tau) + \frac{C}{2}\int_0^T d\tau\, ds\, (x(\tau) - x(s))^2 e^{-W|\tau-s|} \qquad (3.3.192)$$

where C and W are constants to be adjusted to obtain the best upper bound. The choice of the action S_0 can be justified by the following arguments: (i) it is quadratic and hence the corresponding Gaussian path integral can be calculated; (ii) it is retarded potential; (iii) the exponential factor is analogous to that in the initial action S.

Thus, to find the upper bound for the ground-state energy in the case of intermediate values of the coupling constant, we have to calculate

$$\langle S - S_0 \rangle_{S_0} = -\frac{\alpha}{\sqrt{8}}\left\langle \int_0^T d\tau\, ds\, \frac{e^{-|\tau-s|}}{|x(\tau) - x(s)|}\right\rangle_{S_0} - \frac{C}{2}\left\langle \int_0^T d\tau\, ds\, (x(\tau) - x(s))^2 e^{-|\tau-s|}\right\rangle_{S_0}. \qquad (3.3.193)$$

Recall that the averaging is understood with respect to the action S_0 as the ratio of two path integrals:

$$\langle F[x(\tau)]\rangle_{S_0} \stackrel{\text{def}}{\equiv} \frac{\int \mathcal{D}x(\tau)\, e^{-S_0} F[x(\tau)]}{\int \mathcal{D}x(\tau)\, e^{-S_0}}. \qquad (3.3.194)$$

As usual, it is convenient to calculate, first, the generating functional

$$\mathcal{Z}[J] = \left\langle \exp\left\{i\int_0^T d\tau\, J(\tau)x(\tau)\right\}\right\rangle_{S_0}. \qquad (3.3.195)$$

Then the mean value $\langle F[x(\tau)]\rangle_{S_0}$ for polynomial functionals F (in particular, for the second term on the right-hand side of (3.3.193)) can be obtained by functional differentiation, while more general functions are calculated with the help of the Fourier transform. In particular, for the first term in (3.3.193) we may write the representation

$$\left\langle \frac{1}{|\boldsymbol{x}(\tau) - \boldsymbol{x}(s)|} \right\rangle_{S_0} = \left\langle \int \frac{d^3k}{(2\pi)^3} \frac{4\pi}{k^2} e^{ik(\boldsymbol{x}(\tau) - \boldsymbol{x}(s))} \right\rangle_{S_0}$$

$$= \int \frac{d^3k}{(2\pi)^3} \frac{4\pi}{k^2} \mathcal{Z}[J(u)] \bigg|_{J(u) = k[\delta(u-\tau) - \delta(u-s)]} . \quad (3.3.196)$$

The generating functional (3.3.195) can be calculated by any method considered in this book, e.g., discretization, square completion, quadratic 'approximation' (which is exact for quadratic Lagrangians), mode expansion. In problem 3.3.8, we suggest the reader calculates this generating functional by the last method. Using the result of this calculation (see (3.3.227)) we readily obtain for the second term in (3.3.193)

$$\frac{C}{2} \left\langle \int_0^T d\tau\, ds\, |\boldsymbol{x}(\tau) - \boldsymbol{x}(s)|^2 e^{-W|\tau-s|} \right\rangle_{S_0} = \frac{3TC}{UW} \quad (3.3.197)$$

where

$$U^2 \equiv W^2 + \frac{4C}{W}$$

while for the first term in (3.3.193) the calculation of the Fourier transform (cf (3.3.196)) at large values of T gives

$$\int_0^T d\tau\, ds\, \left\langle \frac{e^{-|\tau-s|}}{|\boldsymbol{x}(\tau) - \boldsymbol{x}(s)|} \right\rangle_{S_0} \approx \frac{4TU}{\sqrt{2\pi}} \int_0^\infty du\, \frac{e^{-u}}{\sqrt{W^2 u + [(U^2 - W^2)/U](1 - e^{-uU})}} . \quad (3.3.198)$$

The ground-state energy E_0 corresponding to the test action S_0 can be determined using the general formula (3.3.182) but it is easier to use the following trick. First, we find the derivative

$$\frac{dE_0(C)}{dC} = \lim_{T \to \infty} \frac{1}{-T \int \mathcal{D}\boldsymbol{x}(\tau) \exp\{-S_0\}} \frac{d}{dC} \int \mathcal{D}\boldsymbol{x} \exp\{-S_0\}$$

$$= -\frac{1}{T} \left\langle -\frac{1}{2} \int d\tau\, ds\, |\boldsymbol{x}(\tau) - \boldsymbol{x}(s)|^2 e^{-W|\tau-s|} \right\rangle_{S_0}$$

$$= \frac{3}{UW} = \frac{3}{W\sqrt{W^2 + 4C/W}} . \quad (3.3.199)$$

Now E_0 can be found by the integration of (3.3.199) with the obvious boundary condition $E_0|_{C=0} = 0$. This yields

$$E_0 = \tfrac{3}{2}(U - W). \quad (3.3.200)$$

Collecting all the results, we finally find the upper bound for the polaron ground-state energy E:

$$E \leq E_0 + \frac{1}{T} \langle S - S_0 \rangle_{S_0}$$

$$\xrightarrow[T \to \infty]{} \frac{3}{4U}(U - W)^2 - \frac{\alpha U}{\sqrt{\pi}} \int_0^\infty du\, \frac{e^{-u}}{\sqrt{W^2 u + [(U^2 - W^2)/U](1 - e^{-uU})}} . \quad (3.3.201)$$

The constants W and U (or C) should be adjusted to obtain a lowest upper bound. For extremal values of the coupling constant, this can be done analytically:

(i) For small α, the best values of the constants are:

$$W = 3 \qquad U = 3\left[1 + \frac{2\alpha}{3W}\left(1 - \frac{2}{W}\left[\sqrt{1-W} - 1\right]\right)\right]. \qquad (3.3.202)$$

Then the upper bound turns out to be

$$E \leq -\alpha - 1.23\left(\frac{\alpha}{10}\right). \qquad (3.3.203)$$

This estimation can be compared with the correct result obtained from the perturbation expansion:

$$E \approx -\alpha - 1.26\left(\frac{\alpha}{10}\right). \qquad (3.3.204)$$

(ii) For large α, the best values for W and U are

$$W = 1 \qquad U = \frac{4\alpha^2}{9\pi} - 4\left(\ln 2 + \frac{1}{2}\mathfrak{C}\right) + 1 \qquad (3.3.205)$$

where $\mathfrak{C} = 0.5772\ldots$ is the Euler constant. These values give the following estimation:

$$E \leq -\frac{\alpha^2}{2\pi} - \frac{3}{2}(2\ln 2 + \mathfrak{C}) - \frac{3}{4} + \mathcal{O}\left(\frac{1}{\alpha^2}\right). \qquad (3.3.206)$$

For intermediate values of the coupling constant α, the integration in (3.3.201) cannot be done analytically and (rather simple) numerical calculations should be used. It is necessary to stress that there are no other methods, except the Feynman variational one based on the path-integral technique, which would give reliable results for the intermediate values of the electron–phonon coupling constant α. The reader may find further details and results on the polaron problem in Feynman (1972a), Kittel (1987) and Heeger (1988).

3.3.6 Problems

Problem 3.3.1. Prove the *Furry theorem* (Furry 1937), which can be formulated as follows:

- The determinant (3.3.20) of the Dirac operator in the external field $A_\mu(x)$ is an *even* function of $A_\mu(x)$.

Hint. As we have mentioned, the functional determinant must be regularized. The most natural way is to divide it by the free Dirac determinant (see the regularization of the corresponding determinant for a harmonic oscillator in an external field, section 2.2.2). Thus, we have to prove that the ratio

$$\frac{\det(\gamma_\mu(\partial_\mu + ieA_\mu) - m)}{\det(\gamma_\mu(\partial_\mu - m))}$$

of the Dirac operators is an even function of $A\mu(x)$. The required statement follows from the following chain of equalities:

$$\det\left(1 + \frac{1}{\slashed{\partial} - m}ie\slashed{A}\right) = \det\left(1 + ie\slashed{A}\frac{1}{\slashed{\partial} - m}\right)$$

$$= \det\left[1 + \left(ie\slashed{A}\frac{1}{\slashed{\partial} - m}\right)^\top\right] = \det\left(1 + \frac{1}{\slashed{\partial}^\top - m}ie\slashed{A}^\top\right)$$

$$= \det\left[1 + \left(\gamma_2 \frac{1}{\slashed{\partial} - m}\gamma_2^{-1}\right)\left(\gamma_2 ie\slashed{A}\gamma_2^{-1}\right)\right]$$

$$= \det\left(1 + \frac{1}{\gamma_2\slashed{\partial}\gamma_2^{-1} - m}\, ie\gamma_2\slashed{A}\gamma_2^{-1}\right)$$

$$= \det\left(1 - \frac{1}{\slashed{\partial} - m}\, ie\slashed{A}\right).$$

Note that we use *Euclidean γ-matrices*:

$$\gamma_\mu\gamma_\nu + \gamma_\nu\gamma_\mu = 2\delta_{\mu\nu} \qquad \mu,\nu = 1,2,3,4$$

in the standard representation

$$\gamma_i = \begin{pmatrix} 0 & \sigma_i \\ \sigma_i & 0 \end{pmatrix} \quad i = 1,2,3 \qquad \gamma_4 = \begin{pmatrix} 1 & 0 \\ 0 & -1 \end{pmatrix}$$

for which the following relations are correct:

$$\gamma_\mu^\top = \gamma_\mu \quad \text{for } \mu = 1,3,4, \quad \gamma_2^\top = -\gamma_2.$$

The latter relations, together with the obvious one, $\partial_\mu^\top = -\partial_\mu$, imply

$$\gamma_2\slashed{\partial}^\top\gamma_2^{-1} = \slashed{\partial} \qquad \gamma_2\slashed{A}^\top\gamma_2^{-1} = -\slashed{A}.$$

Problem 3.3.2. Calculate the Jacobian of the functional change of variables (cf (3.3.59)):

$$\widetilde{\pi}(t,\rho) = -f(\rho)\frac{p + \int d\rho\, \varpi(t,\rho)[\partial_x\widetilde{\varphi} - fc]}{\int d\rho\, f\partial_x\widetilde{\varphi}} + \varpi(\tau,\rho)$$

where $f(\rho)$ is such a function that $\int d\rho\, f^2(\rho) = 1$, and p, c are constants.

Hint. We have to calculate the determinant of the operator with the kernel

$$\frac{\delta\widetilde{\pi}(t,\rho)}{\delta\varpi(t,\rho')} = \left[\delta(\rho - \rho') - \frac{1}{\int d\rho''\, f(\rho'')\partial_x\widetilde{\varphi}(\rho')}f(\rho)(\partial_x\widetilde{\varphi}(\rho') - cf(\rho'))\right]\delta(t - t')$$

$$\equiv \left[\mathbb{I} - \frac{1}{\langle f|\partial_x\widetilde{\varphi}\rangle}|f\rangle(\langle\partial_x\widetilde{\varphi}| - c\langle f|)\right]\delta(t - t')$$

where in the second line we have used the Dirac notation for the Hilbert space of functions with the scalar product $\langle f|g\rangle = \int d\rho\, f(\rho)g(\rho)$. Then,

$$\det\left[\frac{\delta\widetilde{\pi}(t,\rho)}{\delta\varpi(t,\rho')}\right] = \prod_t \exp\left\{\operatorname{Tr}\ln\left[\mathbb{I} - \frac{1}{\langle f|\partial_x\widetilde{\varphi}\rangle}|f\rangle(\langle\partial_x\widetilde{\varphi}| - c\langle f|)\right]\right\}$$

$$= \prod_t \exp\left\{\ln\frac{c}{\langle f|\partial_x\widetilde{\varphi}\rangle}\right\} = \prod_t c\left[\int d\rho\, f(\rho)\partial_x\widetilde{\varphi}(t,\rho)\right]^{-1}.$$

The second equality follows from the relation:

$$\operatorname{Tr}\left[\frac{1}{\langle f|\partial_x\widetilde{\varphi}\rangle}|f\rangle(\langle\partial_x\widetilde{\varphi}| - c\langle f|)\right]^n = \frac{(\langle\partial_x\widetilde{\varphi}|f\rangle - c\langle f|f\rangle)^n}{\langle f|\partial_x\widetilde{\varphi}\rangle}$$

$$= \left[\frac{\langle\partial_x\widetilde{\varphi}|f\rangle - c}{\langle f|\partial_x\widetilde{\varphi}\rangle}\right]^n.$$

146   Quantum field theory: the path-integral approach

Problem 3.3.3. Verify that for a 'pure gauge' field configuration (3.3.98), the Chern–Simons topological characteristic (3.3.100) turns into the winding Pontryagin number (3.3.99).

Hint. Write the integrand of (3.3.100) in matrix form, substitute (3.3.98) and convert into (3.3.99) using the antisymmetry in the indices and the unitarity relation $U^\dagger U = 1$.

Problem 3.3.4. Using decomposition (3.3.152) of the Dirac operator, calculate the regularized determinant of the Dirac operator represented in terms of the Fourier transformed eigenfunctions as in (3.3.151).

Hint. Use the fact that the plane waves shift the differential operator (f is an arbitrary smooth function)
$$e^{-ikx} f(\partial_\mu) e^{ikx} = f(\partial_\mu + ik_\mu) \tag{3.3.207}$$

so that after rescaling the integration variable, $k_\mu \to Mk_\mu$, the sum (3.3.151) can be represented as follows:

$$\sum_n \varphi_n^\dagger(x)\gamma_5 \varphi_n(x) = \lim_{M\to\infty} M^4 \frac{1}{(2\pi)^4} \int d^4k \, \text{Tr}\, \gamma_5 \exp\left\{k_\mu k^\mu - \frac{2ik_\mu D^\mu}{M} - \frac{D_\mu D^\mu}{M^2} - \frac{\gamma^\mu \gamma^\nu F_{\mu\nu}}{2M^2}\right\}. \tag{3.3.208}$$

It is important that the properties of γ-matrices, namely,
$$\text{Tr}\, \gamma_5 = \text{Tr}(\gamma_5 \gamma^\mu \gamma^\nu) = 0 \tag{3.3.209}$$
$$\text{Tr}(\gamma_5 \gamma^\mu \gamma^\nu \gamma^\rho \gamma^\sigma) = -4\varepsilon^{\mu\nu\rho\sigma} \tag{3.3.210}$$

leave in the integrand of (3.3.208) only the quadratic term in $\gamma^\mu \gamma^\nu F_{\mu\nu}$, in the limit $M \to \infty$. A subsequent Gaussian integration over k_μ produces expression (3.3.153).

Problem 3.3.5. Calculate the singlet chiral quantum anomaly in a *two*-dimensional Abelian gauge-field theory.

Hint. To calculate of the Jacobian of the chiral transformation (cf (3.3.138)–(3.3.149)), we have to regularize the sum
$$\sum_n \varphi_n^\dagger(x)\gamma_5 \varphi_n(x) = \lim_{M\to\infty} \sum_n \varphi_n^\dagger(x)\gamma_5 \exp\left\{-\frac{\displaystyle{\not{D}}^2}{M^2}\right\} \varphi_n(x). \tag{3.3.211}$$

Decomposing the squared Dirac operator (cf (3.3.152)), shifting the differential operator (see the hint to the preceding problem, equation (3.3.207)) and rescaling the momentum $k_\mu \to Mk_\mu$, we obtain for the sum:

$$\sum_n \varphi_n^\dagger(x)\gamma_5 \varphi_n(x) = \lim_{M\to\infty} M^2 \frac{1}{(2\pi)^4}$$
$$\times \int d^2k \, \text{Tr}\left(\gamma_5 \exp\left\{-k_\mu k^\mu - \frac{2ik_\mu D^\mu}{M} - \frac{D_\mu D^\mu}{M^2} - \frac{i\gamma^\mu \gamma^\nu F_{\mu\nu}}{2M^2}\right\}\right). \tag{3.3.212}$$

In two-dimensional Minkowski spacetime, the γ-matrices and metric have the form
$$\gamma^0 = \sigma_2 \qquad \gamma^1 = i\sigma_1 \qquad \gamma_5 = \gamma^0 \gamma^1 = \sigma_3$$
$$g_{\mu\nu} = \begin{pmatrix} 1 & 0 \\ 0 & -1 \end{pmatrix} \qquad \varepsilon_{01} = 1 \tag{3.3.213}$$

while in two-dimensional Euclidean spacetime, they become

$$\gamma^0 = i\gamma_4 \qquad \gamma_5 = \gamma^0\gamma^1 = i\gamma^4\gamma^1$$
$$g_{\mu\nu} = -\delta_{\mu\nu} \qquad \varepsilon_{41} = i\varepsilon_{01} = i. \qquad (3.3.214)$$

Note that the relations

$$\gamma_\mu\gamma_5 = \varepsilon_{\mu\nu}\gamma^\nu \qquad \text{Tr}\,\gamma_5\gamma_\mu\gamma_\nu = -2\varepsilon_{\mu\nu} \qquad (3.3.215)$$

are valid in both Minkowski and Euclidean spaces. Next, we expand the exponential and take the trace of the Dirac matrices. As a result, we find the regularized sum

$$\sum_n \varphi_n^\dagger(x)\gamma_5\varphi_n(x) = -\frac{i}{4\pi}\varepsilon_{\mu\nu}F^{\mu\nu}. \qquad (3.3.216)$$

Thus the Jacobian of the corresponding path-integral measure reads as

$$J[\beta] = \exp\left\{-\int dx\,\beta(x)\frac{1}{2\pi}\varepsilon_{\mu\nu}F^{\mu\nu}\right\} \qquad (3.3.217)$$

and the anomaly proves to be the following:

$$\mathfrak{A} = \frac{1}{2\pi}\varepsilon_{\mu\nu}F^{\mu\nu}. \qquad (3.3.218)$$

(This result is valid both in Minkowski and Euclidean spaces if the ε-tensor is understood in the appropriate sense, i.e. as in (3.3.213) or in (3.3.214).)

Problem 3.3.6. Prove the regularization independence of the chiral anomaly, i.e. check that the calculation with an arbitrary regularization function $f(x)$ satisfying the conditions (3.3.158) (instead of the damping exponential) leads to the same result (3.3.218).

Hint. Repeating the steps of the calculations performed with exponential damping, we arrive at the expression

$$\sum_n \varphi_n^\dagger(x)\gamma_5\varphi_n(x) = -\frac{1}{2(2\pi^4)}\int d^4k\,f''(k^2)\varepsilon^{\mu\nu\alpha\beta}\,\text{Tr}(F_{\mu\nu}F^{\mu\nu}) \qquad (3.3.219)$$

and the subsequent integration by parts taking into account conditions (3.3.158) gives the same result (3.3.153), as in the case of the exponential regularization function.

Problem 3.3.7. Calculate the expectation value (3.3.161) of the commutator of the gauge-covariant Dirac operator $\slashed{D}$ and the γ_5-matrix.

Hint. Expanding the Dirac spinors $\bar\psi$, ψ in terms of eigenfunctions of $\slashed{D}$ (see equation (3.3.141)), we obtain

$$\langle 0|\bar\psi(x)2i\gamma_5\slashed{D}\psi(x)|0\rangle = \frac{1}{N}\int\prod_i da_i\,d\bar{b}_i\sum_{m,n}\bar{b}_m a_n\varphi_m^\dagger(x)2i\gamma_5\slashed{D}\varphi_n(x)\exp\left\{\sum_k(i\lambda_k - m)\bar{b}_k a_k\right\}$$
$$(3.3.220)$$

where the normalization factor is the usual Dirac determinant

$$N = \det(i\slashed{D} - m). \qquad (3.3.221)$$

148   Quantum field theory: the path-integral approach

This integration gives (recall that $\bar{b}_i$, a_i are Grassmann variables)

$$\langle 0|\bar{\psi}(x)2i\gamma_5 \slashed{D}\psi(x)|0\rangle = 2\sum_n \frac{\varphi_n^\dagger(x)\gamma_5 \slashed{D}\varphi_n(x)}{\lambda_n + im}. \qquad (3.3.222)$$

For the commutator we then find

$$\begin{aligned}\langle 0|\bar{\psi}(x)i[\gamma_5, \slashed{D}]\psi(x)|0\rangle &= \langle 0|\bar{\psi}(x)2i\gamma_5\slashed{D}\psi(x)|0\rangle \\ &= 2\sum_n \frac{\varphi_n^\dagger(x)\gamma_5\slashed{D}\varphi_n(x)}{\lambda_n + im} \\ &= 2\sum_n \frac{\varphi_n^\dagger(x)\gamma_5(\slashed{D}+im)\varphi_n(x)}{\lambda_n + im} - 2m\sum_n \frac{\varphi_n^\dagger(x)i\gamma_5\varphi_n(x)}{\lambda_n + im} \\ &= -\frac{1}{16}\varepsilon^{\mu\nu\alpha\beta}\,\mathrm{Tr}\,F_{\mu\nu}F_{\alpha\beta} + 2m\langle 0|\bar{\psi}(x)\gamma_5\psi(x)|0\rangle \qquad (3.3.223)\end{aligned}$$

where we have used the regularization result (3.3.153). So we have obtained the anomaly and also the mass term which *explicitly* breaks the chiral symmetry (the chiral symmetry exists only in the massless theory).

Problem 3.3.8. Calculate the generating functional (3.3.195) for the test action S_0 (3.3.192), used for estimating the upper bound for the polaron ground-state energy.

Hint. The periodic boundary conditions for the variables $x(\tau)$ implies the following mode decomposition:

$$x(\tau) = x(0) + \sum_{n=1}^\infty a_n \sin\frac{n\pi\tau}{T}. \qquad (3.3.224)$$

The terms of the action S_0 now acquire the form

$$\int_0^T d\tau\,\frac{1}{2}\dot{x}^2 = \frac{1}{4}\sum_{n=1}^\infty a_n^2 \frac{n^2\pi^2}{T}$$

$$\frac{1}{2}C\int_0^T d\tau\,ds\,[x(\tau)-x(s)]^2 e^{-W|\tau-s|} = \frac{1}{2}C\int_0^T d\tau\,ds\left[\sum_{n=1}^\infty a_n\left(\sin\frac{n\pi\tau}{T} - \sin\frac{n\pi s}{T}\right)\right]^2 e^{-W|\tau-s|}$$

$$\underset{\text{large } T}{\approx} \frac{C}{W}\sum_{n=1}^\infty \left(\frac{n^2\pi^2/T}{W^2 + n^2\pi^2/T^2}\right)a_n^2.$$

Introducing also the Fourier transform of the external source:

$$b_n = i\int_0^T d\tau\,J(\tau)\sin\frac{n\pi\tau}{T}$$

we can easily find, after Gaussian integration over the mode variables a_n,

$$\mathcal{Z}[J] = \exp\left\{\sum_{n=1}^\infty \frac{b_n^2}{4A_n}\right\} \qquad (3.3.225)$$

where
$$A_n = \frac{n^2\pi^2}{4T}\left(1 + \frac{4C/W}{W^2 + n^2\pi^2/T^2}\right).$$

For the particular form of the external source
$$\boldsymbol{J}(u) = \boldsymbol{K}[\delta(u-\tau) - \delta(u-s)]$$

expression (3.3.225) becomes

$$\mathcal{Z}[\boldsymbol{K}] = \exp\left\{\sum_{n=1}^{\infty} \frac{-K^2[\sin(n\pi\tau/T) + \sin(n\pi s/T)]^2}{\frac{n^2\pi^2}{T}\left[1 + \frac{4C/W}{W^2+n^2\pi^2/T^2}\right]}\right\} \tag{3.3.226}$$

$$\xrightarrow[T\to\infty]{} \exp\left\{-\frac{K^2}{2}\left[\frac{W^2}{U^2}|\tau-s| + \frac{4C}{WU^3}\left(1 - e^{-|\tau-s|U} + e^{-(\tau+s)U} - \frac{1}{2}e^{-2Ut} - \frac{1}{2}e^{-2Us}\right)\right]\right\}$$

$$\approx \exp\left\{-\frac{K^2}{2}\left[\frac{W^2}{U^2}|\tau-s| + \frac{4C}{WU^3}(1 - e^{-|\tau-s|U})\right]\right\} \tag{3.3.227}$$

where
$$U^2 = W^2 + \frac{4C}{W}.$$

To calculate (3.3.226), we have substituted the sum by the integral over the variable $n\pi/T$ (which is valid at large values of T) and then we have neglected the terms which are exponentially small in almost all the domain of variation of τ and s.

3.4 Path integrals in the theory of gravitation, cosmology and string theory: advanced applications of path integrals

This section contains several rather involved topics on path-integral applications in modern theoretical models such as quantum gravity, (super)strings, cosmology and black holes. The style of this section is necessarily different from the rest of this book: each of these topics deserves a special book for detailed discussion. The brief review in this section is intended only to provide a general understanding of the problems without presenting all the technical details or supplying all the motivations. For discussions of the details we shall refer the reader to the appropriate literature (where further references can be found). Some acquaintance on the part of the reader with the basic facts from the differential geometry of Riemann manifolds (see supplement V) as well as from Einstein's general theory of relativity is assumed.

3.4.1 Path-integral quantization of a gravitational field in an asymptotically flat spacetime and the corresponding perturbation theory

A complete *theory of quantum gravity* is still far from being complete. Moreover, at present, there is the common belief that a complete and self-consistent quantum gravitational theory cannot be constructed within the framework of local field theory (e.g., on the basis of Einstein's *general theory of relativity* or some modification of it) but requires more general theoretical concepts, including the quantum theory of relativistic extended objects such as *strings* and *membranes* (see, e.g., Green *et al* (1987) and Polchinski (1994, 1996)).

However, if we are interested only in phenomena with energies much lower than the natural gravitational scale, namely the *Planck mass*, $M_p \approx 1.2 \times 10^{19}$ GeV $\approx 2.2 \times 10^{-5}$ g, we can use the local field theory as an *effective* theory describing phenomena with relatively low energy. The theory proves to be non-renormalizable and hence cannot be regarded as a fundamental one. But for low-energy processes, the Planck mass M_p serves as a natural ultraviolet cut-off scale. In this subsection, we consider the path-integral quantization of this effective theory, while in the section 3.4.5 we shall briefly discuss string theory.

It is worth stressing that, in both cases, the path-integral technique proves to be crucial for the successful development of the theories.

◇ **Classical action for a gravitational field; gauge invariance and constraints**

A gravitational field can be considered as a type of the gauge fields which we have discussed in most of this chapter. Thus, we may approach the problem of the quantization of the gravitational field in the framework of the general formalism for gauge-field quantization. This time, the gauge transformations are *general coordinate transformations (diffeomorphisms)*

$$x^\mu \longrightarrow x'^\mu = f^\mu(x)$$

(f^μ being an arbitrary differentiable function) of the spacetime manifold under consideration. In the case of an *asymptotically flat* spacetime, i.e. topologically trivial (topologically equivalent to $\mathbb{R}^4$) and with a flat Minkowski metric at infinity, we can clearly separate local (gauge, unphysical) and global spacetime symmetry transformations: the gauge transformations are diffeomorphisms which do not affect space infinity (a flat region), while the global Poincaré group transformations act in the whole space, including the asymptotic flat region, and form the global symmetry group (it is obvious that from the mathematical point of view, the Poincaré transformations are a particular case of diffeomorphisms). These global Poincaré transformations include time shifts, defining thereby the proper time variable and the physical evolution of a gravitational system with asymptotically flat spacetime.

According to Einstein's general relativity theory (see, e.g., Dirac (1975) and Misner *et al* (1973)), a gravitational field is described by a metric $g_{\mu\nu}(x)$, which is a function of the spacetime coordinates x_μ, $-\infty < x_\mu < \infty$, $\mu = 0, 1, 2, 3$ (due to the topological triviality these coordinates can be chosen globally on the whole spacetime manifold). In this section we shall denote the flat Minkowski metric as $\eta_{\mu\nu}$: $\eta_{\mu\nu} \stackrel{\text{def}}{\equiv} \text{diag}\{+1, -1, -1, -1\}$, to distinguish it from the arbitrary non-flat spacetime metric $g_{\mu\nu}(x)$. The condition of asymptotical flatness implies that the coordinates can be chosen so that

$$g_{\mu\nu}(x) \xrightarrow[r\to\infty]{} \eta_{\mu\nu} + \mathcal{O}\left(\frac{1}{r}\right) \qquad (3.4.1)$$

where $r = \sqrt{(x^1)^2 + (x^2)^2 + (x^3)^2}$. The *Einstein action* functional

$$S_{\text{gr}} = \frac{1}{16\pi G_N} \int d^4x \sqrt{-g}\, R(g_{\mu\nu}) \qquad (3.4.2)$$

can be cast into the form

$$S_{\text{gr}} = \frac{1}{16\pi G_N^2} \int d^4x\, [-\Gamma^\rho_{\mu\rho}\partial_\nu(\sqrt{-g}g^{\mu\nu}) + \Gamma^\rho_{\mu\nu}\partial_\rho(\sqrt{-g}g^{\mu\nu}) + \sqrt{-g}g^{\mu\nu}(\Gamma^\rho_{\mu\sigma}\Gamma^\sigma_{\rho\nu} - \Gamma^\rho_{\mu\nu}\Gamma^\sigma_{\rho\sigma})]. \qquad (3.4.3)$$

Here G_N is Newton's constant (which with $\hbar = c = 1$ has the units of [length]^2 or [mass]$^{-2}$: $G_N = M_p^{-2}$), $g \stackrel{\text{def}}{\equiv} \det g_{\mu\nu}$ and $R(g_{\mu\nu})$ is the spacetime curvature corresponding to the metric $g_{\mu\nu}$ (cf supplement V or section 2.5), $\Gamma^\rho_{\mu\nu}$ denote the Christoffel symbols:

$$\Gamma^\rho_{\mu\nu} \stackrel{\text{def}}{\equiv} \tfrac{1}{2}g^{\rho\sigma}(\partial_\mu g_{\nu\sigma} + \partial_\nu g_{\mu\sigma} - \partial_\sigma g_{\mu\nu}) \qquad (3.4.4)$$

and the matrix $g^{\mu\nu}$ is the inverse of $g_{\mu\nu}$: $g^{\mu\sigma} g_{\sigma\nu} = \delta^\mu_\nu$. Under the infinitesimal coordinate transformations

$$\delta x^\mu = \epsilon^\mu(x) \tag{3.4.5}$$

with the local (coordinate-dependent) parameters $\epsilon^\mu(x)$, the basic quantities transform as follows:

$$\delta g^{\mu\nu} = -\epsilon^\lambda \partial_\lambda g^{\mu\nu} + g^{\mu\lambda} \partial_\lambda \epsilon^\nu + g^{\nu\lambda} \partial_\lambda \epsilon^\mu \tag{3.4.6}$$

$$\delta \Gamma^\rho_{\mu\nu} = -\epsilon^\lambda \partial_\lambda \Gamma^\rho_{\mu\nu} - \Gamma^\rho_{\mu\lambda} \partial_\nu \epsilon^\lambda - \Gamma^\rho_{\nu\lambda} \partial_\mu \epsilon^\lambda + \Gamma^\lambda_{\mu\nu} \partial_\lambda \epsilon^\rho. \tag{3.4.7}$$

The variation of the action in the form (3.4.3) over $\Gamma^\rho_{\mu\nu}$, considered as *independent* variables, gives the equation with the solution (3.4.4). Therefore, we are free to work with the so-called *first-order formalism*, where both $g_{\mu\nu}$ and $\Gamma^\rho_{\mu\nu}$ are independent variables in the action (3.4.3), or to make the substitution of expression (3.4.4) for the Christoffel symbols, which yields the action

$$S_{\text{gr}} = \frac{1}{16\pi G_N^2} \int d^4 x \, \mathcal{L}_{\text{gr}}(h^{\mu\nu})$$

$$= \frac{1}{16\pi G_N^2} \int d^4 x \left(h^{\rho\sigma} \partial_\rho h^{\mu\nu} \partial_\nu h_{\sigma\mu} - \frac{1}{2} h^{\rho\sigma} \partial_\rho h^{\mu\nu} \partial_\sigma h_{\mu\nu} + \frac{1}{4} h^{\rho\sigma} \partial_\rho \ln h \, \partial_\sigma \ln h \right) \tag{3.4.8}$$

where we have introduced for compactness the quantity (*covariant density*)

$$h^{\mu\nu} \stackrel{\text{def}}{\equiv} \sqrt{-g} g^{\mu\nu} \qquad h \stackrel{\text{def}}{\equiv} \det h^{\mu\nu}. \tag{3.4.9}$$

The action (3.4.8) corresponds to the *second-order formalism*.

To develop the Hamiltonian formalism, whihc in gravitational theories is also called the *Arnowitt–Deser–Misner (ADM) formalism* (Arnowitt *et al* 1960) (see also, e.g., Misner *et al* (1973)), and to construct the corresponding path integral, it is convenient to start from the first-order formalism, i.e. from action (3.4.3) (similarly to the case of Yang–Mills theories, cf section 3.2.3). Standard constrained system analysis of the Lagrangian in (3.4.3) shows that it contains the non-dynamical variables Γ^0_{i0}, Γ^k_{i0}, Γ^k_{ij} ($i, j, k = 1, 2, 3$) which can be expressed via the dynamical variables $h^{\mu\nu}$, Γ^0_{ik} making use of the following secondary second-class constraints:

$$h^{ik} \Gamma^0_{ik} + h^{00} \Gamma^i_{0i} + \partial_i h^{i0} = 0$$

$$2h^{k0} \Gamma^0_{ik} + h^{00} (\Gamma^0_{i0} - \Gamma^k_{ik}) + \partial_i h^{00} = 0$$

$$\partial_k h^{i0} + h^{in} \Gamma^0_{nk} + h^{00} \Gamma^i_{k0} + h^{0n} \Gamma^i_{nk} - h^{i0} \Gamma^n_{kn} = 0$$

$$\partial_k h^{ij} + h^{iv} \Gamma^j_{\nu k} + h^{j\nu} \Gamma^i_{\nu k} - h^{ij} \Gamma^\nu_{k\nu} = 0. \tag{3.4.10}$$

The natural phase-space variables prove to be

$$q^{ik} = h^{i0} h^{k0} - h^{00} h^{ik} \qquad \pi_{ik} = -\frac{1}{h_{00}} \Gamma^0_{ik} \tag{3.4.11}$$

and the Lagrangian, after substituting the solution of the constraints (3.4.10), takes the canonical form:

$$\mathcal{L}_{\text{gr}}(q^{ik}, \pi_{ik}) = \frac{1}{2G^2} \left[\pi_{ik}(x) \partial_0 q^{ik}(x) - H(x) - \left(\frac{1}{h^{00}(x) - 1} \right) T_0(x) - \frac{h^{i0}}{h^{00}(x)} T_i(x) \right] \tag{3.4.12}$$

with the secondary first-class constraints:

$$T_0(x) \equiv q^{ij}q^{kl}(\pi_{ik}\pi_{jl} - \pi_{ij}\pi_{kl}) + g_3 R_3 = 0 \qquad (3.4.13)$$

$$T_i(x) \equiv 2[\nabla_i(q^{kl}\pi_{kl}) - \nabla_k(q^{kl}\pi_{il})] = 0 \qquad (3.4.14)$$

(the coefficients in front of them in the Lagrangian play the role of Lagrange multipliers) and with a Hamiltonian of the form

$$H(x) = T_0(x) - \partial_i \partial_k q^{ik}(x). \qquad (3.4.15)$$

In these formulae, R_3 denotes the three-dimensional curvature generated by the three-dimensional part g_{ik}, $i, k = 1, 2, 3$ of the metric $g_{\mu\nu}$ and $g_3 = \det g_{ik}$. The symbol ∇_i denotes the covariant derivative with respect to this three-dimensional metric g_{ik}. We can check that the constraints $T_\mu(x)$ are in involution, i.e. they satisfy characteristic property (3.2.57) of first-class constraints. Counting the degrees of freedom gives two possible physical polarizations of the gravitational field: 6 (coordinates q^{ij}) − 4 (constraints) = 2.

◇ Phase-space path integral for the gravitational field in an asymptotically flat space

To construct the phase-space path integral, let us choose the gauge conditions accompanying first-class constraints (3.4.13) and (3.4.14), in the form (see Popov (1983))

$$\ln \det q^{ik} = \Phi(x) \qquad q^{ik} = 0, \; i \neq k \qquad (3.4.16)$$

where $\Phi(x)$ is a function with the appropriate asymptotic behaviour: $\Phi(x) \xrightarrow[r\to\infty]{} \text{constant}/r$. The reader may verify that the necessary condition (3.2.56) is fulfilled for such a choice. Now we are ready to write down the phase-space path integral for the S-matrix:

$$\int \prod_{i\leq k}[\mathcal{D}\pi_{ik}(x)\mathcal{D}q^{ik}(x)]\,\mathcal{D}\lambda^0(x)\,\mathcal{D}\lambda^i(x)\prod_{a=0}^{3}\delta[\chi_a(x)]\det\{T_\mu,\chi_a\}$$

$$\times \exp\left\{i\int d^4x\,[\pi_{ik}\partial_0 q^{ik} - \lambda^i T_i - \lambda^0 T_0 - H(x)]\right\} \qquad (3.4.17)$$

where we have denoted, for compactness,

$$\chi_0 = \ln \det q^{ik} - \Phi(x) \qquad \chi_1 = q^{23} \qquad \chi_2 = q^{31} \qquad \chi_3 = q^{12} \qquad (3.4.18)$$

$$\lambda^0 = \left(\frac{1}{h^{00}} - 1\right) \qquad \lambda^i = \frac{h^{0i}}{h^{00}}. \qquad (3.4.19)$$

We do not discuss the boundary conditions for the fields $\pi_{ik}(x)$, $q^{ik}(x)$: for an asymptotically flat spacetime, the consideration is quite similar to that of Yang–Mills fields, which we discussed in section 3.2.

◇ Transition to the Lagrangian path integral

Gaussian integration over the momentum variables π_{ik} produces, as usual, the configuration path integral (details of the calculation can be found in Popov (1983))

$$\mathfrak{N}^{-1}\int\prod_x\left(h^{-5/2}(x)\prod_{\mu\leq\nu}dh^{\mu\nu}(x)\right)\prod_{a=0}^{3}\delta[\chi_a]\det\widehat{B}\exp\{iS_{\text{gr}}[h^{\mu\nu}]\} \qquad (3.4.20)$$

where the operator $\widehat{B}$ is defined by the Poisson brackets $\{T_\mu, \chi_a\}$ of the constraints and the gauge conditions: in fact $\det \widehat{B}$ is obtained from $\det\{T_\mu, \chi_a\}$, in the process of Gaussian integration, substituting the momenta π_{ik} with their expressions through the metric tensor (cf (3.4.11)) and multiplying by the local factor $1/h^{00}$. Note that the integration measure in the path integral (3.4.20) contains an additional factor $\prod_x h^{-5/2}(x)$. This is the reason for writing it explicitly in (3.4.20) as a product over spacetime points (instead of just the symbolical notations $\mathcal{D}h^{\mu\nu}(x)$). Of course, the product over points is understood in the sense of an appropriate regularization (discretization or truncated mode expansion). It is very important that the local prefactor $h^{-5/2}$ makes the measure in the path-integral gauge invariant.

The reader may explicitly check that operator $\widehat{B}$ satisfies the equality

$$\det \widehat{B} \int \mathcal{D}a(x) \prod_{a=0}^{3} \delta[\chi_a] = 1. \qquad (3.4.21)$$

According to the general theory of quantum gauge fields (see section 3.2), this means that integral (3.4.20) goes over the classes of gauge-invariant fields (recall that in the case of gravitational fields, the role of gauge transformations is played by diffeomorphic transformations of spacetime coordinates; the corresponding infinitesimal transformations of the metric tensor have the form (3.4.6)).

Now, to present an expression for the configuration path integral in an explicitly relativistically covariant form (but, of course, not generally, i.e. diffeomorphically covariant form, because we have already imposed the gauge condition), we can use the Faddeev–Popov trick in order to pass to a covariant gauge condition. The most convenient such gauge is the so-called *harmonicity condition*:

$$\partial_\nu(\sqrt{-g}g^{\mu\nu}) = a^\mu(x) \qquad (3.4.22)$$

where $a^\mu(x)$ is some fixed vector field. To this aim, we introduce as usual the functional $\Delta_h[g^{\mu\nu}]$, such that

$$\Delta_h[g^{\mu\nu}] \int \mathcal{D}f(x) \prod_\mu \delta[\partial_\nu(h^{\mu\nu})^{f(x)} - a^\mu(x)] = 1 \qquad (3.4.23)$$

where $(h^{\mu\nu})^{f(x)}$ denotes the metric density subjected to the diffeomorphic transformation defined by the function $f(x)$. In fact, the integrand does not vanish only in the infinitesimal vicinity of the surface defined by the δ-functional. Therefore, it is enough to use only the infinitesimal form of the diffeomorphic transformations and the integral can be explicitly calculated (as in Yang–Mills fields), with the following result for the functional $\Delta_h[g^{\mu\nu}]$:

$$\Delta_h[g^{\mu\nu}] = \det \widehat{B}_h \qquad (3.4.24)$$

where the operator $\widehat{B}_h$ acts on a field $\varepsilon^\mu(x)$ according to the relation:

$$(\widehat{B}_h \epsilon)^\mu = \partial_\nu(h^{\nu\lambda}\partial_\lambda \epsilon^\mu) - \partial_\lambda(\partial_\nu h^{\mu\nu}\epsilon^\lambda). \qquad (3.4.25)$$

After the transition to the corresponding α-gauge with the help of averaging over the field a^μ (cf section 3.2) and after the introduction of the appropriate ghost fields in order to present the determinant of the operator $\widehat{B}_h$ in exponential form, we obtain the path-integral representation for the S-matrix of the gravitational fields in the form

$$S = \frac{1}{N} \int \prod_x \left(g^{5/2}(x) \prod_{\mu \leq \nu} dg^{\mu\nu}(x)\right) \left(\prod_\mu \mathcal{D}c^\mu \mathcal{D}\bar{c}^\mu\right)$$
$$\times \exp\left\{iS[g^{\mu\nu}] + \frac{i\alpha}{4}\int d^4x\, \partial_\rho h^{\mu\rho}\eta_{\mu\nu}\partial_\sigma h^{\nu\sigma} + i\int d^4x\, \bar{c}^\mu (\widehat{B}_h)_{\mu\nu}c^\nu\right\}. \qquad (3.4.26)$$

◇ **Elements of the perturbation theory**

Expression (3.4.26) serves as a starting point for the development of perturbation theory for processes which involve particles mediating gravitational interactions, called *gravitons*. To develop this theory, we should separate the non-flat part of the metric either additively

$$h^{\mu\nu} = \eta^{\mu\nu} + G_N u^{\mu\nu} \qquad (3.4.27)$$

or multiplicatively

$$h^{\mu\nu} = \eta^{\mu\sigma}[\exp\{G_N\Phi\}]_\sigma{}^\nu. \qquad (3.4.28)$$

Here the matrix fields $u^{\mu\nu}$ or $\Phi^{\mu\nu}$ generate gravitons, while the Minkowski metric $\eta^{\mu\nu}$ plays the role of a classical background. The gravitational action now takes the form

$$S_{gr} = S_2 + \sum_{n=1}^{\infty} G_N^n S_{n+2} \qquad (3.4.29)$$

where S_2 is the quadratic form in the field $u^{\mu\nu}$ or $\Phi^{\mu\nu}$ and S_m, $m > 2$, are terms of the order m in $u^{\mu\nu}$, $\Gamma^{\mu\nu}$ and their first derivatives. Now the reader may easily construct the Feynman diagram technique in either of the previously mentioned parametrizations of the gravitational field. We present here only the free propagators for gravitons:

$$\begin{aligned}G^{\mu\nu,\rho\sigma} =\ & \frac{2}{k^2}[\eta^{\mu\rho}\eta^{\nu\sigma} + \eta^{\mu\sigma}\eta^{\nu\rho} + (\alpha^{-1} - 2)\eta^{\mu\nu}\eta^{\rho\sigma}] \\ & + \frac{2(1-\alpha^{-1})}{k^4}[2k^\mu k^\nu \eta^{\rho\sigma} + 2k^\rho k^\sigma \eta^{\mu\nu} - k^\mu k^\rho \eta^{\nu\sigma} - k^\nu k^\rho \eta^{\mu\sigma} - k^\mu k^\sigma \eta^{\nu\rho} - k^\nu k^\sigma \eta^{\mu\rho}]\end{aligned}$$

$$(3.4.30)$$

and, for the ghost fields,

$$G_{gh}^{\mu\nu} = -\frac{\eta^{\mu\nu}}{k^2}. \qquad (3.4.31)$$

It is clear that, because of the higher powers of the fields and their derivatives in action (3.4.29), the corresponding quantum field theory of gravitation is *non-renormalizable*. This is, in fact, the central problem in constructing a self-consistent field theory of quantum gravity and it has induced persistent attempts to construct more general and self-consistent (renormalizable or even finite) theoretical models. As we have mentioned, it is assumed that the previously discussed quantum theory based on Einstein's general relativity, from the point of view of these more general models, plays the role of an *effective* theory with a restricted range of validity (i.e. at *relatively* low energies $E \ll M_P$). Although we cannot say that we already have such a model which provides a complete theory of quantum gravity, there have been remarkable successes on the way mainly related to the development of the *superstring models*. We shall discuss these very briefly later in this section.

Now we pass to the discussion of some more profound (in comparison with the perturbation theory in the asymptotically flat spacetime) problems of the effective low-energy quantum field theory of gravitation.

3.4.2 Path integrals in spatially homogeneous cosmological models

In the preceding subsection, we discussed the perturbation theory for the quantum field theory of the gravitational interaction based on the Einstein action. Although a complete theory of the $3+1$ quantum

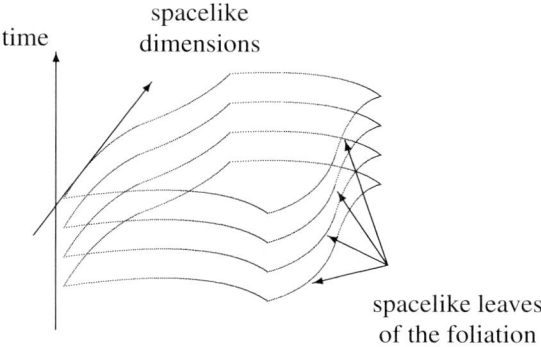

Figure 3.19. A schematic representation of a spacetime admitting a foliation by spacelike leaves.

gravity is not yet available, many interesting *non-perturbative* results have also been obtained. A common tool used to achieve them is the path-integral representation of a quantum gravity transition amplitude and the semiclassical approximation. Quantum general relativity has a number of peculiar features which are not encountered in the quantum theories of non-gravitational interactions: for example, the absence of a background geometry and suitable symmetries to single out the vacuum and to select the Hermitian scalar product of state vectors. It is therefore natural, as the first steps, to apply the study the simpler, truncated models, both to test its viability and to gain insight into the type of technique that will be needed in the full theory. Two main classes of such models exist:

- Four-dimensional 'solvable' spatially homogeneous cosmologies, i.e. cosmologies which admit additional symmetries. In the classical theory, the presence of these symmetries enables us to integrate the field equations completely. We shall see that their presence also simplifies the task of quantization.
- Models based on $(2+1)$-dimensional general relativity (two spatial- and one time-dimension spaces); these models serve to clarify several interesting and important points, both conceptual and technical.

In this subsection, we shall briefly present some applications of the path-integral techniques for the quantization of models with homogeneous cosmologies. In the two subsequent subsections, we shall discuss quantum processes in the $(2 + 1)$-dimensional general relativity with alternating spacetime topologies and the path-integral derivation of the basic quantities in the black hole physics.

◇ **Homogeneous cosmologies: minisuperspace models**

A spacetime is said to be *spatially homogeneous* if it admits a foliation by spacelike submanifolds (see figure 3.19) such that the isometry group of the four-metric acts on each leaf transitively. If the isometry group admits a (not necessarily proper) subgroup which acts simply transitively on each leaf, the spacetime is said to be of *Bianchi* type. In this case, we focus on this subgroup and further classify spacetimes using the properties of the corresponding Lie algebras. If the trace $\sum_a f^a{}_{ba}$ of structure constants $f^a{}_{bc}$ of the Lie algebra vanishes, the spacetime belongs to *Bianchi class A*, while if the trace does not vanish, it belongs to *Bianchi class B*.

156 *Quantum field theory: the path-integral approach*

A spatially homogeneous four-metric is said to be *diagonal* if it can be written in the form:

$$ds^2 = -N^2(t)\,dt^2 + \sum_{i=1}^{3} g_{ii}(t)(\omega^i)^2 \qquad (3.4.32)$$

where $N(t)$ is the *lapse function* and ω^i is a basis of spatial one-forms which are left invariant by the action of the isometry group. We can always change the time coordinate t to the proper time, $t \to t'$: $dt' = N(t)\,dt$, so that the coefficient of the first term becomes simply -1. The diagonal metric is then characterized by the three components $g_{ii}(t)$ which are only functions of time. A key issue, however, is whether the diagonal form of the metric is compatible with the classical field equations. This is the case for models for which the vector (or the diffeomorphism) constraints (3.4.14) are identically satisfied and only the scalar (or the Hamiltonian) constraint (3.4.13) remains to be imposed. We shall restrict ourselves to this class of models which belongs to Bianchi class A, since they admit a Hamiltonian formulation, which is the starting point for canonical quantization. Since the trace of the structure constants f^a_{bc} vanishes for these models, they can be entirely formulated in terms of a symmetric matrix n^{ab}:

$$f^a{}_{bc} = \varepsilon_{dbc} n^{da} \qquad (3.4.33)$$

where ε_{dbc} is the completely anti-symmetric tensor. The signature of n^{ad} can then be used to divide class A models into various types: if n^{ab} vanishes identically, we have Bianchi type I; if it has signature $(0, 0, +)$, we have type II; signature $(+, -, 0)$ corresponds to type VI_0; $(+, +, 0)$ corresponds to VII_0; $(+, +, -)$ to type VIII and $(+, +, +)$ to type IX.

- These types of spacetime manifold are called *minisuperspaces* (actually, the precise definition of minisuperspaces includes important and rather involved refinements; we refer the reader for details to, e.g., Ashtekar (1991) and Ashtekar *et al* (1993)).

A very useful *Misner parametrization* of the diagonal spatial metric exists:

$$g_{ii}(t) = e^{2x^i(t)} \qquad i = 1, 2, 3. \qquad (3.4.34)$$

Here $x^i(t)$ are considered to be arbitrary functions of the time variable, parametrizing the metric (do not confuse them with the spacelike coordinates of the genuine spacetime). We can use the ADM procedure (see preceding subsection) to arrive at the Hamiltonian formulation of minisuperspace models. Since $g_{ii}(t)$ ($i = 1, 2, 3$) and $x^i(t)$ are functions only of time, we actually deal with a quantum-mechanical model with a finite number of degrees of freedom, not with a field theoretical system. The Misner parameters x^i serve as coordinates of the configuration space of this model and take values in the interval $(-\infty, \infty)$, so that the space is topologically trivial. We will denote the momenta conjugate to x^i by p_i. Thus, the fundamental Poisson bracket relations are $\{x^i, p_j\} = \delta^i_j$.

◇ **Quantization of the minisuperspace models**

Thus, we may consider the minisuperspace models as quantum-mechanical systems with Hamiltonians of the form (cf (3.4.13) and (3.4.15))

$$H = g^{ij}(x) p_i p_j + V(x) \qquad i = 1, 2, 3. \qquad (3.4.35)$$

This is the general form of a Hamiltonian for a particle in a curved space, which we have discussed in section 2.5. The essentially new feature of the gravitational minisuperspace models is that expression (3.4.35) must now be considered as a *constraint* (cf (3.4.13)). The fact that a Hamiltonian becomes a constraint (and hence vanishes in the physical sector) is the general feature of gravitational systems which appears due to the reparametrization invariance (i.e. *absence of a unique time variable*). This problem is not applicable for the asymptotically flat spacetimes considered in the preceding subsection, since in the latter case we can use the timelike coordinate in the *flat region* as a proper physical time.

To start the analysis of the minisuperspace models (see, e.g., Marolf (1996)), we first quantize the Poisson brackets for x^i and p_j, while completely ignoring the constraint. This provides an *auxiliary* Hilbert space $\mathcal{H}_{\text{aux}}$. This space is called auxiliary because it contains much more than the *physical states* that satisfy the constraints. In our case, we will take this space to be $\mathcal{H}_{\text{aux}} = L^2(\mathbb{R}^3)$, with the operators $\widehat{x}^i$ (coordinates) and $\widehat{p}_i$ (momenta) acting in the usual way.

The next step in the procedure is to 'quantize' the constraint $H = 0$. For our purposes, this simply means that we choose some self-adjoint operator $\widehat{H}$ on $\mathcal{H}_{\text{aux}}$, which has the function H as its classical limit. The usual ordering ambiguity is present at this level and we make no attempt to give a unique prescription.

Now, if the spectrum of $\widehat{H}$ were entirely discrete, the implementation of the Dirac prescription (Dirac 1964) would be straightforward: those eigenstates of H with zero eigenvalue would become the physical states of our theory and the physical Hilbert space could simply be the $\widehat{H} = 0$ eigenspace of $\mathcal{H}_{\text{aux}}$. However, in typical cases, $\widehat{H}$ also has a continuous spectrum at zero eigenvalue, for which the corresponding eigenstates are not normalizable in the auxiliary Hilbert space but proved to be instead 'generalized eigenstates' of $\widehat{H}$, i.e. distributions. We shall, in fact, assume the spectrum of $\widehat{H}$ to be *entirely* continuous at $\widehat{H} = 0$: many minisuperspace models can be formulated with a constraint having only continuous spectrum at $E = 0$ and we restrict ourselves to this case (e.g., the case of the Bianchi IX model).

In this situation and under a certain technical assumption concerning the operator $\widehat{H}$, the physical Hilbert space is straightforward to construct. What we would really like to do is to project $\mathcal{H}_{\text{aux}}$ onto the (generalized) states which are zero-eigenvalue eigenvectors of H. Of course, since none of these states is normalizable, this is not a projection in the rigorous sense. Instead, it corresponds to an object $\delta(\widehat{H})$, an analog of the Dirac δ-function. Given the previously mentioned assumption on H, the object $\delta(\widehat{H})$ can be shown to exist and to be uniquely defined. It exists not as an operator in the Hilbert space $\mathcal{H}_{\text{aux}}$, but as a map from a dense subspace $\mathcal{S}$ of $\mathcal{H}_{\text{aux}}$ to the space $\mathcal{S}'$ of linear functionals on $\mathcal{S}$ (i.e. to the dual space). The space $\mathcal{S}$ may typically be thought of as a Schwarz space; that is, as the space of smooth rapidly decreasing functions on the configuration space. In this case, $\mathcal{S}'$ is the usual space of tempered distributions.

Then the key idea is the following. Although the generalized eigenstates of $\widehat{H}$ do not lie in $\mathcal{H}_{\text{aux}}$, they can be related to normalizable states through the action of the operator $\delta(\widehat{H})$. That is, the generalized eigenstates $|\psi_{\text{phys}}\rangle$ of $\widehat{H}$ with zero eigenvalue can always be expressed in the form $\delta(\widehat{H})|\psi_0\rangle$, where $|\psi_0\rangle$ is a normalizable state in $\mathcal{S} \subset \mathcal{H}_{\text{aux}}$:

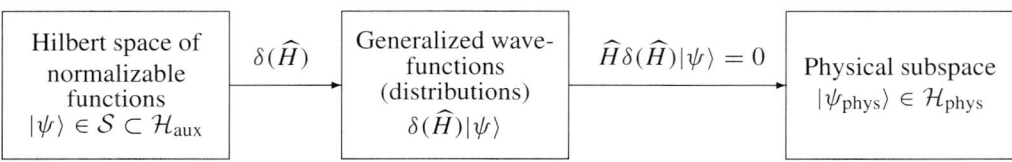

This choice of $|\psi_0\rangle$ is, of course, not unique and, in fact, we associate with a physical state $|\psi_{\text{phys}}\rangle$ the entire *equivalence class* of normalizable states $|\psi\rangle \in \mathcal{S}$, satisfying

$$\delta(\widehat{H})|\psi\rangle = |\psi_{\text{phys}}\rangle. \tag{3.4.36}$$

Each equivalence class of normalizable states will form a *single* state of the physical Hilbert space.

All that is left now is to construct the physical inner product from the auxiliary Hilbert space. Naively, the inner product of two physical states $|\phi_{\text{phys}}\rangle$ and $|\psi_{\text{phys}}\rangle$ may be written as $\langle\phi|\delta(\widehat{H})\delta(\widehat{H})|\psi\rangle$, where $|\phi\rangle$ and $|\psi\rangle$ are normalizable states in the appropriate equivalence classes. This inner product is clearly divergent, as it contains $[\delta(\widehat{H})]^2$. Instead, we define the *physical* inner product to be

$$\langle\phi_{\text{phys}}|\psi_{\text{phys}}\rangle_{\text{phys}} = \langle\phi|\delta(\widehat{H})|\psi\rangle_{\text{aux}} \tag{3.4.37}$$

where the subscripts phys and aux at the brackets indicate the two different inner products. Note that (3.4.37) does not depend on which particular states $|\phi\rangle$, $|\psi\rangle \in \mathcal{S}$ were chosen to represent the physical states $|\phi_{\text{phys}}\rangle$ and $|\psi_{\text{phys}}\rangle$. This construction parallels the case of a purely discrete spectrum: if P_H were a projection onto normalizable zero-eigenvalue eigenstates of $\widehat{H}$, we would have $P_H^2 = P_H$. Note that if $\widehat{H}$ is the Hamiltonian for a free relativistic particle (see below, section 3.4.5), this positive definite inner product corresponds to the Klein–Gordon inner product on the positive-frequency states, but it corresponds to *minus* the Klein–Gordon inner product on the negative-frequency states. The positive- and negative-frequency subspaces are orthogonal as usual.

The algebra of observables commutes with the constraint $\widehat{H}$. These are the analogs of the gauge invariants of classical physics. Each such operator $\widehat{A}$ then *induces* the operator $\widehat{A}_{\text{phys}}$ on $\mathcal{H}_{\text{phys}}$ through

$$\widehat{A}_{\text{phys}}|\psi_{\text{phys}}\rangle \equiv \delta(\widehat{H})\widehat{A}|\psi\rangle \tag{3.4.38}$$

where again $|\psi\rangle$ is any state for which $|\psi_{\text{phys}}\rangle = \delta(\widehat{H})|\psi\rangle$.

◇ What quantity should we derive a path integral for?

As we have learned, path integrals represent transition amplitudes that encode the time evolution of quantum systems. However, for the cases we consider in this subsection, the Hamiltonian explicitly vanishes on the physical Hilbert space. Thus, the operator $e^{-i\widehat{H}t}$ is just the identity. Nevertheless, the physical states contain information that can be called dynamical. Thus, there should be some object, more or less similar to a transition amplitude. It turns out that such an object is just the matrix elements of the operator $\delta(\widehat{H})$ in $\mathcal{H}_{\text{aux}}$ (Marolf 1996). That is, we have to compute $\langle x_f|\delta(H)|x_i\rangle$ where $|x_f\rangle$ and $|x_i\rangle$ are generalized eigenstates of the coordinate operators x^i. Indeed, when one of the coordinates (say x^1) is considered to represent a 'clock' and when this clock behaves semiclassically this object, in a certain sense, describes the amplitude for the 'evolution' of the state $|x_i\rangle$ at the time x_i^1 to the state $|x_f\rangle$ at the time x_f^1. Here x_f and x_i represent the coordinates on the slices through the configuration space of constant values of x^1.

It is now straightforward to represent this object as a path integral. To do so, consider the path-integral expression for the operator $e^{-iN\widehat{H}}$ on $\mathcal{H}_{\text{aux}}$. The new parameter N serves as a formal 'time' variable. We then integrate N from $-\infty$ to ∞ to turn $e^{-iN\widehat{H}}$ into $\delta(\widehat{H})$.

The resulting path integral is then

$$\langle x_f|\delta(\widehat{H})|x_i\rangle = \frac{1}{2\pi}\int_{-\infty}^{\infty} dN \int \mathcal{D}x(t)\,\mathcal{D}p(t)\,\exp\left\{i\int_0^N dt\,[p\dot{x} - H(x(t), p(t))]\right\}$$

$$= \frac{1}{2\pi}\int_{-\infty}^{\infty} dN \int \mathcal{D}x(t)\,\mathcal{D}p(t)\,\exp\left\{i\int_0^1 dt\,[p\dot{x} - N(t)H(x(t), p(t))]\right\}$$

$$= \frac{1}{2\pi} \int \mathcal{D}N(t)\, \mathcal{D}x(t)\, \mathcal{D}p(t) \prod_t \delta(\dot{N}(t)) \Delta[N(t)]$$

$$\times \exp\left\{ i \int_0^1 dt\, [p\dot{x} - n(t)H(x(t), p(t))] \right\} \qquad (3.4.39)$$

where $\int_{-\infty}^{\infty} dN$ denotes an integral over the *single* variable N, while $\mathcal{D}x(t)\mathcal{D}p(t)$ denotes the usual path-integral measure. The second equality is obtained by a simple change of the time variable $t \to Nt$, while the last is obtained by converting the *ordinary* integral over N into the path integral with a simultaneous fixing of all modes of the function $N(t)$, except the constant one, and putting them to zero due to the 'gauge condition' $\dot{N}(t) \equiv \partial N/\partial t = 0$ (represented by the δ-functional in the integrand). The functional $\Delta(G)$ is the associated Faddeev–Popov determinant.

◇ **An example of path-integral representation for the inner product**

Having derived a path integral for $\langle x_f | \delta(\widehat{H}) | x_i \rangle$, it is of interest to see what form this distribution takes in the simple cases where an exact analytic expression can be obtained.

The Bianchi I model is a minisuperspace describing spatially homogeneous spacetimes of the form $\mathcal{M} = T^3 \times \mathbb{R}$ which has a foliation by three-tori with flat Riemannian metrics (so that the tori form spacelike hypersurfaces of $\mathcal{M}$). In the diagonal version of this model, the metric is such that at each spacetime point of $\mathcal{M}$, three mutually orthogonal closed geodesics intersect and each encircle an arm of the torus once. This system may be formulated on the configuration space $\mathcal{Q} = \mathbb{R}^3$ (i.e. on the space of possible metrics (3.4.34) on $\mathcal{M} = T^3 \times \mathbb{R}$), with a constraint of the form

$$H_{\text{BI}} = \tfrac{1}{2}(-p_1^2 + p_2^2 + p_3^2). \qquad (3.4.40)$$

In this case, the coordinate x^1 describes the volume of the three-torus T^3, while the coordinates x^2 and x^3 describe the ratios of the lengths of minimal curves encircling the torus in different directions. It is technically easier to consider a slightly modified model with an additional constant term in the constraint:

$$H = \tfrac{1}{2}(-p_1^2 + p_2^2 + p_3^2 + m^2) \qquad (3.4.41)$$

for $m^2 > 0$. To have an idea about the form of the amplitudes $\langle x_f | \delta(\widehat{H}) | x_i \rangle$ in minisuperspace models, we consider only the case of the Hamiltonian constraint. The Hamiltonian (3.4.40) (or (3.4.41)) is unbounded below. The same is true for the corresponding Euclidean action. This is a general property of quantum general relativity. Since a bounded Euclidean action is required for common arguments involving analytic continuation to Euclidean time, this property has raised the concern about how a path integral for gravity might be defined and analyzed (Gibbons *et al* 1978). The minisuperspace models, being essentially simpler than the complete general relativity, provide a good opportunity for studying this potentially dangerous peculiarity of gravitational systems.

Note that the Hamiltonian (3.4.41) looks exactly like that for the free relativistic particle with mass m (we shall consider a proper relativistic particle in section 3.4.5). However, the *physical* contents relating to the two Hamiltonians are quite different. This follows from the fact that the metric which defines the constraint's 'kinetic term' has a different interpretation in each of the two cases. A free relativistic particle with $p_2 < 0$ is usually interpreted as 'traveling backwards in time', a process which physically corresponds to the creation of an antiparticle. In the Bianchi-like models, a negative p_1 means only that the torus decreases with the proper time, that is, that the universe is *collapsing*.

We now proceed to compute the integral

$$\langle x_f | \delta(\widehat{H}) | x_i \rangle = \frac{1}{2\pi} \int_{-\infty}^{\infty} dN\, \langle x_f | e^{-i\widehat{H}N} | x_i \rangle = \frac{1}{\pi} \operatorname{Re}\left(\int_0^{\infty} dN\, \langle x_f | e^{-i\widehat{H}N} | x_i \rangle \right) \qquad (3.4.42)$$

where Re denotes the real part. The operator $e^{-i\widehat{H}N}$ is just $e^{-im^2N/2}$ times the evolution operator for a free non-relativistic particle (with unit mass). Since we consider matrix elements of the operator with a purely imaginary exponent, the different signs of the 'kinetic terms' in the Hamiltonian (3.4.41) do not lead to any trouble. As a result, its matrix elements are readily seen to be (cf (2.2.59))

$$\langle x_f|e^{-i\widehat{H}N}|x_i\rangle = \frac{1}{(2\pi iN/m)^{3/2}}\exp\left[-\frac{im^2}{2}\left(N - \frac{(x_f-x_i)^2}{m^2N}\right)\right] \qquad (3.4.43)$$

for $N > 0$. The integration over N in (3.4.42) yields for $(x_f - x_i)^2 > 0$ (see Gradshteyn and Ryzhik (1980), formula 3.471)

$$\langle x_f|\delta(H)|x_i\rangle = \frac{2}{\pi(2\pi)^{n/2}}\left[\frac{\sqrt{(x_f-x_i)^2}}{m}\right]^{(1-n/2)} K_{(n/2)-1}\left(m\sqrt{(x_f-x_i)^2}\right) \qquad (3.4.44)$$

where $K_{(n/2)-1}$ is the modified Hankel function of order $(n/2) - 1$. Similarly, for $(x_f - x_i)^2 < 0$, we find

$$\langle x_f|\delta(\widehat{H})|x_i\rangle = -\frac{1}{(2\pi)^{n/2}}\left[\frac{\sqrt{-(x_f-x_i)^2}}{m}\right]^{(1-n/2)} N_{(n/2)-1}\left(m\sqrt{-(x_f-x_i)^2}\right) \qquad (3.4.45)$$

($N_{(n/2)-1}$ is a Bessel function of the second kind). Note that, for $-(x_f-x_i)^2 m^2 \gg 1$, the matrix elements are roughly $\cos(m\sqrt{-(x_f-x_i)^2})$. When $(x_f-x_i)^2 m^2 \gg 1$, the matrix elements contain only the *decreasing* exponential $\exp\{-m\sqrt{(x_f-x_i)^2}\}$. This occurs even though the Euclidean action is unbounded from below.

Recall that the matrix elements (3.4.44) and (3.4.45) describe an evolution (in the sense defined before equation (3.4.39)) of the components $g_{ii}(t)$ of the metric (3.4.32), parametrized with the help of the Misner parameters x_i. This metric is defined on a spacetime without *a priori* assumed asymptotical flatness but which is spatially homogeneous (more precisely, on a minisuperspace).

3.4.3 Path-integral calculation of the topology-change transitions in (2 + 1)-dimensional gravity

The path integral in general relativity is a sum over geometries, and it is natural to ask whether this sum should be extended to include *different* topologies as well. Since realistic four-dimensional quantum gravity is a difficult theory, to study this problem it is natural to look again, as in the preceding subsection, for simpler models that share important features with general relativity. The choice of a simplified model depends on what questions we wish to ask and, as far as the dynamics of spacetime topologies is concerned, a particularly useful model is general relativity in *three* spacetime dimensions. The classical works in this area are by Deser, Jackiw and 't Hooft (Deser *et al* 1984) and Witten (1988) (as a review, see, e.g., Carlip (1995) and further references therein).

The underlying conceptual issues of quantum gravity and some of the technical aspects as well, are identical in 2+1 and 3+1 dimensions. But the elimination of one dimension greatly simplifies the theory, making many computations possible. Moreover, general relativity in 2 + 1 dimensions is renormalizable (it is, in fact, finite), allowing us to avoid the difficult problems of interpreting path integrals in (3 + 1)-dimensional gravity.

◇ **Preliminaries on the (2 + 1)-gravitation theory**

Let us begin by examining the reasons for the simplicity of general relativity in 2 + 1 dimensions. In any spacetime, the curvature tensor may be decomposed into a curvature scalar R, a Ricci tensor $R_{\mu\nu}$, and a remaining trace-free, conformally invariant piece, the Weyl tensor $C_{\mu\nu\rho}{}^\sigma$ (see supplement V). In $2+1$ dimensions, however, the Weyl tensor vanishes identically, and the full curvature tensor is determined algebraically by the curvature scalar and the Ricci tensor:

$$R_{\mu\nu\rho\sigma} = g_{\mu\rho}R_{\nu\sigma} + g_{\nu\sigma}R_{\mu\rho} - g_{\nu\rho}R_{\mu\sigma} - g_{\mu\sigma}R_{\nu\rho} - \tfrac{1}{2}(g_{\mu\rho}g_{\nu\sigma} - g_{\mu\sigma}g_{\nu\rho})R. \tag{3.4.46}$$

In particular, this implies that any solution of the Einstein field equations (without matter fields)

$$R_{\mu\nu} = 0 \tag{3.4.47}$$

is *flat* (i.e. $R_{\mu\nu\rho\sigma} = 0$), and that any solution of the field equations with a cosmological constant,

$$R_{\mu\nu} = 2\Lambda g_{\mu\nu} \tag{3.4.48}$$

has constant curvature. Physically, a $(2+1)$-dimensional spacetime has no local degrees of freedom: there are no gravitational waves in the classical theory, and no gravitons in the quantum theory. The vanishing of the curvature tensor means that any point in a spacetime $\mathcal{M}$ has a neighborhood that is isometric to the Minkowski space. If $\mathcal{M}$ has a trivial topology, a single neighborhood can be extended globally, and the geometry is indeed trivial; but if $\mathcal{M}$ contains non-contractible curves, such an extension may not be possible.

The convenient fundamental variables for a suitable formulation of the $(2+1)$-gravity are now a triad $e_\mu{}^a(x)$ (i.e. components of orthonormal frames) and a spin connection $\omega_\mu{}^a{}_b$. The Einstein–Hilbert action can be written as

$$S_{\text{gr}} = 2\int_{\mathcal{M}} e^a \wedge (d\omega_a + \tfrac{1}{2}\epsilon_{abc}\omega^b \wedge \omega^c) \tag{3.4.49}$$

where $e^a = e_\mu{}^a\,dx^\mu$ and $\omega^a = \tfrac{1}{2}\epsilon^{abc}\omega_{\mu bc}\,dx^\mu$. Technically, it is more convenient to deal with forms and their wedge products (see supplement V) than with their components $e_\mu{}^a$, $\omega_{\mu bc}$. Besides, to simplify formulae, in this subsection we choose such units, that $16\pi G_N = 1$. The action (3.4.49) is invariant under local $SO(2,1)$ transformations (the three-dimensional analog of the Lorentz transformations in the Minkowski spacetime),

$$\begin{aligned}\delta e^a &= \epsilon^{abc}e_b\tau_c \\ \delta\omega^a &= d\tau^a + \epsilon^{abc}\omega_b\tau_c\end{aligned} \tag{3.4.50}$$

as well as 'local translations',

$$\begin{aligned}\delta e^a &= d\sigma^a + \epsilon^{abc}\omega_b\sigma_c \\ \delta\omega^a &= 0.\end{aligned} \tag{3.4.51}$$

Of course, S_{gr} is also invariant under diffeomorphisms of $\mathcal{M}$ but this is not an independent symmetry: it can be shown that when the triad $e_\mu{}^a$ is invertible, diffeomorphisms are equivalent to the transformations (3.4.50)–(3.4.51).

The equations of motion coming from action (3.4.49) are easily derived:

$$T^a[e,\omega] = de^a + \epsilon^{abc}\omega_b \wedge e_c = 0 \tag{3.4.52}$$

and

$$R^a[\omega] = d\omega^a + \tfrac{1}{2}\epsilon^{abc}\omega_b \wedge \omega_c = 0. \tag{3.4.53}$$

The first of these determines ω in terms of e. The second then implies that the connection ω is flat or, equivalently, that the curvature of the metric $g_{\mu\nu} = e_\mu{}^a e_\nu{}^b \eta_{ab}$ vanishes, thus reproducing field equations (3.4.47).

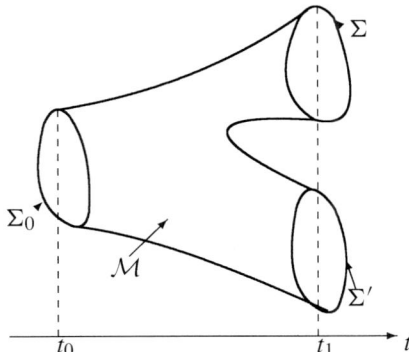

Figure 3.20. The simplest example of a topology-change manifold for gravity in $(1+1)$-dimensional spacetime: a one-dimensional 'universe' Σ_0 is splitting into two disconnected manifolds ('universes') Σ and Σ', thereby changing its topology; the two-dimensional manifold $\mathcal{M}$ corresponds to this topology-change process.

◇ **Path integrals for topology-change transition amplitudes**

We are interested in path integrals of the form

$$K[\omega|_{\partial\mathcal{M}}] = \int \mathcal{D}\omega\, \mathcal{D}e\, \exp\{iS_{\mathrm{gr}}[\mathcal{M}]\} \quad (3.4.54)$$

where $\mathcal{M}$ is a manifold whose boundary

$$\partial\mathcal{M} = \Sigma_1 \cup \Sigma_2 \quad (3.4.55)$$

is the disjoint union of an 'initial' surface Σ_1 and a 'final' surface Σ_2. (Σ_1 and Σ_2 need not be connected surfaces.) In figure 3.20 we illustrate this with the help of the much simpler case of the $(1+1)$-gravity. The reader may find concrete examples of topology-change manifolds in $(2+1)$-dimensional gravity in, e.g., Carlip and Cosgrove (1994). Visually, they are more complicated and we do not present them here.

The quantity $K[\omega|_{\partial\mathcal{M}}]$ represented by path integral (3.4.54) depends on the values of the dynamical variables on the boundary surfaces Σ_1 and Σ_2 and has the meaning of transition amplitude (similarly to the ordinary quantum-mechanical amplitude $K(x, t|x_0, t_0)$).

◇ **Boundary conditions for the connection and triad for topology-change processes**

Since we are dealing with manifolds with a boundary, we must first determine the appropriate boundary conditions. The canonical quantization of the $(2+1)$-gravity on a manifold $\mathbb{R}\times\Sigma$ shows that the states are gauge-invariant functionals $\Psi[\omega_i{}^a]$ of the spatial part of the connection, subject to the constraint that ω be flat on Σ. The corresponding boundary conditions for the path integral therefore require us to fix a flat connection $\omega_i{}^a$ on $\partial\mathcal{M}$. Recall that, if there exists a map from a manifold $\mathcal{X}$ to a manifold $\mathcal{Y}$, $f: \mathcal{X} \to \mathcal{Y}$, then any form Λ on $\mathcal{Y}$ induces the form $f^*\Lambda$ on $\mathcal{X}$. For example, for a one-form $\Lambda = \Lambda_\alpha\, dy^\alpha$, the form $f^*\Lambda$ is $f^*\Lambda = \Lambda_\alpha \frac{\partial f^\alpha}{\partial x^i} dx^i$, where $y^\alpha = f^\alpha(x)$ is the function locally defining the map $\mathcal{X} \to \mathcal{Y}$. Since $\partial\mathcal{M}$ is a boundary of $\mathcal{M}$, there exists an inclusion map $I: \partial\mathcal{M} \to \mathcal{M}$. We can then freely specify the induced connection one-form $I^*\omega$ on $\partial\mathcal{M}$, as long as the induced curvature I^*R vanishes. The $SO(2,1)$ gauge invariance of the resulting amplitude is formally guaranteed by the functional integral over the normal component of ω: at $\partial\mathcal{M}$, $\omega_{\perp a}$ is a Lagrange multiplier for the constraint

$$N^a = \tfrac{1}{2}\epsilon^{ij}(\partial_i e_j{}^a - \partial_j e_i{}^a + \epsilon^{abc}(\omega_{ib}e_{jc} - \omega_{ic}e_{jb})) \quad (3.4.56)$$

that generates the $SO(2, 1)$ transformations of $I^*\omega$. Observe that we must integrate over $\omega_\perp$ at the boundary to enforce this constraint, i.e. we must not fix $\omega_\perp$ as part of the boundary data. This is in accordance with canonical theory, in which the wavefunctionals depend only on the tangential (i.e. spatial) components of ω.

The specification of $I^*\omega$ is not quite sufficient to give us a well-defined path integral. As usual, it is useful to decompose the fields to be integrated, in particular ω, into a classical field $\omega^{(cl)}$ that satisfies the classical field equations, and a fluctuation Ω:

$$\omega = \omega^{(cl)} + \Omega \qquad d\omega^{(cl)a} + \tfrac{1}{2}\epsilon^{abc}\omega^{(cl)}_b \wedge \omega^{(cl)}_c = 0. \qquad (3.4.57)$$

Assuming now that $\omega^{(cl)}$ exists, the boundary condition

$$I^*\Omega = 0$$

(the usual boundary condition for quantum fluctuations) can be recognized as *part* of the standard Dirichlet, or *relative*, boundary conditions for a one-form. In order to impose complete Dirichlet conditions, this should be accompanied by the relation

$$*\bar{D} * \Omega = 0. \qquad (3.4.58)$$

Here, $*$ is the Hodge-star operator with respect to an *auxiliary* Riemannian metric h, which we introduce in order to define a direction normal to the boundary, while $\bar{D}$ is the covariant exterior derivative coupled to the background connection $\omega^{(cl)}$,

$$\bar{D}\beta^a = d\beta^a + \epsilon^{abc}\omega^{(cl)}{}_b \wedge \beta_c.$$

Since (3.4.58) depends on the non-physical metric h, we must check that the final transition amplitudes are independent of h.

In order to impose the boundary conditions for the triad e^a, which are consistent with those for ω^a, we observe that $e_{\perp a}$ acts as a Lagrange multiplier for the constraint

$$\tilde{N}^a = \tfrac{1}{2}\epsilon^{ij}(\partial_i\omega_j{}^a - \partial_j\omega_i{}^a + \epsilon^{abc}\omega_{ib}\omega_{jc}) \qquad (3.4.59)$$

and so, if we integrated over it at the boundaries, this would lead to a delta-functional $\delta[\tilde{N}^a] = \delta[I^*R^a]$ at the boundary. But we have already required that $I^*\omega$ be flat, therefore such a delta-functional would diverge. We avoid this redundancy by fixing $e_\perp$ at $\partial\mathcal{M}$. Eventually, we have to prove that transition amplitudes do not depend on the specific value of $e_\perp$, so this does not contradict the canonical picture (states depend only on ω^a).

As with ω, we can obtain additional boundary conditions by decomposing e into a classical background field and a fluctuation

$$e = e^{(cl)} + E \qquad de^{(cl)a} + \epsilon^{abc}\omega^{(cl)}_b \wedge e^{(cl)}_c = 0 \qquad (3.4.60)$$

where $E_\perp$ vanishes, i.e. $I^*(*E) = 0$. This restriction on E is a part of the standard Neumann, or *absolute*, boundary conditions for a one-form,

$$I^*(*E) = 0 \qquad I^*(*\bar{D}E) = 0. \qquad (3.4.61)$$

⋄ Calculation of the path integral

The next step in constructing the path-integral representation for topology-change amplitudes is to choose gauge conditions to fix the transformations (3.4.50)–(3.4.51). In order to do this, we employ the auxiliary Riemannian metric h introduced earlier and impose the Lorentz gauge conditions

$$*D * E^a = *D * \Omega^a = 0 \qquad (3.4.62)$$

(the Hodge-star operation is defined with respect to the auxiliary metric h). For later convenience, we use the covariant derivative D coupled to the full connection ω rather than $\omega^{(cl)}$ in our gauge-fixing condition. D and $\bar{D}$ agree at the boundary, however, so the gauge condition on Ω reduces to the second equation of (3.4.58) on $\partial \mathcal{M}$.

To impose (3.4.62) in the path integral, it is convenient to introduce a pair of three-form Lagrange multipliers u_a and v_a, and add the term

$$S_{\text{gauge}} = -\int_{\mathcal{M}} (u_a \wedge *D * E^a + v_a \wedge *D * \Omega^a) \qquad (3.4.63)$$

to the action. It is not difficult to see that for the path integral to be well defined, u should obey relative boundary conditions ($I^*(*D * u) = 0$), while v should obey absolute boundary conditions ($*v = 0$ on $\partial \mathcal{M}$). The latter restriction has again a rather straightforward interpretation: since we are already imposing the gauge condition (3.4.58) on Ω at the boundary, we do not need the added delta-functional $\delta[*D * \Omega]$ that would come from integrating over v at $\partial \mathcal{M}$.

As usual, the gauge-fixing process leads to a Faddeev–Popov determinant, which can be incorporated by adding a ghost term

$$S_{\text{gh}} = -\int_{\mathcal{M}} (\bar{f} \wedge *D * Df + \bar{g} \wedge *D * Dg) \qquad (3.4.64)$$

where f, $\bar{f}$, g and $\bar{g}$ are anticommuting ghost fields. We must be careful again about the boundary conditions: corresponding to restrictions (3.4.58) and (3.4.61) on Ω and E, we choose f and $\bar{f}$ to satisfy relative boundary conditions and g and $\bar{g}$ to satisfy the absolute boundary conditions. The full gauge-fixed action is then

$$\begin{aligned} S &= S_{\text{gr}} + S_{\text{gauge}} + S_{\text{gh}} \\ &= \int_{\mathcal{M}} [E^a \wedge (\bar{D}\Omega_a + \tfrac{1}{2}\epsilon_{abc}\Omega^b \wedge \Omega^c + *D * u_a) \\ &\quad + \tfrac{1}{2}\epsilon_{abc} e^{(cl)a} \wedge \Omega^b \wedge \Omega^c - v^a \wedge *D * \Omega_a - \bar{f} \wedge *D * Df - \bar{g} \wedge *D * Dg]. \end{aligned} \qquad (3.4.65)$$

E and v occur linearly in (3.4.65), so we can first integrate over these fields to obtain delta-functionals. There is one subtlety here: certain modes of E do not contribute to the action. These are nothing but the familiar zero modes (see section 3.3.2) and, as usual, they must be treated separately in the integration measure. The integral over the 'non-zero modes' of E will give a delta-functional of $\bar{D}\Omega_a + \tfrac{1}{2}\epsilon_{abc}\Omega^b \wedge \Omega^c + *D * u_a$. The zeros of this expression form a surface $(\tilde{\Omega}(s), \tilde{u}(s))$ in the field space, and if we expand the action around these zeros, only those fields infinitesimally close to this surface should contribute to the path integral. Writing $\Omega = \tilde{\Omega} + \delta\Omega$, we find that the relevant zero modes of E are those $\tilde{E}$ for which

$$D_{\omega^{(cl)}+\tilde{\Omega}} \tilde{E} = 0. \qquad (3.4.66)$$

(Note that $\tilde{E}$ depends on $\tilde{\Omega}$, so the order of integration below cannot be changed.) Performing the integration over E and v, we obtain

$$\int \mathcal{D}\Omega \, \mathcal{D}u \, \mathcal{D}E \, \mathcal{D}v \, e^{iS} = \int \mathcal{D}\Omega \, \mathcal{D}u \, \mathcal{D}\tilde{E} \, \delta[\bar{D}\Omega_a + \tfrac{1}{2}\epsilon_{abc}\Omega^b \wedge \Omega^c + *D * u_a] \delta[*D * \Omega_a]. \qquad (3.4.67)$$

The argument of the first delta-functional vanishes only when $D*D*u_a = 0$; assuming that the connection ω is irreducible, this implies that $u_a = 0$. The delta-functional then imposes the condition

$$\bar{D}\Omega_a + \tfrac{1}{2}\epsilon_{abc}\Omega^b \wedge \Omega^c = 0 \tag{3.4.68}$$

which can be recognized as the requirement that $\omega = \omega^{(\mathrm{cl})} + \Omega$ be a flat connection. This, in turn, allows us to eliminate the term

$$\int_{\mathcal{M}} \tfrac{1}{2}\epsilon_{abc}\, e^{(\mathrm{cl})a} \wedge \Omega^b \wedge \Omega^c = -\int_{\mathrm{M}} e^{(\mathrm{cl})a} \wedge \bar{D}\Omega_a = \int_{\mathrm{M}} \bar{D}e^{(\mathrm{cl})a} \wedge \Omega_a = 0$$

in (3.4.65).

We can now use the delta-functionals to perform the remaining integration over Ω. By a straightforward calculation, we can show that

$$\mathcal{D}\Omega\, \delta[\bar{D}\Omega_a + \tfrac{1}{2}\epsilon_{abc}\Omega^b \wedge \Omega^c + *D*u_a]\delta[*D*\Omega_a] = \mathcal{D}\tilde{\omega}|\det'\tilde{L}^{\mathrm{rel}}_{-}|^{-1} \tag{3.4.69}$$

where $\tilde{\omega} = \omega^{(\mathrm{cl})} + \tilde{\Omega}$ ranges over the flat connections with our specified boundary values and the operation $\tilde{L}^{\mathrm{rel}}_{-} = *D_{\tilde{\omega}} + D_{\tilde{\omega}}*$ maps a one-form plus a three-form (α, β) obeying relative boundary conditions to a one-form plus a three-form $(*D_{\tilde{\omega}}\alpha + D_{\tilde{\omega}}*\beta, D_{\tilde{\omega}}*\alpha)$ obeying absolute boundary conditions. Performing the ghost integrals, we finally obtain

$$K[\omega|_{\partial\mathcal{M}}] = \int [d\tilde{\omega}][d\tilde{E}] \frac{\det'\tilde{\Delta}^{\mathrm{rel}}_{(0)}\det'\tilde{\Delta}^{\mathrm{abs}}_{(0)}}{|\det'\tilde{L}^{\mathrm{rel}}_{-}|} \tag{3.4.70}$$

where $\tilde{\Delta}_{(k)}$ is the Laplacian $*D_{\tilde{\omega}}*D_{\tilde{\omega}} + D_{\tilde{\omega}}*D_{\tilde{\omega}}*$ acting on k-forms and the superscripts 'rel' and 'abs' indicate the function space (with relative or absolute boundary conditions) in which the respective Laplacians act.

Now, by expanding one-forms and three-forms in modes of $L^{\dagger}_{-}L_{-}$, we may prove that

$$|(\det'\tilde{L}_{-})(\det'\tilde{L}^{\dagger}_{-})| = \det'\tilde{\Delta}^{\mathrm{rel}}_{(1)}\det'\tilde{\Delta}^{\mathrm{rel}}_{(3)} = \det'\tilde{\Delta}^{\mathrm{abs}}_{(1)}\det'\tilde{\Delta}^{\mathrm{abs}}_{(3)}. \tag{3.4.71}$$

Moreover, $\det'\tilde{\Delta}^{\mathrm{rel}}_{(k)} = \det'\tilde{\Delta}^{\mathrm{abs}}_{(3-k)}$, since the Hodge-star operator maps any eigenfunction α of $\tilde{\Delta}^{\mathrm{rel}}_{(k)}$ to an eigenfunction $*\alpha$ of $\tilde{\Delta}^{\mathrm{abs}}_{(3-k)}$ with the same eigenvalue. Similar manipulations then show (Carlip and Cosgrove 1994) that

$$K[\omega|_{\partial\mathcal{M}}] = \int \mathcal{D}\tilde{\omega}\, \mathcal{D}\tilde{E}\, T[\tilde{\omega}],$$

$$T[\tilde{\omega}] = \frac{(\det'\tilde{\Delta}^{\mathrm{rel}}_{(3)})^{3/2}(\det'\tilde{\Delta}^{\mathrm{rel}}_{(1)})^{1/2}}{(\det'\tilde{\Delta}^{\mathrm{rel}}_{(2)})} = \frac{(\det'\tilde{\Delta}^{\mathrm{abs}}_{(3)})^{3/2}(\det'\tilde{\Delta}^{\mathrm{abs}}_{(1)})^{1/2}}{(\det'\tilde{\Delta}^{\mathrm{abs}}_{(2)})}. \tag{3.4.72}$$

In principle, integral (3.4.72) determines the transition amplitude for an arbitrary topology change in $2+1$ dimensions. In practice, however, the evaluation of the determinants is a rather complicated problem. Since it is not directly related to a path-integral technique, we refer the reader to the original papers (see Carlip and Cosgrove (1994), Carlip (1995) and references therein). This calculation shows that path integrals representing spatial topology change in $(2+1)$-dimensional general relativity need not vanish, but such topology-changing amplitudes may diverge, thanks to the existence of zero modes $\tilde{E}^a$ of the triad e^a. These divergences presumably reflect the appearance of 'classical' spacetimes in which the distances measured with the metric $g_{\mu\nu} = e_\mu{}^a e_{\nu a}$ become arbitrarily large.

Clearly, no firm conclusions about topology change can be drawn without a much better understanding of the overall normalization of amplitudes in $(2+1)$-dimensional gravity, which would remove these divergences without breaking the symmetries of the original theory. This is a difficult problem which still has to be solved.

3.4.4 Hawking's path-integral derivation of the partition function for black holes

In this subsection we return to the physical four-dimensional spacetime.

A spacetime domain with such a strong gravitational field that even light cannot leave it is called a *black hole* (see, e.g., Misner *et al* (1973), Novikov and Frolov (1989) and Chandrasekhar (1983)). Black holes are justly considered the most exotic objects in the Universe and their study is related to the most fundamental problems of spacetime physics.

The content of this subsection could justly pertain to chapter 4, since we shall discuss mainly the *statistical* and *thermodynamical* properties of black holes, but we prefer to collect together the applications of path-integral techniques in gravitational theory.

John Wheeler was the first to point out that the existence of black holes contradicts the basic thermodynamical law, namely, the increasing entropy law, unless we attribute entropy to black holes themselves. In the early 1970s, Bekenstein argued that black holes indeed have entropy (Bekenstein 1973). The argumentation is based on Stephen Hawking's theorem which states that the square A_H of a black hole does not decrease with any classical processes, i.e. A_H behaves similarly to entropy. Two years later, Hawking showed that black holes have temperature (Hawking 1975) and a few years later these conclusions were confirmed and further developed by Gibbons and Hawking (1977), who used path-integral methods to evaluate the black-hole partition function.

A number of authors have speculated that the large entropy of a black hole should be associated with a large number of internal states, hidden by the *horizon* (boundary of the black hole), that are consistent with the few external parameters (mass, angular momentum and electric charge) that characterize the black hole. Others have argued that the black-hole entropy can be associated with a large number of possible initial states that can collapse to form a given black hole. Path-integral analysis does not support either of these views in an obvious way. Essential progress in understanding the origin of black-hole entropy and its relation with microscopic quantum properties has only been reached very recently in the framework of string–membrane theories and the fruitful idea of fundamental dualities in these models (see, e.g., Strominger and Vafa (1996), Horowitz (1996), Maldacena *et al* (1997) and references therein). We shall not discuss any further this involved and still intensively developing topic but confine ourselves to a short presentation of the path-integral derivation of the black-hole partition function, following Gibbons and Hawking (1977) and Brown and York (1994).

◇ A short tour into black-hole thermodynamics and statistical mechanics

Since we are going to discuss the black-hole partition function before chapter 4 which is devoted to path integrals in statistical physics and since black-hole thermodynamics and statistical mechanics have some unusual peculiarities, we start from a short presentation of the basic facts needed for the subsequent derivation of the partition function.

Hawking's analysis shows that the temperature of a black hole, as measured at spatial infinity, equals the surface gravity divided by 2π. For a Schwarzschild black hole of mass M, it follows that the inverse temperature β at infinity is $8\pi G_N M$ (recall that the Schwarzschild horizon is a sphere of radius $R_g = 2G_N M$). On the other hand, the standard thermodynamical definition of inverse temperature is $\beta = \partial S(E)/\partial E$, where $S(E)$ is the *entropy* function and E is the thermodynamical internal energy. If the mass M at infinity and the internal energy E are identified, then the relationship $\partial S(E)/\partial E = 8\pi G_N M$ can be integrated to yield $S(E) = 4\pi G_N^2 E^2$ (plus an additive constant). This result is in complete agreement with the prediction made by Bekenstein that a black hole has an entropy proportional to the area of its event horizon.

The black-hole entropy $S(E) = 4\pi G_N^2 E^2$ is a convex function of E. This is characteristic for an unstable thermodynamical system: the instability arises because energy and temperature are inversely related for black holes. Thus, if fluctuations cause a black hole to absorb an extra amount of thermal radiation from its environment, its mass will increase and its temperature will decrease. The tendency then is for a cooler black hole to absorb even more radiation from its hotter environment, causing the black hole to grow without bound.

These results can be reformulated within the context of statistical mechanics. First, consider the canonical partition function $\mathcal{Z}_\beta$ for an arbitrary system. In general, $\mathcal{Z}_\beta$ is a sum over quantum states weighted by the Boltzmann factor $e^{-\beta E}$. If $\nu(E)$ is the density of quantum states with energy E, then

$$\mathcal{Z}_\beta = \int dE\, \nu(E) e^{-\beta E} \qquad (3.4.73)$$

(see also section 4.1). The partition function can also be expressed as

$$\mathcal{Z}_\beta = \int dE\, e^{-I(E)} \qquad (3.4.74)$$

where the 'action' is defined by $I(E) \equiv \beta E - \mathcal{S}(E)$ and the entropy function $\mathcal{S}(E)$ is the logarithm of the density of states: $\mathcal{S}(E) \equiv \ln \nu(E)$. As usual, the integral over E can be evaluated in the stationary-phase approximation by expanding the action $I(E)$ to quadratic order around the stationary points $E_c(\beta)$, which satisfy

$$0 = \left.\frac{\partial I}{\partial E}\right|_{E_c} = \beta - \left.\frac{\partial \mathcal{S}}{\partial E}\right|_{E_c}. \qquad (3.4.75)$$

The Gaussian integral associated with a stationary point E_c will converge if the second derivative of the action at E_c is positive:

$$\left.\frac{\partial^2 I}{\partial E^2}\right|_{E_c} = -\left.\frac{\partial^2 \mathcal{S}}{\partial E^2}\right|_{E_c} > 0. \qquad (3.4.76)$$

This condition shows that the entropy $\mathcal{S}(E)$ should be a concave function at the extremum E_c, in order for the Gaussian integral to converge.

A further significance of condition (3.4.76) can be seen as follows. In the stationary-phase approximation, the expectation value of the energy is $\langle E \rangle \equiv -\partial \ln \mathcal{Z}_\beta/\partial \beta \approx E_c$ and the *heat capacity* is $C \equiv \partial \langle E \rangle/\partial \beta^{-1} \approx \partial E_c/\partial \beta^{-1}$. By differentiating (3.4.75) with respect to β, we find

$$1 = \frac{\partial E}{\partial \beta} \left.\frac{\partial^2 \mathcal{S}}{\partial E^2}\right|_{E_c}. \qquad (3.4.77)$$

Therefore, the heat capacity is given by

$$C \approx -\beta^2 \left(\left.\frac{\partial^2 \mathcal{S}}{\partial E^2}\right|_{E_c}\right)^{-1} = \beta^2 \left(\left.\frac{\partial^2 I}{\partial E^2}\right|_{E_c}\right)^{-1}. \qquad (3.4.78)$$

Thus, we see that in the stationary-phase approximation, the convergence of the integral for the canonical partition function is equivalent to the thermodynamical stability of the system (the concavity of the entropy), which in turn is equivalent to the positivity of the heat capacity.

For a black hole in particular, the entropy $\mathcal{S}(E) = 4\pi G_N^2 E^2$ is not a concave function of the internal energy and the integral for the partition function diverges.

The canonical partition function $\mathcal{Z}_\beta$ characterizes the thermal properties of thermodynamically *stable* systems. For unstable systems, $\mathcal{Z}_\beta$ can give information concerning the rate of decay from a quasi-stable configuration (such as a 'hot flat space' in the black-hole example), but it cannot be used to define thermodynamical properties such as expectation values, fluctuations, response functions, etc. Thus, before the partition function can be used as a probe of black-hole thermodynamics, it is first necessary to stabilize the black hole. It was recognized (see Brown and York (1994) and references therein) that a black hole is rendered thermodynamically stable by enclosing it in a spatially finite 'box' or boundary, with walls maintained at a finite temperature. In this case, the energy and the temperature at the boundary are not inversely related because of the *blueshift effect* for the temperature in a stationary gravitational field.

The stabilizing effect of a finite box can be explained as follows. Consider a Schwarzschild black hole of mass M surrounded by a spherical boundary of radius R (do not confuse the boundary radius R with the Schwarzschild radius $R_g = 2G_N M$). The inverse temperature at infinity is $8\pi G_N M$, while the inverse temperature at the boundary is blueshifted to $\beta = 8\pi G_N M \sqrt{1 - (2G_N M)/R}$. On the other hand, the inverse temperature is defined by $\beta = \partial S(E)/\partial E$, where again $S(E)$ is the entropy as a function of the internal energy E. The entropy of the black hole depends only on the black-hole size and is unaffected by the presence of a finite box. Thus, the entropy is given by $S(E) = 4\pi G_N^2 M^2$ as before, so that equating the two expressions for inverse temperature, we find that

$$8\pi M \sqrt{1 - (2G_N M)/R} = \frac{\partial (4\pi G_N^2 M^2)}{\partial E}. \qquad (3.4.79)$$

In this case, the energy E and mass M as measured at infinity do not coincide. Equation (3.4.79) can be integrated to yield

$$E = G_N^{-1} R \left(1 - \sqrt{1 - (2G_N M)/R}\right) \qquad (3.4.80)$$

where, for convenience, the integration constant has been chosen so that $E \to M$ in the limit $R \to \infty$ with M fixed. The significance of this expression can be seen by expanding E in powers of $G_N M/R$ with the result $E = M + M^2/(2R) + \cdots$. This shows that the internal energy inside the box equals the energy at infinity M *minus* the binding energy $-G_N M^2/(2R)$ of a shell of mass M and radius R. The binding energy $-G_N M^2/(2R)$ is the energy associated with the gravitational field outside the box. Observe also that the internal energy takes values in the range $0 \le E \le R/G_N$.

By solving (3.4.80) for M as a function of E, we obtain the entropy function

$$S(E) = 4\pi G_N^2 E^2 (1 - G_N E/(2R))^2. \qquad (3.4.81)$$

Note that the derivative $\partial S/\partial E$ is a concave function of E (schematically depicted in figure 3.21) that vanishes at the extreme values $E = 0$ and $E = R/G_N$. It follows that $\partial S/\partial E$ has a maximum β_{cr}. For $\beta > \beta_{\text{cr}}$, the equation $\beta = \partial S/\partial E$ has no solutions for E. On the other hand, for $\beta < \beta_{\text{cr}}$, there are two solutions, E_1 and E_2. At the larger of these two solutions, say, E_2, the second derivative $\partial^2 S/\partial E^2$ is negative and the stability criterion (3.4.76) is satisfied. At the smaller of these two solutions, E_1, the second derivative $\partial^2 S/\partial E^2$ is positive and the stability criterion (3.4.76) is violated. These considerations indicate that for a small box at low temperature ($\beta > \beta_{\text{cr}}$), the equilibrium configuration consists of a flat space. For a large box at high temperature ($\beta < \beta_{\text{cr}}$), the stable equilibrium configuration consists of a large black hole with the energy E_2. The unstable black hole with the energy E_1 is an *instanton* that governs the nucleation of black holes from flat space. In the limit $R \to \infty$, the stable black-hole configuration is lost and only the instanton solution survives.

◇ **Gravitational action with boundary terms**

Since the black-hole stability condition requires the restriction of the spacetime by a finite box, we have to take care about the appropriate boundary conditions and complete the gravitational action with boundary terms.

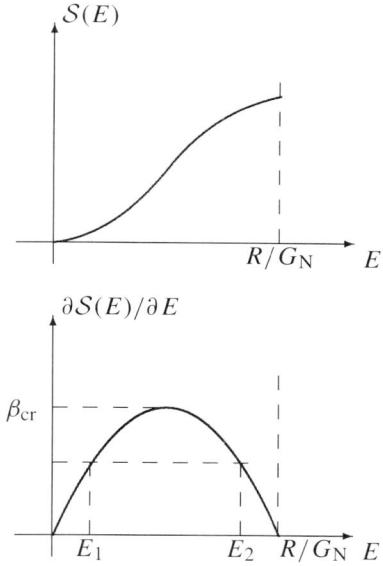

Figure 3.21. Plots of the black-hole entropy $\mathcal{S}(E)$ and its derivative $\partial \mathcal{S}(E)/\partial E$, in the presence of a finite box of radius R.

Assume that the spacetime manifold $\mathcal{M}$ is topologically the product of a spacelike hypersurface and a real (time) line interval, $\Sigma \times I$. The boundary of Σ (the finite 'box') is denoted $\partial \Sigma \equiv B$; $I = [t', t''] \in \mathbb{R}$. The spacetime metric is $g_{\mu\nu}$, with the associated curvature tensor $R_{\mu\nu\sigma\rho}$ and derivative operator ∇_μ. The boundary $\partial \mathcal{M}$ of $\mathcal{M}$ consists of the initial and final spacelike hypersurfaces Σ' and Σ'' (cf figure 3.22) at t' and t'', respectively, and a timelike hypersurface $^3B = B \times I$ joining them. The induced metric on the spacelike hypersurfaces at t' and t'' is denoted by h_{ij}, and the induced metric on ^3B is denoted by γ_{ij}.

Consider the gravitational action

$$S = \frac{1}{2\kappa} \int_\mathcal{M} d^4x \sqrt{-g}(\mathcal{R} - 2\Lambda) + \frac{1}{\kappa} \int_{t'}^{t''} d^3x \sqrt{h} K - \frac{1}{\kappa} \int_{^3B} d^3x \sqrt{-\gamma} \Theta. \qquad (3.4.82)$$

Here, $\kappa = 8\pi G_N$ and Λ is the cosmological constant. The symbol $\int_{t'}^{t''} d^3x$ denotes an integral over the boundary Σ'' minus an integral over the boundary surface Σ'. The function K is the trace of the extrinsic curvature K_{ij} (see supplement V) for the boundary surfaces Σ' and Σ'', defined with respect to the future pointing unit normal u^μ. Likewise, Θ is the trace of the extrinsic curvature Θ_{ij} of the boundary element ^3B, defined with respect to the outward pointing unit normal n^μ.

Under variations of the metric, the action (3.4.82) varies according to

$$\delta S = \text{(terms that vanish when the equations of motion hold)}$$
$$+ \int_{t'}^{t''} d^3x \, P^{ij} \delta h_{ij} + \int_{^3B} d^3x \, \pi^{ij} \delta \gamma_{ij} - \frac{1}{\kappa} \int_{B'}^{B''} d^2x \sqrt{\sigma} \delta \alpha. \qquad (3.4.83)$$

The coefficient of δh_{ij} in the boundary terms at t' and t'' is the gravitational momentum

$$P^{ij} = \frac{1}{2\kappa} \sqrt{h}(K h^{ij} - K^{ij}). \qquad (3.4.84)$$

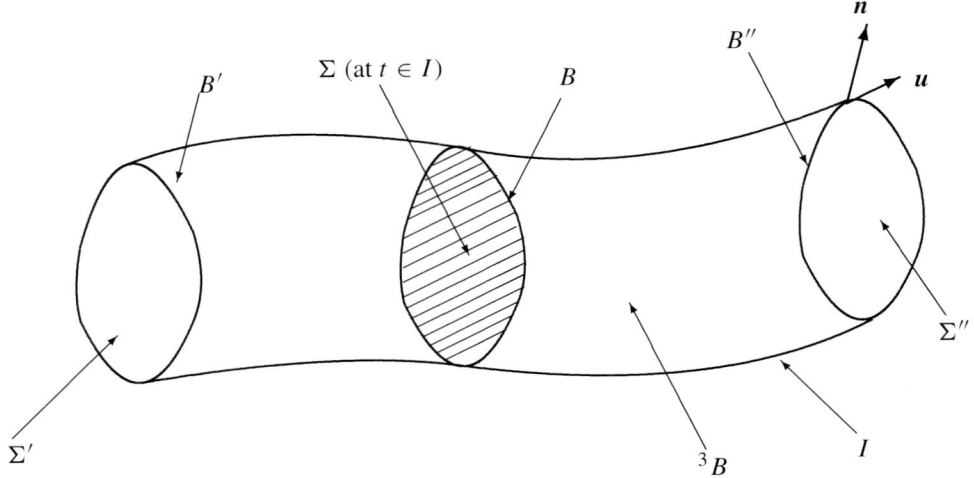

Figure 3.22. Schematic presentation of the spacetime manifold $\mathcal{M}$ and its boundary $\partial\mathcal{M}$ consisting of the initial Σ' and final Σ'' spacelike hypersurfaces and the timelike hypersurface $^3B = B \times I$ joining them. The unit normal $\boldsymbol{n}$ to ^3B and the normal $\boldsymbol{u}$ to Σ'' at the boundary B'' are also depicted.

Likewise, the coefficient of $\delta\gamma_{ij}$ in the boundary term at ^3B is

$$\pi^{ij} = -\frac{1}{2\kappa}\sqrt{-\gamma}(\Theta\gamma^{ij} - \Theta^{ij}). \tag{3.4.85}$$

Equation (3.4.83) also includes integrals over the 'corners' $B'' = $ (hypersurface Σ'' at $t'') \cap {}^3B$ and $B' = $ (hypersurface at $t') \cap {}^3B$, whose integrands are proportional to the variation of the 'angle' $\alpha = \sinh^{-1}(u \cdot n)$ between the unit normals u^μ of the hypersurfaces at t'' and t' and the unit normal n^μ of ^3B (see figure 3.22). The determinant of the two-metric on B' or B'' is denoted by σ.

Let us foliate the boundary element ^3B into two-dimensional surfaces B with induced two-metrics σ_{ab}. The three-metric γ_{ij} can be written according to the so-called *Arnowitt–Deser–Misner decomposition* as

$$\gamma_{ij}\,dx^i\,dx^j = -N^2\,dt^2 + \sigma_{ab}(dx^a + V^a\,dt)(dx^b + V^b\,dt) \tag{3.4.86}$$

where N is the lapse function and V^a is the shift vector.

◇ The action and related quantities in Lorentzian and Euclidean spaces

In the preceding sections and chapters we used both Lorentz (i.e. real time) and Euclidean (imaginary time) forms of the path integrals, the latter being obtained by analytical continuation in the time variable. In the case of a (gravitational) theory with a *dynamical* metric, the path integral, in general, includes integration over the variety of all metrics with different signatures and strictly speaking, there is no distinction between the 'Lorentzian action' and the 'Euclidean action', or between the 'Lorentzian equations of motion' and the 'Euclidean equations of motion'. Of course, a particular *solution* of the classical equations of motion can be Lorentzian or Euclidean. But for the action functional itself, the only distinction between Lorentzian and Euclidean is simply one of *notation*. We have already used what might be called Lorentzian notation: the action S is defined with the convention that $\exp\{iS\}$ is the phase in the path integral; the volume elements for $\mathcal{M}$ and ^3B are written as $\sqrt{-g}$ and $\sqrt{-\gamma}$, respectively; the lapse function associated with the foliation of $\mathcal{M}$ into hypersurfaces Σ is defined by $N \equiv \sqrt{-1/g^{tt}}$. (The lapse function that appears in (3.4.86) is the restriction of this spacetime lapse to the boundary element ^3B. It is defined by $N \equiv \sqrt{-1/\gamma^{tt}}$.) Therefore S, $\sqrt{-g}$, $\sqrt{-\gamma}$ and N are real for Lorentzian metrics and imaginary for Euclidean metrics.

We can re-express the action in Euclidean *notation* by making the following changes. Define a new action functional by $S_E[g] \equiv -iS[g]$ so that the phase in the path integral is given by $\exp\{-S_E\}$. We also rewrite the volume elements for $\mathcal{M}$ and ^3B as $\sqrt{g} \equiv i\sqrt{-g}$ and $\sqrt{\gamma} \equiv i\sqrt{-\gamma}$, respectively, and define a new lapse function by $\bar{N} \equiv \sqrt{1/g^{tt}} \equiv i\sqrt{-1/g^{tt}} \equiv iN$. A bit of care is required in defining the square roots. For example, the appropriate definition of $\sqrt{-g}$ is obtained by taking the branch cut in the upper half complex plane, say, along the positive imaginary axis. Then the imaginary part of $\sqrt{-g}$ is negative. Correspondingly, the appropriate definition of $\sqrt{g}$ is obtained by taking the branch cut along the negative imaginary axis. Then the imaginary part of $\sqrt{g}$ is positive.

It is also convenient to redefine the timelike unit normal of the slices Σ. In Lorentzian notation, the unit normal is defined by $u_\mu \equiv -N\delta_\mu^t$ and satisfies $u \cdot u = -1$. A new unit normal is defined by $\bar{u}_\mu \equiv \bar{N}\delta_\mu^t \equiv iN\delta_\mu^t \equiv -iu_\mu$ and satisfies $\bar{u} \cdot \bar{u} = +1$. In some contexts, it is also useful to define a new extrinsic curvature $\bar{K}_{\mu\nu}$ in terms of the normal $\bar{u}_\mu$. $\bar{K}_{\mu\nu}$ is related to the old extrinsic curvature $K_{\mu\nu}$ by $\bar{K}_{\mu\nu} \equiv -(\delta_\mu^\sigma - \bar{u}^\sigma \bar{u}_\mu)\nabla_\sigma \bar{u}_\nu \equiv i(\delta_\mu^\sigma + u^\sigma u_\mu)\nabla_\sigma u_\nu \equiv -iK_{\mu\nu}$. In turn, $\bar{K}_{\mu\nu}$ can be used to define a new gravitational momentum $\bar{P}^{ij}$ that is related to the momentum of (3.4.84) by $\bar{P}^{ij} \equiv -iP^{ij}$. We will, however, continue to use the old notation K_{ij} and P^{ij}.

In addition to the notational changes described here, we will also define a new shift vector by $\bar{V}_i \equiv ig_{ti} \equiv iV_i$. This notation is different from the standard Euclidean notation in the sense that $\bar{V}_i$ *is imaginary for Euclidean metrics*. One of the motivations for this change is the following. Apart from surface terms, the gravitational Hamiltonian is a linear combination of constraints built from the gravitational canonical data with the lapse function and shift vector as coefficients. In conjunction with the new notation $\bar{N}$, $\bar{V}_i$ for the lapse and shift, we choose to continue to denote the gravitational canonical data by h_{ij}, P^{ij}, as previously mentioned. Then the constraints are unaffected by the change in notation, and the Hamiltonian can be written as $H[N, V] \equiv -iH[\bar{N}, \bar{V}]$. The overall factor $(-i)$ that appears in this relationship is precisely what is required for the connection between the evolution operator ($e^{-i\hat{H}t}$ in particle mechanics) and the density operator ($e^{-\hat{H}\beta}$ in ordinary statistical mechanics). When the gravitational field is coupled to other gauge fields, such as the Yang–Mills or electromagnetic ones, it is natural to redefine the Lagrange multipliers associated with the gauge constraints as well.

With our new notation, (3.4.82) becomes

$$S_E = -\frac{1}{2\kappa} \int_\mathcal{M} d^4x \sqrt{g}(\mathcal{R} - 2\Lambda) - \frac{i}{\kappa} \int_{t'}^{t''} d^3x \sqrt{h} K + \frac{1}{\kappa} \int_{^3B} d^3x \sqrt{\gamma}\Theta \qquad (3.4.87)$$

and (3.4.83) becomes

$$\delta S_E = \text{(terms that vanish when the equations of motion hold)}$$
$$- i\int_{t'}^{t''} d^3x \, P^{ij} \delta h_{ij} - i\int_{^3B} d^3x \, \pi^{ij} \delta\gamma_{ij} + \frac{1}{\kappa} \int_{B'}^{B''} d^2x \sqrt{\sigma}\delta\bar{\alpha}. \qquad (3.4.88)$$

Here, we have defined $\bar{\alpha} \equiv \cos^{-1}(\bar{u} \cdot n)$ so that $\delta\bar{\alpha} \equiv i\delta\alpha$. Thus, $\bar{\alpha}$ is the angle between the unit normals $\bar{u}$ and n of the boundary elements t'' (or t') and ^3B.

172   *Quantum field theory: the path-integral approach*

◇ **Functional integral**

A path integral constructed from an action S_E is a functional of the quantities that are held fixed in the variational principle $\delta S_E = 0$. What are held fixed in the variational principle are the quantities that appear to be varied in the boundary terms of δS_E. The fixed boundary data for the action (3.4.87) are the metric h_{ij} on the surfaces Σ', Σ'' at t' and t'', the angle $\bar{\alpha}$ at the corners B' and B'', and the lapse function $\bar{N}$, the shift vector $\bar{V}^a$ and the two-metric σ_{ab} on ^3B. In the path integral, the gauge-invariant part of the data on ^3B corresponds to the inverse temperature β, the chemical potential ω^a and the two-geometry of the boundary B. A detailed analysis (Brown and York 1994) shows that these boundary data correspond to the thermodynamical description of the corresponding black hole as a grand canonical ensemble:

- The inverse temperature is defined in terms of the boundary data on ^3B by

$$\beta = \int dt\, \bar{N}|_B. \qquad (3.4.89)$$

In geometrical terms, this is the proper distance between t' and t'' as measured along the curves in ^3B that are orthogonal to the slices B.
- The chemical potential is defined in terms of the boundary data on ^3B by

$$\omega^a = \frac{\int dt\, \bar{V}^a|_B}{\int dt\, \bar{N}|_B} = \frac{\int dt\, V^a|_B}{\int dt\, N|_B}. \qquad (3.4.90)$$

The physical meaning of the 'chemical potential' ω^a is the proper velocity of the physical system as measured with respect to observers who are at rest at the system boundary B.

The path integral constructed from the action S_E is

$$\rho[h'', h'; \bar{\alpha}'', \bar{\alpha}'; \beta, \omega, \sigma] = \int \mathcal{D}g\, e^{-S_E[g]} \qquad (3.4.91)$$

where h'' and h' denote the metrics on Σ'' and Σ', while $\bar{\alpha}''$ and $\bar{\alpha}'$ denote the angles at the corners B'' and B'. This path integral is the grand canonical density matrix for the gravitational field in a box B. The grand canonical partition function, denoted $\Xi[\beta, \omega, \sigma]$, is obtained by tracing over the initial and final configurations. In path-integral language, this amounts to performing a periodic identification, so that the manifold topology becomes $\mathcal{M} = \Sigma \times S^1$. In addition, $\bar{\alpha}''$ and $\bar{\alpha}'$ should be chosen so that the total angle $\bar{\alpha}'' + \bar{\alpha}'$ equals π. This insures that the boundary $\partial \mathcal{M}$ is smooth when the initial and final hypersurfaces are joined together. Thus, the grand canonical partition function can be written as

$$\Xi[\beta, \omega, \sigma] = \int Dh\, \rho[h, h; \bar{\alpha}'', \bar{\alpha}'; \beta, \omega, \sigma]\bigg|_{\bar{\alpha}'' + \bar{\alpha}' = \pi}. \qquad (3.4.92)$$

The right-hand side of this expression apparently depends on the angle difference $\bar{\alpha}'' - \bar{\alpha}'$. However, we expect that with periodic identification, $\bar{\alpha}'' - \bar{\alpha}'$ is a pure gauge and in a more detailed analysis would be absent from the path integral.

One can consider various density matrices and partition functions corresponding to different combinations of thermodynamical variables, where one variable is selected from each of the conjugate pairs. For example, in ordinary statistical mechanics the thermodynamically conjugate pairs might consist of the inverse temperature and energy $\{\beta, E\}$, and the chemical potential and particle number $\{\mu, N\}$. Then the grand canonical partition function is $\Xi(\beta, \mu)$, the canonical partition function is $\mathcal{Z}(\beta, N)$ and the microcanonical partition function (the density of states) is $\nu(E, N)$. These partition functions are related to each other by Laplace and inverse Laplace transforms, where each transform has the effect of switching the functional dependence from some thermodynamical variable (such as β) to its conjugate (such as E).

When the gravitational field is included in the description of the system, all of the thermodynamical data can be expressed as boundary data. In the path-integral formalism, the effect of the Laplace and inverse Laplace transforms is simply to add or subtract certain boundary terms from the action.

◇ Black-hole entropy

So far the partition function $\Xi[\beta, \omega, \sigma]$ and the density of states $\nu[\varepsilon, j, \sigma]$ were constructed as functional integrals over the gravitational field on manifolds whose topologies are necessarily $\Sigma \times S^1$. This would seem to be an unavoidable consequence of deriving $\Xi[\beta, \omega, \sigma]$ and $\nu[\varepsilon, j, \sigma]$ from traces of density matrices because the density matrices ρ are defined in terms of functional integrals on manifolds $\mathcal{M}$, with the product topology $\Sigma \times I$. However, experience has shown that for a black hole, the functional integrals for the partition function and density of states are extremized by a metric on the manifold $R^2 \times S^2$ (S^2 corresponds to the topology of the black-hole horizon). Thus, we would expect the black-hole contribution to the density of states to come from a path integral that is defined on a manifold with the topology $R^2 \times S^2$. Let us discuss how the black-hole density of states can be related to the microcanonical density matrix.

We begin by considering the manifold $\mathcal{M} = \Sigma \times I$, where Σ is topologically a thick spherical shell ($S^2 \times I$). The boundary $\partial \Sigma = B$ consists of two disconnected surfaces, an inner sphere B_i and an outer sphere B_o. The boundary element ^3B consists of disconnected surfaces as well, $^3B_i = B_i \times I$ and $^3B_o = B_o \times I$. The results of the previous sections can be applied in constructing various density matrices for the gravitational field on Σ. We wish to consider the particular density matrix ρ_* that is defined through the path integral with the action

$$S_E^{(BH)} = S_E - \int_{^3B_o} d^3x \sqrt{\sigma}(\bar{N}\varepsilon - \bar{V}^a j_a) + \int_{^3B_i} d^3x \sqrt{\sigma} \bar{N} s_a^a/2 \qquad (3.4.93)$$

where ε is identified as an energy surface density for the system. Likewise, we identify j_i as the momentum surface density and s^{ab} as the spatial stress. $S_E^{(BH)}$ differs from the action S_E by boundary terms which are not the same for the two disconnected parts of ^3B. The contributions to the variation $\delta S_E^{(BH)}$ from ^3B_i and ^3B_o are

$$\delta S_E^{(BH)}|_{^3B_i} = \int_{^3B_i} d^3x \, ((\sqrt{\sigma}\varepsilon)\delta\bar{N} - (\sqrt{\sigma} j_a)\delta\bar{V}^a + \sigma_{ab}\delta(\bar{N}\sqrt{\sigma}s^{ab}/2)) \qquad (3.4.94)$$

$$\delta S_E^{(BH)}|_{^3B_o} = \int_{^3B_o} d^3x \, (-\bar{N}\delta(\sqrt{\sigma}\varepsilon) + \bar{V}^a \delta(\sqrt{\sigma} j_a) - (\bar{N}\sqrt{\sigma}s^{ab}/2)\delta\sigma_{ab}). \qquad (3.4.95)$$

The choice of the boundary terms at the outer boundary element ^3B_o corresponds to the microcanonical boundary conditions. At the inner boundary element ^3B_i none of the traditional 'conserved' quantities like energy, angular momentum or area is fixed. Thus, they are allowed to fluctuate on the inner boundary element while their conjugates, the inverse temperature, chemical potential and spatial stress are held fixed.

We can show (Brown and York 1994) that the black-hole density of states $\nu_*[\varepsilon, j, \sigma]$ is obtained from the trace of the density matrix $\rho_* = \int \mathcal{D}g \, \exp\{-S_E^{(BH)}[g]\}$ along with the following special choice of data on the inner boundary element ^3B_i:

$$\bar{N} = 0 \qquad (3.4.96)$$

$$\bar{V}^a = 0 \tag{3.4.97}$$

$$\bar{N}\sqrt{\sigma}\eta^{ab} = 0 \tag{3.4.98}$$

$$\bar{N}\theta = -\frac{4\pi}{\kappa(t''-t')}. \tag{3.4.99}$$

From the geometrical point of view, this follows from the fact that conditions (3.4.96)–(3.4.99) effectively transform the topology $\Sigma \times S^1$ into the required one, namely, $R^2 \times S^2$. In (3.4.99), $(t''-t')$ is just the range of the coordinate time t.

The correctness of this prescription can be confirmed by considering the evaluation of the functional integral for $\nu_*[\varepsilon, j, \sigma]$, where the data ε, j_a, σ_{ab} on the outer boundary $^3B_{\rm o}$ correspond to a stationary black hole. That is, let ε, j_a, σ_{ab} be the stress–energy–momentum for a topologically spherical two-surface $B_{\rm o}$ within a time slice of a stationary Lorentzian black-hole solution $g_{\rm L}$ of the Einstein equations. In the path integral for $\nu_*[\varepsilon, j, \sigma]$, let us fix these data on each slice $B_{\rm o}$ of the outer boundary $^3B_{\rm o}$. The path integral can be evaluated semiclassically by searching for metrics that extremize the action $S_E^{\rm (BH)}$ and satisfy the conditions at both $^3B_{\rm o}$ and $^3B_{\rm i}$. One such metric will be the complex metric g_C that is obtained by substituting $t \to -it$ in the Lorentzian black-hole solution $g_{\rm L}$. In the zero-loop approximation for the path integral, the density of states becomes

$$\nu_*[\varepsilon, j, \sigma] \approx \exp\{-S_E^{\rm (BH)}[g_C]\}. \tag{3.4.100}$$

The calculation of the extremal action yields

$$S_E^{\rm (BH)}[g_C] = -\frac{2\pi}{\kappa}\int_{B_{\rm i}} d^2x\,\sqrt{\sigma}. \tag{3.4.101}$$

The integral that remains is just the area $A_{\rm H}$ of the black-hole event horizon. Thus, in the approximation (3.4.100), the entropy is

$$\mathcal{S}[\varepsilon, j, \sigma] = \ln \nu_*[\varepsilon, j, \sigma] \approx \frac{2\pi}{\kappa} A_{\rm H}. \tag{3.4.102}$$

With $\kappa = 8\pi G_{\rm N}$, this is the standard result $\mathcal{S} = A_{\rm H}/(4G_{\rm N})$ for the black-hole entropy.

It is worth noting that, as in quantum cosmology, a very promising investigative direction in black-hole physics is the study of gravitation theories in low-dimensional, especially two-dimensional, spacetime. Due to their relative simplicity, such models allow us to develop and probe new theoretical ideas and methods in the theory of black holes. Besides, a strong motivation to study a particular class of such theories, namely *two-dimensional dilaton gravity*, appears from the fact that the spherical reduction (i.e. the assumption that all fields under consideration depend only on time and radial coordinates) of four-dimensional Einstein gravity precisely produces a theory of this type. Path integrals prove to be a very powerful tool for the consideration of this model. In particular, in some cases (even for gravitation with matter fields) the path-integral method allows us to obtain exact non-perturbative results. For a review, we refer the reader to Kummer and Vassilevich (1999).

3.4.5 Path integrals for relativistic point particles and in the string theory

Path integrals have found one of their most impressive and successful applications in the theory of (super)strings. The latter nowadays is the most viable candidate for the realization of the old dream and, in a sense, the ultimate aim of physicists: the construction of a theory of 'everything'. More precisely, this is a candidate for a theory describing in a unified way all fundamental interactions, including gravitation. It is important that string theory should fit very nicely into the pre-existing picture of what physics beyond the standard model might look like. Besides gravity, string theory necessarily incorporates a number of previous unifying ideas (though sometimes in a transmuted form): grand unification, Kaluza–Klein theory (unification via extra dimensions), supersymmetry and extended supersymmetry. Moreover, it unifies these ideas in an elegant way, and resolves some of the problems which previously arose—most notably, difficulties in obtaining chiral (parity-violating) gauge interactions and the renormalizability problem of the Kaluza–Klein theory, which is even more severe than for four-dimensional gravity. Furthermore, some of the simplest string theories give rise to precisely the gauge groups and matter representations which previously arose in grand unification (that is, unification of strong, weak and electromagnetic interactions on the basis of a single simple Lie group). Thus, we can justly say that string theory is at least a step toward the unification of gravity, quantum mechanics and particle physics. However, it is worth mentioning that at present, it is clear that even the original superstring theory (see the book by Green *et al* (1987) and references therein) is not general enough to serve as such a unification theory: this ambitious aim requires a more general theory (sometimes called *M-theory*), including extended objects of higher dimensions, such as membranes, as well as a more general formulation in which different extended objects would appear as particular excitations. At the moment, only isolated results and ideas concerning such a theory exist (for a review see, e.g., Duff (1999)) and we shall not even slightly touch it in our book. Moreover, the 'original' (super)string theory is also too extensive to be presented here even in a short version. We shall only be able to introduce the reader to the very basic ideas and some results of this theory, stressing the advantage of the application of path integrals for their derivations.

To make this introduction easier, we shall start by rederiving the relativistic particle propagator (Green function) with the help of a path integral, in the *first-quantized formalism*. Then the basic techniques and the starting point in the string theory becomes a natural generalization of the result for relativistic particles.

◇ Propagator for a relativistic point particle in the first-quantized formalism

There are several forms of the action for a relativistic point particle in flat Minkowski space, all of which lead to the same answer. The most straightforward one is (see, e.g., Landau and Lifshitz (1987))

$$S'[x(\tau)] = -m \int_{x_0}^{x_f} ds = -m \int_0^1 du \sqrt{\eta_{ab}\dot{x}^a(u)\dot{x}^b(u)} \qquad (3.4.103)$$

where the integral is taken over the world-line of the particle with the endpoints x_0, x_f and u is a parameter (proper time) on the world-line, such that $x^a(0) = x_0^a$, $x^a(1) = x_f^a$. This form of the action has some disadvantages, in particular, it is non-polynomial and highly inconvenient for a consideration of the massless limit, $m \to 0$. Therefore, we shall use a more suitable form of the action:

$$S[x; N] = \frac{1}{2} \int_0^1 du \left(\frac{\dot{x}^a \eta_{ab} \dot{x}^b}{N} - m^2 N \right). \qquad (3.4.104)$$

The reader may easily verify that both actions S' and S lead to equivalent equations of motion (i.e. the same classical physics). We use the indices $a, b, \ldots$ (and not $\mu, \nu, \ldots$) to stress that the dimension of the spacetime is not necessarily equal to four. In order to prevent the determinant of the metric from depending on the spacetime dimensionality, in this subsection we choose another signature of the Minkowski metric: $\eta_{ab} = \text{diag}\{-1, +1, \ldots, +1\}$. The action (3.4.104) possesses the reparametrization gauge invariance with respect to the group $\mathcal{G}$ of one-dimensional diffeomorphic transformations

$$x^a(u) \to x^a(f(u))$$

$$N(u) \to \frac{df}{du} N(f(u)) \qquad (3.4.105)$$

for any differentiable function $f(u)$ that leaves the endpoints of the interval $[0, 1]$ fixed, i.e. provided that

$$f(0) = 0 \qquad f(1) = 1. \tag{3.4.106}$$

The existence of the reparametrization invariance (3.4.105) shows that the action (3.4.104) can be considered as an action for a one-dimensional gravity, the field $N(u)$ playing the role of a lapse function (one-dimensional metric) and $x^a(u)$ that of the matter fields. The appearance of the determinant of the reparametrization, $u \to f(u)$, in the transformation of N in (3.4.105) characterizes it as a density of weight 1. The coordinate x^a is a density of weight zero, i.e. a scalar under coordinate reparametrizations.

The momenta conjugate to x^a and N are:

$$p_a \equiv \frac{\delta S}{\delta \dot{x}^a} = \frac{1}{N} \eta_{ab} \dot{x}^b$$

$$p_N \equiv \frac{\delta S}{\delta \dot{N}} = 0$$

respectively. Since the momentum conjugate to N vanishes identically, the Hamiltonian system is a degenerate one with a primary first-class constraint. The secondary first-class constraint following from the primary one $p_N = 0$ is

$$\dot{p}_N = -\frac{\delta S}{\delta N} = p_a \eta^{ab} p_b + m^2 \equiv H = 0. \tag{3.4.107}$$

Thus, the Hamiltonian H for a relativistic particle vanishes at the surface defined by the constraints. Because of this degeneracy, the phase-space Hamiltonian path-integral quantization should be based on the phase-space path integral with constraints. However, in this section rather than entering into that discussion of the phase-space integral, let us follow a different route and construct the configuration-space path integral by geometric reparametrization-invariance considerations alone (see, e.g., Cohen *et al* (1986) and Mottola (1995)).

The space $\mathcal{X}$ of all configurations of the functions $(x^a(u), N(u))$ may be treated by methods borrowed from Riemannian geometry. At every 'point' $X^i = (x^a(u), N(u))$ in the function space we introduce the cotangent space $\delta \mathcal{X}$, labeled by the basis vectors $\delta X^i = (\delta x^a(u), \delta N(u))$. In the Riemannian geometry, we can introduce a metric on a space by defining a quadratic form, the line element, that maps $\delta \mathcal{X} \times \delta \mathcal{X}$ to the real numbers,

$$(\delta s)^2 = G_{ij}(X) \delta X^i \delta X^j. \tag{3.4.108}$$

The (infinitesimal) invariant volume measure of integration on the cotangent space, $\delta \mathcal{X}$

$$d(\delta \mathcal{V}) = \sqrt{\det G_{ij}(X)} \, d(\delta X^1) \times d(\delta X^2) \times \cdots \tag{3.4.109}$$

may be chosen to satisfy the Gaussian normalization condition,

$$\int d(\delta \mathcal{V}) \exp\{-\tfrac{1}{2} \delta X^i G_{ij}(X) \delta X^j\} = 1 \tag{3.4.110}$$

and immediately induces an invariant volume measure on the full space $\mathcal{X}$,

$$d\mathcal{V} = \sqrt{\det G_{ij}(X)} \, dX^1 \times dX^2 \times \cdots . \tag{3.4.111}$$

All the construction can be justified by a suitable finite-dimensional regularization (via discretization or Fourier mode cutoff).

To define the path integral for a system possessing a reparametrization gauge invariance, we have only to define a quadratic inner product on the cotangent space $\delta \mathcal{X}$. Then the construction of the invariant measure on the function space of all configurations proceeds exactly along the lines of (3.4.108)–(3.4.111). Since the path integral is specified by an invariant action functional and an invariant integration measure, this procedure preserves the classical reparametrization invariance under quantization. Then, by identifying configurations which differ only by a reparametrization of coordinates, we may integrate only over *equivalence classes* of coordinates, $\mathcal{X}/\mathcal{G}$, in a manifestly gauge or coordinate invariant way.

Since both N and δN have weight one under (3.4.105), and

$$\int_0^1 N(u)\, du \equiv \tau \tag{3.4.112}$$

is invariant under reparametrizations (acting from the right), we can define an invariant inner product on $\mathcal{X}$ by

$$\langle \delta N | \delta N \rangle_N \stackrel{\text{def}}{\equiv} \int_0^1 \left(\frac{\delta N}{N}\right)^2 N\, du = \int_0^1 \frac{(\delta N)^2}{N}\, du$$

$$\langle \delta x | \delta x \rangle_N \stackrel{\text{def}}{\equiv} \int_0^1 \delta x^a \eta_{ab} \delta x^b N\, du. \tag{3.4.113}$$

These invariant products are unique up to a multiplicative constant which may be absorbed into the normalization of the measure by the condition

$$\int \mathcal{D}(\delta N) \exp\{-\tfrac{1}{2}\langle \delta N, \delta N \rangle_N\} = 1 \tag{3.4.114}$$

$$\int \mathcal{D}(\delta x) \exp\left\{-\frac{i}{2}\langle \delta x, \delta x \rangle_N\right\} = 1. \tag{3.4.115}$$

A factor of i is inserted into the second Gaussian because the Minkowski metric is pseudo-Riemannian with one negative eigenvalue in the timelike direction, so that the volume form in (3.4.111) remains real. In Euclidean signature metrics, this factor would be absent. These definitions generate an invariant functional measure on the full space $\mathcal{X}$ by equations (3.4.108)–(3.4.111). In order to pull this measure back to an invariant measure on the quotient space of equivalence classes $\mathcal{X}/\mathcal{G}$, let us parametrize the gauge orbits $\mathcal{G}$ by the set of differentiable functions $f(u)$ satisfying the endpoint conditions (3.4.106). Then an arbitrary lapse function $N(u)$ may be written in the form

$$N(u) = \tau \frac{df}{du} \qquad f(u) = \frac{1}{\tau} \int_0^u N(u)\, du. \tag{3.4.116}$$

In other words, the gauge equivalence class $\mathcal{M}/\mathcal{G}$ of all the functions $N(u)$ is characterized completely by the single parameter τ defined in (3.4.112), while the gauge fiber $\mathcal{G}$ is coordinatized by $f(u)$. Hence, the integration measure on the quotient space is given by

$$\frac{[\mathcal{D}N]}{[\mathcal{D}f]}[\mathcal{D}x] = J(\tau)\, d\tau\, [\mathcal{D}x] \tag{3.4.117}$$

where J is the Jacobian of the change of variables from N to (τ, f). The invariant path integral is then

$$K_{\text{rel}}(x, 1; x_0, 0) = \int J(\tau)\, d\tau \int_{\mathcal{C}\{x(0)=x_0, 0; x(1)=x, 1\}} \mathcal{D}x(\tau)\, e^{iS[x;\tau]}. \tag{3.4.118}$$

To avoid confusion, it is necessary to stress that here we just factor out the volume of the gauge orbits, while in the case of Yang–Mills theories (section 3.2) we integrated over chosen representatives from each orbit (this aim is achieved by the imposition of gauge-fixing conditions).

Our task now is to determine the Jacobian J. As follows from (3.4.116), a variation of the lapse function $N(u)$ can be written as

$$\delta N = (\delta\tau)\frac{df}{du} + \tau\frac{d(\delta f)}{du}. \tag{3.4.119}$$

After substituting this form into the inner product definition (3.4.113), the cross term between $\delta\tau$ and δf vanishes by using the endpoint conditions (3.4.106). Changing variables in the last term from u to $v \equiv f(u)$ and defining

$$\xi(v) \equiv (\delta f)(u)|_{u=f^{-1}(v)} \tag{3.4.120}$$

we obtain

$$\langle \delta N | \delta N \rangle_N = \frac{(\delta\tau)^2}{\tau} + \tau \int_0^1 dv\, \xi(v)\left[-\frac{d^2}{dv^2}\right]\xi(v) \tag{3.4.121}$$

after an integration by parts. Now, it is straightforward to verify that the quantity $\xi(v)$ (which parametrizes infinitesimal diffeomorphisms) is invariant under diffeomorphism group transformations operating from the *right*, i.e.

$$\xi(v) \to \xi(v) \qquad f(u) \to f(\alpha(u)) \tag{3.4.122}$$

but that it transforms as a density of weight -1 under the *inverse* diffeomorphism group transformations operating from the *left*, i.e.

$$\xi(w) \to \frac{1}{\left(\frac{d\beta^{-1}(w)}{dw}\right)}\xi(\beta^{-1}(w)) \qquad f(u) \to \beta(f(u)) \equiv w. \tag{3.4.123}$$

The only quadratic form in ξ that is invariant under both of these two transformations is (see Polyakov (1981, 1987))

$$\langle \xi | \xi \rangle \equiv \int_0^1 \tau dv\, (\tau\xi)^2(v) \tag{3.4.124}$$

provided that τ remains invariant under the first transformation, but

$$\tau \to \tau\left(\frac{d\beta^{-1}(w)}{dw}\right) \tag{3.4.125}$$

under the second.

With the fully invariant inner product (3.4.124), we may define the integration measure on $\mathcal{G}$ by the Gaussian normalization condition

$$\int \mathcal{D}\xi(\tau) \exp\left\{-\frac{i}{2}\langle \xi | \xi \rangle\right\} = 1. \tag{3.4.126}$$

Returning then to our problem of evaluating the Jacobian J in (3.4.117), we find:

$$1 = \int \mathcal{D}(\delta N) \exp\left\{-\frac{i}{2}\langle \delta N | \delta N \rangle_N\right\}$$

$$= \int d(\delta\tau)\, J(\tau) \int \mathcal{D}\xi\, \exp\left\{-\frac{i(\delta\tau)^2}{2\tau} - \frac{i\langle \xi|(-d^2/dv^2)\xi\rangle}{2\tau^2}\right\}$$

$$= J\left(\frac{2\pi\tau}{i}\right)^{\frac{1}{2}}\left[\det\left(-\tau^{-2}\frac{d^2}{dv^2}\right)\right]^{-\frac{1}{2}}$$
$$= \text{constant} \times J \qquad (3.4.127)$$

where the last result follows by (1.2.227). Hence, the Jacobian J is a constant independent of τ.

Evaluating the propagator in (3.4.118) is now straightforward. The functional integration over $x^a(u)$ with fixed endpoint conditions is the same as that for the free non-relativistic particle, and we are left with only a simple integral over the proper time τ to perform:

$$K_{\text{rel}}(x,1;x_0,0) = J\int_0^\infty d\tau\, (2\pi i\tau)^{-\frac{d}{2}} \exp\left\{\frac{i}{2\tau}(x-x_0)^a\eta_{ab}(x-x_0)^b - \frac{i}{2}\tau m^2\right\}$$
$$= J\int_0^\infty d\tau \int \frac{d^d p}{(2\pi)^d} \exp\left\{ip_a(x-x_0)^a - \frac{i\tau}{2}(p_a\eta^{ab}p_b + m^2)\right\}$$
$$= \int \frac{d^d p}{(2\pi)^d} \frac{e^{ip\cdot(x-x_0)}}{(p^2+m^2-i0)} \qquad (3.4.128)$$

provided that the constant $J = i/2$ and an infinitesimal negative imaginary part is added to m^2 to define the τ integral. With this normalization, expression (3.4.128) is recognized as the Feynman propagator for the free relativistic scalar field, that is, $K_{\text{rel}}(x,1;x_0,0) = D_c(x-x_0)$ (cf (3.1.93)), which we have obtained here by the path-integral treatment of the reparametrization invariant *first-quantized* particle action (3.4.104).

Thus, relativistic particles, as well as non-relativistic ones, can be treated in the framework of the first-quantized formalism. In order to describe particle interactions, we should include, together with the free particle trajectories as depicted in figure 3.23(a), the merging and splitting trajectories as depicted in figure 3.23(b). It is seen, that with an appropriate choice of vertex operator, the set of this type of trajectory corresponds to the first-order contribution to the S-matrix (or four-point Green function) of the scalar φ^4-theory. This observation can be generalized, and the whole perturbation expansion for any field theory can be reconstructed in the first-quantized formalism. However, the approach has obvious shortcomings:

- it is not clear how to consider non-perturbative phenomena (e.g., solitons, instantons, strong coupling interactions, etc);
- we have to postulate a specific choice of vertices and merging–splitting rules for trajectories which can be justified only by comparison with the corresponding field theory and
- in the vicinity of the vertices, the trajectories (for example, of the type presented in figure 3.23(b)) are not a one-dimensional manifold (this fact complicates the geometrical description of such 'branching' trajectories).

The situation drastically changes in the case of string theory, where attempts to construct straightforwardly a second-quantized (analog of quantum field) theory meet essential difficulties, while the first-quantized theory proves to be well defined, self-consistent and powerful.

◇ **String basics**

The most natural action describing the dynamics of strings, i.e. one-dimensional extended objects, has the following form

$$S = -T \times (\text{Area of two-dimensional world-sheet})$$
$$= -T\int d^2x\, \sqrt{-\det \partial_\alpha X^a \partial_\beta X_a} \qquad \alpha,\beta = 0,1 \qquad (3.4.129)$$

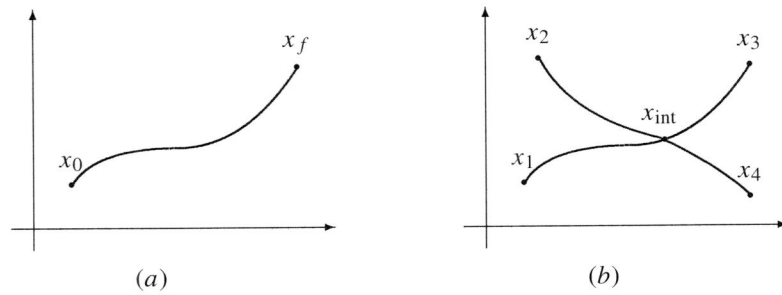

Figure 3.23. A sample of a free particle trajectory (*a*) and a sample of trajectories of interacting particles (*b*).

and is called the *Nambu–Goto action* (the indices $a, b = 0, 1, \ldots, d - 1$ are raised and lowered with the flat-space metric $\eta_{ab} = \text{diag}(-, +, +, \ldots, +)$) This generalizes the relativistic action (3.4.103) for a point particle, which is minus the mass times the invariant length of the world-line. For a static string, this action reduces to minus the length of the string times the time interval times T, so the latter is the *string tension*. Note that in the second line we are describing the world-sheet by $X^\mu(x^0, x^1)$, using a parametrization x^α of the world-sheet, but the action is independent of the choice of parametrization (world-sheet coordinate invariant). So the two-dimensional spacetime is the string world-sheet, while the spacetime is the field space where the X^μ live, the *target space* of the map X^μ : world-sheet $\to$ spacetime.

As for relativistic particles, it is useful to rewrite action (3.4.129) in a form which removes the square root from the derivatives. Let us add a world-sheet metric $g_{\alpha\beta}(x)$ ($\alpha, \beta = 0, 1$) and let

$$S_\text{P} = T \int d^2x \, \sqrt{g} \, g^{\alpha\beta} \partial_\alpha X^a \partial_\beta X_a \tag{3.4.130}$$

where $g = \det g_{\alpha\beta}$. This is commonly known as the *Polyakov action* because he emphasized its virtues for quantization. The equation of motion for the metric determines it up to a position-dependent normalization

$$g_{\alpha\beta} \sim \partial_\alpha X^a \partial_\beta X_a; \tag{3.4.131}$$

inserting this back into the Polyakov action gives the Nambu–Goto action. The Polyakov action makes sense for either a Lorentzian metric on a world-sheet, with signature $(-, +)$, or a Euclidean metric, with signature $(+, +)$. Much of the development can be carried out in either case. We shall use the Euclidean formalism.

In addition to the two-dimensional coordinate invariance mentioned earlier

$$X'(x') = X(x) \qquad \frac{\partial x'^\alpha}{\partial x^\gamma} \frac{\partial x'^\beta}{\partial x^\delta} g'_{\alpha\beta}(x') = g_{\gamma\delta}(x) \tag{3.4.132}$$

the Polyakov action has another local symmetry, namely, the *Weyl invariance*, i.e. position-dependent rescalings of the metric:

$$g'_{\alpha\beta}(x) = e^{2\sigma(x)} g_{\alpha\beta}(x). \tag{3.4.133}$$

To proceed with the quantization, we need to remove the redundant degrees of freedom from the local symmetries. Noting that the metric has three components and there are three local symmetries (two coordinate reparametrizations and the scale of the metric), it is natural to do this by imposing the following conditions on the metric:

$$g_{\alpha\beta}(x) = \delta_{\alpha\beta}. \tag{3.4.134}$$

This is always possible, at least locally.

After quantization, a string is represented by an infinite set of *normal modes*, i.e. the set of massive quantum-mechanical states. The mass gaps Δm^2 in this set are proportional to the string tension T. The spectrum of a bosonic string starts from a *tachyon state* (a state with imaginary mass) and hence the purely bosonic string theory is not fully self-consistent. The situation is improved in the theory of *superstrings*, that is, after the inclusion of fermionic degrees of freedom (introduction of two-dimensional fermionic fields on the world-sheet of a string). These fermionic degrees of freedom are added in a supersymmetric way and this provides the ground state with zero mass (cf supplement VI), so that the undesirable tachyon disappears from the spectrum.

Strings can be open or closed. The zero modes of the open strings consist of spin-1 particles corresponding to *Yang–Mills fields*, while closed strings in the massless sector contain *gravitons* (spin-2 particles) and, in the case of superstrings, *gravitinos* (spin-$\frac{3}{2}$ particles). At relatively low energies (essentially less than the Planck mass scale $\sim 10^{19}$ GeV), the massive modes effectively decouple and the low-energy physics can be described by the effective local field theory for the zero modes (that is, by supergravity or supersymmetric Yang–Mills theory).

In the first-quantized formalism, string interactions are described similarly to the case of point particles, i.e. by path integration over *world-sheets* with appropriate topologies. Examples of such world-sheets in the case of a four-string interaction for open (*b*) and closed (*c*) strings are depicted in figure 3.24. As we have already mentioned, attempts to construct a second-quantized string theory, i.e. a 'string field theory', have met critical problems which have not been successfully overcome so far. On the other hand, the shortcomings of the first-quantized formulation pointed out earlier for relativistic particles, become milder in the case of strings. In particular, it is seen in figure 3.24 that the world-sheets corresponding to the string interactions are still two-dimensional manifolds (in contrast to the case of point particles) and the arbitrariness in the definition of the interaction terms in the string perturbation theory is strongly restricted by the symmetries on the world-sheets, namely by the two-dimensional diffeomorphisms and Weyl symmetries. Of course, this by no means removes the necessity of a non-perturbative formulation of the (super)string theory. But the current hopes for such a construction are connected with the so-called *M-theory* (see, e.g., Duff (1999)) rather than with a straightforward generalization of a local field theory to the string case.

We have already pointed out that superstring theory is too extensive and versatile a subject to present it in this book even briefly. Instead, to give the reader an idea about the typical calculations and peculiarities of this theory, we shall discuss the so-called *Weyl anomaly* in the simplest model.

⋄ **The Weyl anomaly in two dimensions**

A simple application of path integrals over string world-sheets is the calculation of the *Weyl anomaly* in two dimensions. For this aim, let us consider a single free massless scalar field with the classical action,

$$S_{\text{cl}}[g, \phi] = \int d^2x \sqrt{-g} g^{\alpha\beta} \partial_\alpha \phi \partial_\beta \phi. \tag{3.4.135}$$

This is the simplest variant of the string action (3.4.130) with only one space or time coordinate ($X^a \to \varphi$). Action (3.4.135) is clearly invariant under general coordinate transformations. In addition, it is also invariant under Weyl rescalings, since when writing

$$g_{\alpha\beta} = e^{2\sigma} \eta_{\alpha\beta} \qquad \sqrt{-g} = e^{2\sigma} \tag{3.4.136}$$

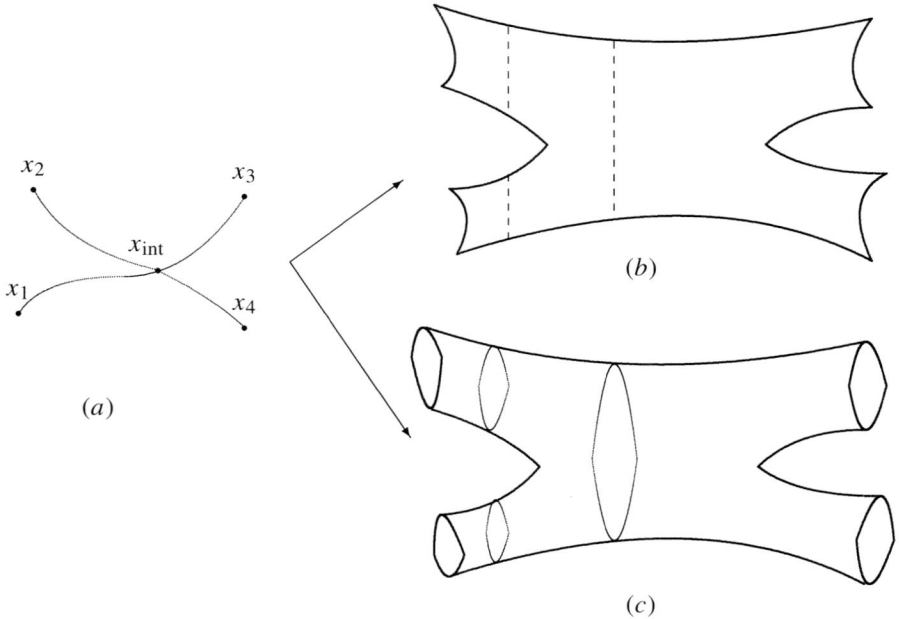

Figure 3.24. Generalization of the four-particle interaction world-lines (*a*) to the case of the four-string interaction world-sheet for open (*b*) and closed (*c*) strings.

we observe that $S[\phi]$ is independent of σ. The general coordinate invariance implies that the energy–momentum tensor derived from $S[\phi]$ is covariantly conserved,

$$\nabla_\alpha T^{\alpha\beta}[g,\phi] = 0 \qquad T^{\alpha\beta}[g,\phi] = \frac{2}{\sqrt{-g}}\frac{\delta S[g,\phi]}{\delta g_{\alpha\beta}} = \partial^\alpha\phi\partial^\beta\phi - \frac{1}{2}(\partial\phi)^2 g^{\alpha\beta} \qquad (3.4.137)$$

while the Weyl invariance guarantees that this classical energy–momentum tensor is traceless,

$$g_{\alpha\beta}T^{\alpha\beta}[\phi] = e^{-2\sigma}\frac{\delta S}{\delta\sigma}[g = e^{2\sigma}\eta,\phi] = 0. \qquad (3.4.138)$$

We define the quantum effective action (in Euclidean time) by the covariant path integral

$$\exp(-S_{\text{eff}}[g]) = \int \mathcal{D}\phi \, \exp\{-S_{\text{cl}}[g,\phi]\} \qquad (3.4.139)$$

where the generally covariant integration measure over scalar fields must be defined such that

$$\int \mathcal{D}\phi \, \exp\{-\tfrac{1}{2}\langle\phi|\phi\rangle\} = 1 \qquad \langle\phi|\phi\rangle \equiv \int d^2x \, \sqrt{-g}\phi^2. \qquad (3.4.140)$$

Now, the point is that this inner product and the corresponding integration measure over scalar fields $\mathcal{D}\phi$ are invariant under general coordinate transformations but *not* under Weyl rescalings. Hence, we must expect the energy–momentum tensor of the quantized theory to remain conserved but have non-zero trace (in contrast to its classical counterpart). Thus the so-called *trace anomaly* appears. Although this had to be discovered by laborious calculations in the operator quantization method, it is actually obvious from the Weyl non-invariance of the covariant integration measure (cf section 3.3.4). Note that it is logically possible to define a Weyl invariant scalar inner product and integration measure by leaving out the $\sqrt{+g}$ in (3.4.140), at the price of making it not generally coordinate invariant. In this case, the quantum energy–momentum tensor would remain traceless, but it would no longer be conserved (this is an analog of the competition between the chiral and gauge symmetries; see section 3.3.4). If coordinate invariance is assumed to be a more fundamental symmetry of nature than Weyl invariance, this possibility must be rejected.

In order to calculate the Weyl trace anomaly, we perform the Gaussian integration in (3.4.139) and obtain

$$S_{\text{eff}}[g] = \frac{1}{2}\operatorname{Tr}\ln(-\Box) \to -\frac{1}{2}\int_\epsilon^\infty \frac{dt}{t}\operatorname{Tr}\exp\{-t\Box\} \qquad (3.4.141)$$

where the second equality is the regularized definition (the so-called heat kernel definition) of 'Tr ln', which introduces a cutoff on the lower limit of integration. This regulated form is most convenient for evaluating the trace anomaly by varying $S_{\text{eff}}[g]$ with respect to σ. Using

$$\Box = \frac{1}{\sqrt{-g}}\partial_\alpha[\sqrt{g}g^{\alpha\beta}\partial_\beta] = e^{-2\sigma}\overline{\Box} \qquad (3.4.142)$$

with $\overline{\Box}$ evaluated in the *flat* Euclidean metric, we find

$$\begin{aligned}
g_{\alpha\beta}T^{\alpha\beta} &= e^{-2\sigma}\frac{\delta S_{\text{eff}}[g]}{\delta\sigma(x)} \\
&= -\int_\epsilon^\infty dt\,\langle x|\Box\exp\{-t\Box\}|x\rangle \\
&= \langle x|\exp\{-\epsilon\Box\}|x\rangle \\
&= \frac{1}{4\pi}\left[\frac{1}{\epsilon} + \frac{R}{6} + \mathcal{O}(\epsilon)\right]
\end{aligned} \qquad (3.4.143)$$

where the expansion of the heat kernel for $\exp\{-\epsilon\Box\}$ has been used in the last step as $\epsilon \to 0$, and we have assumed that the operator $\Box$ has no zero modes, so that the upper limit of the t integral does not contribute to the trace. The background metric-independent, divergent first term is associated with the infinite energy density of the vacuum in flat space, which can be regulated by the full ζ-function method (see section 1.2.7), or simply subtracted from the definition of the energy–momentum tensor by a normal-ordering procedure. The finite second term proportional to the Ricci scalar curvature of the background metric $g_{\alpha\beta}$ is the trace anomaly. Note that it indeed comes from the Weyl non-invariance of the integration measure (3.4.140), since if we used the Weyl invariant measure omitting $\sqrt{g}$ from the inner product defined there, $\Box$ in (3.4.142) would be replaced by $\sqrt{g}\Box = e^{2\sigma}\Box = \overline{\Box}$ which is independent of σ, so that the variation in (3.4.143) would then give zero identically. Therefore, the trace anomaly is a necessary and immediate consequence of the covariant definition of the path-integral measure in (3.4.140).

In the conclusion of this section, we shall briefly discuss the case of an action with an arbitrary number of matter fields. The general form of the trace anomaly of the energy–momentum tensor for classically Weyl invariant matter in a background gravitational field is

$$T_\alpha^\alpha \text{ (matter)} = \frac{c_m}{24\pi}R \qquad (3.4.144)$$

$$= \frac{c_m}{24\pi}e^{-2\sigma}(\overline{R} - 2\overline{\Box}\sigma) \qquad (3.4.145)$$

in the decomposition $g_{\alpha\beta} = e^{2\sigma}\bar{g}_{\alpha\beta}$. The coefficient c_m is defined by the number of matter fields: $c_m = (N_S + N_F)$ for N_S scalar and N_F (Dirac) fermion fields. From (3.4.143), this implies that there exists an effective anomalous quantum action, such that

$$\frac{\delta S_{\text{anom}}[g]}{\delta \sigma(x)} = \frac{c_m}{24\pi}(\overline{R} - 2\overline{\Box}\sigma). \tag{3.4.146}$$

Since the right-hand side of this equation is linear in σ, we may integrate both sides immediately with respect to σ to obtain the anomalous action:

$$S_{\text{anom}}[g = e^{2\sigma}\bar{g}] = S_{\text{anom}}[\bar{g}] + \frac{c_m}{24\pi}\int d^2x\,\sqrt{-\bar{g}}[-\sigma\overline{\Box}\sigma + \overline{R}\sigma]. \tag{3.4.147}$$

The action $S_{\text{anom}}[g]$ must be a scalar under general coordinate transformations and a functional of only the full $g_{\alpha\beta}$, so that we may use this information to determine the σ-independent integration constant $S_{\text{anom}}[\bar{g}]$ and write down the fully covariant but non-local form of the anomalous action:

$$S_{\text{anom}} = -\frac{c_m}{96\pi}\int d^2x\sqrt{-g}\int d^2x'\,\sqrt{-g'}R(x)\Box^{-1}(x,x')R(x'). \tag{3.4.148}$$

The contribution of the world-sheets with a given topology to the string amplitudes is proportional to the path integral (after the factorization of the volume of the gauge group of two-dimensional diffeomorphisms)

$$\int \mathcal{D}\sigma\, \mathcal{D}g^{\perp}_{\mu\nu} J(g = e^{2\sigma}g^{\perp})\exp\{-S_{\text{P}} - S_{\text{anom}}\} \tag{3.4.149}$$

where $g^{\perp}_{\mu\nu}$ parametrize orbits of the gauge group of two-dimensional diffeomorphisms in the set of the metrics (similarly to the parameter τ in the case of the relativistic point particle, cf (3.4.112)). The number of these parameters is finite and depends on the topology of the considered world sheet. The Jacobian J appears, analogously to the case of point particles, because of the change of variables: $g_{\mu\nu} \to \{\xi_{\mu}, g^{\perp}_{\mu\nu}\}$ (ξ_{μ} are parameters of diffeomorphisms). Finally, S_{P} is the classical action (3.4.130) or its supersymmetric extension and S_{anom} is given by (3.4.148). After extracting the Weyl parameter σ from the Jacobian J, i.e. the transition from $J(g = e^{2\sigma}g^{\perp})$ to $\bar{J}(g^{\perp})$, the coefficient c_m in front of the anomalous action changes as follows:

$$c_m \longrightarrow c = c_m - 26 = N_{\text{S}} + N_{\text{F}} - 26$$

so that the path integral (3.4.149) becomes

$$\int \mathcal{D}\sigma\, \mathcal{D}g^{\perp}_{\mu\nu}\, \bar{J}(g^{\perp})\exp\left\{-S_{\text{P}} - \frac{c}{c_m}S_{\text{anom}}\right\}.$$

All the dependence of the integrand on the gauge parameter $\sigma(x)$ is concentrated in S_{anom}. Thus, if we want the quantum system to have the same number of physical degrees of freedom as the classical prototype (that is, the quantum effective action does not depend on $\sigma(x)$), we have to require that the parameter c, called the *central charge*, be equal to zero

$$c = N_{\text{S}} + N_{\text{F}} - 26 = 0.$$

In the case of a purely bosonic string ($N_{\text{F}} = 0$), this means that a self-consistent quantization can be carried out only in a 26-dimensional spacetime (i.e. the number of the coordinates X^a, $a = 0,\ldots,25$ is equal to 26). In superstrings, fermionic degrees of freedom appear and the supersymmetry fixes their number N_{F}, so that the central charge is equal to zero at $N_{\text{S}} = 10$. Thus, non-anomalous superstrings exist only in a ten-dimensional spacetime.

3.4.6 Quantum field theory on non-commutative spacetimes and path integrals

In the last subsection of this chapter, we shall discuss applications of path integrals in a recently emerged branch of quantum field theory, namely, in the field theory of non-commutative spacetimes.

The standard concept of a geometric space is based on the notion of a manifold $\mathcal{M}$ with points $x \in \mathcal{M}$ locally labeled by a finite number of real coordinates $x^\mu \in \mathbb{R}^4$. However, it is generally believed that this picture of spacetime as a manifold $\mathcal{M}$ would break down at very short distances of the order of the *Planck length* $\lambda_P \approx 1.6 \times 10^{-33}$ cm. This implies that the mathematical concepts for high-energy (small-distance) physics have to be changed or, more precisely, our classical geometrical concepts may not be well suited for describing physical phenomena at small distances. No convincing alternative description of physics at very short distances is known, though different routes to progress have been proposed. One such direction is to try to formulate physics on some non-commutative spacetime. There appear to be too many possibilities to do this, and it is difficult to see what the right choice is. There have been investigations in the context of Connes' formulation of the non-commutative geometry (Connes 1994) and his approach to construction of the standard model of electroweak and strong interactions (Connes and Lott 1990). Another approach is based on the relation between measurements at very small distances and black-hole formations (Doplicher *et al* 1994, 1995). One more possibility is based on quantum group theory (see, e.g., Chaichian and Demichev (1996)). As shown by Seiberg and Witten (1999), the non-commutative geometry naturally appears in string theory. This result provides us with a solid background for the study of field theories on non-commutative spacetimes which supposedly correspond to the low-energy limit of such strings theories. It is worth noting that the generalization of commutation relations for the canonical operators (coordinate–momentum or creation–annihilation operators) was suggested long ago by Heisenberg (1954) in attempts to achieve regularization for his (non-renormalizable) nonlinear spinor field theory.

The essence of non-commutative geometry consists in reformulating, first, the geometry in terms of *commutative* algebras of smooth functions, and then generalizing them to their *non-commutative* analogs. One of the main motivations for studying QFT on non-commutative spacetimes is that the notion of points as elementary geometrical entities is lost and we might expect an ultraviolet cutoff to appear. The simplest model of this kind is the fuzzy sphere (see Berezin (1975), Hoppe (1989), Grosse *et al* (1997) and references therein), i.e. the non-commutative analog of a two-dimensional sphere. As is known from standard quantum mechanics, a quantization of any *compact* space, in particular a sphere, leads to finite-dimensional representations of the corresponding operators, so that in this case any calculation is reduced to manipulations with finite-dimensional matrices and thus there is simply no place for ultraviolet divergences. Things are not so easy in the case of non-compact manifolds. Quantization leads to infinite-dimensional representations and we have no guarantee that non-commutativity of the spacetime coordinates removes the ultraviolet divergences.

In order to illustrate the main peculiarities of QFT on non-commutative spacetimes, we shall use a relatively simple example of non-commutative geometry, namely, the Euclidean non-compact plane with Heisenberg-like commutation relations among coordinates (Chaichian *et al* 2000). In discussing this example, we shall see that physically meaningful quantities in non-commutative QFT are the correlation functions (Green functions) of mean values of quantum fields on a non-commutative spacetime in *localized states* from the Hilbert space of a representation of the corresponding coordinate algebra. These localized states are the best counterparts of points on an ordinary commutative space, which 'label' the (infinite number of) degrees of freedom of the QFT. Thus we must consider a quantum field in non-commutative QFT as a map from the *set of states* on the corresponding non-commutative space into the algebra of secondary quantized operators. We also use this example for the study of symmetry transformations of non-commutative spacetimes with Lie algebra commutation relations for coordinates. The non-commutative coordinates prove to be tensor operators, and we consider concrete examples of the corresponding transformations of localized states (an analog of spacetime point transformations).

◇ Two-dimensional quantum field theory on non-commutative spacetime with Heisenberg-like commutation relations

A complex scalar field $\varphi(x)$ on a Euclidean plane $P^{(2)} = \mathbb{R}^2$ is a prescription

$$x = (x_1, x_2) \in P^{(2)} \to \varphi(x) \in \mathbb{C}$$

which assigns to any point x of the plane the complex number $\varphi(x)$. In order to pass to a non-commutative plane and to the corresponding 'fields', we introduce in the two-dimensional Euclidean *coordinate* plane $P^{(2)}$ the following Poisson bracket:

$$\{x_i, x_j\} = \varepsilon_{ij} \qquad i, j = 1, 2 \tag{3.4.150}$$

and extend it, by the Leibniz rule, to all smooth functions on $P^{(2)}$ (here ε_{ij} is the antisymmetric tensor, $\varepsilon_{12} = 1$). The brackets are invariant with respect to the canonical transformations $x_i \to M_{ij}x_j + a_i$, where M_{ij} is an unimodular (i.e. det M = 1) matrix; a_1, a_2 are arbitrary constants. In particular, the bracket is invariant with respect to the two-dimensional group $E(2)$ of isometries of $P^{(2)}$ formed by

(i) rotations: $\quad x_1 \to x_1 \cos\phi + x_2 \sin\phi \qquad x_2 \to x_2 \cos\phi - x_1 \sin\phi$

(ii) translations: $\quad x_1 \to x_1 + a_1 \qquad x_2 \to x_2 + a_2.$ $\tag{3.4.151}$

In the non-commutative version $P^{(2)}_\lambda$ of the plane, we replace the commuting parameters by the Hermitian operators $\hat{x}_i$, $(i, j = 1, 2)$ satisfying the commutation relations

$$[\hat{x}_i, \hat{x}_j] = i\lambda^2 \varepsilon_{ij} \qquad i, j = 1, 2 \tag{3.4.152}$$

where λ is a positive constant of the dimension of *length*. We realize the operators $\hat{x}_i$, $i, j = 1, 2$ in a suitable Fock space $\mathcal{F}$ introducing the annihilation and creation operators

$$\hat{\alpha} = \frac{1}{\lambda\sqrt{2}}(\hat{x}_1 + i\hat{x}_2) \qquad \hat{\alpha}^\dagger = \frac{1}{\lambda\sqrt{2}}(\hat{x}_1 - i\hat{x}_2) \tag{3.4.153}$$

and putting

$$\mathcal{F} = \left\{ |n\rangle = \frac{1}{\sqrt{n!}} \hat{\alpha}^{*n} |0\rangle; n = 0, 1, \ldots \right\}.$$

Here $|0\rangle$ is a normalized state satisfying $\hat{\alpha}|0\rangle = 0$.

For all operators of the form

$$\hat{f} = \frac{\lambda^2}{(2\pi)^2} \int d^2k \, \tilde{f}(k) e^{ik\hat{x}} \tag{3.4.154}$$

(with a suitable smooth decreasing $\tilde{f}(k)$), we introduce the 'integral' (linear functional) $I_\lambda[\hat{f}]$ as follows:

$$I_\lambda[\hat{f}] \stackrel{\text{def}}{\equiv} \text{Tr}\, \hat{f} = \tilde{f}(0). \tag{3.4.155}$$

Here Tr denotes the trace in the Fock space and $k\hat{x} = k_1\hat{x}_1 + k_2\hat{x}_2$. The non-commutative analogs of field derivatives $\partial_i \hat{\varphi}$, $i = 1, 2$ are defined as

$$\partial_i \hat{\varphi} = \varepsilon_{ij} \frac{i}{\lambda^2}[\hat{x}_j, \hat{\varphi}] \qquad i = 1, 2. \tag{3.4.156}$$

They satisfy the Leibniz rule and reduce to the usual derivatives in the commutative limit.

In the non-commutative case, the Euclidean action of self-interacting scalar non-commutative quantum field theory reads as

$$S^{(\lambda)}[\hat{\varphi}, \hat{\varphi}^\dagger] = S_0^{(\lambda)}[\hat{\varphi}, \hat{\varphi}^\dagger] + S_{\text{int}}^{(\lambda)}[\hat{\varphi}, \hat{\varphi}^\dagger] \qquad (3.4.157)$$

with the free part having the form

$$S_0^{(\lambda)}[\hat{\varphi}, \hat{\varphi}^\dagger] = I_\lambda[(\partial_i \hat{\varphi})^\dagger (\partial_i \hat{\varphi}) + m^2 \hat{\varphi}^\dagger \hat{\varphi}]. \qquad (3.4.158)$$

The interaction part S_{int} of the action we shall discuss later (see (3.4.170) and below).

◇ **Calculations by the method of operator symbols**

The calculation of Green functions and other quantities in non-commutative QFT can be carried out in a simple and natural way by the use of operator symbols (see section 2.3.1). In fact, transition to the momentum representation $\varphi(\hat{x}) \to \widetilde{\varphi}(k)$ in (3.4.154) is the first step in the construction of the corresponding *Weyl symbol*. If, in addition, we now make the inverse *ordinary* Fourier transform:

$$\varphi_{\text{w}}(x) = \frac{\lambda^2}{(2\pi)^2} \int d^2 k\, \widetilde{\varphi}(k) e^{ikx} \qquad (3.4.159)$$

we just obtain the Weyl symbol $\varphi_{\text{w}}(x)$ of the operator $\varphi(\hat{x})$. Thus, the Weyl symbols or their Fourier transforms (which plays the role of the momentum representation for a field on the non-commutative plane $P_\lambda^{(2)}$) are in one-to-one correspondence with the set of fields (operators) $\varphi(\hat{x})$ on the non-commutative space. This correspondence is based on the relation

$$\text{Tr}\exp\{ik_i \hat{x}_i\} = 2\pi \lambda^{-2} \delta^{(2)}(k).$$

As we mentioned in section 2.3.1, the trace of an operator $f(\hat{x})$ it is expressed via its Weyl symbol as follows:

$$\text{Tr}\, f(\hat{x}) = \frac{1}{2\pi \lambda^2} \int d^2 x\, f_{\text{w}}(x) = I_\lambda[f(\hat{x})].$$

Now any action for non-commutative QFT can be obtained from the corresponding classical action by the substitution of the ordinary point-wise function multiplication by the $\star$-product. For example,

$$\text{Tr}\sum_i [\hat{x}_i, \varphi(\hat{x})]^2 = \int d^2 x\, \{x_i, \varphi_{\text{w}}(x)\}_M \star \{x_i, \varphi_{\text{w}}(x)\}_M$$

$$= \int d^2 x\, (\partial_i \varphi_{\text{w}} \star \partial_i \varphi_{\text{w}})(x)$$

where $\{\cdot, \cdot\}_M$ is the *Moyal bracket*:

$$\{\varphi, \psi\}_M \stackrel{\text{def}}{\equiv} \frac{1}{\lambda^2}(\varphi \star \psi - \psi \star \varphi).$$

Equivalently, we may use the Fourier transform $\widetilde{\varphi}(p)$ of the Weyl symbol (momentum representation). The existence of this field, depending on the *commutative* variables p_1 and p_2, corresponds to the commutativity of the momentum operators of the system considered.

The 'second quantization' in the Euclidean case amounts to calculating the path integral over a *set of operator symbols* which gives the generating functional $\mathcal{Z}[J]$ for Green functions:

$$\mathcal{Z}[J] = \mathfrak{N}^{-1} \int \mathcal{D}\varphi_{\text{w}}(x) \exp\{-S[\varphi_{\text{w}}, J, \star]\} \qquad (3.4.160)$$

where $S[\varphi_w, J, \star]$ is the operator action (3.4.157) on the non-commutative space expressed in terms of symbols or, in other words, the usual classical action in which the ordinary point-wise multiplication of the fields is substituted by the star-product.

The Weyl symbol has some special properties which makes it convenient for the calculations. In particular, the explicit form of the $\star$-product which makes the algebra of Weyl symbols isomorphic to the operator algebra is defined by the expression

$$\begin{aligned}(\varphi_w \star \psi_w)(x) &= \varphi_w(x) \exp\left\{i\frac{\lambda^2}{2} \overleftarrow{\partial}_i \varepsilon^{ij} \overrightarrow{\partial}_j\right\} \psi_w(x) \\ &= \sum_{m=0}^{\infty} \frac{1}{m!} \left(\frac{i\lambda^2}{2}\right)^m \varepsilon^{i_1 j_1} \cdots \varepsilon^{i_m j_m} (\partial_{i_1} \cdots \partial_{i_m} \varphi_w)(\partial_{j_1} \cdots \partial_{j_m} \psi_w) \\ &= \varphi_w(x) \psi_w(x) + \mathcal{O}(\lambda^2).\end{aligned} \qquad (3.4.161)$$

This immediately shows that a quadratic term in the non-commutative QFT action, written in terms of the Weyl symbols, has the same form as that on the classical space:

$$\begin{aligned}\int d^2 x \, (\varphi_w \star \psi_w)(x) &= \int d^2 x \, \varphi_w(x) \exp\left\{-i\frac{\lambda^2}{2} \overrightarrow{\partial}_i \varepsilon^{ij} \overrightarrow{\partial}_j\right\} \psi_w(x) \\ &= \int d^2 x \, \varphi_w(x) \psi_w(x)\end{aligned} \qquad (3.4.162)$$

because ε^{ij} is antisymmetric. Therefore, the free action of the non-commutative QFT in terms of the Weyl symbols has the same form as the usual QFT on commutative space. Higher-order (interaction) terms contain non-locality, but the analysis in Filk (1996) shows that they do not remove ultraviolet divergences.

One more property of the Weyl symbols is their nice behaviour with respect to linear canonical transformations: if we consider transformations (cf (3.4.151))

$$\widehat{x}'_i = M_{ij} \widehat{x}_j + b_i$$

the corresponding Weyl symbol transforms as follows

$$\varphi'_w(x) = \varphi_w(Mx + b)$$

i.e. it transforms as an ordinary scalar field. This essentially simplifies our study of the invariance properties of non-commutative quantum field theory.

When considering the Green functions $G_w(x, y) \equiv \langle \varphi_w(x) \varphi_w(y) \rangle$, we should take into account the fact that the value of an operator symbol at a point on the classical counterpart of a non-commutative space has no direct physical or even mathematical meaning. Only the total function can be considered as a symbol and this defines the corresponding operator. Thus the function $G_w(x, y)$ has the meaning of an operator symbol acting in the direct product $\mathcal{H} \otimes \mathcal{H}$ of two copies of the Hilbert space in which a representation of the coordinate algebra is realized. Now let us recall that in standard QFT, the points of a commutative space (labeled by values of the coordinates x_1, x_2) are considered to 'enumerate' different degrees of freedom of the field system. In non-commutative geometry there are no longer any points but there are states in the Hilbert space of representations of a coordinate algebra instead. Thus we must consider a quantum field in non-commutative QFT as a map from a set of states on the corresponding non-commutative space into the algebra of secondary quantized operators, so that the physically meaningful object in non-commutative QFT is the mean values of the field operators: $\langle \Psi | \hat{\varphi} | \Psi \rangle$, $|\Psi\rangle \in \mathcal{H}$. Of course, we can choose any complete set of states, but for clear physical interpretation and comparison with the commutative limit, the set should satisfy the following requirements:

(i) the states must be localized in spacetime and
(ii) as the parameter of non-commutativity λ goes to zero, the states must shrink to a point.

This consideration shows that in order to convert the Green function $G_w(x, y)$ into a physically meaningful object, we must average it over some localized state. States which correspond to optimal rotationally invariant localization around the point $x = (x_1, x_2)$ of the plane are uniquely (up to a phase factor) given as the coherent states $|\xi\rangle$ for the operators (3.4.153): $|\xi\rangle = \exp\{\xi\hat{a}^\dagger - \xi^*\hat{a}\}|0\rangle$ ($|0\rangle$ is the vacuum state in the Fock space $\mathcal{F}$; $\xi = (x_1 + ix_2)/(\sqrt{2}\lambda)$); see (2.3.107).

It can be shown that

$$\varphi_N(x) \equiv \langle\xi|\hat{\varphi}|\xi\rangle = \int d^2x'\, \omega_\lambda(x - x')\varphi_w(x') \tag{3.4.163}$$

with the smearing function

$$\omega_\lambda(x - x') = \frac{1}{\pi\lambda^2}\exp\left\{-\frac{(x - x')^2}{\lambda^2}\right\}.$$

In fact, $\varphi_N(x)$ is the *normal symbol* (see section 2.3.1) of the operator $\hat{\varphi}$. Therefore, the physical Green function $G_\lambda(x, y)$ is given as

$$G_\lambda(x, y) = \langle\varphi_N(x)\varphi_N(y)\rangle = \int d^2x\, d^2y\, \omega_\lambda(x - x')\omega_\lambda(y - y')G_w(x', y') \tag{3.4.164}$$

and it represents a quantum average of the true field functional $\varphi_N(x)\varphi_N(y) = \langle\xi|\hat{\varphi}|\xi\rangle\langle\zeta|\hat{\varphi}|\zeta\rangle$ (where $\xi = (x_1 + ix_2)/\lambda\sqrt{2}$, $\zeta = (y_1 + iy_2)/\lambda\sqrt{2}$). Similarly, any higher Green functions $G_\lambda(x_1, \ldots, x_n)$ are obtained by smearing the corresponding Green functions $G_w(x_1, \ldots, x_n)$.

The formal Green functions $G_w(x_1, \ldots, x_n)$ are, as a matter of rule, singular if some arguments coincide. However, the physical Green functions $G_\lambda(x_1, \ldots, x_n)$ are regular due to *intrinsic* effective smearing induced by the non-commutativity of the coordinates. In a free field, the formal Green function $G_w^{(0)}(x, y) = \langle\varphi_w(x)\varphi_w(y)\rangle_0$ is given by the standard formula

$$G_w^{(0)}(x, y) = \frac{1}{(2\pi)^2}\int d^2k\, \frac{e^{ik(x-y)}}{k^2 + m^2}. \tag{3.4.165}$$

According to (3.4.164) the corresponding physical Green function $G_\lambda^{(0)} = \langle\varphi_N(x)\varphi_N(y)\rangle_0$ can be straightforwardly calculated with the result

$$G_\lambda^{(0)}(x, y) = \frac{1}{(2\pi)^2}\int d^2k\, \frac{e^{ik(x-y)-\lambda^2 k^2/2}}{k^2 + m^2}. \tag{3.4.166}$$

This can easily be derived by use of the normal symbols. In this case the star-product for normal symbols has the form

$$\varphi_N(\bar{\xi}, \xi) \star \varphi_N(\bar{\xi}, \xi) = \varphi_N(\bar{\xi}, \xi)\exp\{\lambda^2 \overleftarrow{\partial}_\xi \overrightarrow{\partial}_{\bar{\xi}}\}\varphi_N(\bar{\xi}, \xi). \tag{3.4.167}$$

The free action in terms of the normal symbols takes the form

$$S_0^{(N)} = \int d^2\xi\, [\partial_i \varphi_N(\bar{\xi}, \xi)\exp\{\lambda^2 \overleftarrow{\partial}_\xi \overrightarrow{\partial}_{\bar{\xi}}\}\partial_i\varphi_N(\bar{\xi}, \xi) + m^2\varphi_N(\bar{\xi}, \xi)\exp\{\lambda^2 \overleftarrow{\partial}_\xi \overrightarrow{\partial}_{\bar{\xi}}\}\varphi_N(\bar{\xi}, \xi)]$$

$$= \frac{1}{(2\pi)^2}\int d^2\kappa\, \widetilde{\varphi}_N(-\kappa)(2\bar{\kappa}\kappa + m^2)e^{\lambda^2\bar{\kappa}\kappa}\widetilde{\varphi}_N(\kappa). \tag{3.4.168}$$

Here $\widetilde{\varphi}_N(\kappa)$ is the Fourier transform of the normal symbol

$$\varphi_N(\bar{\xi}, \xi) = \frac{1}{(2\pi)^2} \int d^2\kappa \, e^{i(\kappa\bar{\xi} + \bar{\kappa}\xi)} \widetilde{\varphi}_N(\kappa)$$

and $\kappa = \lambda(k_1 + ik_2)/\sqrt{2}$.

Whereas $G_w^{(0)}(x, y)$ is logarithmically divergent for $x \to y$, the physical Green function is finite:

$$|G_\lambda^{(0)}(x, y)| \leq G_\lambda^{(0)}(x, x) = \frac{1}{(2\pi)^2} \int d^2k \, \frac{e^{-\lambda^2 k^2/2}}{k^2 + m^2} \quad (3.4.169)$$

depending only on a dimensionless parameter $a = \lambda m$ characterizing the non-commutativity.

If the interaction is switched on, the problem of a perturbative determination of the full Green function $G_\lambda = \langle \varphi_N(x) \varphi_N(y) \rangle$ naturally appears. Within perturbation theory the problem is reduced to calculating the free-field averages of the type $\langle \varphi_N(x) \varphi_N(y) S_{\text{int}}^n \rangle_0$. However, now the problem of a non-commutative generalization of the interaction term arises. If we choose, as a commutative prototype, the $(\varphi^* \varphi)^2$-interaction, the most direct non-commutative generalization is

$$S_{\text{int}}^\lambda[\hat{\varphi}, \hat{\varphi}^\dagger] = g \int d^2x \, \varphi_N^*(x) \star \varphi_N(x) \star \varphi_N^*(x) \star \varphi_N(x). \quad (3.4.170)$$

This action produces vertices containing factors $e^{\lambda^2 k^2/2}$ on each leg with the momentum $k_i, i = 1, 2, 3, 4$, plus additional phase factors $\exp\{\pm i\lambda^2(k_1 \times k_2 + k_3 \times k_4)/2\}$ (here $k \times p \stackrel{\text{def}}{=} \varepsilon_{ij} k_i p_j$). The Gaussian factors $e^{-\lambda^2 k^2/2}$ from the propagators are canceled in Feynman diagrams and the ultraviolet divergences appear. Of course, calculations with different types of operator symbol, being different at intermediate steps, give the same physical results. Note, however, that the normal symbols of the field operators on the non-commutative plane have a much clearer physical interpretation since they are related (in fact, equal) to the mean values over localized coherent states.

However, this is not the only possibility. Insisting only on a commutative limit condition, $\lim_{\lambda \to 0} S_{\text{int}}^\lambda[\hat{\varphi}, \hat{\varphi}^\dagger] = S_{\text{int}}[\varphi, \varphi^*]$, the integrand in the non-commutative integral $I[\hat{\varphi}^\dagger \hat{\varphi} \hat{\varphi}^\dagger \hat{\varphi}]$ is defined up to the *operator ordering*. There is no problem in modifying the operator ordering of the generators $\widehat{x}_1$ and $\widehat{x}_2$ in the integrand $\hat{\varphi}^\dagger \hat{\varphi} \hat{\varphi}^\dagger \hat{\varphi}$ in such a way that the vertices will not contain the exponential factors $\exp\{\lambda^2 k_i^2/2\}$ on legs. For example, we can use the normal symbols to construct the free action but the Weyl symbols to construct the interaction part. The resulting action will lead to ultraviolet-regular Feynman diagrams. However, besides this pragmatic point of view, any deeper principle preferring such a different ordering is not known so far.

◇ Symmetry transformations on the quantum plane

Some subgroup of the group of the canonical transformations of the commutation relations for the coordinate operators can be considered as a group of spacetime symmetry for non-commutative QFT. As we discussed earlier, the degrees of freedom of non-commutative QFT correspond to a set of localized (e.g., coherent) states. Thus a natural question about the behaviour of such states under the quantum spacetime symmetry transformations arises.

The fact that a linear transformation preserves commutation relations for a set of some operators means that the latter are *tensor operators* (see supplement IV). It is worth separating the commuting from the non-commuting operators:

(1) *A set of commutative operators.* For a general linear transformation of commutative operators $x_i \to x'_i = M_{ij} x_j + b_i$, where M_{ij}, b_i are ordinary c-number group parameters, the vector $|\psi_x\rangle$ remains an eigenvector of the transformed operator x' but with a shifted eigenvalue $x'_i = M_{ij} x_j + b_i$.

(2) *A set of non-commutative operators: tensor operator.* A tensor operator $\widehat{A}_i$ acting in some Hilbert space $\mathcal{H}$, has, by the definition, the property

$$\widehat{A}'_i \equiv M_{ij}(g)\widehat{A}_j = \widehat{U}(g)\widehat{A}_i\widehat{U}^{-1}(g) \qquad (3.4.171)$$

where $M_{ij}(g)$ $(i, j = 1, \ldots, d)$ is a matrix finite-dimensional representation of a Lie group G, $g \in G$ and $\widehat{U}(g)$ is a unitary operator in the Hilbert space $\mathcal{H}$. In general, the components $\widehat{A}_i$ $(i = 1, \ldots, d)$ of a tensor operator do not commute with each other. Consider an eigenvector $|\lambda\rangle_A$ of one component, say $\widehat{A}_d$, of the tensor operator. After the transformation, the eigenvector $|\lambda\rangle_{A'}$ of the transformed component A'_d is related to $|\lambda\rangle_A$ by the operator $\widehat{U}(g)$:

$$|\lambda\rangle_{A'} = \widehat{U}(g)|\lambda\rangle_A = \sum_{\lambda'} {}_A\langle\lambda|\widehat{U}(g)|\lambda'\rangle^*_A |\lambda'\rangle_A. \qquad (3.4.172)$$

Then considering the action of transformed component $\widehat{A}'_d$ on the initial eigenstate $|\lambda\rangle_A$, we obtain

$$\begin{aligned}\widehat{A}'_d|\lambda\rangle_A &= \widehat{U}(g)\widehat{A}_d\widehat{U}^{-1}(g)|\lambda\rangle_A \\ &= \sum_{\lambda'} \lambda' {}_A\langle\lambda'|\widehat{U}(g)|\lambda\rangle_A |\lambda'\rangle_{A'}.\end{aligned} \qquad (3.4.173)$$

Let us apply this consideration (well known in standard quantum mechanics) to the examples of Euclidean and pseudo-Euclidean quantum planes. While ultraviolet behaviour in these cases is the same, their properties with respect to the symmetry transformations are quite different.

We shall consider only the homogeneous part of the transformations. In the Euclidean plane, these are rotations (3.4.151) (a one-dimensional subgroup of the group $Sp(2) \sim SL(2, \mathbb{R})$ of the canonical transformations). The corresponding creation and annihilation operators (3.4.153) are transformed separately

$$\hat{\alpha} \to e^{i\phi}\hat{\alpha} \qquad \hat{\alpha}^\dagger \to e^{-i\phi}\hat{\alpha}^\dagger$$

so that the corresponding localized (coherent) states $|\xi\rangle$ are transformed in very simple way:

$$|\xi\rangle \longrightarrow |e^{i\phi}\xi\rangle. \qquad (3.4.174)$$

Thus the localized coherent states are transformed in the simple and physically transparent way. On the contrary, coordinate eigenstates are transformed non-locally according to (3.4.173). Indeed, the coordinates are transformed under Euclidean rotations by the formula

$$\begin{aligned}\widehat{x}_1 &\to \widehat{x}'_1 = (\cos\phi)\widehat{x}_1 + (\sin\phi)\widehat{x}_2 = \widehat{U}_\phi \widehat{x}_1 \widehat{U}_\phi^{-1} \\ \widehat{x}_2 &\to \widehat{x}'_2 = -(\sin\phi)\widehat{x}_1 + (\cos\phi)\widehat{x}_2 = \widehat{U}_\phi \widehat{x}_2 \widehat{U}_\phi^{-1}.\end{aligned} \qquad (3.4.175)$$

The explicit form of the operator $\widehat{U}_\phi$ is easily found and proves to be

$$\widehat{U}_\phi = \exp\left\{-\frac{i}{2}\phi(\widehat{x}_1^2 + \widehat{x}_2^2)\right\}.$$

Formally, this operator coincides with the evolution operator for a particle in the harmonic potential. We have already calculated it by the path-integral method (see (2.2.79))

$$\langle x'_1|\widehat{U}_\phi|x_1\rangle = \frac{1}{\sqrt{2\pi i\lambda^2}}\sqrt{\frac{1}{\sin\phi}}\exp\left\{\frac{1}{4\lambda^2}\frac{1}{\sin\phi}[((x'_1)^2 + x_1^2)\cos\phi - 2x'_1 x_1]\right\}. \qquad (3.4.176)$$

Insertion of this kernel into the formulae (3.4.172) and (3.4.173) leads to the non-local transformation of eigenstates of the operator $\hat{x}_1$ (eigenstates of $\hat{x}_2$ are transformed quite similarly).

The situation is opposite in the case of the pseudo-Euclidean (Minkowski) plane. Now we have to use another subgroup of the canonical group $SL(2, \mathbb{R})$: two-dimensional Lorentz group $SO(1, 1)$

$$\begin{aligned} x_0 \to x_0' &= (\cosh \eta) x_0 + (\sinh \eta) x_1 \\ x_1 \to x_1' &= (\sinh \eta) x_0 + (\cosh \eta) x_1. \end{aligned} \qquad (3.4.177)$$

It is convenient to use the light-front variables

$$x_\pm = \frac{1}{\sqrt{2}}(x_0 \pm x_1).$$

The boosts (3.4.177) now have the simple form

$$x_\pm \to e^{\pm \eta} x_\pm$$

(η has the meaning of rapidity). On the non-commutative plane the coordinates satisfy the commutation relations

$$[\hat{x}_+, \hat{x}_-] = i\lambda^2. \qquad (3.4.178)$$

The corresponding annihilation and creation operators

$$\hat{\alpha} = \frac{1}{\lambda\sqrt{2}}(\hat{x}_+ + i\hat{x}_-) \qquad \hat{\alpha}^\dagger = \frac{1}{\lambda\sqrt{2}}(\hat{x}_+ - i\hat{x}_-) \qquad (3.4.179)$$

are transformed non-trivially

$$\begin{aligned} \hat{\alpha} \to \hat{\alpha}_\eta &= (\cosh \eta)\hat{\alpha} + (\sinh \eta)\hat{\alpha}^\dagger = \widehat{U}_\eta \hat{\alpha} \widehat{U}_\eta^{-1} \\ \hat{\alpha}^\dagger \to \hat{\alpha}_\eta^\dagger &= (\sinh \eta)\hat{\alpha} + (\cosh \eta)\hat{\alpha}^\dagger = \widehat{U}_\eta \hat{\alpha}^\dagger \widehat{U}_\eta^{-1}. \end{aligned} \qquad (3.4.180)$$

The explicit form of the operator $\widehat{U}_\eta$ is easily found and proves to be

$$\widehat{U}_\eta = \exp\{-\tfrac{1}{2}\eta((\hat{\alpha}^\dagger)^2 - \hat{\alpha}^2)\}.$$

Thus now the corresponding localized coherent states are transformed as

$$|\xi\rangle \longrightarrow |\xi_\eta\rangle = \widehat{U}_\eta |\xi\rangle.$$

The most convenient way of calculating the matrix elements $\langle \zeta | \widehat{U}_\eta | \xi \rangle$ uses the path integral

$$\langle \zeta | \widehat{U}_\eta | \xi \rangle = \int \prod_\tau \frac{d\bar{z}(\tau)\, dz(\tau)}{2\pi} \exp\left\{ \bar{\zeta} z(\eta) + \int_0^\eta d\tau\, [-\bar{z}(\tau)\dot{z}(\tau) - \tfrac{1}{2}(\bar{z}^2(\tau) - z^2(\tau))] \right\}.$$

This is a Gaussian integral and, as we know, its value is given by the integrand at the extremum trajectory of the exponent with the boundary conditions: $z(0) = \xi$, $\bar{z}(\eta) = \bar{\zeta}$. The result is:

$$\langle \zeta | \widehat{U}_\eta | \xi \rangle = \exp\left\{ -\frac{|\zeta|^2}{2} - \frac{|\xi|^2}{2} + \frac{\bar{\zeta}\xi}{\cosh \eta} - \frac{\bar{\zeta}^2 - \xi^2}{2} \tanh \eta \right\}.$$

Now to realize the properties of the transformed state $|\xi_\eta\rangle = \widehat{U}_\eta|\xi\rangle$ we can calculate it in the coordinate representation (either x_+ or x_-). For example,

$$|\langle x_+|\xi_\eta\rangle|^2 \equiv |\langle x_+|\widehat{U}_\eta\xi\rangle|^2 = \frac{1}{\sqrt{\pi}\lambda e^{-\eta}} \exp\left\{-\frac{(x_+ - \lambda e^{-\eta}\xi_1)^2}{\lambda^2 e^{-2\eta}}\right\}$$

(here $\xi = (\xi_1 + i\xi_2)/\sqrt{2}$). This expression shows that $|\xi_\eta\rangle$ is also a localized state and with respect to the coordinate x_+ it is located around the point $e^{-\eta}\sqrt{2}\lambda \operatorname{Re}\xi$ with the dispersion $(\lambda e^{-\eta})^2$ (while $|\xi\rangle$ is located around $\sqrt{2}\lambda \operatorname{Re}\xi$ with the dispersion λ^2). Similarly, with respect to the coordinate x_-, the state $|\xi_\eta\rangle$ is located around the point $e^{\eta}\sqrt{2}\lambda \operatorname{Im}\xi$ with the dispersion $(\lambda e^{\eta})^2$.

◇ **Remarks on the relation between an ultraviolet behaviour of QFT on non-commutative spacetimes and topology**

As we have mentioned, the transition to a non-commutative spacetime does not necessarily lead to an ultraviolet regularization of the quantum field theory constructed in this space, at least in the most natural way of introducing non-commutativity as presented above. In particular, QFT on non-commutative planes with Heisenberg-like commutation relations for coordinates and a deformed plane with quantum $E_q(2)$-symmetry still contain divergent tadpoles (Filk 1996, Chaichian *et al* 2000). However, in general, theories which have the same ultraviolet behaviour on classical spaces may acquire essentially different properties after the quantization. The reason is that quantization procedure is highly sensitive to the topology of the manifold under consideration. Thus, while in classical spacetime the theories on a sphere, cylinder or plane have ultraviolet divergences, in non-commutative spacetime the two-dimensional theories on the fuzzy sphere and on the quantum cylinder do not have divergences at all. This can be traced to the compactness properties of the spacetime in question:

- In the case of a fuzzy sphere, models contain a finite number of modes and thus all the usual integrations are replaced by final sums and, consequently, no ultraviolet divergences can appear.
- In the case of a cylinder, we cannot *a priori* claim whether the quantum field theory is finite. However, the non-commutativity of the spacetime together with the compactness of the space (circle) lead to the intrinsic cutoff in the energy modes. This guarantees the removal of ultraviolet divergences in the two-dimensional case (Chaichian *et al* 2000).
- On a non-commutative plane (with a commutative limit which is non-compact in both directions) with Heisenberg-like or even with commutation relations induced by quantum groups, the non-commutativity of the spacetime does not lead to an ultraviolet-regular theory (Filk 1996, Chaichian *et al* 2000).

Thus, the non-commutativity itself does not guarantee the removal of ultraviolet divergences: in addition, global topological restrictions are needed—namely, at most one dimension (time) is allowed to be non-compact, in order to achieve the removal of ultraviolet divergences of a quantum field theory formulated in a non-commutative spacetime of arbitrary dimensions.

Chapter 4

Path integrals in statistical physics

In the first three chapters we have considered problems related to the physical behaviour of one or a few particles (or some other physical objects which can be effectively described by one or a few (quasi)particles as, e.g., the random walk model in polymer physics considered in chapter 1). Though quantum field theory describes systems with an arbitrary number of degrees of freedom (an arbitrary number of (quasi)particles), in chapter 3 we actually applied it to systems with a restricted and small number of particles (e.g., for the description of the scattering process for a few particles). Another way to express this fact is to say that in chapter 3 we have considered field theories at *zero temperature*. However, the majority of realistic systems contain many identical (indistinguishable) particles such as atoms, electrons, photons etc. An attempt to describe these systems in terms of the individual trajectories of *all* particles is absolutely hopeless. Instead, we are interested in the collective behaviour of systems and describe them in terms of *partition functions, mean values, correlation functions*, etc. The methods of derivation, analysis and calculation of such collective characteristics constitute the subject of *classical and quantum statistical mechanics*.

In fact, the statistical properties of indistinguishable particles play an important role in the quantum mechanics of a *few* particles as well. As is known from standard courses (and as we have previously mentioned in this book), the Schrödinger wavefunction can be classified according to the irreducible representations of the permutation group (corresponding to permutations of the quantum numbers of different but indistinguishable particles). In more than two space dimensions, the representations which definitely occur in nature are:

- completely symmetrical under the permutations;
- completely antisymmetrical.

The particles which always appear with symmetric wavefunctions are called *bosons* and those with antisymmetric wavefunctions are called *fermions*, the names which we have already used in this book many times. These features with respect to permutations of the two sorts of particle are in one-to-one correspondence with the specific forms of the partition functions and, hence, with their different collective behaviour: *Bose* or *Fermi* statistics. Note that, in general, systems of particles with more general symmetry properties under a permutation group (e.g., obeying the so-called *parastatistics*, see section 4.3.3) may exist. This possibility is especially important for the two-dimensional physics, where it can be responsible for some interesting phenomena (in particular, *the fractional Hall effect*).

The problem to be discussed in this chapter is how to incorporate the statistical properties into the path-integral formalism for the study of many-particle systems. As in the preceding chapters, we start from a short review of the basic notions of classical and quantum statistical mechanics. Then, in section 4.2, we discuss some applications of the path-integral formalism in classical statistical physics. These applications are mainly related to a convenient representation of the so-called *configuration integral* (entering the classical *partition function*) for easier calculation. In section 4.3, we pass to quantum systems and in order to establish a 'bridge' to what we considered in chapter 2, we introduce, at first, a path-integral representation for an arbitrary but *fixed* number of indistinguishable particles obeying the Bose or Fermi statistics. As we shall see, this problem is mathematically equivalent to the construction of path integrals in a restricted (bounded) domain of the $3N$-dimensional space (N is the number of particles, bosons or fermions, in the three-dimensional space) with appropriate conditions depending of the type of statistics. We shall also discuss the generalization to the case of particles with parastatistics.

The next step (section 4.4) is the transition to the case of an *arbitrary number* of particles, which requires the use of the second quantization, and hence, field theoretical methods. The consideration of path-integral methods in quantum field theory in chapter 3 will be highly useful in the derivation of the path-integral representation for partition functions of statistical systems with an arbitrary number of particles. Moreover, this path-integral representation reveals similarity (at least formal) between the basic objects of classical or quantum statistics and those of the Euclidean quantum field theory (we have stressed this similarity in the introduction).

A part of section 4.4 and two subsequent sections 4.5, 4.6 are devoted to some of the most fruitful applications of the path-integral techniques to the study of fundamental problems of quantum statistical physics, such as the analysis of critical phenomena (phase transitions), calculations in field theory at finite (non-zero) temperature and in field theory at finite energy (describing systems with the microcanonical distribution) as well as to the study of non-equilibrium systems and the phenomena of superfluidity and superconductivity. One subsection is devoted to the presentation of basic elements of the method of *stochastic quantization*, which non-trivially combines ideas borrowed from the theory of stochastic processes (chapter 1), quantum mechanics (chapter 2) and quantum field theory (chapter 3), as well as methods of non-equilibrium statistical mechanics (present chapter). The last section of this chapter is devoted to systems defined on lattices. Of course, there are no continuous trajectories on a lattice and, hence, no true path integrals in this case. On the other hand, we have learned that the path integrals in quantum mechanics are defined through their discrete approximations. Therefore, the partition function or generating functional for a system on a lattice are, in fact, very close to those for continuous systems and may serve as a regularization for the latter. Thus, the discussion of some principal ideas of the lattice field theory in a book devoted to path integrals seems to be quite relevant.

4.1 Basic concepts of statistical physics

As in all preceding chapters, we start from a short review of the main facts and an introduction of the main objects of statistical physics (see, e.g., Balescu (1975), Kittel (1987) and Feynman (1972a)). The reader well acquainted with the standard formulae and statements of this subject may use this section only for checking the notational conventions.

- The principal aim of statistical physics is to express the properties of macroscopic objects, i.e. systems consisting of a huge number of identical particles (molecules, atoms, electrons etc), through the properties of these constituents and their mutual interactions.

The existence of a *large* number of particles leads to specific *statistical* laws. The most important of them is that a system in an arbitrary state and being in contact with a thermal reservoir tends to turn into some *equilibrium state*. The properties of the latter are defined by such general characteristics of the initial state as the number of particles, their total energy etc. The process of transition of a system into its equilibrium state is called *relaxation* and the characteristic time of this process is called the *relaxation time*.

◇ Classical statistical systems

Let us consider a system of N particles assuming, for simplicity, that they do not have internal degrees of freedom (as, e.g., spin). All statistical properties of this system are encoded in its phase-space *partition function* $w(x, p)$, $x = \{x_1, x_2, \ldots, x_{3N}\}$; $p = \{p_1, p_2, \ldots, p_{3N}\}$, the quantity $w(x, p) d^{3N}x\, d^{3N}p$ being the probability to find the coordinates and momenta of the particles of the system in the vicinity $d^{3N}x\, d^{3N}p$ of the values x, p. If a system is not in the equilibrium state, the partition function depends also on the time t.

A statistical description starts with the fact that the possible forms of the partition function of a system in the equilibrium state can be determined on the basis of a general consideration without going into the details of the system's behaviour (i.e. without solving the equations of motion of the system). There are three basic cases:

- *Microcanonical distribution.* In a closed system, the total energy is conserved and the points in the corresponding phase space characterizing the states of the system are *uniformly* distributed over a surface of a given value of the energy. This leads to *the microcanonical distribution*:

$$w_E(x, p) = A\, \delta(H(x, p) - E) \qquad (4.1.1)$$

where $H(x, p)$ is the energy of the system expressed in terms of the phase variables (i.e. the Hamiltonian) and E is some value of it (A is a normalization constant).

- *Canonical distribution.* In reality, we mainly deal with some *small* subsystems of bigger systems (strictly speaking, there are no absolutely closed (isolated) systems at all, except, perhaps, the whole Universe). The distribution function of a subsystem is different from (4.1.1) but does not depend on the concrete properties of the rest of the entire system, called the *thermostat*. To obtain the distribution function of a subsystem from (4.1.1), we should integrate over the coordinates and momenta of particles of the thermostat. This can be done by taking into account the *smallness* of the energy of the subsystem in comparison with that of the thermostat. As a result, we arrive at the *canonical distribution*

$$w_C(x, p) = \exp\left\{\frac{F - H(x, p)}{kT}\right\} \equiv \exp\{\beta(F - H(x, p))\}. \qquad (4.1.2)$$

The quantity T in this expression has the physical meaning of the *temperature* of the system. It is convenient to introduce the inverse temperature $\beta \equiv 1/k_B T$ (in units of the Boltzmann constant k_B). The normalization coefficient for the distribution (4.1.2),

$$\mathcal{Z}^{(\text{cl})} \stackrel{\text{def}}{\equiv} \exp\{-\beta F\} = \int dx\, dp\, \exp\{-\beta H(x, p)\} \qquad (4.1.3)$$

is called the *partition function* (it is also called the *statistical integral* or *statistical sum*) and is defined by the condition

$$\int d^{3N}x\, dp^{3N} \exp\{\beta(F - H(x, p))\} = 1. \qquad (4.1.4)$$

In distinction from the microcanonical distribution, the energy of a system obeying the canonical distribution is not fixed but distributed in a thin interval around its mean value (physically, this corresponds to the possibility of an energy exchange with the thermostat).

- *Grand canonical distribution.* If the particles of a subsystem may leave it and return through a surface bounding the subsystem, the probability for the subsystem to be in a state with the energy $H(x, p)$ and the number of particles N is given by the *grand canonical distribution*

$$w_G(x, p) = \exp\{\beta(\Omega - H(x, p) + \mu N)\} \tag{4.1.5}$$

where μ is the so-called *chemical potential* related to an average number of particles in the subsystem and $\exp\{\beta\Omega\}$ is defined by the normalization condition

$$\int d^{3N}x\, d^{3N}p\, w_G(x, p) = 1$$

that is

$$\Xi^{(\text{cl})} \stackrel{\text{def}}{\equiv} \exp\{-\beta\Omega\} = \int d^{3N}x\, d^{3N}p\, \exp\{-\beta(H(x, p) - \mu N)\}. \tag{4.1.6}$$

◇ **Quantum statistical systems**

In quantum mechanics, the distribution functions are substituted by the *density operator* (also called the *density matrix* or *statistical operator*) $\widehat{\rho}$. The mean value of a physical quantity represented by an operator $\widehat{f}$ is given by the expression:

$$\langle f \rangle = \text{Tr}[\widehat{\rho}\widehat{f}]. \tag{4.1.7}$$

The density operator in an equilibrium state and in the coordinate representation has the form

$$\widehat{\rho}(x, x') = \sum_n w_n \psi_n(x) \psi_n^*(x') \tag{4.1.8}$$

where $\psi_n(x)$ are the eigenfunctions of the Hamiltonian operator of the system under consideration and w_n is the distribution of probabilities that the quantum system is in the state with the energy E_n. The exact form of this distribution depends again on the general properties of the considered system:

- For closed systems with fixed total energy, volume and number of particles, the density operator has the form

$$\widehat{\rho}_M = A\delta(\widehat{H}(\widehat{x}, \widehat{p}) - E) \tag{4.1.9}$$

and the distribution is given by

$$w_n = A(E_n, V, N) \tag{4.1.10}$$

where $A(E_n, V, N)$ is the statistical weight, i.e. the number of quantum states in the vicinity of the energy E_n.
- For *subsystems* with fixed number of particles (canonical ensemble), the density operator has the form:

$$\widehat{\rho}_C = \mathcal{Z}^{-1} e^{-\beta \widehat{H}} \tag{4.1.11}$$

the distribution being given by an expression analogous to (4.1.2):

$$w_n = \exp\{\beta(F - E_n)\} \tag{4.1.12}$$

with the *partition function* (normalization factor):

$$\mathcal{Z} \equiv \exp\{-\beta F\} = \sum_n \exp\{-\beta E_n\}. \tag{4.1.13}$$

In the operator form, this formula can be rewritten as

$$\mathcal{Z} = \operatorname{Tr} \exp\{-\beta \widehat{H}\} \qquad (4.1.14)$$

the operator $\widehat{H}$ being the Hamiltonian of a system.

- For a *grand canonical ensemble* of quantum particles (i.e. for systems with varying number of particles), the density operator takes the form:

$$\hat{\rho}_{\mathrm{G}} = \Xi^{-1} e^{-\beta(\widehat{H} - \mu \widehat{N})} \qquad (4.1.15)$$

($\widehat{N}$ is the particle number operator) and the distribution w_n reads as

$$w_n = \exp\{\beta(\Omega - E_n - \mu N_n)\} \qquad (4.1.16)$$

with the corresponding normalization factor

$$\Xi \equiv \exp\{-\beta\Omega\} = \sum_n \exp\{-\beta(E_n - \mu N_n)\} \qquad (4.1.17)$$

which is the *grand canonical partition function*.

◇ **Thermodynamical quantities and fluctuations**

One of the main results of statistical mechanics is the clarification and statistical interpretation of thermodynamical quantities. In particular, the exponent F of the canonical partition function $\mathcal{Z} = \exp\{-\beta F\}$ has the thermodynamical meaning of the *free energy* of a system, while its derivative with respect to the temperature (at fixed external conditions, e.g., volume of the system) gives the *entropy* S:

$$S = -\frac{\partial F}{\partial T}. \qquad (4.1.18)$$

Simple manipulations allow us to derive from (4.1.18) a more general relation:

$$S = k_{\mathrm{B}} \ln \Delta n_{\mathrm{cont}} \qquad (4.1.19)$$

where Δn_{cont} is the number of states which give *essential* contribution to the partition function. The relation (4.1.19) is even valid for non-equilibrium states.

The cornerstone of statistical physics is the fact that the physical quantities X_i characterizing a *macroscopic* body are equal, approximately but with high precision, to their mean values. However, due to the approximate nature of these equalities, the quantities X_i have small stochastic *fluctuations* around their mean values, characterized by the *dispersion*

$$\mathbb{D} X_i \stackrel{\mathrm{def}}{\equiv} \langle (X_i - \langle X_i \rangle)^2 \rangle. \qquad (4.1.20)$$

The correlation between two different quantities X_i and X_j is characterized by the function

$$\langle (X_i - \langle X_i \rangle)(X_j - \langle X_j \rangle) \rangle.$$

If X_i and X_j are the values of the *same* physical quantity at the different space points labeled i and j, the corresponding characteristic

$$\langle (X_i - \langle X \rangle)(X_j - \langle X \rangle) \rangle \qquad (4.1.21)$$

Basic concepts of statistical physics 199

is called the (space) *correlation function*. When the distance between two points grows, the correlation tends to zero (usually exponentially, except the case of phase transitions), since the fluctuations at distant points are independent.

The applications of statistical mechanics to the equilibrium behaviour of different macroscopic systems are reduced to calculations of the partition function. Then, the characteristics of a system can be easily derived in the same way as the different vacuum expectation values in quantum mechanics and quantum field theory are derived from the generating functionals.

◇ **Classical limit and configuration integral**

In the quantum-mechanical formulae (4.1.11)–(4.1.17), we assumed that the spectrum of a Hamiltonian is purely discrete. This actually occurs in all systems in finite volumes. However, because of the large number of particles, these spectra are very dense and it is technically reasonable to pass to an (approximate) integration instead of the summation. Then, the partition function takes the form

$$\mathcal{Z} = \int_0^\infty dE\, g(E) \exp\{-\beta E\} \quad (4.1.22)$$

where $g(E)$ is the density of states at the energy E. In the classical limit, we can pass to the integration in the formula (4.1.13) using the standard substitution $\sum_n \to \int d^{3N}x\, d^{3N}p/(2\pi\hbar)^{3N}$ and dividing the result by $N!$, because a quantum-mechanical state is not changed under the permutation of identical particles. In the *classical limit* this yields

$$\mathcal{Z}^{(\text{cl})} = \frac{1}{(2\pi\hbar)^{3N} N!} \int d^{3N}x\, d^{3N}p\, \exp\{-\beta H(x, p)\}. \quad (4.1.23)$$

Note that this expression differs by a factor from the purely classical counterpart given by (4.1.3).

The simplest many-body system is the *ideal gas*, a collection of a large number of non-interacting particles. Due to the absence of the interaction the partition function can be computed exactly and the expression for the free energy reads

$$F = -\frac{1}{\beta} \ln \frac{L^{3N}}{\lambda_B^{3N}}, \quad (4.1.24)$$

where L^3 is the volume of the ideal gas and

$$\lambda_B \stackrel{\text{def}}{\equiv} \sqrt{2\pi\hbar^2\beta/m} \quad (4.1.25)$$

is the so-called *Boltzmann wavelength*.

For more realistic models the problem of the calculation of the partition function is very complicated. Hence, different approximation methods are required and for this aim, the path-integral approach proves to be very fruitful and powerful.

The standard Hamiltonian for a system of *classical* particles with pair-wise interaction has the form

$$H(x, p) = \sum_{i=1}^{N} \left(\frac{p_i^2}{2m} + V_1(x_i) \right) + \sum_{i<j} V_2(x_i, x_j). \quad (4.1.26)$$

Here the function $V_1(x_i)$ represents a potential in an external field, while the symmetric function $V_2(x_i, x_j)$ is a potential of pair-wise particle interactions. After the Gaussian integration over the particle momenta, the partition function becomes

$$\mathcal{Z}^{(cl)} = \frac{Q}{\lambda_B^{3N} N!} \qquad (4.1.27)$$

where again $\lambda_B = \sqrt{2\pi \hbar^2 \beta / m}$ and

$$Q \stackrel{\text{def}}{\equiv} \int d^{3N}x \, \exp\left\{-\beta\left(\sum_i V_1(x_i) + \sum_{i<j} V_2(x_i, x_j)\right)\right\}. \qquad (4.1.28)$$

This quantity Q is called the *configuration integral*; its calculation is the main technical problem in classical statistical mechanics. As a matter of rule, it is impossible to calculate this integral exactly in the case of a non-trivial interaction between particles.

The grand canonical partition function Ξ can be written as

$$\Xi = \sum_{N=0}^{\infty} \mathcal{Z}_N^{(cl)} \exp\{\beta \mu N\}$$

$$= \sum_{N=0}^{\infty} \frac{z^N}{N!} \int d^{3N}x \prod_i \exp\{-\beta V_1(x_i)\} \prod_{i<j}(1 + g_{ij}) \qquad (4.1.29)$$

where $\mathcal{Z}_N^{(cl)}$ is the canonical partition function for N particles (it is assumed that $\mathcal{Z}_0^{(cl)} = 1$) and

$$g_{ij} \equiv -1 + \exp\{-\beta V_2(x_i, x_j)\}. \qquad (4.1.30)$$

The factor at the configuration integrals in the series (4.1.29) proves to be $z^N/N!$, where the quantity

$$z \stackrel{\text{def}}{\equiv} \frac{1}{\lambda_B^3} e^{\beta \mu} \qquad (4.1.31)$$

is called the *activity*. An arbitrary term of (4.1.29) can be represented by a diagram with N vertices and lines corresponding to the so-called *superpropagator* g_{ij}. This expansion and the corresponding diagrams are called the *Mayer expansion* and *Mayer diagrams*. They are analogous to the expansion in quantum field theory and the Feynman diagrams but for *non-polynomial*, exponential interaction.

In the next section, we shall show that the use of a path-integral representation for the configuration integrals allows us to present the latter in a form more similar to that encountered in *polynomial* quantum field theory and, hence, to develop an expansion which is more convenient in many cases.

4.2 Path integrals in classical statistical mechanics

The calculation of the free energy F (see (4.1.13) and (4.1.24)) for the classical ideal gas can be carried out explicitly and straightforwardly. This is not the case for the more general problem of evaluating the partition sum for a non-ideal system. The main difficulty is the calculation of the configuration integral (4.1.28). There are several approaches to analyze the thermodynamics of non-ideal systems with short-range interactions: Mayer's method, the correlation function method, the integral equation method, the renormalization group method, etc (see, e.g., Balescu (1975)). One more possibility is to represent the classical partition function in terms of a path integral and then to use for its calculation one of the methods discussed in this book.

Consider a system of N identical particles in a volume L^3 interacting pairwise through a two-body potential. Denoting the positions of particles i and j by $\boldsymbol{x}_i$ and $\boldsymbol{x}_j$, their interaction energy $V(\boldsymbol{x}_i - \boldsymbol{x}_j)$ can often be decomposed in a natural way into the sum of two terms:

$$V(\boldsymbol{x}_i - \boldsymbol{x}_j) = V_0(\boldsymbol{x}_i - \boldsymbol{x}_j) + V_1(\boldsymbol{x}_i - \boldsymbol{x}_j). \tag{4.2.1}$$

For example, in many models, V_0 denotes a repulsive hard core potential

$$V_0(\boldsymbol{x}_i - \boldsymbol{x}_j) = \begin{cases} +\infty & \text{if } |\boldsymbol{x}_i - \boldsymbol{x}_j| \leq \sigma \\ 0 & \text{if } |\boldsymbol{x}_i - \boldsymbol{x}_j| > \sigma \end{cases}$$

($\sigma \in \mathbb{R}$ is the size of particles) and V_1 denotes an attractive potential. If the latter is a weak (i.e. its depth is small compared to $k_B T$) and a long-distance one (the range of interaction is large compared to σ), the system is called a *van der Waals gas*. The classical canonical partition function $\mathcal{Z}^{(\text{cl})}$ of such N particles is given by a multiple integral of the type (4.1.27), (4.1.28):

$$\mathcal{Z}^{(\text{cl})}(N, \beta, L^3) = \frac{1}{\lambda_B^{3N} N!} \int_{L^3} d^3 x_1 \cdots d^3 x_N \, \exp\left\{ -\beta \sum_{i<j} V_0(\boldsymbol{x}_i - \boldsymbol{x}_j) - \beta \sum_{i<j} V_1(\boldsymbol{x}_i - \boldsymbol{x}_j) \right\}. \tag{4.2.2}$$

The explicit calculation of this integral for large N and a realistic form of $V_1(\boldsymbol{x})$ is very complicated. As a step towards its solution, we may try to rewrite (4.2.2) in terms of a path integral and then either calculate it exactly (if this is possible) or use some approximation method.

The representation of the classical partition function in terms of path integrals is based on the parametrization of the potential term V_1 by means of the auxiliary Gaussian *random functions* $\varphi(\boldsymbol{x})$ with zero mean value and with the correlation function

$$\langle \varphi(\boldsymbol{x}) \varphi(\boldsymbol{x}') \rangle = -\beta V_1(\boldsymbol{x} - \boldsymbol{x}'). \tag{4.2.3}$$

We have discussed such Gaussian random functions (fields) in section 1.2.8. Making use of the results of this discussion, namely formula (1.2.239), we can present the exponential of the potential V_1 in terms of path integrals (Wiegel (1986) and references therein).

Note that the configuration integral with the potential V_0 is simple and can be calculated straightforwardly. Therefore, for brevity, we put $V_0 = 0$. We also drop the subscript of potential V_1, because in the case of a vanishing V_0 we have $V_1 = V$.

◇ **Path-integral representation for the configuration integral**

Let us consider an auxiliary functional

$$I[\eta] = \mathfrak{N}^{-1} \int \mathcal{D}\varphi(\boldsymbol{x}) \exp\left\{ -\frac{\beta}{2} \int d^3x \, d^3x' \, \varphi(\boldsymbol{x}) H(\boldsymbol{x} - \boldsymbol{x}') \varphi(\boldsymbol{x}') + i\beta \int d^3x \, \eta(\boldsymbol{x}) \varphi(\boldsymbol{x}) \right\} \tag{4.2.4}$$

where

$$\mathfrak{N} = \int \mathcal{D}\varphi(\boldsymbol{x}) \exp\left\{ -\frac{\beta}{2} \int d^3x \, d^3x' \, \varphi(\boldsymbol{x}) H(\boldsymbol{x} - \boldsymbol{x}') \varphi(\boldsymbol{x}') \right\}.$$

$H(\boldsymbol{x})$ is the kernel of some linear operator. We can choose the function $H(\boldsymbol{x})$ so that it satisfies the condition

$$\int d^3x' \, V(\boldsymbol{x} - \boldsymbol{x}') H(\boldsymbol{x}' - \boldsymbol{x}'') = \delta(\boldsymbol{x} - \boldsymbol{x}''). \tag{4.2.5}$$

This condition means that V is the inverse operator with respect to H. Then, the usual calculation of the Gaussian path integral (4.2.4) yields

$$\exp\left\{-\beta/2 \int d^3x\, d^3x'\, \eta(x) V(x-x') \eta(x')\right\} = I[\eta(x)]$$

$$= \mathfrak{N}^{-1} \int \mathcal{D}\varphi \exp\left\{\beta/2 \int d^3x\, d^3x'\, \varphi(x) H(x-x')\varphi(x') + i\beta \int d^3x\, \eta(x)\varphi(x)\right\}. \quad (4.2.6)$$

The main aim of this section is to associate (4.2.6) with the statistical mechanics of a classical system. For this purpose, we consider a system consisting of $N = N_+ + N_-$ charged pointlike particles (N_+ and N_- are the numbers of positively and negatively charged particles, respectively). The microscopic charge density of the system can be written as

$$\eta(x) = \sum_{i=1}^{N_+} \delta(x - x_i) + \sum_{j=1}^{N_-} (-1)\delta(x - x_j) \quad (4.2.7)$$

so that we obtain for the full energy

$$U(x_1 \ldots x_N) = \tfrac{1}{2}\sum_{i \neq j=1}^{N_+} V(x_i - x_j) + \tfrac{1}{2}\sum_{i \neq j=1}^{N_-} V(x_i - x_j) - \sum_{i=1}^{N_+}\sum_{j=1}^{N_-} V(x_i - x_j)$$

$$= \int d^3x\, d^3x'\, \eta(x) V(x-x') \eta(x') - (N_+ + N_-)V(0). \quad (4.2.8)$$

Using (4.2.5) and (4.2.6), we can write the exponential $\exp\{-\beta U(x_1 \ldots x_N)\}$ of the system through the path integral:

$$\exp\left\{-\frac{\beta}{2}\left(\sum_{i\neq j=1}^{N_+} V(x_i - x_j) + \sum_{i\neq j=1}^{N_-} V(x_i - x_j) - 2\sum_{i=1}^{N_+}\sum_{j=1}^{N_-} V(x_i - x_j)\right)\right\}$$

$$= e^{N\beta V(0)/2} \mathfrak{N}^{-1} \int \mathcal{D}\varphi \exp\left\{-\frac{\beta}{2}\langle \varphi, |H\varphi\rangle + i\beta\langle\eta|\varphi\rangle\right\} \quad (4.2.9)$$

where we have introduced the scalar products

$$\langle\varphi|H\varphi\rangle \stackrel{\text{def}}{\equiv} \int d^3x\, d^3x'\, \varphi(x) H(x-x')\varphi(x')$$

$$\langle\varphi|\eta\rangle \stackrel{\text{def}}{\equiv} \int d^3x\, \varphi(x)\eta(x). \quad (4.2.10)$$

◇ **Example: Kac–Uhlenbeck–Hemmer model**

For a particular form of the potential V_1 for a *one-dimensional* van der Waals gas of the type

$$V_1(y - y') = -\tfrac{1}{2} w\gamma \exp\{-\gamma|y - y'|\} \quad (4.2.11)$$

(the so-called *Kac–Uhlenbeck–Hemmer model*; w and γ are the parameters of the model), the outlined procedure leads to a Wiener-like path integral.

Note that the inverse of the correlation function of the auxiliary stochastic field, according to the representation (4.2.3), proves to be

$$\langle\varphi(y)\varphi(y')\rangle^{-1} = \frac{1}{\beta w_0 \gamma^2}\left(\frac{d^2}{dy^2} - \gamma^2\right)\delta(y-y')$$

(here we understand the correlation function to be the operator (4.2.5)). After substituting this formula into (4.2.9), the derivative d^2/dy^2 produces in the exponent the term

$$\sim \int dy \left(\frac{d\varphi}{dy}\right)^2$$

so that the resulting path integral has a Wiener–Feynman–Kac-like form, the variable y playing a role analogous to the time variable τ in a genuine Wiener integral. For more details on this application of path integrals and generating functionals, as well as for further references, we refer the reader to Wiegel (1986).

◇ Diagrammatic expansions

In order to illustrate the usefulness of the path-integral representation for the configuration integral, let us consider the grand partition function for a plasma of charged particles at a temperature $1/\beta$:

$$\Xi^{(\mathrm{cl})} = \sum_{N_+,N_-=0}^{\infty} \frac{z_+^{N_+} z_-^{N_-}}{N_+! N_-!} \int d^{3N}x \, \exp\left\{\frac{\beta}{2}\sum_{i\neq j} V(\boldsymbol{x}_i - \boldsymbol{x}_j)\right\} \qquad (4.2.12)$$

where z_+ and z_- are the activities of positively and negatively charged particles, respectively. Using (4.2.6) (or (4.2.9)), we can rewrite this equation as follows:

$$\Xi^{(\mathrm{cl})} = \sum_{N_+,N_-=0}^{\infty} \frac{z_+^{N_+} z_-^{N_-}}{N_+! N_-!} \int\int d^{3N}x \, e^{\beta N V(0)/2} \mathfrak{N}^{-1} \int \mathcal{D}\varphi \, \exp\{-\beta/2\langle\varphi|H\varphi\rangle + i\beta\langle\varphi|\eta\rangle\}. \qquad (4.2.13)$$

We assume that the system under consideration is 'neutral', $N_+ = N_-$ and such that $z_+ = z_- = z$. Keeping in mind (4.2.7), we obtain:

$$\Xi^{(\mathrm{cl})} = \mathfrak{N}^{-1} \int \mathcal{D}\varphi \, \exp\left(-\beta/2\langle\varphi|H\varphi\rangle + 2z_V \int d^3x \, \cos(\beta\varphi(\boldsymbol{x}))\right) \qquad (4.2.14)$$

where $z_V = z e^{\beta V(0)/2}$.

In order to calculate Ξ approximately, we can use the usual perturbation theory for non-Gaussian path integrals. To this aim, we introduce the auxiliary external source J and define

$$\Xi_0^{(\mathrm{cl})}[J] = \mathfrak{N}^{-1} \int \mathcal{D}\varphi \, \exp\left\{-\beta/2\langle\varphi|H\varphi\rangle + \int d^3x \, J\varphi\right\}$$

$$= \exp\left\{\frac{1}{2\beta} \int d^3x \, d^3x' \, J(\boldsymbol{x}) V(\boldsymbol{x}-\boldsymbol{x}') J(\boldsymbol{x}')\right\} \qquad (4.2.15)$$

so that the mean value of any combination $f(\varphi(\boldsymbol{x}))$ of the fields $\varphi(\boldsymbol{x})$ is given by the functional derivative

$$\langle f(\varphi(\boldsymbol{x}))\rangle_H \equiv \mathfrak{N}^{-1} \int \mathcal{D}\varphi f(\varphi(\boldsymbol{x})) e^{-\beta/2\langle\varphi|H\varphi\rangle} = f\left(\frac{\delta}{\delta J(\boldsymbol{x})}\right) \Xi_0^{(\mathrm{cl})}[J]\bigg|_{J=0}. \qquad (4.2.16)$$

Thus, for $\Xi^{(cl)}$ we obtain the perturbation series

$$\Xi^{(cl)} = \sum_{n=1}^{\infty} \frac{1}{n!} \left[2z_V \int d^3x \sum_{k=0}^{\infty} \frac{(-1)^k}{(2k)!} \left(\beta \frac{\delta}{\delta J(x)} \right)^{2k} \right]^n \Xi_0^{(cl)}[J] \bigg|_{J=0} \qquad (4.2.17)$$

with the modified activity z_V playing the role of an expansion parameter (an analog of the coupling constant in quantum field theory).

Following the method used in the preceding chapter for quantum field theory, we can now represent each term of the sum in (4.2.17) as a diagram with n vertices connected by k lines. Each vertex contributes an integration over d^3x, while the factor $-\beta V(r_i - r_j)$ corresponds to each line connecting the vertices i and j.

Let us consider a vertex with l lines which begins and ends at the same vertex (recall that such diagrams are called 'tadpoles'; the corresponding factor is proportional to $V(0)$) and j non-tadpole lines going out from it (external lines with respect to the vertex). The combinatorial prefactor in the integral, corresponding to such a diagram, reads as

$$\frac{(2k)!}{l!\, j!\, 2^l}$$

where $2k = 2l + j$ is the total number of lines going out from the vertex.

We can show that the summation over the number l of tadpole lines for a given vertex with a fixed number j of external lines, is equivalent to the renormalization of z_V to its original value:

$$z_V \equiv e^{\beta V(0)/2} z \longrightarrow e^{-\beta V(0)/2} z_V = z \qquad (4.2.18)$$

and the singular (local) factor $\exp\{N\beta V(0)\}$ in (4.2.9) just disappears.

Moreover, the summation of multiple lines connecting two vertices leads to the substitution of each set of diagrams with multiple lines by one diagram with one effective line, where the effective line is associated with Mayer's superpropagator $f_{ij} = e^{-\beta V(r_i - r_j)} - 1$ (cf (4.1.30)). Thus, the perturbation theory derived from the path-integral representation of the grand partition function (4.2.14) is equivalent to the Mayer series derived directly from (4.2.12) (cf (4.1.29)).

◇ The expansion of the canonical partition function in powers of density

For a system with two sorts of particle with the densities ρ_+ and ρ_-, the partition sum can be written as

$$\mathcal{Z}^{(cl)} = \frac{1}{\lambda_B^{3N} N_+! N_-!} \int d^{3N}x \, \exp\left\{ -\frac{\beta}{2} \sum_{ij} V_{ij}(r_i - r_j) \right\} \qquad (4.2.19)$$

(λ_B is defined in (4.1.25)) and for the free energy F this yields

$$-\beta F = \ln \mathcal{Z}^{(cl)} = N_+[1 - \ln \lambda_B \rho_+] + N_-[1 - \ln \lambda_B \rho_-] - \beta F_{int}$$

where F_{int} is the non-ideal part of the free energy:

$$F_{int}(N_+, N_-, \beta, L^3) = \ln \int \frac{d^3x_1}{L^3} \cdots \frac{d^3x_N}{L^3} e^{-\beta U} \qquad (4.2.20)$$

and L^3 is the d-dimensional volume of the system.

Now we rewrite $\exp\{-\beta U\}$ using (4.2.9):

$$e^{-\beta F_{\text{int}}} = \exp\left\{\frac{N\beta}{2}V(0)\right\}\mathfrak{N}^{-1}\int \mathcal{D}\varphi\, \exp\{-\beta/2\langle\varphi|H\varphi\rangle\}e^{S_1} \qquad (4.2.21)$$

$$S_1 = N_+\ln\int\frac{d^3x_+}{L^3}e^{i\varphi(x_+)} + N_-\ln\int\frac{d^3x_-}{L^3}e^{-i\varphi(x_-)}.$$

Since the path integral (4.2.21) is non-Gaussian, it can be calculated only in some approximation. For example, for a dilute gas we can use the expansion in powers of the densities ρ_+, ρ_- or the quadratic approximation (i.e. the expansion of S_1 up to the second order in φ). In the latter case, we have, for S_1 (we assume φ is oscillating around $\varphi_c = 0$),

$$S_1 \approx \frac{\rho_+ + \rho_-}{L^3}|\widetilde{\varphi}(0)|^2 - \frac{\rho_+ + \rho_-}{2}\int\frac{d^3q}{(2\pi)^3}|\widetilde{\varphi}(q)|^2 \qquad (4.2.22)$$

($\widetilde{\varphi}(q)$ is the Fourier transform of $\varphi(x)$), and for the non-ideal part of the free energy we obtain the so-called *Debye–Hückel approximation*:

$$\frac{\beta F_{\text{int}}}{L^3} \approx \frac{1}{2}\int\frac{d^3q}{(2\pi)^3}[\ln(1+\beta(\rho_+ + \rho_-)\widetilde{V}(q)) - \beta(\rho_+ + \rho_-)\widetilde{V}(q)]. \qquad (4.2.23)$$

4.3 Path integrals for indistinguishable particles in quantum mechanics

Statistical methods are applicable to an ensemble of a large number of identical (or several types of identical) particles. Before we develop path-integral methods for the derivation of statistical characteristics (partition functions) in the framework of quantum statistical mechanics, let us discuss peculiarities of path-integral representation for quantum-mechanical transition amplitudes in the case of a few (a fixed number) *indistinguishable* particles. As we have learned in chapter 2, the calculation of a quantum-mechanical transition amplitude for *distinguishable particles* can be carried out by using the Feynman–Kac formula (for a general class of scalar potentials). In order to treat identical particles, we can exploit the fact that this method separates the problem of the potential, dealt with by the Feynman–Kac functional, from the problem of a correct choice of a set of paths to be integrated over. This allows us to consider the latter problem for a non-interacting system.

The consideration here is applicable both to real and imaginary time formulations of quantum-mechanical processes. For definiteness, we shall use the imaginary-time formalism which leads to the Wiener path integrals and Brownian-like particle trajectories. The propagator over the configuration space can be obtained by the application of permutations to a linear combination of standard Brownian processes (quantum particle motion in imaginary time). The boson and fermion diffusion processes are fundamental to this approach (Lemmens *et al* 1996) and their relation to the standard Brownian motion is settled by restricting the configuration space of N particles to a specific domain, as well as by the appropriate boundary conditions: absorption for the fermion diffusion process and reflection for the boson diffusion process. In combination with the Feynman–Kac functional, this approach allows us to write the propagator of the many-body Schrödinger equation as path integration with such boundary conditions (section 4.3.1).

In section 4.3.2, we shall apply this consideration to derive the partition function for fermionic and bosonic particles in the oscillator potential. In section 4.3.3, we shall expand our consideration to the case of the *parastatistics*.

4.3.1 Permutations and transition amplitudes

The basic idea of this subsection is that for indistinguishable particles, the order in which the position values are measured is irrelevant. This means that any permutation of the observed values should have the same probability, so that we can restrict the domain of possible values to an ordered set of positions $x_1 \geq x_2 \geq \cdots \geq x_N$.

We shall first illustrate the technique for two particles on a line and then generalize the result to N particles in the three-dimensional space.

◇ Two particles on a line

Let x_1 and x_2 be the coordinates of the first and second particle, respectively. The configuration space is two-dimensional: $(x_1, x_2) \in \mathbb{R}^2$. If the particles are identical, the configuration (x_1, x_2) and the configuration (x_2, x_1) should indicate the same state. For *fermions* with parallel spin, the anti-symmetry under interchange of the two particles is taken into account by the propagator

$$\langle x_1, x_2 | e^{-t\widehat{H}/\hbar} | x_1', x_2' \rangle = K(x_1, t | x_1', 0) K(x_2, t | x_2', 0) - K(x_1, t | x_2', 0) K(x_1, t | x_2', 0) \quad (4.3.1)$$

where $K(x, t | x_0, t_0)$ is the propagator for a single particle. For convenience, we use here the Euclidean-time formalism because later we shall be interested mainly in the corresponding partition function (cf 4.1.4). Let D_N define a domain on a line satisfying the condition $x_1 \geq x_2 \geq \cdots \geq x_n$, where $x_1, x_2, \ldots, x_n$ denote the possible components of the positions of the particles on a line. We consider formula (4.3.1) for $x_1 \geq x_2$ and $x_1' \geq x_2'$, so that $(x_1, x_2) \in D_2$ and $(x_1', x_2') \in D_2$. The boundary of D_2 is defined by $x_1 = x_2$ and denoted by ∂D_2 and the propagator (4.3.1) has the *absorption* boundary condition

$$\langle x_1', x_2' | \exp\{-t\widehat{H}/\hbar\} | x_1, x_2 \rangle |_{x_1 = x_2} = \langle x_1', x_2' | \exp\{-t\widehat{H}/\hbar\} | x_1, x_2 \rangle |_{x_1' = x_2'} = 0$$

on this boundary.

Similarly, the symmetry of *bosons* under permutations leads to the propagator

$$\langle x_1, x_2 | e^{-tH/\hbar} | x_1', x_2' \rangle = K(x_1, t | x_1', 0) K(x_2, t | x_2', 0) + K(x_1, t | x_2', 0) K(x_1, t | x_2', 0). \quad (4.3.2)$$

Again, propagator (4.3.2) is a transition amplitude of a two-dimensional diffusion process on D_2, but now with *reflecting* boundary conditions on ∂D_2. As for fermions, the required conditions for such a transition amplitude are easily verified. It is clear that (4.3.1) and (4.3.2) can be written respectively as a determinant and a permanent:

$$\langle x_1, x_2 | e^{-tH/\hbar} | x_1', x_2' \rangle = \begin{vmatrix} K(x_1, t | x_1', 0) & K(x_1, t | x_2', 0) \\ K(x_1, t | x_2', 0) & K(x_2, t | x_2', 0) \end{vmatrix}_\xi \quad (4.3.3)$$

where $|\mathbf{a}|_\xi$ with $\xi = +1$ refers to the permanent of a matrix $\mathbf{a}$ (for bosons) and $|\mathbf{a}|_\xi$ with $\xi = -1$ means the determinant of $\mathbf{a}$ (fermions). (Recall that the *permanent* of a matrix $\mathbf{a}$ is defined by the following sum over all permutations P:

$$\text{perm}(\mathbf{a}) \equiv |\mathbf{a}|_+ = \sum_P a_{1P(1)} a_{2P(2)} \cdots a_{NP(N)}$$

see, e.g., Ryser (1963).) This observation allows us to generalize the process for two identical particles to a process for N identical particles moving in one dimension: we need only substitute the 2×2 permanent or determinant by a $N \times N$ analog. In this case, the form of the transition amplitude will automatically take the boundary conditions into account.

◇ Transition amplitudes for an arbitrary number of identical particles in a three-dimensional space

The starting point of the construction is the projection of the transition amplitudes for distinguishable particles on a transition amplitude which has the correct symmetry properties under permutation of the particle positions:

$$K_{\rm I}(\bar{X}, t|\bar{X}', 0) = \frac{1}{N!} \sum_P \xi^P K(\bar{X}_P, t|\bar{X}', 0) \qquad (4.3.4)$$

($K_{\rm I}$ is the propagator for *indistinguishable* particles). This projection is a weighted average over all elements P of the permutation group. The weight is the character ξ^P of the representation; i.e. $\xi^P = 1$ for bosons, whereas for fermions $\xi^{P_+} = 1$ for even permutations P_+ and $\xi^{P_-} = -1$ for odd permutations P_-. Here a vector $\bar{X}$ belongs to the configuration space $\mathbb{R}^{3N}$, with the x, y and z components of the jth particle as the $(3j)$th, $(3j+1)$th and $(3j+2)$th components of $\bar{X}$. Its permutations $\bar{X}_P$ can be represented as

$$\bar{X}_P = \mathsf{P}\bar{X} \qquad (4.3.5)$$

where P is a $3N \times 3N$-dimensional matrix with one 3×3 identity matrix on each block row and block column, corresponding to each particle. For instance, for two particles, P can take one of the two forms

$$\begin{bmatrix} \mathbb{I}_3 & \bigcirc \\ \bigcirc & \mathbb{I}_3 \end{bmatrix} \quad \text{or} \quad \begin{bmatrix} \bigcirc & \mathbb{I}_3 \\ \mathbb{I}_3 & \bigcirc \end{bmatrix}$$

with

$$\mathbb{I}_3 = \begin{bmatrix} 1 & 0 & 0 \\ 0 & 1 & 0 \\ 0 & 0 & 1 \end{bmatrix} \quad \text{and} \quad \bigcirc = \begin{bmatrix} 0 & 0 & 0 \\ 0 & 0 & 0 \\ 0 & 0 & 0 \end{bmatrix}.$$

The transition amplitude for the non-interacting identical non-relativistic particles takes the form (cf (2.2.41))

$$K_{\rm I}(\bar{X}, t|\bar{X}', 0) = \left(\frac{m}{2\pi\hbar t}\right)^{3N/2} \frac{1}{N!} \sum_P \xi^P \exp\left\{-\frac{m}{2\hbar t}[\mathsf{P}\bar{X} - \bar{X}']^\top[\mathsf{P}\bar{X} - \bar{X}']\right\} \qquad (4.3.6)$$

which can readily be rewritten as

$$K_{\rm I}(\bar{X}, t|\bar{X}', 0) = \left(\frac{m}{2\pi\hbar t}\right)^{3N/2} \exp\left\{-\frac{m}{2\hbar t}(\bar{X} \cdot \bar{X} + \bar{X}' \cdot \bar{X}')\right\} \left(\frac{1}{N!} \sum_P \xi^P \exp\left\{\frac{m}{\hbar t}\mathsf{P}\bar{X} \cdot \bar{X}'\right\}\right). \qquad (4.3.7)$$

◇ Projection on even permutations

We now separate the even permutations P_+ from the odd permutations P_-, which can be written as $P_- = rP_+$, where r is an element of the permutation group which interchanges two particles (i.e. r is a transposition). Without loss of generality, we take the first and second particles. Using the fact that $\xi^{P_+} = 1$ for both fermions and bosons, we obtain

$$\sum_P \xi^P \exp\left\{\frac{m}{\hbar t}\mathsf{P}\bar{X} \cdot \bar{X}'\right\} = \sum_{P_+} \left(\exp\left\{\frac{m}{\hbar t}\mathsf{P}_+\bar{X} \cdot \bar{X}'\right\} + \xi^r \exp\left\{\frac{m}{\hbar t}\mathsf{r}\mathsf{P}_+\bar{X} \cdot \bar{X}'\right\}\right) \qquad (4.3.8)$$

where r is a $3N \times 3N$ matrix whose action is to interchange the coordinates of the first and the second particle. Hence, r only differs from the identity matrix in the block column and the block row corresponding to these particles:

$$r = \begin{bmatrix} O & \mathbb{I}_3 & O & \cdots & O \\ \mathbb{I}_3 & O & O & \cdots & O \\ O & O & \mathbb{I}_3 & \cdots & O \\ \vdots & \vdots & \vdots & \ddots & \vdots \\ O & O & O & \cdots & \mathbb{I}_3 \end{bmatrix}. \tag{4.3.9}$$

Note that $\xi^r = -1$ for fermions (since r is an odd permutation) and $\xi^r = 1$ for bosons. Some elementary algebra then gives

$$\sum_P \xi^P \exp\left\{\frac{m}{\hbar t} P \bar{X} \cdot \bar{X}'\right\} = \sum_{P_+} \exp\left\{\frac{1}{2}\frac{m}{\hbar t}[\mathbb{I}_{3N} + r]P_+ \bar{X} \cdot \bar{X}'\right\}$$
$$\times \left(\exp\left\{\frac{1}{2}\frac{m}{\hbar t}[\mathbb{I}_{3N} - r]P_+ \bar{X} \cdot \bar{X}'\right\}\right.$$
$$\left.+ \xi^r \exp\left\{-\frac{1}{2}\frac{m}{\hbar t}[\mathbb{I}_{3N} - r]P_+ \bar{X} \cdot \bar{X}'\right\}\right) \tag{4.3.10}$$

where $\mathbb{I}_{3N}$ denotes the $3N \times 3N$ identity matrix (not to be confused with the 3×3 identity matrix, $\mathbb{I}_3$). Since

$$\mathbb{I}_{3N} - r = \begin{bmatrix} \mathbb{I}_3 & -\mathbb{I}_3 & O & \cdots & O \\ -\mathbb{I}_3 & \mathbb{I}_3 & O & \cdots & O \\ O & O & O & \cdots & O \\ \vdots & \vdots & \vdots & \ddots & \vdots \\ O & O & O & \cdots & O \end{bmatrix} \tag{4.3.11}$$

we readily obtain

$$\mathbb{I}_{3N} - rP_+ \bar{X} \cdot \bar{X}' = (x_{P_+,1} - x_{P_+,2}) \cdot (x'_1 - x'_2) \tag{4.3.12}$$

where $x_{P_+,1}$ and $x_{P_+,2}$ are the coordinates of the first and second particles in $P_+\bar{X}$. The vector x_j denotes the usual three-dimensional position vector, in contrast to $\bar{X}$, which is a vector of dimension $3N$, as previously described.

◇ **The parity of $K_I(\bar{X}, t|\bar{X}', 0)$ and of its components**

The transition amplitude for N three-dimensional non-interacting identical particles is given by

$$K_I(\bar{X}, t|\bar{X}', 0) = \left(\frac{m}{2\pi \hbar t}\right)^{3N/2} \exp\left\{-\frac{m}{2\hbar t}(\bar{X} \cdot \bar{X} + \bar{X}' \cdot \bar{X}')\right\}$$
$$\times \frac{1}{N!} \sum_{P_+} \exp\left\{\frac{m}{2\hbar \tau}[\mathbb{I}_{3N} + r]P_+ \bar{X} \cdot \bar{X}'\right\}$$
$$\times \begin{cases} 2\cosh\left(\frac{1}{2}\frac{m}{\hbar t}(x_{P_+,1} - x_{P_+,2}) \cdot (x'_1 - x'_2)\right) & \text{for bosons} \\ 2\sinh\left(\frac{1}{2}\frac{m}{\hbar t}(x_{P_+,1} - x_{P_+,2}) \cdot (x'_1 - x'_2)\right) & \text{for fermions.} \end{cases} \tag{4.3.13}$$

This form of the amplitude allows us to answer the questions about the state space and boundary conditions. Indeed, the decompositions

$$\cosh \boldsymbol{a} \cdot \boldsymbol{b} = \cosh a_x b_x \cosh a_y b_y \cosh a_z b_z + \cosh a_x b_x \sinh a_y b_y \sinh a_z b_z$$

$$+ \sinh a_x b_x \cosh a_y b_y \sinh a_z b_z + \sinh a_x b_x \sinh a_y b_y \cosh a_z b_z \quad (4.3.14)$$

$$\sinh \boldsymbol{a} \cdot \boldsymbol{b} = \sinh a_x b_x \sinh a_y b_y \sinh a_z b_z + \sinh a_x b_x \cosh a_y b_y \cosh a_z b_z$$
$$+ \cosh a_x b_x \sinh a_y b_y \cosh a_z b_z + \cosh a_x b_x \cosh a_y b_y \sinh a_z b_z \quad (4.3.15)$$

allow us to rewrite the transition amplitude as a sum of four terms $K_{\mathrm{I}}(\bar{X}, t|\bar{X}', 0; \ell); \ell = 0 \ldots 3$:

$$K_{\mathrm{I}}(\bar{X}, t|\bar{X}', 0) = \sum_{l=0}^{3} K_{\mathrm{I}}(\bar{X}, t|\bar{X}', 0; \ell). \quad (4.3.16)$$

Here the summation index ℓ is associated with combinations of given parities (with respect to the interchange of the indicated coordinates of the three-dimensional space):

Parity of $K_{\mathrm{I}}(\bar{X}, t|\bar{X}', 0; \ell)$ for bosons

index:	$\ell = 0$	$\ell = 1$	$\ell = 2$	$\ell = 3$
x-coordinate	even	even	odd	odd
y-coordinate	even	odd	even	odd
z-coordinate	even	odd	odd	even

(4.3.17)

Parity of $K_{\mathrm{I}}(\bar{X}, t|\bar{X}', 0; \ell)$ for fermions

index:	$\ell = 0$	$\ell = 1$	$\ell = 2$	$\ell = 3$
x-coordinate	odd	odd	even	even
y-coordinate	odd	even	odd	even
z-coordinate	odd	even	even	odd

(4.3.18)

In this way we have reduced the problem of construction of the boundary conditions for three-dimensional particles to the same problem for particles on a line, which we have discussed earlier. The important consequence is that it is sufficient to analyze each component $K_{\mathrm{I}}(\bar{X}, t|\bar{X}', 0; \ell)$ with a given parity individually, with respect to the interchange of particles. For a given value of ℓ, this function, defined on the configuration space, can be obtained from a transition amplitude defined on the state space $D_N^3 \equiv D_N \otimes D_N \otimes D_N$, because it is a product of the transition probabilities of three independent processes, each defined on a D_N with the appropriate (bosonic or fermionic) boundary conditions.

Let $\{X_{\mathrm{F}\ell}(t); t \geq 0\}$ be the set of paths for identical fermions moving in $\mathbb{R}^3$. Then, this set is given according to the following rule

$$X_{\mathrm{F}\ell}(t) = \begin{cases} \ell = 0 \\ X_{\mathrm{F}}(t) \\ Y_{\mathrm{F}}(t) \\ Z_{\mathrm{F}}(t) \end{cases} \begin{cases} \ell = 1 \\ X_{\mathrm{F}}(t) \\ Y_{\mathrm{B}}(t) \\ Z_{\mathrm{B}}(t) \end{cases} \begin{cases} \ell = 2 \\ X_{\mathrm{B}}(t) \\ Y_{\mathrm{F}}(t) \\ Z_{\mathrm{B}}(t) \end{cases} \begin{cases} \ell = 3 \\ X_{\mathrm{B}}(t) \\ Y_{\mathrm{B}}(t) \\ Z_{\mathrm{F}}(t) \end{cases} \quad (4.3.19)$$

where $X_{\mathrm{F}}(t)$, $Y_{\mathrm{F}}(t)$ and $Z_{\mathrm{F}}(t)$ denote the set of paths for fermions in the x-, y- and z-directions. Similarly, $X_{\mathrm{B}}(t)$, $Y_{\mathrm{B}}(t)$ and $Z_{\mathrm{B}}(t)$ are the set of paths for bosons in the x-, y- and z-directions. For bosons the decomposition is as follows

$$X_{\mathrm{B}\ell}(t) = \begin{cases} \ell = 0 \\ X_{\mathrm{B}}(t) \\ Y_{\mathrm{B}}(t) \\ Z_{\mathrm{B}}(t) \end{cases} \begin{cases} \ell = 1 \\ X_{\mathrm{B}}(t) \\ Y_{\mathrm{F}}(t) \\ Z_{\mathrm{F}}(t) \end{cases} \begin{cases} \ell = 2 \\ X_{\mathrm{F}}(t) \\ Y_{\mathrm{B}}(t) \\ Z_{\mathrm{F}}(t) \end{cases} \begin{cases} \ell = 3 \\ X_{\mathrm{F}}(t) \\ Y_{\mathrm{F}}(t) \\ Z_{\mathrm{B}}(t). \end{cases} \quad (4.3.20)$$

For example, the fermion case with $\ell = 1$ is invariant under the even permutations P_+ of the particle coordinates. Furthermore, under r (interchange of two particles) it is antisymmetric in the x-direction and symmetric in the y- and z-directions. These symmetry properties allow us to restrict the transitions $\bar{X}' \to \bar{X}$ to a domain $D_N^3 \equiv D_N \otimes D_N \otimes D_N$, simultaneously satisfying the conditions

$$\bar{X} \in D_N^3 \iff \begin{cases} x_1 \geq x_2 \geq \cdots \geq x_N \\ y_1 \geq y_2 \geq \cdots \geq y_N \\ z_1 \geq z_2 \geq \cdots \geq z_N \end{cases} \tag{4.3.21}$$

with the boundary condition that $K_I(\bar{X}, \bar{X}'; \tau; \ell = 1)$ is zero if, during the transition process, the boundary ∂D_N is hit in the x-direction, being at the same time symmetric with respect to the boundary ∂D_N in the y- and the z-direction.

The transition amplitude for particles with interactions can be constructed now with the help of the Feynman–Kac formula. Of course, for a straightforward application of the permutation symmetries, the potential term must not spoil the indistinguishability of the particles. In other words, it should possess appropriate symmetry properties under the permutation of particle positions.

The spin states are left out of the picture by assuming that there are no spin-dependent interactions involved and therefore the spin, as an additional degree of freedom, does not have to be considered explicitly. Of course, the spin degrees of freedom are implicitly present because two identical particles are only considered indistinguishable if they are in the same spin state.

4.3.2 Path-integral formalism for coupled identical oscillators

Now we proceed to study the path-integral approach to the calculation of partition functions and generating functionals in *quantum* statistical mechanics. First, we consider the direct approach to such calculations in the familiar case of a harmonic potential. In the next sections, we shall study a general approach in the framework of field theories at non-zero temperature and at finite energy.

The case of identical particles in a parabolic confinement potential with either harmonic interactions between the particles, or with an anisotropy induced by a homogeneous magnetic field on top of the parabolic confinement gives rise to repetitive Gaussian integrals and allows us to derive an explicit expression for the generating function of the canonical partition function (Brosens *et al* 1997). For an ideal gas of non-interacting particles in a parabolic well, this generating function coincides with the grand canonical partition function. With interactions, the calculation of this generating function (instead of the partition function itself) circumvents the constraints on the summation over the *cycles* of the permutation group and, because of this fact, allows us to calculate the canonical partition function recursively, for the system with harmonic two-body interactions.

Note that the model of N identical particles in a parabolic well, in the presence of a magnetic field and with harmonic repulsive or attractive two-body interactions, has its intrinsic value since it constitutes an exactly soluble idealization of atoms in a magnetic trap.

◇ Harmonically interacting identical particles in a parabolic well

We shall calculate the partition function for N identical particles with the following Lagrangian, including one-body and two-body potentials:

$$L = \frac{1}{2} \sum_{j=1}^{N} \dot{r}_j^2 - V_1 - V_2$$

$$V_1 = \frac{\Omega^2}{2} \sum_{j=1}^{N} r_j^2 \qquad V_2 = -\frac{\omega^2}{4} \sum_{j,l=1}^{N} (r_j - r_l)^2.$$

The potentials can be rewritten in terms of the centre-of-mass coordinate r and the coordinates u_j describing the positions of the particles measured from the centre of mass:

$$R = \frac{1}{N} \sum_{j=1}^{N} r_j \qquad u_j = r_j - R \qquad (4.3.22)$$

so that

$$V_1 + V_2 = V_{\text{CM}} + V$$

$$V_{\text{CM}} = \frac{1}{2} N \Omega^2 R^2 \qquad V = W^2 \sum_{j=1}^{N} u_j^2$$

with

$$W = \sqrt{\Omega^2 - N\omega^2}. \qquad (4.3.23)$$

The requirement that W be real (i.e. $\Omega^2 \geq N\omega^2$) expresses the stability condition that the confining potential be strong enough to overcome the repulsion between the particles. If a harmonic interparticle *attraction* is considered, the eigenfrequency W would become $W = \sqrt{\Omega^2 + N\omega^2}$, and no stability condition has to be imposed on the confining potential.

Since the system consists in each direction of one degree of freedom with the frequency Ω and $(N - 1)$ degrees of freedom with the frequency W, the propagator

$$K_{\text{D}}(r_1'' \cdots r_N'', \beta | r_1' \cdots r_N', 0) \equiv \langle r_1'' \cdots r_N'' | e^{-\beta H} | r_1' \cdots r_N' \rangle_{\text{D}} \qquad (4.3.24)$$

for *distinguishable* particles (indicated by the subscript D for three dimensions and d in one dimension) can be calculated from the action expressed in the imaginary-time variable, and it is of course a product of the propagators K_{d} per component:

$$K_{\text{D}}(r_1'' \cdots r_N'', \beta | r_1' \cdots r_N', 0) = K_{\text{d}}(\bar{x}'', \beta | \bar{x}', 0) K_{\text{d}}(\bar{y}'', \beta | \bar{y}', 0) K_{\text{d}}(\bar{z}'', \beta | \bar{z}', 0) \qquad (4.3.25)$$

where the column vector $\bar{x}$ contains the x-components of the particles, i.e. $\bar{x}^{\text{T}} = (x_1, \ldots, x_N)$ and similarly for $\bar{y}$ and $\bar{z}$. Knowing the propagator $K(x'', \beta | x', 0)$ of a single harmonic oscillator (cf (2.2.77)), we find for the one-dimensional propagator K_{d} of the N distinguishable oscillators in the interacting system that

$$K_{\text{d}}(\bar{x}'', \beta | \bar{x}', 0) = \frac{K(\sqrt{N} X'', \beta | \sqrt{N} X', 0)_\Omega}{K(\sqrt{N} X'', \beta | \sqrt{N} X', 0)_W} \prod_{j=1}^{N} K(x_j'', \beta | x_j', 0)_W \qquad (4.3.26)$$

where the factor $\sqrt{N}$ in $\sqrt{N} X''$, $\sqrt{N} X'$ accounts for the additional factor N in V_{CM}. The denominator in (4.3.26) compensates for the fact that $(N - 1)$ instead of N degrees of freedom of frequency W are available. The three-dimensional propagator K_{D} (4.3.24) for N distinguishable oscillators of the interacting system is, according to (4.3.25) and (4.3.26), given by

$$K_{\text{D}}(\bar{r}'', \beta | \bar{r}', 0) = \frac{K(\sqrt{N} R'', \beta | \sqrt{N} R', 0)_\Omega}{K(\sqrt{N} R'', \beta | \sqrt{N} R', 0)_W} \prod_{j=1}^{N} K(r_j'', \beta | r_j', 0)_W \qquad (4.3.27)$$

$$K(r_j'', \beta | r_j', 0)_W = K(x_j'', \beta | x_j', 0)_W K(y_j'', \beta | y_j', 0)_W K(z_j'', \beta | z_j', 0)_W \qquad (4.3.28)$$

where the vectors with bar $\bar{r}$ denote (as in the preceding section) the points in the configuration space $\mathbb{R}^{3N}$, i.e. $\bar{r} \equiv \{(x_1, y_1, z_1), \ldots, (x_N, y_N, z_N)\}$. Similarly to the case of free particles considered in preceding section, the symmetrized density matrix K_I for three-dimensional identical particles (indicated by the subscript I) can be obtained by using the following projection, with P denoting the permutation matrix:

$$K_I(\bar{r}'', \beta|\bar{r}', 0) = \frac{1}{N!} \sum_P \xi^P K_D(P\bar{r}'', \beta|\bar{r}', 0) \qquad (4.3.29)$$

where $\xi = +1$ for bosons and $\xi = -1$ for fermions. It should be emphasized that P acts on the particle indices, not on the components of r separately. The partition function $\mathcal{Z}_I \equiv \text{Tr}\exp\{-\beta\widehat{H}\}$ is then readily obtained by integrating over the configuration space

$$\mathcal{Z}_I = \int d^{3N}\bar{r}\, K_I(\bar{r}, \beta|\bar{r}, 0) = \int d^{3N}\bar{r}\, \frac{1}{N!} \sum_P \xi^P K_D(P\bar{r}, \beta|\bar{r}, 0). \qquad (4.3.30)$$

The integration proceeds in three steps:

(i) the first stage deals with the centre-of-mass treatment;
(ii) the second concerns the *cyclic decomposition* of the permutations in (4.3.30) and
(iii) at the third step, the summation over the cycles will be performed.

◇ Step 1: The centre-of-mass decoupling

Making use of the δ-function to separate the centre-of-mass variable and the Fourier transform, we obtain

$$\mathcal{Z}_I = \int d^3r \int \frac{d^3k}{(2\pi)^3} e^{i\mathbf{k}\cdot\mathbf{R}} \frac{K(\sqrt{N}\mathbf{R}, \beta|\sqrt{N}\mathbf{R}, 0)_\Omega}{K(\sqrt{N}\mathbf{R}, \beta|\sqrt{N}\mathbf{R}, 0)_W}$$
$$\times \int d^{3N}\bar{r}\, \frac{1}{N!} \sum_P \xi^P \prod_{j=1}^N K((P\mathbf{r})_j, \beta|\mathbf{r}_j, 0)_W e^{-i\mathbf{k}\cdot\mathbf{r}_j/N}. \qquad (4.3.31)$$

This transformation makes $\mathbf{R}$ independent of the particle positions relative to the centre of mass. The real dependence on the relative positions is reintroduced by the Fourier transform. It should be noted that the explicit dependence of propagator (4.3.27) on $\mathbf{R}$, and the presence of the factor $\exp\{-i\mathbf{k}\cdot\mathbf{r}_j/N\}$ are consequences of the two-body interactions.

◇ Step 2: Cyclic decomposition

Any permutation can be broken up into *cycles* (see, e.g., Hamermesh (1964)). Recall that a cycle $P_l^{(c)}$ of length ℓ, in this context, is a special type of general permutations P. Acting on a subset of ℓ elements x_i ($i = 1, \ldots, \ell$), it produces the permutation: $P_l^{(c)}(x_i) = x_{i+1}$ ($i = 1, \ldots, \ell-1$), $P_l^{(c)}(x_l) = x_1$. Suppose that the cyclic decomposition of a particular permutation contains M_ℓ cycles of length ℓ. It is known that the positive integers M_ℓ and ℓ then have to satisfy the constraint

$$\sum_\ell \ell M_\ell = N. \qquad (4.3.32)$$

Furthermore, the number $M(M_1, \ldots M_N)$ of cyclic decompositions with M_1 cycles of length $1, \ldots, M_\ell$ cycles of length $\ell, \ldots$ is known to be

$$M(M_1, \ldots, M_N) = \frac{N!}{\prod_\ell M_\ell! \ell^{M_\ell}}. \qquad (4.3.33)$$

A cycle of length ℓ can be obtained from $(\ell - 1)$ transpositions. Therefore, the sign factor ξ^P can be decomposed as

$$\xi^P = \prod_\ell \xi^{(\ell-1)M_\ell}. \tag{4.3.34}$$

Combining these results, we obtain

$$\mathcal{Z}_I = \int d^3R \int \frac{d^3k}{(2\pi)^3} e^{i\mathbf{k}\cdot\mathbf{R}} \frac{K(\sqrt{N}\mathbf{R}, \beta|\sqrt{N}\mathbf{R}, 0)_\Omega}{K(\sqrt{N}\mathbf{R}, \beta|\sqrt{N}\mathbf{R}, 0)_W} \sum_{M_1\cdots M_N} \prod_\ell \frac{\xi^{(\ell-1)M_\ell}}{M_\ell! \ell^{M_\ell}} (\mathcal{K}_\ell(\mathbf{k}))^{M_\ell} \tag{4.3.35}$$

$$\mathcal{K}_\ell(\mathbf{k}) = \int d^3r_{\ell+1} \int d^3r_\ell \cdots \int d^3r_1 \, \delta(\mathbf{r}_{\ell+1} - \mathbf{r}_1) \prod_{j=1}^N K(\mathbf{r}_{j+1}, \beta|\mathbf{r}_j, 0)_W e^{-i\mathbf{k}\cdot\mathbf{r}_j/N}. \tag{4.3.36}$$

The δ-function expresses the fact that the decomposition is cyclic. It is obvious that

$$\mathcal{K}_\ell(\mathbf{k}) = \mathcal{K}_\ell^{(1D)}(k_x)\mathcal{K}_\ell^{(1D)}(k_y)\mathcal{K}_\ell^{(1D)}(k_z) \tag{4.3.37}$$

which allows us to analyze $\mathcal{K}_\ell(\mathbf{k})$ from its one-dimensional constituents:

$$\mathcal{K}_\ell^{(1D)}(k_x) = \int dx_{\ell+1} \int dx_\ell \cdots \int dx_1 \, \delta(x_{\ell+1} - x_1) \prod_{j=1}^N K(x_{j+1}, \beta|x_j, 0)_W e^{-ik_x x_j/N}. \tag{4.3.38}$$

Using the semigroup property of the harmonic oscillator propagator $K(x_{j+1}, \beta|x_j, 0)_W$, all integrations but one can be performed

$$\mathcal{K}_\ell^{(1D)}(k_x) = \int dx \, K(x, \ell\beta|x, 0)_W \exp\left\{-\int_0^{\ell\beta} d\tau \, f_x(\tau) x(\tau)\right\} \tag{4.3.39}$$

where

$$f_x(\tau) = i\frac{k_x}{N} \sum_{j=0}^{\ell-1} \delta(\tau - j\beta). \tag{4.3.40}$$

The integral (4.3.39) is the propagator $K_{W,f}$ of a driven harmonic oscillator with the Lagrangian

$$L_{W,f_x} = \tfrac{1}{2}\dot{x}^2 - \tfrac{1}{2}W^2 x^2 + f_x(\tau)x \tag{4.3.41}$$

studied in sections 1.2.7 and 1.2.8 (see (1.2.262) and also problem 2.2.14, page 198, volume I). It should be noted that without the two-body interactions, the driving force (4.3.40) is absent. Taking the result from (1.2.262) and integrating over the configuration space, we obtain

$$\mathcal{Z}_{W,f_x}(\beta) = \int dx \, K_{W,f_x}(x, \beta|x, 0)$$

$$= \frac{1}{2\sinh\tfrac{1}{2}\beta W} \exp\left\{\frac{1}{2}\int_0^\beta d\tau \int_0^\beta d\sigma \, \frac{f_x(\tau)f_x(\sigma)}{2W} \frac{\cosh((\tfrac{\beta}{2} - |\tau - \sigma|)W)}{\sinh\tfrac{1}{2}\beta W}\right\}. \tag{4.3.42}$$

After some straightforward algebra, we obtain for the one-dimensional function $\mathcal{K}_\ell^{(1D)}(k_x)$:

$$\mathcal{K}_\ell^{(1D)}(k_x) = \frac{1}{2\sinh\tfrac{1}{2}\ell\beta W} \exp\left(-\frac{\ell}{4N^2}\frac{k_x^2}{W}\frac{1 + e^{-\beta W}}{1 - e^{-\beta W}}\right) \tag{4.3.43}$$

214   Path integrals in statistical physics

and for its three-dimensional extension

$$\mathcal{K}_\ell(\vec{k}) = \left(\frac{1}{2\sinh\frac{1}{2}\ell\beta W}\right)^3 \exp\left(-\frac{\ell}{4N^2}\frac{k^2}{W}\frac{1+e^{-\beta W}}{1-e^{-\beta W}}\right). \tag{4.3.44}$$

Using (4.3.32), we are then left with a sixfold integral for the partition function

$$\mathcal{Z}_\mathrm{I} = \int d^3R \int \frac{d^3k}{(2\pi)^3} e^{i\mathbf{k}\cdot\mathbf{R}} \frac{K(\sqrt{N}\mathbf{R},\beta|\sqrt{N}\mathbf{R},0)_\Omega}{K(\sqrt{N}\mathbf{R},\beta|\sqrt{N}\mathbf{R},0)_\mathrm{W}} \exp\left\{-\frac{1}{4N}\frac{k^2}{W}\frac{1+e^{-\beta W}}{1-e^{-\beta W}}\right\}$$

$$\times \sum_{M_1\cdots M_N} \prod_\ell \frac{\xi^{(\ell-1)M_\ell}}{M_\ell!\ell^{M_\ell}} \left(\frac{1}{2\sinh\frac{1}{2}\ell\beta W}\right)^{3M_\ell}. \tag{4.3.45}$$

Both the integrations over k and R are Gaussian, leading to the following series for $\mathcal{Z}_\mathrm{I}$:

$$\mathcal{Z}_\mathrm{I} = \left(\frac{\sinh\frac{1}{2}\beta W}{\sinh\frac{1}{2}\beta\Omega}\right)^3 \mathcal{Z}^{(0)}_\mathrm{I}(N)$$

$$\mathcal{Z}^{(0)}_\mathrm{I}(N) \equiv \sum_{M_1\cdots M_N} \prod_\ell \frac{\xi^{(\ell-1)M_\ell}}{M_\ell!\ell^{M_\ell}} \left(\frac{e^{-\frac{1}{2}\ell\beta W}}{1-e^{-\ell\beta W}}\right)^{3M_\ell}. \tag{4.3.46}$$

Without the two-body interactions ($W = \Omega$, that is $\omega = 0$, cf (4.3.23)), $\mathcal{Z}^{(0)}_\mathrm{I}(N)$ is the partition function of a set of identical oscillators. The partition function $\mathcal{Z}_\mathrm{I}$ only differs from it by a centre-of-mass correction and the actual values of W.

◇ Step 3: The generating function

The remaining summation over the cycles involves the constraint (4.3.32), which, however, can be removed by the use of the *generating function technique*. Concentrating on the explicit dependence of $\mathcal{Z}^{(0)}_\mathrm{I}(N)$ on N (with W considered as a parameter), we can construct the following *generating function*

$$\Xi(u) \stackrel{\text{def}}{\equiv} \sum_{N=0}^{\infty} \mathcal{Z}^{(0)}_\mathrm{I}(N) u^N \tag{4.3.47}$$

with $\mathcal{Z}^{(0)}_\mathrm{I}(0) = 1$ by definition. The partition function $\mathcal{Z}^{(0)}_\mathrm{I}(N)$ can then be obtained by taking the appropriate derivatives of $\Xi(u)$ with respect to u, assuming that the series for $\Xi(u)$ is convergent near $u = 0$:

$$\mathcal{Z}^{(0)}_\mathrm{I}(N) = \frac{1}{N!}\frac{d^N}{du^N}\Xi(u)\bigg|_{u=0}. \tag{4.3.48}$$

The summation over the number of cycles with the length ℓ is now unrestricted and can easily be performed:

$$\Xi_\mathrm{I}(u) = \exp\left\{\sum_{\ell=1}^{\infty}\xi^{\ell-1}\frac{e^{-\frac{3}{2}\ell\beta W}u^\ell}{\ell(1-e^{-\ell\beta W})^3}\right\}. \tag{4.3.49}$$

This series can be rewritten into the more convenient form

$$\Xi_\mathrm{I}(u) = \exp\left\{-\xi\sum_{\nu=0}^{\infty}\frac{1}{2}(\nu+1)(\nu+2)\ln(1-\xi u e^{-\beta W(\frac{3}{2}+\nu)})\right\}. \tag{4.3.50}$$

It should be noted that, in the case of a model without two-body interaction ($W = 0$), the function $\Xi_I(u)$ coincides with the *grand canonical partition function* of a set of identical particles in a parabolic well.

Considering the differentiation in (4.3.48) step by step, i.e.

$$\mathcal{Z}^{(0)}{}_I(N) = \frac{1}{N!} \frac{d^{N-1}}{du^{N-1}} \frac{d}{du} \Xi(u) \bigg|_{u=0}$$

together with the product rule and an elementary binomial expansion, we can find the following reccurence relation (Brosens *et al* 1997):

$$\mathcal{Z}^{(0)}{}_I(N) = \frac{1}{N} \sum_{m=0}^{N-1} \xi^{N-m-1} \left(\frac{b^{\frac{1}{2}(N-m)}}{1 - b^{N-m}} \right)^3 \mathcal{Z}^{(0)}{}_I(m) \qquad (4.3.51)$$

where

$$b \equiv e^{-\beta W}. \qquad (4.3.52)$$

The corresponding one-dimensional version of this recurrence relation becomes

$$\mathcal{Z}^{(0)}{}_{I,(1D)}(N) = \frac{1}{N} \sum_{m=0}^{N-1} \xi^{N-m-1} \frac{b^{\frac{1}{2}(N-m)}}{1 - b^{N-m}} \mathcal{Z}^{(0)}{}_{I,(1D)}(m) \qquad (4.3.53)$$

leading to the following explicit expression for the one-dimensional boson $\mathcal{Z}^{(0)}{}_b$ and one-dimensional fermion $\mathcal{Z}^{(0)}{}_f$ partition functions:

$$\mathcal{Z}^{(0)}{}_b = \frac{b^{\frac{1}{2}N}}{\prod_{j=1}^{N}(1 - b^j)}$$

$$\mathcal{Z}^{(0)}{}_f = \frac{b^{\frac{1}{2}N^2}}{\prod_{j=1}^{N}(1 - b^j)}. \qquad (4.3.54)$$

It is easy to check that these partition functions are the solution of the recurrence relation for $\mathcal{Z}^{(0)}{}_{I,(1D)}(N)$ with $\xi = 1$ for bosons and $\xi = -1$ for fermions. Unfortunately, for the three-dimensional case an analytic solution of (4.3.51) has not been found and we have to rely on numerical schemes.

Note that the same techniques is applicable to the calculation of the partition function for N identical oscillators in a constant magnetic field with the Lagrangian:

$$L_{\omega_c} = \tfrac{1}{2} \sum_{j=1}^{N} (\dot{r}_j - 2\omega_c x_j \dot{y}_j)^2 - \tfrac{1}{2} \Omega^2 \sum_{j=1}^{N} r_j^2 \qquad (4.3.55)$$

where ω_c is the cyclotron frequency (Brosens *et al* 1997).

Having at our disposal expressions for the partition functions (explicit in the one-dimensional case or obtained numerically from the recurrence relations in the three-dimensional case), we can find all the thermodynamical characteristics of systems of bosons or fermions in the harmonic potential and in the magnetic field by using the standard formulae, for example, the free energy $F = -\beta^{-1} \ln \mathcal{Z}$, the internal energy $U = \partial(\beta F)/\partial \beta$ and the specific heat $C = \partial U/\partial T$.

4.3.3 Path integrals and parastatistics

In the preceding subsections, we discussed the construction of path integrals for identical bosons and fermions. In this subsection, we shall generalize our consideration to the more general case of particles obeying *parastatistics*. Parastatistics, invented by H S Green (1953), is the first ever consistent extension of fundamental statistics. In this, the standard bosonic or fermionic fields which would create identical particles are replaced by composite fields whose components commute with themselves and anticommute with each other for parabosons, or vice versa for parafermions. The number p of components of the fields defines the 'order' of the parastatistics. In general, we can put, at most, p parafermions in a totally symmetric wavefunction and, at most, p parabosons in a totally antisymmetric one.

Although direct physical applications of parastatistics are absent, it is quite instructive to learn the path-integral techniques needed for such a generalization.

◇ Basics of the parastatistics

Green noted that the commutator of the number operator with the annihilation and creation operators is the same for both bosons and fermions:

$$[\widehat{n}_k, \widehat{a}_l^\dagger] = \delta_{kl} \widehat{a}_l^\dagger. \qquad (4.3.56)$$

The number operator can be written as

$$\widehat{n}_k = \tfrac{1}{2}[\widehat{a}_k^\dagger, \widehat{a}_k]_\pm + \text{constant} \qquad (4.3.57)$$

with an anticommutator ($[\cdot,\cdot]_+ \equiv \{\cdot,\cdot\}$) in the case of bosons and a commutator ($[\cdot,\cdot]_- \equiv [\cdot,\cdot]$) in the case of fermions. If these expressions are inserted in the commutation relation (4.3.56), the resulting relation is *trilinear* in the annihilation and creation operators (Green's trilinear commutation relation for his parabose and parafermi statistics):

$$[[\widehat{a}_k^\dagger, \widehat{a}_l]_\pm, \widehat{a}_m^\dagger]_- = 2\delta_{lm}\widehat{a}_k^\dagger. \qquad (4.3.58)$$

Since these rules are trilinear, the usual vacuum condition

$$\widehat{a}_k|0\rangle = 0 \qquad (4.3.59)$$

does not suffice to allow the calculation of matrix elements of the as and $a^\dagger$s. Hence, a condition on the one-particle states must be added:

$$\widehat{a}_k \widehat{a}_l^\dagger |0\rangle = \delta_{kl}|0\rangle. \qquad (4.3.60)$$

Green found an infinite set of solutions of his commutation rules, one for each integer, by giving an ansatz which he expressed in terms of Bose and Fermi operators. Let

$$\widehat{a}_k^\dagger = \sum_{\alpha=1}^p \widehat{b}_k^{(\alpha)\dagger} \qquad \widehat{a}_k = \sum_{\alpha=1}^p \widehat{b}_k^{(\alpha)} \qquad (4.3.61)$$

and let $\widehat{b}_k^{(\alpha)}$ and $\widehat{b}_k^{(\beta)\dagger}$ be Bose (Fermi) operators for $\alpha = \beta$, but anticommute (commute) for $\alpha \neq \beta$, for the 'parabose' ('parafermi') cases. This ansatz clearly satisfies Green's relation. The integer p is the order of the parastatistics. The physical interpretation of p is that, for parabosons, p is the maximum number of particles that can occupy an antisymmetric state, while for parafermions, p is the maximum number of particles that can occupy a symmetric state (in particular, the maximum number which can occupy the same state). The case $p = 1$ corresponds to the usual Bose or Fermi statistics. From Green's ansatz, it is clear that the squares of all norms of states are positive, since the sums of Bose or Fermi operators give positive norms. Thus, parastatistics gives a set of self-consistent theories. The violations of statistics provided by parastatistics are gross. Parafermi statistics of order 2 has up to two particles in each quantum state. High-precision experiments are not necessary to rule this out for all particles we think are fermions. It is important to note that the parastatistics of order p is related by the so-called Klein transformation to models with usual bosons or fermions *and* with exact $SO(p)$ or $SU(p)$ internal symmetry.

◇ The permutation group S_N and quantization of many-body systems

As in the case of fermions or bosons, we can deal with a parastatistical system with a fixed number of particles in a first-quantized formalism. In this approach, the N-body Hilbert space is decomposed into irreducible representations of the particle permutation group S_N (for basic facts about S_N see, e.g., in Hamermesh (1964)). Since the particles are indistinguishable, this group should be viewed as a 'gauge' symmetry of the system, and the states transforming in the same representation have to be identified. Moreover, since all physical operators are required to commute with the permutation group, each irreducible component is a superselection sector. Therefore, we can project the Hilbert space to only *some* of the irreducible representations of S_N. Further, only one state in each irreducible representation need be kept as a representative of the multiplet of physically equivalent states. The resulting reduced space constitutes a consistent quantization of N indistinguishable particles. The choice of included irreducible representations constitutes a choice of quantum statistics.

This description relies on a canonical quantization of the many-body system. It is of interest to have also a path-integral formulation of a quantum system, since this complements and completes the conceptual framework and usually offers orthogonal intuition in several cases. For ordinary statistics, this question was studied in sections 4.3.1 and 4.3.2. Here, we present such a realization for parastatistics (Polychronakos 1996).

The starting point, as in the ordinary bosonic statistics in section 4.3.1, will be the coordinate representation of the full (unprojected) Hilbert space, spanned by the position eigenstates $|x_1, \ldots, x_N\rangle \equiv |x\rangle$ (where x_i can be in a space of any dimension). The collection of such states for a set of distinct x_i transforms in the $N!$-dimensional defining representation of S_N

$$\widehat{P}|x\rangle \equiv |Px\rangle = |x_{P^{-1}(1)}, \ldots, x_{P^{-1}(N)}\rangle \tag{4.3.62}$$

where P is a permutation (the appearance of P^{-1} is necessary to represent the products of permutations in the right order).

◇ Projection of states to irreducible representations of S_N

Projecting the Hilbert space to an irreducible representation R of S_N amounts to keeping only linear combinations of states within this multiplet transforming in R, that is,

$$|a; x\rangle = \sum_P C_a(P)\widehat{P}|x\rangle \qquad a = 1, \ldots, d_R \equiv \dim(R) \tag{4.3.63}$$

where the sum is taken over all the elements of the permutation group and $C_a(P)$ are appropriately chosen coefficients. Let us denote by $R_{ab}(\widehat{P})$ the matrix elements of the permutation $\widehat{P}$ in the representations R, so that

$$\widehat{P}|a, x\rangle = \sum_b R_{ab}(P^{-1})|b, x\rangle. \tag{4.3.64}$$

The defining representation decomposes into irreducible components, classified by the *Young tableaux*, each appearing with a certain multiplicity. Thus, we need to clarify whether we have to keep only one irreducible representation out of each multiplicity or the whole set of a given irreducible representations. To do this, we note that if instead of the base state $|x\rangle$ for the construction of the states $|a, x\rangle$ in (4.3.63) we choose a different permutation $P|x\rangle$, then although the new states $|a, Px\rangle$ constructed through (4.3.63) still transform in the irreducible representation R, in general they are *not* linear combinations of $|a, x\rangle$, but rather span a different copy of R. Since we can continuously move in the configuration space from $|x\rangle$ to $P|x\rangle$, we conclude that we must keep *all* irreducible representations R in the decomposition of the defining representation according to a given multiplicity. To realize this explicitly, we construct the states

$$|ab, x\rangle = \sqrt{\frac{d_R}{N!}} \sum_P R_{ab}(P) P|x\rangle. \qquad (4.3.65)$$

Using the group property of the representation $R(P_1)R(P_2) = R(P_1 P_2)$, we deduce that under the action of the group S_N and under a change of the base point x, these states transform as

$$\widehat{P}|ab, x\rangle = \sum_c R_{ac}(P^{-1})|cb, x\rangle$$
$$|ab, Px\rangle = \sum_c R_{cb}(P^{-1})|ac, x\rangle. \qquad (4.3.66)$$

Thus, we see that the first index in these states labels the different elements of a single irreducible representation R, while the second index labels the different equivalent irreducible representations in the multiplet. Since both indices take d_R values, we recover the standard result that each irreducible representation of S_N is embedded in the defining representation a number of times equal to its dimension.

Consider now the matrix element $\langle ab, x|\widehat{A}|cd, y\rangle$, where $\widehat{A}$ is any physical operator, that is, any operator commuting with all the elements $\widehat{P}$ of S_N. Substituting the definition (4.3.65) and using the unitarity of $\widehat{P}$ ($\widehat{P}^\dagger = \widehat{P}^{-1}$) and of R ($R^*_{ab}(P) = R_{ba}(P^{-1})$) we obtain, after a change of summation variable,

$$\langle ab, x|\widehat{A}|cd, y\rangle = \frac{d_R}{N!} \sum_{P,P',e} R_{be}(P') R_{ea}(P^{-1}) R_{cd}(P) \langle x|\widehat{A}\widehat{P}'|y\rangle. \qquad (4.3.67)$$

Using further the *Schur orthogonality relation* for the representations, i.e.

$$\sum_P R_{ab}(P) R_{cd}(P^{-1}) = \frac{N!}{d_R} \delta_{ad} \delta_{bc} \qquad (4.3.68)$$

we finally obtain

$$\langle ab, x|\widehat{A}|cd, y\rangle = \sum_P \delta_{ac} R_{bd}(P) \langle x|\widehat{A}|Py\rangle. \qquad (4.3.69)$$

Let us first choose $\widehat{A} = \mathbb{1}$. Then this provides the overlap between the states:

$$\langle ab, x|cd, y\rangle = \sum_P \delta_{ac} R_{bd}(P) \delta(x - Py). \qquad (4.3.70)$$

For x in the neighborhood of y, it is $P = 1$ which contributes to the normalization, for which $R_{bd}(1) = \delta_{bd}$ and we recover the standard normalization for the states.

◇ Propagator for particles with parastatistics defined by an irreducible representation of the permutation group S_N

Now we can choose $\widehat{A} = e^{-it\widehat{H}}$, where $\widehat{H}$ is the Hamiltonian, and thus find the propagator $K(ab; x, t|cd; y, 0)$ between the states of the system. It is clear from (4.3.69) that the first index a in the state $|ab, x\rangle$ propagates trivially. Since this is the index that corresponds to the different but physically equivalent states within each irreducible representation R, we conclude that the required projection of the Hilbert space to the physical subspace amounts to simply omitting this index from all states (that is, freeze this index to the same fixed value for all states of the theory; no physical quantity will ever depend on the choice of this value). On the other hand, the second index, corresponding to different equivalent irreducible representations, does *not* propagate trivially and must, as argued before, be kept. We are led therefore to the physical states $|ba, x\rangle \to |a, x\rangle$ and the propagator

$$K_R(a; x, t|b; y, 0) = \sum_P R_{ab}(P) K_D(x, t|Py, 0) \qquad (4.3.71)$$

where $K_D(x, t|Py, 0) = \langle x| \exp\{-it\widehat{H}\} \widehat{P} |y\rangle$ is the usual many-body propagator for *distinguishable* particles. Expressing the latter in terms of the standard path integral and using (4.3.71), we obtain the path-integral form of the propagator for N particles obeying the parastatistics. We note that, due to the transformation property (4.3.66), the states $|a, Px\rangle$ are linear combinations of the states $|a, x\rangle$. Therefore, projecting down to the physical subspace corresponding to R amounts to trading the original $N!$ copies of physically equivalent states $|Px\rangle$ for a number d_R of *global internal degrees of freedom* for the system, labeled by the index a.

It is now easy to write down the path integral corresponding to identical particles quantized in the R-irreducible representation of S_N. $K_D(x, t|Py, 0)$ can be expressed as an N-body path integral in the standard way, with particles starting from the positions $Py_i = y_{P^{-1}(i)}$ and ending in the positions x_i. Since all permutations of particle positions are physically equivalent, (4.3.71) instructs us to sum over *all* sectors where particles end up in such permuted positions, weighted with the factors $R_{ab}(P)$ depending on the internal degrees of freedom of the initial and final states. From (4.3.65) and (4.3.70) we can write the completeness relation within the physical subspace

$$\mathbb{1}_R = \int \frac{d^N x}{N!} \sum_a |a, x\rangle\langle a, x| \qquad (4.3.72)$$

and with the use of (4.3.72) it is easy to prove that this transition amplitude satisfies the standard semigroup property

$$\int \frac{d^N y}{N!} \sum_b K_R(a; x, t|b; y, t') K_R(b; y, t'|c; z, 0) = K_R(a; x, t|c; z, 0) \qquad 0 < t' < t. \qquad (4.3.73)$$

◇ Extension to parabosons and parafermions

The extension to parabosons, parafermions or any similar statistics is immediate. Let $S = \{R_1, \ldots, R_n\}$ be the set of allowed irreducible representations of S_N in the Hilbert space. The internal degree of freedom now takes the values $A = (R, a)$, where $R \in S$ and $a = 1, \ldots, d_R$ labels the internal degrees of freedom within each irreducible representation. So, overall, A takes $d_{R_1} + d_{R_2} + \cdots + d_{R_n}$ different values. The propagator (and corresponding path integral) is obviously

$$K_S(A; x, t|B; y, 0) = \sum_P S(P)_{AB} K_D(x, t|Py, 0) \qquad (4.3.74)$$

where $S(P)_{AB} = \delta_{R_A,R_B}(R_A)_{ab}(P)$. For parabosons (parafermions) of the order p, S is the set of Young tableaux with up to p rows (columns). We note that the irreducible representations for parafermions are the duals of those for parabosons (the dual of a tableau is the tableau with the rows and columns interchanged). In an appropriate basis, the representation matrices of the dual irreducible representations R, $\tilde{R}$ are real and satisfy

$$\tilde{R}_{ab}(P) = (-1)^P R_{ab}(P) \tag{4.3.75}$$

where $(-1)^P$ is the parity of the permutation. We arrive then at the relation between the weights for parabosons and parafermions of the order p:

$$S_{pF}(P)_{AB} = (-1)^P S_{pB}(P)_{AB}. \tag{4.3.76}$$

This extends a similar relation for ordinary fermions and bosons, for which there are no internal degrees of freedom and $S_B(P) = 1$.

From the path integral we can evaluate the partition function, by simply shifting to Euclidean time $t \to -i\beta$ and summing over all initial and final states, with the measure implied by (4.3.72). Given that

$$\sum_a R_{aa}(P) = \operatorname{Tr} R(P) = \chi_R(P) \tag{4.3.77}$$

we get an expression in terms of the *characters* χ_R of S_N:

$$\mathcal{Z}_S(\beta) = \int \frac{d^N x}{N!} \sum_P S(P) \langle x | e^{-\beta \hat{H}} | P x \rangle \quad \text{where } S(P) = \sum_{R \in S} \chi_R(P). \tag{4.3.78}$$

The interpretation in terms of a periodic Euclidean path integral is obvious. The characters $\chi_R(P)$ are a set of integers, and thus the 'statistical factors' $S(P)$ weighing each topological sector of the path integral are (positive or negative) integers. In the case of parabosons of any order p, however, we note that the statistical weights are *positive* (or zero) integers. The ones for parafermions can be either positive or negative, as given by

$$S_{pF}(P) = (-1)^P S_{pB}(P) \qquad S_{pB}(P) \geq 0. \tag{4.3.79}$$

A general formula for $S_{pB}(P)$ for an arbitrary p is absent.

◇ **Partition function for an ideal parabosonic gas**

From these results we can derive the partition function for a gas of parastatistical particles, as well as the allowed occupancy of single-particle states. Consider a collection of non-interacting particles, for which the Hamiltonian is separable into a sum of one-body Hamiltonians $H = \sum_i H(x_i)$. Let the energy eigenvalues of the one-body problem be ϵ_i and the corresponding one-body Boltzmann factors $Z_i = e^{-\beta \epsilon_i}$. Consider now a sector of the Euclidean path integral characterized by the permutation of final points P. It is clear that this path integral $\mathcal{Z}_P$ decomposes into a product of disconnected components, characterized by the fact that the particle world-lines in each component mix particles only within the same component. Similarly to the cyclic decomposition in the preceding subsection (see step 2 of the derivation of the partition function for identical particles), we can decompose the path integral into cyclic permutations:

$$\mathcal{Z}_P = \prod_{\ell \in \text{cycles}(P)} \mathcal{Z}_\ell. \tag{4.3.80}$$

The path integral $\mathcal{Z}_\ell$ for a cyclic permutation of ℓ particles taking into account the periodic boundary conditions for partition functions can be thought of as the path integral of a single particle winding ℓ times around the Euclidean time β. This means that

$$\mathcal{Z}_\ell(\beta) = \mathcal{Z}_1(\ell\beta) = \sum_i z_i^\ell \qquad (4.3.81)$$

and the corresponding expression for $\mathcal{Z}_P$ becomes

$$\mathcal{Z}_P = \prod_{\ell \in \text{cycles}(P)} \sum_i z_i^\ell. \qquad (4.3.82)$$

The expression for the full partition function then becomes

$$\mathcal{Z}_S = \sum_{R \in S} \sum_P \frac{1}{N!} \chi_R(P) \prod_{n \in \text{cycles}(P)} \sum_i z_i^\ell. \qquad (4.3.83)$$

Expression (4.3.83) can be presented in a more explicit form via the so-called Schur functions. This representation does not deal with a path-integral technique and for further details we refer the reader to the original papers by Suranyi (1990) and Chaturvedi (1996).

4.3.4 Problems

Problem 4.3.1. In chapter 2, we learned that in imaginary time a quantum-mechanical particle formally looks like a Brownian particle. Check that transition *amplitude* (4.3.3) for indistinguishable bosons can be interpreted as a transition *probability* amplitude for a Brownian particle (boson diffusion process).

Hint. Let $\bar{X}$ and $\bar{Y}$ be two elements of D_N^3 (see definition after (4.3.1)) and construct the following permanent

$$K_{\text{IB}}(\bar{X}, t|\bar{Y}, 0) = \text{perm}|K(x_i, t|y_j, 0)|. \qquad (4.3.84)$$

It is clear that K_{IB} (the subscript IB means 'indistinguishable bosons') is positive for all $(\bar{X}, \bar{Y})$ pairs and that it also satisfies the required initial condition

$$\lim_{t \to 0} K_{\text{IB}}(\bar{X}, t|\bar{Y}, 0) = \delta(\bar{X} - \bar{Y}). \qquad (4.3.85)$$

Furthermore, in order that K_{IB} can be used as a transition probability density, it has to satisfy the conservation of probability (normalization) and the semigroup property

$$\int_{D_N^3} d^{3N}\bar{Y} \, K_{\text{IB}}(\bar{X}, t|\bar{Y}, 0) = 1 \qquad (4.3.86)$$

$$\int_{D_N^3} d^{3N}\bar{Y} \, K_{\text{IB}}(\bar{X}, t|\bar{Y}, 0) K_{\text{IB}}(\bar{Y}, s|\bar{z}, 0) = K_{\text{IB}}(\bar{X}, t+s|\bar{z}, 0). \qquad (4.3.87)$$

The conservation of probability can be derived using the property that a permanent is invariant under an interchange of two rows or columns. Hence

$$\int_{D_N^3} d^{3N}\bar{Y} \, K_{\text{IB}}(\bar{X}, t|\bar{Y}, 0) = \int_{D_N^3} d^{3N}\bar{Y} \frac{1}{N!} \sum_P K_{\text{IB}}(\bar{X}, t|\bar{Y}_p, 0)$$

$$= \frac{1}{N!} \int_{\mathbb{R}^{3N}} d^{3N}\bar{Y} \, \text{perm}|K(x_i, t|y_j, 0)| = 1 \qquad (4.3.88)$$

where use has been made of the fact that $K(x_i, t|y_j, 0)$ conserves probability.

The semigroup property follows with an analogous procedure by extending the integration domain D_N to $\mathbb{R}^N$ using the permutation symmetry and subsequently using the semigroup property of the single-particle propagators:

$$\int_{D_N^3} d^{3N}\bar{Y}\, K_{\text{IB}}(\bar{X}, t|\bar{Y}, s) K_{\text{IB}}(\bar{Y}, s|\bar{Z}, 0) = \frac{1}{N!}\int_{D_N^3} d^{3N}\bar{Y} \sum_p K_{\text{IB}}(\bar{X}, t|\bar{Y}_p, s) K_{\text{IB}}(\bar{Y}_p, s|\bar{Z}, 0)$$

$$= \frac{1}{N!}\int_{\mathbb{R}^{3N}} d^{3N}\bar{Y}\, \text{perm}|K(x_i, t|y_j, s)| \times \text{perm}|K(y_j, s|z_k, 0)| = \text{perm}|K(x_i, t|z_k, 0)|.$$
(4.3.89)

In the last step, the semigroup property of the one-particle propagators gives rise to $N!$ identical contributions.

In order to see how the integration over two permanents leads again to a permanent, the following argument might be useful. Denote by $|\bar{Y}\rangle$ a fully symmetrized and properly normalized solution of the Schrödinger equation for free bosons. The resolution of unity is then given by

$$1 = \frac{1}{N!}\int_{\mathbb{R}^{3N}} d^{3N}\bar{Y}\, |\bar{Y}\rangle\langle\bar{Y}|.$$
(4.3.90)

Denoting by H_0^i the Hamiltonian for the ith free particle, and by $H_0 = \sum_{i=1}^N H_0^i$ the Hamiltonian for N free non-interacting bosons, a diffusion from $\bar{z} \in D_N$ to $\bar{X} \in D_N$ is given by

$$K_{\text{IB}}(\bar{X}, t|\bar{Z}, 0) = \langle\bar{X}|e^{-\widehat{H}_0 t/\hbar}|\bar{Z}\rangle = \frac{1}{N!}\int_{\mathbb{R}^{3N}} d^{3N}\bar{Y}\, \langle\bar{X}|e^{-\widehat{H}_0(t-s)/\hbar}|\bar{Y}\rangle\langle\bar{Y}|e^{-\widehat{H}_0 s/\hbar}|\bar{Z}\rangle.$$
(4.3.91)

The reduction of all identical contributions to the preceding integral by permutation symmetry then leads to

$$K_{\text{IB}}(\bar{X}, t+s|\bar{Z}, 0) = \int_{D_N^3} d^{3N}\bar{Y}\, \langle\bar{X}|e^{-\widehat{H}_0 t/\hbar}|\bar{Y}\rangle\langle\bar{Y}|e^{-\widehat{H}_0 s/\hbar}|\bar{Z}\rangle$$

$$= \int_{D_N^3} d^{3N}\bar{Y}\, K_{\text{IB}}(\bar{X}, t|\bar{Y}, 0) K_{\text{IB}}(\bar{Y}, s|\bar{Z}, 0).$$
(4.3.92)

Therefore, $K_{\text{IB}}(\bar{X}, t|\bar{Y}, 0)$ is a transition probability density to go from $\bar{Y}$ to $\bar{X}$ in a time lapse t for a system of non-interacting identical particles with Bose–Einstein statistics. The boundary conditions for this process are determined by the behaviour of $K_{\text{IB}}(\bar{X}, t|\bar{Y}, 0)$ at the boundary ∂D_N. Because $\bar{\nabla}K_{\text{IB}}(\bar{X}, t|\bar{Y}, 0)$ is zero for $\bar{X} \in \partial D_N$, K_{IB} satisfies Neumann boundary conditions, leading to reflection for the process at the boundary.

Problem 4.3.2. Transform the initial expression (4.3.30) for the partition function of identical particles into the form (4.3.31), with separated centre-of-mass variables.

Hint. The centre-of-mass coordinate $\boldsymbol{R}$ does not only depend on the coordinates of all the particles, but it also has its own propagator. Therefore, substituting $\boldsymbol{R}$ by its expression in terms of the particle positions and then performing the integration does not seem to be the most adequate way to deal with the integration over the configuration space. Instead, the following identity is used for the formal treatment of $\boldsymbol{R}$ as an independent coordinate, at the expense of additional integrations:

$$\int d^{3N}\bar{r}\, f\!\left(\bar{r}, \frac{1}{N}\sum_{j=1}^N \boldsymbol{r}_j\right) = \int d^3R \int d^{3N}\bar{r}\, f(\bar{r}, \boldsymbol{R})\delta\!\left(\boldsymbol{R} - \frac{1}{N}\sum_{j=1}^N \boldsymbol{r}_j\right).$$
(4.3.93)

The Fourier transformation of the δ-function then leads to

$$\int d^{3N}\bar{r}\, f\left(\bar{r}, \frac{1}{N}\sum_{j=1}^{N} r_j\right) = \int d^3R \int \frac{d^3k}{(2\pi)^3} e^{i k \cdot R} \int d^{3N}\bar{r}\, f(\bar{r}, R) e^{-i\bar{k}\cdot\bar{r}} \qquad (4.3.94)$$

where $\bar{k} = \frac{k}{N}\{(1,1,1),\ldots,(1,1,1)\}$ is a $3N$-dimensional vector. Applying this transformation to the partition function $\mathcal{Z}_I$ and rearranging the factors, we obtain (4.3.31).

4.4 Field theory at non-zero temperature

This section is devoted to field theories describing *open* systems with exchange of energy between systems and surrounding thermal reservoirs. In other words, we shall consider canonical quantum statistical ensembles in the second-quantized formalism, that is *quantum field theory at non-zero temperature*.

The systems can be non-relativistic (section 4.4.1) or relativistic (sections 4.4.2 and 4.4.3). In either case, the basic quantities to be calculated are the corresponding canonical density operator (4.1.11), the partition function (4.1.13) and mean values (4.1.7) of operators constructed from the quantum fields. In fact, if we are interested only in the static characteristics of thermal systems in equilibrium, the calculations reduce to expressing the trace of the density operator $\hat{\rho} = \mathcal{Z}_\beta^{-1} \exp\{-\beta\hat{H}\}$ through the path integral (which we have already considered several times in this book; cf sections 2.2.1 and 4.3.2) and to the subsequent calculation of the path integral (exact or approximate). Therefore, sections 4.4.1 and 4.4.2, devoted to static characteristics, contain only some peculiarities of the trace calculation for field theoretical systems (diagram techniques, the method of the effective potential). In contrast, if we wish to study *dynamical* processes for thermal field systems, we need an essential modification of the path-integral representation for the partition function (in particular, the *doubling* of field variables). We shall consider this topic in section 4.4.3.

4.4.1 Non-relativistic field theory at non-zero temperature and the diagram technique

The non-relativistic field theory is the quantum mechanics of systems with an *arbitrary* number of identical particles in the formalism of the second quantization. In classical theory, such systems are described by the complex fields φ, φ^*, which are ordinary or anticommuting functions (depending on whether the particle statistics are bosonic or fermionic). The quantization is carried out with the help of the usual canonical commutation relations in the bosonic case or anticommutation relations in the fermionic case.

The free action functional has the form

$$S_0[\varphi^*, \varphi] = \int d^4x\, \varphi^*(x)[i\partial_t - \widehat{H}_1]\varphi(x) \qquad (4.4.1)$$

where $x = (t, \boldsymbol{x})$ and $\widehat{H}_1$ is a *one-particle* Hamiltonian, i.e. a linear operator acting only on $\boldsymbol{x}$ and having the meaning of the usual quantum-mechanical Hamiltonian for one particle from the system under consideration:

$$\widehat{H}_1 = \frac{\widehat{p}^2}{2m} + V_1(\boldsymbol{x}) - \mu. \qquad (4.4.2)$$

Here the first term is the kinetic energy of the particle, the second is the potential of an external field (e.g., a crystal lattice potential for electrons in a solid body) and μ has the meaning of the *chemical potential*. If the particles have non-zero spin, the fields φ^*, φ carry the spin index (hidden in our formulae).

224   *Path integrals in statistical physics*

It is clear from (4.4.1) that $i\varphi^*$ is the canonically conjugate momentum for the field φ. Thus, the canonical (anti)commutation relations have the form

$$\hat{\varphi}(\boldsymbol{x})\hat{\varphi}^\dagger(\boldsymbol{x}') \pm \hat{\varphi}^\dagger(\boldsymbol{x}')\hat{\varphi}(\boldsymbol{x}) = \hbar\delta(\boldsymbol{x}-\boldsymbol{x}'). \tag{4.4.3}$$

If the fields φ^*, φ are presented as a series expansion over some orthonormal complete set of eigenfunctions Φ_α of the operator $\widehat{H}_1$ or of the kinetic energy part $\widehat{p}^2/(2m)$,

$$\hat{\varphi}(\boldsymbol{x}) = \sum_\alpha \hat{a}_\alpha \Phi_\alpha \qquad \hat{\varphi}^\dagger(\boldsymbol{x}) = \sum_\alpha \hat{a}_\alpha^\dagger \Phi_\alpha^* \tag{4.4.4}$$

the coefficients $\hat{a}_\alpha, \hat{a}_\alpha^\dagger$ obey the standard commutation relations for creation and annihilation operators:

$$\hat{a}_\alpha \hat{a}_\alpha^\dagger \pm \hat{a}_\alpha^\dagger \hat{a}_\alpha = \delta_{\alpha\beta}$$

$$\hat{a}_\alpha \hat{a}_\beta \pm \hat{a}_\beta \hat{a}_\alpha = \hat{a}_\alpha^\dagger \hat{a}_\beta^\dagger \pm \hat{a}_\beta^\dagger \hat{a}_\alpha^\dagger = 0.$$

In order to take into account the interaction between the particles of the system, we have to add higher-order terms into the Hamiltonian. Usually, we consider only pairwise interactions:

$$H_{\text{int}} = \int d^3x\, d^3y\, u(\boldsymbol{x}-\boldsymbol{y})\varphi^*(t,\boldsymbol{x})\varphi(t,\boldsymbol{x})\varphi^*(t,\boldsymbol{y})\varphi(t,\boldsymbol{y}) \tag{4.4.5}$$

where $u(\boldsymbol{x}-\boldsymbol{y})$ is the two-point potential.

If we are interested in calculating the thermodynamical mean values (4.4.16) and (4.4.17), the variable t should be converted into the Euclidean one, $t \to \tau = -it$, and τ plays the role of the temperature: $\tau \in [0, \beta \equiv (k_B T)^{-1}]$. The subsequent formulae are slightly different for the bosonic and fermionic cases.

◇ **Generating functional and diagram technique for the bosonic non-relativistic field theory**

In the bosonic case, the path integral is defined over the space of *periodic* functions

$$\varphi(\tau+\beta, \boldsymbol{x}) = \varphi(\tau, \boldsymbol{x}) \qquad \varphi^*(\tau+\beta, \boldsymbol{x}) = \varphi^*(\tau, \boldsymbol{x}) \tag{4.4.6}$$

(because we are calculating the *trace* (4.1.14)). Assuming that the particles are confined in a box L^3 and taking into account the periodicity (4.4.6), we can make the Fourier transform (i.e. the expansion (4.4.4) over the eigenfunctions of the one-particle kinetic energy):

$$\varphi(\tau, \boldsymbol{x}) = \frac{1}{\sqrt{\beta L^3}} \sum_{\omega_n, \boldsymbol{k}} a(\omega_n, \boldsymbol{k}) e^{i(\omega_n \tau - \boldsymbol{k}\boldsymbol{x})}$$

$$\varphi^*(\tau, \boldsymbol{x}) = \frac{1}{\sqrt{\beta L^3}} \sum_{\omega_n, \boldsymbol{k}} a^*(\omega_n, \boldsymbol{k}) e^{-i(\omega_n \tau - \boldsymbol{k}\boldsymbol{x})} \tag{4.4.7}$$

where

$$\omega_n = 2\pi n/\beta \qquad k_i = 2\pi n_I/L \qquad n, n_I \in \mathbb{Z}. \tag{4.4.8}$$

In terms of these Fourier components, the complete action takes the form

$$S = \sum_{\omega,\boldsymbol{k}} \left(\frac{k^2}{2m} - i\omega - \mu\right) a^*(\omega,\boldsymbol{k}) a(\omega,\boldsymbol{k})$$

$$+ \frac{1}{2\beta L^3} \sum_{\substack{\boldsymbol{k}_1+\boldsymbol{k}_2=\boldsymbol{k}_3+\boldsymbol{k}_4 \\ \omega_1+\omega_2=\omega_3+\omega_4}} \widetilde{u}(\boldsymbol{k}_1-\boldsymbol{k}_3) a^*(\omega_1,\boldsymbol{k}_1) a^*(\omega_2,\boldsymbol{k}_2) a(\omega_3,\boldsymbol{k}_3) a(\omega_4,\boldsymbol{k}_4). \tag{4.4.9}$$

Here $\widetilde{u}(k)$ is the Fourier transform of the two-particle potential $u(x)$,

$$u(x) = \frac{1}{L^3} \sum_k e^{ikx} \widetilde{u}(k) \qquad (4.4.10)$$

and we have used the shorthand notation $\omega_j \equiv \omega_{n_j}$ ($j = 1, 2, 3, 4$).

The basic objects in a field theory at non-zero temperature are the *thermal Green functions*, i.e. the mean values of the products of field variables:

$$\langle\!\langle \mathbf{T}(\hat{\varphi}(\tau_1, x_1) \cdots \hat{\varphi}(\tau_n, x_n))\rangle\!\rangle_\beta \equiv \text{Tr}[\hat{\rho}_C(\beta)(\hat{\varphi}(\tau_1, x_1) \cdots \hat{\varphi}(\tau_n, x_n))] \qquad (4.4.11)$$

($\hat{\rho}_C(\beta)$ is the canonical density operator (4.1.11) at the inverse temperature β). The usual steps which we have carried out several times in this book yield the following path-integral representation for their generating functional:

$$\mathcal{Z}_\beta[j, j^*] = \mathfrak{N}^{-1} \int \mathcal{D}\varphi\, \mathcal{D}\varphi^* \exp\left\{-\frac{1}{\hbar}\left[S + \int d^4x\, (j^*(\tau, x)\varphi(\tau, x) + j(\tau, x)\varphi^*(\tau, x))\right]\right\}. \qquad (4.4.12)$$

The perturbation expansion of this path integral leads to the diagram technique with the following basic elements:

Free two-point Green function: $\xrightarrow{\omega,\, k}$ $G_0 = \left(i\omega - \frac{k^2}{2m} + \mu\right)^{-1}$

Two-particle vertex: [diagram with legs ω_1, k_1; ω_2, k_2; ω_3, k_3; ω_4, k_4] $\widetilde{u}(k_1 - k_3) + \widetilde{u}(k_2 - k_4)$.

◇ **Generating functional and diagram technique for the fermionic non-relativistic field theory**

The quantization of fermionic systems is carried out by integration over Grassmann elements. To obtain correct statistics, we have to input *antiperiodic* boundary conditions in τ (problem 4.4.3, page 254):

$$\varphi(\tau + \beta, x) = -\varphi(\tau, x) \qquad \varphi^*(\tau + \beta, x) = -\varphi^*(\tau, x). \qquad (4.4.13)$$

As a result, the fermionic functions are expanded in the following Fourier series:

$$\varphi(\tau, x) = \frac{1}{\sqrt{\beta L^3}} \sum_{\omega_n, k} a(\omega_n, k) e^{i(\omega_n \tau - kx)}$$
$$\varphi^*(\tau, x) = \frac{1}{\sqrt{\beta L^3}} \sum_{\omega_n, k} a^*(\omega_n, k) e^{-i(\omega_n \tau - kx)} \qquad (4.4.14)$$

where

$$\omega_n = 2\pi(n + 1/2)/\beta \qquad k_i = 2\pi n_\text{I}/L \qquad n, n_\text{I} \in \mathbb{Z}. \qquad (4.4.15)$$

Formally, the action functional for fermionic systems looks the same as for bosonic systems (4.4.9). The diagram technique is also quite similar and has the following basic elements:

Free two-point Green function: $\xrightarrow{\omega,\ \boldsymbol{k}}$ $\qquad G_0 = \left(i\omega - \frac{k^2}{2m} + \mu\right)^{-1}$

Two-particle vertex: (diagram with legs $\omega_1,\boldsymbol{k}_1$; $\omega_2,\boldsymbol{k}_2$; $\omega_3,\boldsymbol{k}_3$; $\omega_4,\boldsymbol{k}_4$) $\qquad \widetilde{u}(\boldsymbol{k}_1 - \boldsymbol{k}_3) - \widetilde{u}(\boldsymbol{k}_2 - \boldsymbol{k}_4)$.

The only difference is that the frequencies $\omega\beta/(2\pi)$ take *half-integer* values and that the two-particle potential is antisymmetrized.

4.4.2 Euclidean-time relativistic field theory at non-zero temperature

In this subsection, we consider the calculation of temperature-dependent quantum effects in relativistic field theories at non-zero temperature. To illustrate the theoretical techniques, we use, at first, the theory of a single scalar field φ. Without any loss of generality we assume that we work in the rest frame of the system so that a Hamiltonian approach is adequate. Further, in this subsection we shall assume that the system under consideration is in a thermal equilibrium. In this case, its Green functions are given by the conventional thermodynamical averaging:

$$\begin{aligned}G_\beta(x_1,\ldots,x_N) &= \langle\!\langle \mathbf{T}(\hat{\varphi}(x_1)\cdots\hat{\varphi}(x_N))\rangle\!\rangle_\beta \\ &\stackrel{\text{def}}{\equiv} \text{Tr}[\hat{\rho}_C(\beta)\mathbf{T}(\hat{\varphi}(x_1)\cdots\hat{\varphi}(x_N))] \\ &= \sum_n \langle n|\mathbf{T}(\hat{\varphi}(x_1)\cdots\hat{\varphi}(x_N))|n\rangle \frac{e^{-\beta E_n}}{\sum_m e^{-\beta E_m}}.\end{aligned} \qquad (4.4.16)$$

In (4.4.16), $|n\rangle$ denote a complete orthonormal set of energy eigenstates of the Hamiltonian with the eigenvalues E_n. As $\beta \equiv 1/(\kappa_B T)$ goes to infinity, we recover the usual expression for the Green functions of the scalar field, defined as the mean value of the time-ordered product in the ground state $|0\rangle$.

The generating functional of these *thermal Green functions* has the form

$$\begin{aligned}\mathcal{Z}_\beta[j] &= \left\langle\!\!\left\langle \mathbf{T}\left(\exp\left\{\frac{i}{\hbar}\int dx\, j(x)\varphi(x)\right\}\right)\right\rangle\!\!\right\rangle_\beta \\ &= \frac{1}{\text{Tr}\exp\{-\beta\widehat{H}\}} \text{Tr}\left[e^{-\beta\widehat{H}}\mathbf{T}\left(\exp\left\{\frac{i}{\hbar}\int dx\, j(x)\varphi(x)\right\}\right)\right] \end{aligned} \qquad (4.4.17)$$

where the trace is taken over any complete set of states.

In the case of zero-temperature quantum field theory, there are two formalisms: in real- and imaginary-time variables. At non-zero temperature, the two techniques are essentially different and the choice of one of them crucially depends on the problem to be solved. In this subsection we shall discuss the Euclidean-time formalism introduced by Feynman (see Feynman (1972a) and references therein) and developed by Matsubara (1955) for non-relativistic systems, and later extended to field theory (see Abrikosov *et al* (1965) and references therein).

◇ Path-integral representation for the generating functional of thermal Green functions

The usual steps allow us to represent the trace (4.4.17) in terms of the path integral:

$$\mathcal{Z}_\beta[j] = \mathfrak{N}^{-1} \int \mathcal{D}\varphi(x) \exp\left\{-\left[S_{E,\beta} - \int_0^\beta d\tau \int d^3x\, j\varphi\right]\right\}$$

$$= \mathfrak{N}^{-1} \int \mathcal{D}\varphi(x) \exp\left\{-\int_0^\beta d\tau \int d^3x \left[\frac{1}{2}(\partial_i\varphi)^2 + \frac{1}{2}m^2\varphi^2 + V(\varphi) - j\varphi\right]\right\} \quad (4.4.18)$$

where $S_{E,\beta}$ denotes the Euclidean action on the finite 'time' interval $[0, \beta]$ for the scalar field φ. Note that $(\partial_i\varphi)^2 \equiv \sum_{i=0}^3(\partial_i\varphi)^2$. It is seen that (4.4.18) is quite similar to the generating functional for ordinary (Euclidean) field theory, which we considered in chapter 3, the only specific feature of $\mathcal{Z}_\beta$ consisting in the periodic boundary condition

$$\varphi(0, \boldsymbol{x}) = \varphi(\beta, \boldsymbol{x}) \quad (4.4.19)$$

on the finite 'time' interval. The periodicity in the 'time' variable τ implies the Fourier decomposition

$$\varphi(\tau, \boldsymbol{x}) = \beta^{-1} \sum_{m=-\infty}^\infty \int \frac{d^3k}{(2\pi)^3} \varphi_m(\boldsymbol{k}) \exp\{i(\boldsymbol{k}\boldsymbol{x} + \omega_m\tau)\} \quad (4.4.20)$$

where $\omega_m = 2\pi m/\beta$.

Let us first consider the free theory. In terms of the Fourier transform, the free action $S_{0,\beta}$ reads

$$S_{0,\beta}[\varphi] = \frac{1}{2\beta} \sum_m \int \frac{d^3k}{(2\pi)^3} (\omega_m^2 + \boldsymbol{k}^2 + m_0^2) \varphi_m(\boldsymbol{k})\varphi_{-m}(\boldsymbol{k}) \quad (4.4.21)$$

where $\varphi_m^* = \varphi_{-m}$, so that the propagator in the momentum space has the form

$$D_\beta(\omega_m, \boldsymbol{k}) = \frac{1}{\omega_m^2 + \boldsymbol{k}^2 + m_0^2}. \quad (4.4.22)$$

Its Fourier transform, i.e. the propagator in the coordinate space,

$$D_\beta(\tau - \tau', \boldsymbol{x} - \boldsymbol{x}') = \frac{1}{\beta} \sum_m \int \frac{d^3k}{(2\pi)^3} \frac{\exp\{i[\omega_m(\tau - \tau') + \boldsymbol{k}(\boldsymbol{x} - \boldsymbol{x}')]\}}{\omega_m^2 + \boldsymbol{k}^2 + m_0^2} \quad (4.4.23)$$

is periodic in the time variable:

$$D_\beta(\tau + \beta, \boldsymbol{x}) = D_\beta(\tau, \boldsymbol{x}) \quad (4.4.24)$$

(the *Kubo–Martin–Schwinger propagator relation*). This is a consequence of the field periodicity condition. As usual, the generating functional for a theory with interaction can be written formally as

$$\mathcal{Z}_\beta[j] = \exp\left\{-\int_0^\beta d\tau \int d^3x\, V\left(\frac{\delta}{\delta j}\right)\right\} \mathcal{Z}_{0,\beta}[j]. \quad (4.4.25)$$

The power expansion of the exponent in (4.4.25) generates the perturbation series for the Green functions of the thermal scalar field theory.

228   *Path integrals in statistical physics*

◇ **Effective potential and critical temperature**

To study the behaviour of a system with a variation of temperature, it is convenient to use the so-called *effective potential*. In general, this can be defined in the zero-temperature quantum field theory as well.

- Consider $\Gamma[\phi]$, the functional Legendre transform of the generating functional $W[J]$ of connected Green functions. As we have explained in section 3.1.5, $\Gamma[\phi]$ generates the one-particle irreducible (OPI) Green functions. Besides its purely technical merit, the functional $\Gamma[\phi]$ has a direct physical meaning, having the interpretation of the quantum generalization of the classical action of the model under consideration. For this reason, it is termed the *effective action* of the theory.

To explain this, recall that the initial generating functional $\mathcal{Z}_\beta[J]$ satisfies the Schwinger equation (cf section 3.1.5, equation (3.1.143)) which can be rewritten as follows:

$$\left[\left(\frac{\delta S[\phi]}{\delta \phi(x)}\right)\bigg|_{\phi=\hat{\phi}} + J(x)\right]\mathcal{Z}_\beta[J] = 0 \qquad (4.4.26)$$

where $\phi(x)$ is substituted by the operator (we explicitly recover the Planck constant $\hbar$ needed for further discussion)

$$\hat{\phi} = -i\hbar \frac{\delta}{\delta J(x)}.$$

Using the relation $\mathcal{Z}_\beta[J] = \exp\{W[J]\}$, the Schwinger equation can be presented as

$$\left[\left(\frac{\delta S[\phi]}{\delta \phi(x)}\right)\bigg|_{\phi=\widetilde{\phi}} + J(x)\right]\mathbb{I} = 0 \qquad (4.4.27)$$

where

$$\begin{aligned}\widetilde{\phi}(x) &= -i\hbar\left(\frac{\delta W}{\delta J(x)} - \frac{\delta}{\delta J(x)}\right) \\ &= \phi(x) - i\hbar\int dy\, \frac{\delta^2 W}{\delta J(x)\delta J(y)}\frac{\delta}{\delta \phi(y)} \\ &= \phi(x) + i\hbar\int dy\, \left(\frac{\delta^2 \Gamma}{\delta \phi(x)\delta \phi(y)}\right)^{-1}\frac{\delta}{\delta \phi(y)} \end{aligned} \qquad (4.4.28)$$

and $\mathbb{I}$ is the unit (trivial) functional (i.e. $(\delta/\delta J)\mathbb{I} = 0$). On the other hand, the properties of the Legendre transformation imply (cf section 3.1.5) that

$$J(x) = -\frac{\delta \Gamma}{\delta \phi(x)}. \qquad (4.4.29)$$

The comparison of (4.4.28) and (4.4.29) yields a rather cumbersome equation for the functional $\Gamma[\phi]$, which in the classical limit gives

$$\lim_{\hbar \to 0} \Gamma[\phi] = S[\phi]. \qquad (4.4.30)$$

This limit justifies the name *effective action* for the generating functional $\Gamma[\phi]$ for OPI-Green functions. Note, however, that while $S[\phi]$ is the integral of a *local* Lagrangian density, the quantum effective action is highly non-local:

$$\Gamma[\phi] = \sum_{m=0}^{\infty} \frac{1}{m!}\int d^4x_1 \cdots d^4x_m\, \phi(x_1)\cdots\phi(x_m)\Gamma_m(x_1,\ldots,x_m) \qquad (4.4.31)$$

(here $\Gamma_m(x_1, \ldots, x_m)$ are m-point OPI-Green functions). We can give a quasi-local form to $\Gamma[\phi]$ by expanding each field $\phi(x_j)$, $j \neq 1$ about the point $x \equiv x_1$ and then by integration over all but $x = x_1$ variables:

$$\Gamma[\phi] = \int d^4x \left[-U(\phi) + \tfrac{1}{2} Z(\phi) \partial_\mu \phi \partial^\mu \phi + \text{terms containing derivatives of the order} \geq 4 \right] \quad (4.4.32)$$

where $U(\phi)$, $Z(\phi)$ are functions of ϕ. The limit of $U(\phi)$ is the total *classical potential* $U_{\text{cl}}(\phi)$:

$$\lim_{\hbar \to 0} U(\phi) = U_{\text{cl}}(\phi) \equiv \tfrac{1}{2} m^2 \phi^2 + V(\phi) \quad (4.4.33)$$

where $V(\varphi)$ defines a self-interaction of the scalar field.

- $U(\phi)$ is the quantum generalization of this classical potential and is known as the *effective potential*.

The effective potential can be isolated in the full quantum theory by taking the mean field ϕ to be *constant* in spacetime (recall that ϕ denotes the mean field $\phi = \delta W[J]/\delta J$, in distinction from the initial dynamical field φ, cf (3.1.154) and (3.1.155)). For such ϕ, only the effective potential remains in the series (4.4.32), irrespective of the magnitude of $\hbar$. For a system in a spacetime box of volume βL^3 and in the case of constant ϕ, we have

$$\Gamma[\phi] = -\beta L^3 V(\phi). \quad (4.4.34)$$

Note that the effective potential is numerically the same quantity in both Euclidean and Minkowski (real-time) variants of quantum field theory. For finite-temperature Euclidean field theory, the effective potential is identical to the conventional thermodynamic *free energy*.

In general, the calculations of most interest are of an effective potential U_β (the subscript β indicates that we consider a theory at *non-zero* temperature) for systems that possess symmetry-breaking at zero temperature. Then, we expect a restoration of the symmetry as temperature increases. As we shall show, a finite-order calculation in the loop expansion is sufficient to show that this expectation is correct.

Let us consider the simplest scalar φ^4-theory with the classical potential term

$$U_{\text{cl}}(\varphi) = \frac{1}{2} m^2 \varphi^2 + \frac{\lambda}{4!} \varphi^4. \quad (4.4.35)$$

As we learned in section 3.2.8, at zero temperature and for $m^2 < 0$ the reflection symmetry $\varphi \to -\varphi$ is broken. We anticipate that, at high temperature, this discrete symmetry is restored.

The stationary-phase method gives the following *one-loop* approximation expression for the effective potential:

$$U_\beta^{(1)}(\phi) = U_{\text{cl}}(\phi) + \frac{1}{2\beta} \sum_n \int \frac{d^3k}{(2\pi)^3} \ln(\mathbf{k}^2 + \omega_n^2 + M^2(\phi)) + U_R^{(1)} \quad (4.4.36)$$

where $M^2(\phi) = U_{\text{cl}}''(\phi)$ and $U_R^{(1)}$ is the one-loop renormalization counterterm. The sum over n in (4.4.36) has the general form

$$f(E) = \sum_n \ln(E^2 + \omega_n^2) \quad (4.4.37)$$

and diverges. To extract its finite part, let us first find its derivative

$$\frac{\partial f(E)}{\partial E} = 2 \sum_n \frac{E}{E^2 + \omega_n^2}$$

$$= 2 \sum_n \frac{E}{E^2 + (2\pi n/\beta)^2}$$

$$= 2\beta \left(\frac{1}{2} + \frac{1}{e^{\beta E} + 1} \right). \quad (4.4.38)$$

Here we have used the result (Gradshteyn and Ryzhik (1980), formula 1.421.4)

$$\sum_{n=1}^{\infty} \frac{x}{x^2 + n^2} = -\frac{1}{2x} + \frac{1}{2}\pi \coth(\pi x). \qquad (4.4.39)$$

The integration of (4.4.38) yields

$$f(E) = 2\beta[\tfrac{1}{2}E + \beta^{-1}\ln(1 - e^{-\beta E})] + \text{constant} \qquad (4.4.40)$$

and inserting this result in (4.4.36), we obtain finally

$$U_\beta^{(1)}(\phi) = U_{\text{cl}}(\phi) + \{U_{\beta=0}^{(1)}(\phi) + U_R^{(1)}(\phi)\} + W_\beta^{(1)}(\phi). \qquad (4.4.41)$$

In this formula, the terms in the curly brackets give a diverging contribution, independent of temperature (and hence it can be calculated in the usual zero-temperature quantum field theory) together with the corresponding counterterms. The temperature-dependent contribution is given by the integral

$$W_\beta^{(1)}(\phi) = \frac{1}{\beta} \int \frac{d^3k}{(2\pi)^3} \ln(1 - e^{-\beta E(k)}). \qquad (4.4.42)$$

Note that this temperature-dependent addition to the one-loop effective potential is ultraviolet finite. For small β (high T), the integrand in (4.4.42) can be expanded in the power series (*high temperature expansion*); this gives

$$W_\beta^{(1)}(\phi) = -\pi^2 90 (k_B T)^4 + \frac{M^2(\phi)}{24}(k_B T)^2 + \mathcal{O}(T). \qquad (4.4.43)$$

The net result for the complete one-loop effective potential (after an additional shifting of the origin to the point $U_\beta^{(1)}(\phi = 0)$) can be written in the form

$$U_\beta^{(1)}(\phi) = \frac{1}{2}m^2\left(1 - \frac{T^2}{T_c^2}\right)\phi^2 + \frac{\lambda}{4!}\phi^4 \qquad (4.4.44)$$

where the *critical temperature* T_c is given by

$$T_c^2 = -\frac{24m^2}{\lambda k_B^2}. \qquad (4.4.45)$$

Recall that we study the model with spontaneous symmetry-breaking *at zero temperature* (and, hence, with $m^2 < 0$). At temperatures $T < T_c$ the one-loop potential retains its degenerate minima. As T increases to T_c these minima move *continuously* to the origin $\phi = 0$, becoming coincident at $T = T_c$. The restoration of symmetry, with a single minimum at $\phi = 0$, takes place as the temperature becomes higher than the critical temperature. This behaviour corresponds to a *second-order* phase transition (in contrast to a *first-order* phase transition in which the minima jump *discontinuously* to the origin at $T = T_c$).

Unfortunately, this picture of the phase transitions as being very illuminative is a bit oversimplified. The signal of this oversimplification comes from the fact that the obtained effective potential contradicts the *general* property of effective potentials, i.e. the *convexity*.

Theorem 4.1 (Symanzik). The effective potential is *convex* even if the classical potential V_{cl} is non-convex (e.g., corresponds to the spontaneous breaking, see figure 4.1) (Symanzik 1964).

Proof. To prove this statement, let us start from the generating functional (in the Euclidean space) with the constant source j and in a finite volume L^3:

$$\mathcal{Z}_\beta[j] = \int \mathcal{D}\varphi \exp\left\{-\frac{1}{\hbar}\left[S_{E,\beta}[\varphi] - j\int_0^\beta d\tau \int_{L^3} dx \varphi(x)\right]\right\}. \tag{4.4.46}$$

Defining $w(j)$ as

$$\mathcal{Z}_\beta[j] \equiv \exp\left\{-\frac{\beta}{\hbar} w(j) L^3\right\} \tag{4.4.47}$$

we obtain

$$-\hbar L^3 \frac{\partial^2 w}{\partial j^2} = \left\langle\left(\int_0^\beta d\tau \int_{L^3} dx\, \varphi\right)^2\right\rangle - \left\langle\left(\int_0^\beta d\tau \int_{L^3} dx\, \varphi\right)\right\rangle^2 \tag{4.4.48}$$

where the averaging $\langle\cdots\rangle$ is defined with respect to the Euclidean (Wiener) functional measure:

$$\langle F[\varphi]\rangle = \frac{\int \mathcal{D}\varphi(x) \exp\left\{-\frac{1}{\hbar}(S_{E,\beta}[\varphi] - j\int_0^\beta d\tau \int_{L^3} dx\, \varphi)\right\} F[\varphi]}{\int \mathcal{D}\varphi(x) \exp\left\{-\frac{1}{\hbar}(S_{E,\beta}[\varphi] - j\int_0^\beta d\tau \int_{L^3} dx\, \varphi)\right\}}. \tag{4.4.49}$$

Applying the Cauchy–Schwarz–Bunyakovskii inequality, $\langle(\int \varphi)^2\rangle \geq \langle(\int \varphi)\rangle^2$, to equation (4.4.48), we find

$$\frac{\partial^2 w}{\partial j^2} \leq 0 \tag{4.4.50}$$

and using the particular case of the general properties of two-point OPI functions (cf (3.1.157)), i.e.

$$\frac{\partial^2 U}{\partial \phi^2} \frac{\partial^2 w}{\partial j^2} = -1 \tag{4.4.51}$$

we finally obtain the desired result:

$$\frac{\partial^2 U}{\partial \phi^2} \geq 0 \tag{4.4.52}$$

(for any finite L^3).

———o———

Note that this proof and the implication of the statement to quantum field theory are rather formal (we did not take into account possible divergences and did not discuss the $L^3 \to \infty$ limit). However, this 'naive' formalism leads to the correct result (Griffiths 1972). The characteristic form of the effective one-loop potential for a classical potential with two minima is depicted in figure 4.1.

To resolve the apparent contradiction between 'naive' one-loop calculation, spontaneous symmetry-breaking phenomena and the convexity of the effective potential, we should take into account *both* extrema of the double-well classical potential. Indeed, if $m^2 < 0$, the action possesses two extremal points $\phi_\pm(j)$ at constant j, which are solutions of the equation

$$\lambda \phi_\pm (\phi_\pm^2 - \phi_0^2) = 6j \tag{4.4.53}$$

where $\phi_0 = 6|m^2|/\lambda$. In the $j \to 0$ limit, $S_E[\phi_+] = S_E[\phi_-]$ and the extremal points contribute equally. To calculate the corresponding path integral we first have to isolate the zero-frequency mode (cf sections 3.3.2 and 3.3.3). Using the separation

$$\phi(x) = a + \sqrt{\hbar}\xi(x) \tag{4.4.54}$$

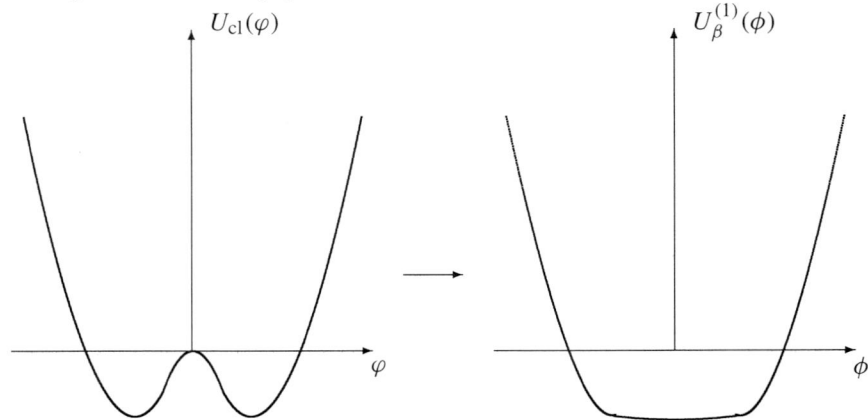

Figure 4.1. The effective potential $U_\beta^{(1)}(\phi)$ for a scalar theory with a double-well classical potential $V_{\text{cl}}(\varphi)$. The effective potential is convex and coincides with the classical potential only outside the domain of its extrema.

and the decomposition of unity

$$1 = \int_{-\infty}^{\infty} da\, \delta\left(a - \frac{1}{\beta L^3}\int_0^\beta d\tau \int_{L^3} dx\, \varphi(x)\right)$$

$$= \int_{-\infty}^{\infty} da\, d\alpha\, \exp\left\{i\alpha\left(a - \frac{1}{\beta L^3}\int_0^\beta d\tau \int_{L^3} dx\, \varphi(x)\right)\right\}$$

we can write the generating functional in the form

$$\mathcal{Z}_\beta(j) = \int da\, d\alpha\, \mathcal{D}\xi(x)\, \exp\left\{-\frac{1}{\hbar}(S_{E,\beta}[a+\sqrt{\hbar}\xi] - ja\beta L^3) + i\alpha \int_0^\beta d\tau \int_{L^3} \delta x\, \xi\right\}$$

$$= \int da\, K(a)\, \exp\left\{-\frac{\beta L^3}{\hbar}(U_{\text{cl}}(a) + U_R^{(1)}(a) - ja)\right\} \qquad (4.4.55)$$

where

$$K(a) = \int d\alpha\, \mathcal{D}\xi\, \exp\left\{\int_0^\beta d\tau \int_{L^3} dx\left[i\alpha\xi(x) - \frac{1}{2}\xi\left(-\nabla^2 + m^2 + \frac{1}{2}\lambda a^2\right)\xi\right.\right.$$

$$\left.\left. - \frac{1}{6}\sqrt{\hbar}\lambda a\xi^3 - \frac{1}{4!}\lambda\xi^4\right]\right\}. \qquad (4.4.56)$$

Using the semiclassical approximation, the behaviour of the mean field $\phi(j)$ in the one-loop approximation as a function of the source j can be found:

$$\phi(j) \approx \phi_0 \tanh\left(\frac{L^3 j\phi}{\hbar}\right). \qquad (4.4.57)$$

It is easy to verify that the behaviour of $\phi(j)$ is different from that for the classical case (i.e. determined from the classical Lagrangian with the external source) but becomes coincident with the classical behaviour in the region $\phi > \phi_0$ (i.e. outside the domain of the extrema of the classical potential). The most important point which we have been aiming to illustrate is that despite the convexity of the one-loop effective potential, *we still have the symmetry-breaking* since $\phi(j) \to \pm\phi_0$ as $j \to \pm 0$, as it is seen from (4.4.57).

At non-zero and increasing temperature the potential has the flat-bottomed 'bucket' profile as in figure 4.1, the base of which gets narrower as T increases and the disappearance of the flatness signals about the restoration of the symmetry. Note that there are arguments that a more rigorous description of this phenomenon requires the consideration of more complicated (than just a constant) field configurations, in particular of the so-called *domain walls* (de Carvalho *et al* 1985).

4.4.3 Real-time formulation of field theory at non-zero temperature

In the preceding subsection, we considered the *static* characteristics of thermal systems in *equilibrium*. The equilibrium is achieved as a result of a *dynamical* process of energy exchange between the fields and the reservoir in which they are immersed. In these calculations, the underlying dynamics was hidden by the imaginary-time formalism. Although the results of the Euclidean-time approach are correct, in order to obtain more detailed information about systems at non-zero temperature, it is desirable to consider the *real-time* formulation.

Any approach to a real-time description leads to the *doubling* of fields, as well as of the corresponding Hilbert spaces of states. To understand this point qualitatively, let us consider a thermal reservoir maintaining a certain number of excited quanta in a system. An exchange of energy can come about by two processes: energy is absorbed by the system

- either by exciting new quanta or
- by annihilating the (Dirac) 'holes' of particles maintained by the reservoir.

The appearance of the new possibility of energy absorption in the case of non-zero temperature is schematically depicted in figure 4.2. The two reverse processes are responsible for the emission of energy by the system. While in the imaginary-time formulation these two types of process are inseparable, in the real-time case they lead to a doubling of the fields. This formalism has been developed mainly by Umezawa with co-authors (see Umezawa *et al* (1982) and references therein).

◇ **Introductory example: the harmonic oscillator with doubling**

To illustrate the technique, let us consider the much simpler problem of a free harmonic oscillator. The doubling implies that we should consider two pairs of mutually commuting creation and annihilation operators $\widehat{a}^\dagger, \widehat{a}$ and $\widehat{b}^\dagger, \widehat{b}$ which act in the direct product Hilbert space spanned by the states

$$|n\rangle|\nu\rangle = (n!)^{-1/2}(\nu!)^{-1/2}(\widehat{a}^\dagger)^n(\widehat{b}^\dagger)^\nu|0\rangle|0\rangle.$$

Then the thermal averaging $\langle\!\langle \widehat{A}(\widehat{a}^\dagger, \widehat{a})\rangle\!\rangle$ (cf (4.4.16)) for any operator $\widehat{A}$ constructed from $\widehat{a}^\dagger$ and $\widehat{a}$ can be presented as the ordinary quantum-mechanical averaging:

$$\langle\!\langle \widehat{A}(\widehat{a}^\dagger, \widehat{a})\rangle\!\rangle_\beta = \langle\beta|\widehat{A}(\widehat{a}^\dagger, \widehat{a})|\beta\rangle \qquad (4.4.58)$$

with respect to the state $|\beta\rangle$

$$|\beta\rangle = \frac{\sum_n \exp\{-\beta E_n/2\}|n\rangle|n\rangle}{\sqrt{\sum_n \exp\{-\beta E_n\}}}$$

$$= \frac{\sum_n \exp\{-\beta\omega n/2\}(n!)^{-1}(\widehat{a}^\dagger)^n(\widehat{b}^\dagger)^n|0\rangle|0\rangle}{\sqrt{\sum_n \exp\{-\beta\omega n\}}}$$

$$= (1 - e^{-\beta\omega})^{1/2} \exp\{e^{-\beta\omega/2}\widehat{a}^\dagger\widehat{b}^\dagger\}|0\rangle|0\rangle \qquad (4.4.59)$$

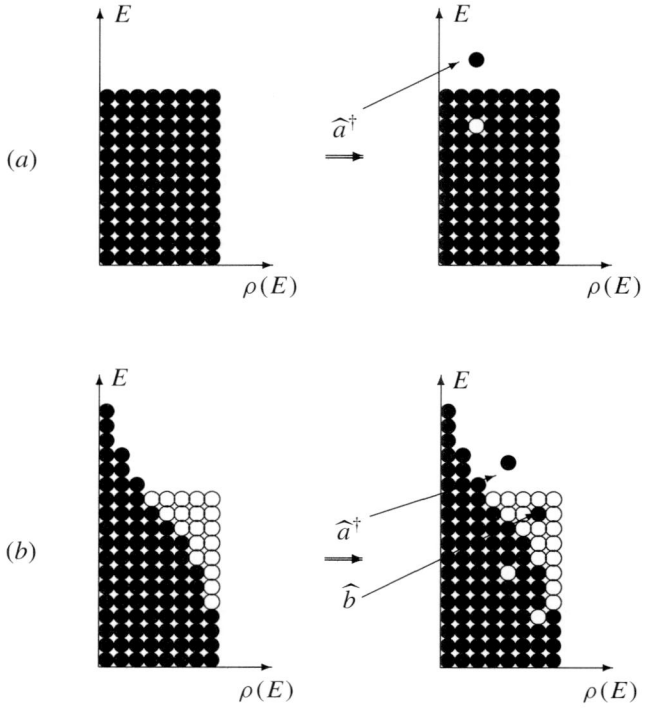

Figure 4.2. Schematic presentation of energy absorption by a system at zero temperature (a) and the two ways of absorption at non-zero temperature (b).

(E_n is the harmonic oscillator energy: $E_n = (n + \frac{1}{2})\omega$).

In the product Hilbert space, time translation is supposed to be generated by the Hamiltonian

$$\widehat{H} = \omega\widehat{a}^\dagger\widehat{a} - \omega\widehat{b}^\dagger\widehat{b} \qquad (4.4.60)$$

which is invariant with respect to the canonical transformations generated by the operator

$$L = \mathrm{i}\theta(\widehat{a}\widehat{b} - \widehat{a}^\dagger\widehat{b}^\dagger) \qquad (4.4.61)$$

i.e.

$$\begin{aligned}\widehat{a}_\beta &= \mathrm{e}^{\mathrm{i}L}\widehat{a}\mathrm{e}^{-\mathrm{i}L} = (\cosh\theta)\widehat{a} + (\sinh\theta)\widehat{b}^\dagger \\ \widehat{a}_\beta^\dagger &= \mathrm{e}^{\mathrm{i}L}\widehat{a}\mathrm{e}^{-\mathrm{i}L} = (\sinh\theta)\widehat{b} + (\cosh\theta)\widehat{a}^\dagger.\end{aligned} \qquad (4.4.62)$$

Such canonical transformations are well known in the literature as the *Bogoliubov transformations*. The state $|\beta\rangle$ can be considered as a result of the corresponding unitary transformation:

$$|\beta\rangle = \mathrm{e}^{\mathrm{i}\widehat{L}}|0\rangle|0\rangle \qquad (4.4.63)$$

with the appropriately chosen parameter θ. The latter can most easily be defined from the equality

$$\langle\!\langle\widehat{a}\widehat{a}^\dagger\rangle\!\rangle = \langle\beta|\widehat{a}\widehat{a}^\dagger|\beta\rangle$$

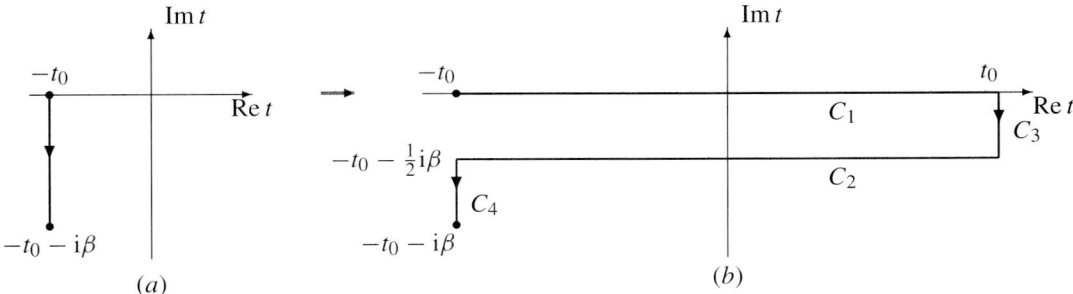

Figure 4.3. The contour in the complex-time plane for non-zero temperature in the imaginary-time (a) and in the real-time (b) formalisms.

which, after simple calculation, gives the relation

$$\cosh^2 \theta = (1 - e^{-\beta \omega})^{-1}.$$

The doublets $\widehat{a}, \widehat{b}^\dagger$ and $\widehat{b}, \widehat{a}^\dagger$ allow us to describe the processes of energy exchange in the thermal bath mentioned earlier: the oscillator system absorbs energy either by the excitation of additional quanta ($\widehat{a}^\dagger$) or by the annihilation of holes of particles maintained by the thermal reservoir ($\widehat{b}$); the energy emission similarly involves $\widehat{a}$ and $\widehat{b}^\dagger$. The proper creation and annihilation operators that describe the excitation and de-excitation of the system are the operators $\widehat{a}_\beta^\dagger, \widehat{a}_\beta$, given by the Bogoliubov transformation.

◇ **Generalization to field theory: path-integral approach**

The generalization to field theory is straightforward, though it requires some caution related to the transition to infinite volume. We shall not discuss the operator thermal formalism for field theory (see Umezava *et al* (1982)) using instead a shorter way based on the path-integral approach and a generalization of the Euclidean (imaginary-time) thermal theory (Niemi and Semenoff 1984).

In the Euclidean formalism of the preceding subsection, we performed the integration in the complex plane of the time variable from $t = 0$ to $t = -i\beta$. The periodicity of the fields enables us to generalize this interval to $[-t_0, -t_0 - i\beta]$ for any real t_0. To obtain the real-time theory, we need to choose a different contour $C = C_1 + C_2 + C_3 + C_4$ (see figure 4.3) with the same endpoints but which includes the *real-time* axis (or at least a very large part of it). The causality condition imposes the constraint $\text{Im}\, t < 0$ (cf the $i\varepsilon$-prescription which we discussed in chapter 3), but there is still considerable freedom in the choice of possible contour. The choice depicted in figure 4.3(b) is technically most convenient. At the end of the calculation, we should take the limit $t_0 \to \infty$. We shall see that the field defined on the piece C_2 can be interpreted as the *second* field $\varphi_2(t, x) = \varphi(t - i\beta/2, x)$. As the temperature $T \to 0$, the contour C_2 retreats to infinitely negative imaginary time and the φ_2 field completely decouples from the theory.

The generating functional formally has the usual form

$$\mathcal{Z}_\beta = \int \mathcal{D}\varphi(\tau, x)\, \mathcal{D}\pi(\tau, x) \exp\left\{ i \int_C d\tau \int d^3x \, (\pi \partial_c \varphi - \mathcal{H} + j\varphi) \right\}$$
$$= \mathfrak{N}^{-1} \int \mathcal{D}\varphi(\tau, x) \exp\left\{ i \int_C d\tau \int d^3x \, (\mathcal{L}(\varphi) + j\varphi) \right\}. \quad (4.4.64)$$

Here ∂_c denotes the derivative in the direction of the contour C. The field φ is integrated over all configurations periodic on C. The Lagrangian density on C can be written as

$$\mathcal{L}(\varphi) = -\tfrac{1}{2}\varphi(\Box_c + m^2)\varphi + V(\varphi) \quad (4.4.65)$$

where $\Box_c = \partial_c^2 - \nabla^2$. The path integral (4.4.64) contains the fields with non-physical arguments. This is not convenient for practical use and the next step is to recast it as an integral over fields at real times.

First, we note that the Gaussian integration for the free-field Lagrangian in (4.4.64) can be performed in the usual way, with the result

$$\mathcal{Z}_{0,\beta} = \exp\left\{-\tfrac{1}{2}i\int_C d^4x\, d^4y\, j(x) D_\beta(x-y) j(y)\right\} \quad (4.4.66)$$

where $D_\beta(x-y)$ is the *thermal* propagator on the contour, satisfying

$$(\Box_c + m^2) D_\beta(x-y) = -\delta_c(x-y). \quad (4.4.67)$$

Here $\delta_c(x) \stackrel{\text{def}}{\equiv} \delta_c(\tau)\delta(\boldsymbol{x})$, the variable τ being defined on the contour C. To find the explicit form of $D_\beta(x)$, let us make the partial Fourier transform

$$\widetilde{D}_\beta(\tau, \boldsymbol{k}) = \int \frac{d^3x}{(2\pi)^3} e^{-i\boldsymbol{k}\boldsymbol{x}} D_\beta(\tau, \boldsymbol{x}) \quad (4.4.68)$$

so that $\widetilde{D}_\beta(\tau, \boldsymbol{k})$ satisfies the equation

$$(\partial_c^2 + E^2(\boldsymbol{k}))\widetilde{D}_\beta(\tau, \boldsymbol{k}) = -\delta_c(\tau) \quad (4.4.69)$$

$$E^2(\boldsymbol{k}) = \boldsymbol{k}^2 + m^2. \quad (4.4.70)$$

It is also helpful to decompose the Green function into retarded $D_\beta^{(r)}$ and advanced $D_\beta^{(a)}$ components:

$$D_\beta(x-y) = \theta_c(\tau_x - \tau_y) D_\beta^{(r)}(x-y) + \theta_c(\tau_y - \tau_x) D_\beta^{(a)}(x-y) \quad (4.4.71)$$

whose partial Fourier transforms satisfy the equations

$$(\partial_c^2 + E^2(\boldsymbol{k}))\widetilde{D}_\beta^{(a,r)}(\tau, \boldsymbol{k}) = 0. \quad (4.4.72)$$

These equations show that $\widetilde{D}_\beta^{(a,r)}(\tau, \boldsymbol{k})$ are linear combinations of $e^{iE\tau}$ and $e^{-iE\tau}$. The coefficients of these linear combinations are found from the condition

$$D_\beta^{(a)}(\tau_x - \tau_y - i\beta, \boldsymbol{x} - \boldsymbol{y}) = D_\beta^{(r)}(\tau_x - \tau_y, \boldsymbol{x} - \boldsymbol{y}) \quad (4.4.73)$$

which, in turn, follows from the periodicity of φ and the Kubo–Martin–Schwinger relation (4.4.24). Equations (4.4.72), (4.4.69) and (4.4.73) imply

$$\widetilde{D}_\beta^{(r)}(\tau, \boldsymbol{k}) = f(E)[e^{-iE\tau} + e^{iE(\tau+i\beta)}]$$
$$\widetilde{D}_\beta^{(a)}(\tau, \boldsymbol{k}) = f(E)[e^{iE\tau} + e^{-iE(\tau-i\beta)}] \quad (4.4.74)$$

$$f(E) = \frac{i}{2E}\frac{1}{(1-e^{-\beta E})}. \quad (4.4.75)$$

The key observation for further development is that $D_\beta(\tau_1 - \tau_2, \boldsymbol{k}) \to 0$ as $t_0 \to \infty$ if τ_1 lies on C_1 or C_2 and τ_2 lies on C_3 or C_4. This means that $\mathcal{Z}_{0,\beta}$ becomes separable in this limit:

$$\mathcal{Z}_{0,\beta}[j] = \mathcal{Z}_{0,\beta}[j; C_1 C_2] \mathcal{Z}_{0,\beta}[j; C_3 C_4] \quad (4.4.76)$$

in which the time integral in the individual factors is restricted to the appropriate parts of the contour (x_0, y_0 lie on either C_1, C_2 or C_3, C_4). For sources localized in time, it can be absorbed into the normalization as $t_0 \to \infty$, and we can identify $\mathcal{Z}_{0,\beta}[j]$ with $\mathcal{Z}_{0,\beta}[j; C_1 C_2]$ alone.

The next step is to define $j_1(t, \mathbf{x}) = j(t, \mathbf{x})$, $j_2(t, \mathbf{x}) = j_2(t - i\beta/2, \mathbf{x})$, and to rewrite the exponent in (4.4.66):

$$\int_{C_1 C_2} dx\, dy\, j(x) D_\beta(x - y) j(y) = \int_{C_1} dx\, dy\, j_a(x) D_{ab}(x - y) j_b(y). \qquad (4.4.77)$$

In the integral (4.4.77), as distinct from (4.4.66), all the functions are defined for *real* time. The matrix propagator has the components:

$$\begin{aligned} D_{11}(x - y) &= D_\beta(x - y) \\ D_{22}(x - y) &= D_\beta(y - x) \\ D_{12}(x - y) &= -D_\beta^{(a)}(x_0 - y_0 + i\beta/2, \mathbf{x} - \mathbf{y}) \\ D_{12}(x - y) &= -D_\beta^{(r)}(x_0 - y_0 - i\beta/2, \mathbf{x} - \mathbf{y}) \end{aligned} \qquad (4.4.78)$$

(x_0 and y_0 are real). The components of the propagator are determined from these definitions and (4.4.71)–(4.4.74). The full momentum-space propagator $D_\beta(k)$ reads

$$D_\beta(k) = \begin{pmatrix} \widetilde{D}_c(k) & 0 \\ 0 & -\widetilde{D}_c^*(k) \end{pmatrix} - \frac{2\pi i \delta(k^2 - m^2)}{e^{\beta E} - 1} \begin{pmatrix} 1 & e^{\beta E/2} \\ e^{\beta E/2} & 1 \end{pmatrix} \qquad (4.4.79)$$

where $\widetilde{D}_c(k)$ is the usual zero-temperature propagator (3.1.93) for a scalar field (in momentum representation). The finite-temperature effect is only felt on the mass-shell. Although such terms may seem surprising, they are the only way in which the defining relation of the Green function, $(\Box_x + m^2) D_{11}(x) = -\delta(x)$, can be sustained.

To recast the path integral in a *double-field* way, let us first write $\mathcal{Z}_{0,\beta}$ as the path integral

$$+ P F_{0,\beta}[j_1, j_2] = \int \mathcal{D}\varphi_1 \mathcal{D}\varphi_2 \exp\left\{ -i \int dx \left[\tfrac{1}{2}\varphi_a D_{ab}^{-1} \varphi_b + j_b \varphi_b \right] \right\} \qquad (4.4.80)$$

(D_{ab}^{-1} is the inverse operator with respect to the matrix propagator). For an interacting theory in the large-t_0 limit we can write

$$\begin{aligned} \mathcal{Z}_\beta[j_1, j_2] &= \exp\left\{ -i \int_C dx\, V\left(-i\frac{\delta}{\delta j}\right) \right\} \mathcal{Z}_{0,\beta}[j] \\ &= \exp\left\{ -i \int_{\text{real time}} dx \left[V\left(-i\frac{\delta}{\delta j_1}\right) - V\left(-i\frac{\delta}{\delta j_2}\right) \right] \right\} \mathcal{Z}_{0,\beta}[j_1, j_2] \\ &= \int \mathcal{D}\varphi_1 \mathcal{D}\varphi_2 \exp\left\{ -i \int dx \left[\tfrac{1}{2}\varphi_a D_{ab}^{-1} \varphi_b - V(\varphi_1) + V(\varphi_2) + j_b \varphi_b \right] \right\}. \qquad (4.4.81) \end{aligned}$$

The minus sign of the second terms in the exponents occurs because of the reverse direction of the contour C_2.

Note that only those diagrams in which the field φ_1, but not φ_2, appears on the external legs have a physical meaning. Hence, the φ_2-field is a kind of ghost field, only occurring in the interior of diagrams.

Thus, we have arrived at the double-field formulation of the scalar field theory at non-zero temperature using the path-integral approach. A similar analysis can be carried out for fermions. The doubling that comes from the deformation of the contour is again inevitable.

4.4.4 Path integrals in the theory of critical phenomena

For a finite system, the partition function $\mathcal{Z}_\beta$ is an entire function of $\beta = 1/(k_B T)$. However, once the infinite volume limit is taken the free energy can have singularities. When the singularities occur for real positive temperature T, the system has phase transitions. The phase transitions and the related spontaneous symmetry-breakings abound in modern condensed matter physics.

If there are other external parameters in the problem, such as a magnetic field, the location of the points of phase transition can depend on these parameters. A plot of the location of the points of phase transition is called a phase diagram. One of the important tasks of statistical mechanics is the derivation of phase diagrams for realistic systems. Phase diagrams are quite dependent on the specific details of the system under consideration. However, once the locations of the points of phase transition are known, there is a remarkable universality in the behaviour of the system near the critical point.

There are many phenomena which occur at an isolated critical temperature T_c and the theory of critical phenomena relates them together. Some of the principal phenomena are:

- the singularities in the free energy;
- the existence of spontaneous symmetry-breaking;
- the behaviour of correlation functions at long distances.

These are related through the construction of the scaling limit and scaling laws. Before a discussion of the path-integral technique, we shall briefly recall some general characteristics of phase transitions.

◇ **Singularities in the free energy**

To discuss singularities in the free energy, it is convenient to define the specific heat c as

$$c = -T \frac{\partial^2 f}{\partial T^2} \qquad (4.4.82)$$

where $f \stackrel{\text{def}}{\equiv} F/L^d$ (F is the free energy). Then the simplest generic singularity the specific heat can have at a critical temperature T_c is

$$c \sim A_\alpha |T - T_c|^{-\alpha}. \qquad (4.4.83)$$

The exponent α is referred to as a *critical exponent*.

◇ **Spontaneous symmetry-breaking**

According to our discussion in section 4.4.2, we consider the mean value

$$\phi(j) = \frac{\delta W[j]}{\delta j} = \lim_{L \to \infty} \langle \varphi(x) \rangle_\beta \qquad (4.4.84)$$

(cf (4.4.57)). In lattice spin systems, the external source j may have the direct physical meaning of the external magnetic field h. Then the mean value $\phi(T) \stackrel{\text{def}}{\equiv} \lim_{h \to 0} \phi(h)$ is termed the *magnetization*.

If $j = 0$, interaction (4.4.35) is invariant under $\varphi \to -\varphi$ and thus, if $\phi(j)$ is continuous at $j = 0$, it follows that $\phi(0) = 0$. If $T > T_c$, this is indeed the case. However, because we are considering the $L \to \infty$ limit in definition (4.4.84), there is no reason that $\phi(j)$ has to be continuous at $j = 0$ and, indeed, for $T < T_c$ the continuity fails.

As we have learned in the preceding subsection, typically as $T \to T_c$, the mean value $\phi(T) \stackrel{\text{def}}{\equiv} \lim_{j \to 0} \phi(j)$ (in particular, spontaneous magnetization in the case of a spin system) vanishes. Thus we define a second critical exponent β as

$$\phi(T) \sim A_\beta (T_c - T)^\beta \qquad \text{as } T \to T_c, \ T < T_c. \tag{4.4.85}$$

Another quantity related to the mean value $\phi(j)$ is the *susceptibility* χ:

$$\chi \stackrel{\text{def}}{\equiv} \left. \frac{\partial \phi(j)}{\partial j} \right|_{j=0}. \tag{4.4.86}$$

This susceptibility also has a singular behaviour at $T = T_c$ and we parametrize this in terms of the exponent γ as

$$\chi \sim A_\gamma |T - T_c|^{-\gamma} \qquad T \to T_c. \tag{4.4.87}$$

In the case of a magnetic external field, χ is the usual *magnetic susceptibility*.

◇ **Correlations**

Not only the bulk thermal properties of the system have singularities at T_c, but also the corresponding phenomena in the correlation functions. Consider, for example, the two-point correlation function (Green function) in the infinite volume limit:

$$G(X) = \langle \varphi(0) \varphi(X) \rangle_\beta. \tag{4.4.88}$$

When $T < T_c$, we find that as $R \equiv \sum_{i=1}^d X_i^2 \to \infty$, the correlation approaches the limiting value of ϕ^2 exponentially as

$$G(X) \sim \phi^2 \left(1 + \frac{C(\theta, T)}{R^p} \exp\{-R/\xi(\theta, T)\} + \cdots \right) \tag{4.4.89}$$

where $\xi(\theta, T)$ is called the *correlation length* (θ denotes the set of angular variables in a polar coordinate frame in the d-dimensional space). Similarly, when $T > T_c$,

$$G(X) \sim \frac{C'(\theta, T)}{R^k} \exp\{-R/\xi'(\theta, T)\} + \cdots. \tag{4.4.90}$$

The correlation lengths depend on T and diverge as $T \to T_c$, and thus we define the exponent ν as

$$\xi(\theta, T) \sim A_\nu(\theta) |T - T_c|^{-\nu} \qquad (T \to T_c). \tag{4.4.91}$$

The divergence of the correlation length as $T \to T_c$ is a signal that expressions (4.4.89) and (4.4.90) break down at T_c and instead it is found that for $T = T_c$ the correlations decay as a power law

$$G(X) \sim \frac{A_c(\theta)}{R^{d-2+\eta}} \qquad (R \to \infty) \tag{4.4.92}$$

where d is the dimensionality of the system and η is called the *anomalous dimension*.

240   *Path integrals in statistical physics*

◇ **Scaling limit and scaling functions**

Of all the phenomena discussed earlier that happen at an isolated critical temperature T_c the most important is the divergence of the correlation length ξ. The physical meaning of this divergence is that at the critical temperature the physical scale is infinitely large compared to scales that appeared at the stage of the initial formulation of a problem (e.g., of interatomic spacing in a crystal body) and thus it is most natural to renormalize our length scale from the initial characteristic length (e.g., the atomic one) to the observed physical length:

$$x_i = \frac{X_i}{\xi(\theta^{(i)}, T)} \qquad (4.4.93)$$

($\theta^{(i)}$ denotes the set of values of the angular variables in the ith coordinate direction). The limit

$$T \to T_c \qquad X_i \to \infty \qquad x_i \text{ fixed} \qquad (4.4.94)$$

is the analog of the mass renormalization (chapter 3) and is part of what is called the *scaling limit*.

If the limit (4.4.94) were the only step, (4.4.89) would vanish because the factor ϕ^2 (the spontaneous 'magnetization') vanishes and the factor $C(\theta, T)$ is found to go to a constant independent of T and θ. Consequently, we also have to divide the correlation function $G(X)$ by ϕ^2 and define the renormalized Green function as

$$G_R(r) = \lim_{\text{scaling}} \phi^{-2} G(X) \qquad (4.4.95)$$

where $r^2 = \sum_i x_i^2$ and by $\lim_{\text{scaling}}$ we mean (4.4.94). The process of dividing $G(X)$ by ϕ^2 is called *wavefunction renormalization*.

◇ **Scaling laws**

Thus far, the theory discussed may be considered to be descriptive and all the exponents and functions introduced may be considered to be independent, subject only to the general requirements of thermodynamic stability. However, if we make an additional assumption that there are no other length scales in the problem other than the atomic length scale of definition and the physical length scale of the correlation length and that these two scales join together smoothly, we find that this theory makes predictions about the relation between critical exponents. These relations are known as *scaling laws*.

Some more assumptions and reasonable approximations even allow us to obtain numerical values for the critical exponents on the basis of the scaling limit. We shall present such a calculation using the path-integral method. In this case, it is more convenient to carry out the scaling transformations of *momentum* variables (after the Fourier transform of the fields under consideration), because this method is based on separate integration over the lower and higher modes of the fields (cf section 3.3.1).

◇ **The scaling transformation in the formalism of path integration**

Let us consider a statistical system in a box L^3 described by the action

$$S = \int d\tau\, d^3x \left(\frac{1}{2}(\partial_\tau \varphi)^2 + \frac{1}{2}(\partial_i \varphi)^2 - \frac{\mu}{2}\varphi^2 + \frac{g}{4!}\varphi^4 \right) \qquad (4.4.96)$$

where $x \in L^3$ and $0 \le \tau \le \beta$. This standard φ^4-functional describes a real scalar field at finite temperature T. If the coefficient μ in (4.4.96) is negative, the action is negatively defined and a phase transition is impossible. If $\mu > 0$, the system may undergo a phase transition at some critical temperature T_c, below which anomalous mean values $\langle \varphi(\tau, x) \rangle_\beta \neq 0$ appear. To study the critical exponents for the phase transitions, we shall again use the method of the separate integration over higher and lower modes (Popov 1983) (cf section 3.3.1) in the Fourier expansion of the field φ:

$$\varphi(\tau, \boldsymbol{x}) = \frac{1}{\sqrt{\beta L^3}} \sum_{k,w} e^{i(\omega\tau - \boldsymbol{k}\boldsymbol{x})} \widetilde{\varphi}(\omega, \boldsymbol{k}). \tag{4.4.97}$$

Now, we define the functional $S_{(0)}^{(\text{eff})}$ by the integration over the higher modes $\widetilde{\varphi}(\omega, \boldsymbol{k})$ with $\omega \neq 0$ and with $\omega = 0$, $k > k_0$:

$$S_{(0)}^{(\text{eff})} = -\ln \int \prod_{\substack{\omega \neq 0 \\ \text{all } k_i}} \prod_{\substack{\omega = 0 \\ k > k_0}} d\widetilde{\varphi}(\omega, \boldsymbol{k}) \, e^{-S}. \tag{4.4.98}$$

The general form of the functional $S_{(0)}^{(\text{eff})}$ reads as

$$S_{(0)}^{(\text{eff})} = c_0 + \frac{1}{2} \int_{k<k_0} d^d k \, u_2(\boldsymbol{k}) \varphi(\boldsymbol{k}) \varphi(-\boldsymbol{k})$$

$$+ \sum_{n=2}^{\infty} \frac{1}{2n!} \int_{k_a < k_0} \left(\prod_{a=1}^{2n} d^d k_a \, \varphi(\boldsymbol{k}_a) \right) u_{2n}(\boldsymbol{k}_1, \ldots, \boldsymbol{k}_{2n}) \delta\left(\sum_{b=1}^{2n} \boldsymbol{k}_b \right) \tag{4.4.99}$$

that is, as the sum over the even powers of the field $\varphi(\boldsymbol{k}) \equiv \varphi(0, \boldsymbol{k})$ and with the cutoff of the integrals at the upper limits. The constant c_0 is not essential for further calculation.

We are going to calculate the asymptotics of the two-point correlator on the basis of the scaling hypothesis. This aim is achieved through the following steps.

(i) Consider the coefficient function $u_2(\boldsymbol{k})$. Let this function be of the form

$$u_2(\boldsymbol{k}) = u_{20} + u_{22} k^2 + \cdots \tag{4.4.100}$$

in the vicinity of $\boldsymbol{k} = 0$ (i.e. it can be expanded in even powers of the momentum variable). After the scaling transformation

$$\varphi(\boldsymbol{k}) \to \zeta \varphi(\boldsymbol{k}) \tag{4.4.101}$$

where the parameter ζ is chosen so that

$$\zeta^2 u_{22} = 1 \tag{4.4.102}$$

we arrive at the functional $S^{(\text{eff})}$ of the form (4.4.99) in which

$$u_2(\boldsymbol{k}) = K + k^2 + \cdots \tag{4.4.103}$$

(K is some new constant). Now, let us integrate the functional $e^{-S_{(0)}^{(\text{eff})}}$ over $\varphi(\boldsymbol{k})$, with $k_0/2 < k < k_0$:

$$S_{(1)}^{(\text{eff})} = \ln \int \prod_{k_0/2 < k < k_0} d\widetilde{\varphi}(\omega, \boldsymbol{k}) \, \exp\{-S_{(0)}^{(\text{eff})}\}. \tag{4.4.104}$$

The functional $S_{(1)}^{(\text{eff})}$ differs from $S_{(0)}^{(\text{eff})}$ by the value of the constants c_0 and by the actual form of the coefficient functions u_{2n}, as well as by the upper limit of the integrals (they are cut off at $k_0/2$ instead of k_0, as in $S_{(0)}^{(\text{eff})}$). Next, we make the transformation

$$\boldsymbol{k} \to 2\boldsymbol{k} \qquad \varphi(\boldsymbol{k}) \to \varphi(2\boldsymbol{k}) \tag{4.4.105}$$

converting the domain of the momenta $k < k_0/2$ back into $k < k_0$ and again make the transformation (4.4.101), so that the lowest non-trivial coefficient function in $S_{(1)}^{(\text{eff})}$ obtains a form analogous to (4.4.103):

$$u_2^{(1)}(k) = K_1 + k^2 + \cdots. \tag{4.4.106}$$

Thus, the path integration (4.4.104), together with the subsequent change of variables (4.4.101) and (4.4.105) convert the functional $(S_{(0)}^{(\text{eff})} - c_0)$ into the functional $(S_{(1)}^{(\text{eff})} - c_1)$ of the same form, but with modified coefficient functions.

(ii) We can expand these transformations to other coefficient functions u_{2n} and consider them as a definition of the nonlinear transformations

$$u^{(k+1)} = \widehat{M}[u^{(k)}] \tag{4.4.107}$$

of the sets of coefficient functions:

$$u^{(k)} = \{u_2^{(k)}, u_4^{(k)}, u_6^{(k)}, \ldots\}. \tag{4.4.108}$$

The *scaling hypothesis* (see page 240) implies the following natural assumption: along the line of a phase transition, the multiple $\widehat{M}$-transformations have, as a limit, a stationary set u_{st} of coefficient functions:

$$u_{\text{st}} = \lim_{n\to\infty} \widehat{M}^n[u^{(0)}] \qquad u_{\text{st}} = \widehat{M}[u_{\text{st}}]. \tag{4.4.109}$$

Indeed, the stationarity means that the rescaling of the momenta, $k \to 2k$, is equivalent to the scaling transformation (4.4.101) (cf also (4.4.95)).

(iii) Consider the two-point correlator $D(k)$:

$$D(k)\delta(k + k') = \langle \varphi(k)\varphi(k') \rangle_\beta. \tag{4.4.110}$$

Let the stationarity condition (4.4.109) be true at $n \geq n_0$ for some n_0, i.e. for the momenta

$$k < 2^{-n_0} k_0 \tag{4.4.111}$$

($k = |\boldsymbol{k}|$, $k_0 = |\boldsymbol{k}_0|$). For an arbitrarily small $\boldsymbol{k}$ and an integer n, such that the momentum $\boldsymbol{k}_1 = 2^n \boldsymbol{k}$ belongs to the interval

$$2^{-n_0-1} k_0 < k_1 < 2^{-n_0} k_0 \tag{4.4.112}$$

we have (due to the stationarity condition (4.4.109))

$$\langle \varphi(k)\varphi(k') \rangle_\beta = \zeta^{2n} \langle \varphi(k_1)\varphi(k_1') \rangle_\beta \tag{4.4.113}$$

or, for the function $D(k)$,

$$D(k) = \zeta^{2n} 2^{-nd} D(k_1) \tag{4.4.114}$$

(d is the space dimensionality). Rewriting (4.4.114) as

$$D(k)k^{(2\ln \zeta/\ln 2 - d)} = D(k_1)k_1^{(2\ln \zeta/\ln 2 - d)} \tag{4.4.115}$$

we finally obtain the asymptotics

$$D(k) \xrightarrow[k\to 0]{} c k^{(d - 2\ln \zeta/\ln 2)}. \tag{4.4.116}$$

The Fourier transform gives the corresponding asymptotics in the coordinate space:

$$D(r) \xrightarrow[r\to\infty]{} r^{(2\ln \zeta/\ln 2 - 2d)}. \tag{4.4.117}$$

A comparison with (4.4.92) immediately gives an expression for the anomalous dimension:

$$\eta = 2 + d - \frac{2\ln\zeta}{\ln 2}. \qquad (4.4.118)$$

Qualitative arguments based on the linearized form of the $\widehat{M}$-transformation, that is

$$u^{(k+1)} - u_{st} = \widehat{A}(u^{(k+1)} - u_{st}) \qquad (4.4.119)$$

($\widehat{A}$ is the *linear* operator obtained by the linearization of $\widehat{M}$), allow us to derive the expression for the correlation length defined in (4.4.89) or (4.4.90) (see problem 4.4.2, page 253):

$$\xi(T) \sim (T - T_c)^{-\ln 2/\ln\lambda_1} \qquad (4.4.120)$$

where λ_1 is the largest eigenvalue of the operator $\widehat{A}$. Thus, the critical exponent ν is given by

$$\nu = \frac{\ln 2}{\ln\lambda_1}. \qquad (4.4.121)$$

Finally, a similar consideration gives the expression for the critical exponent γ, which defines the behaviour of the susceptibility (cf (4.4.87)):

$$\gamma = \frac{2\ln\zeta}{\ln\lambda_1} - d\frac{\ln 2}{\ln\lambda_1}. \qquad (4.4.122)$$

Thus, the separate integration over the higher Fourier modes, together with the hypothesis about the relation between the phase transition of stationary points of the $\widehat{M}$-transformations, allow us to express the critical exponents through the parameters of these transformations. For an explicit calculation of the critical exponents, we need a further approximation.

Let us neglect all the coefficient functions but u_2 and u_4 and take $u_2^{(n)}$, $u_4^{(n)}$ in the form

$$u_2^{(n)} = k^2 + K_n \qquad u_4^{(n)} = Q_n \qquad (4.4.123)$$

(K_n and Q_n are some constants). Then the $\widehat{M}$-transformation for $u_2^{(n)}$, $u_4^{(n)}$ can be written as the following series:

$$u_2^{(n+1)}(k) = \zeta^2 2^{-d}\left(u_2^{(n)}(k/2) + \underset{u_4^{(n)}}{\text{(a)}} + \cdots\right)$$

$$u_4^{(n+1)}(k) \approx u_4^{(n+1)}(0) = \zeta^4 2^{-3d}\left(u_4^{(n)} + \underset{u_4^{(n)}\ u_4^{(n)}}{\text{(b)}} + \cdots\right). \qquad (4.4.124)$$

These series are obtained as a result of integration over the selected set of modes as in (4.4.104). We restricted the perturbation series only to diagrams of types (a) and (b). In the approximation (4.4.123), i.e. with the constant $u_4^{(n)}$, diagram (a) does not depend on the external momentum. Requiring that the coefficient at k^2 be equal to unity, we find the scaling parameter ζ which enters the formulae (4.4.118) and (4.4.122) for the critical exponents:

$$\zeta = 2^{1+d/2}. \qquad (4.4.125)$$

After this choice, equations (4.4.124) for the coefficient functions (4.4.123) acquire the following explicit form:

$$u_2^{(n+1)}(\boldsymbol{k}) = k^2 + K_{n+1}$$
$$= k^2 + 4\left(K_n + \frac{Q_n}{2(2\pi)^d}\int_{(k_0/2)<k<k_0}\frac{dk}{k^2 + K_n}\right)$$
$$u_4^{(n+1)}(\boldsymbol{k}) \approx Q_{n+1}$$
$$= k^2 + 4\left(Q_n - \frac{3}{2}\frac{Q_n^2}{(2\pi)^d}\right)\int_{(k_0/2)<k<k_0}\frac{dk}{(k^2 + K_n)^2} \quad (4.4.126)$$

(note that the factor 3 in the second equation appears due to three diagrams of type (b) with different positions of the external momenta, while the factor $\frac{1}{2}$ is the symmetry factor of the diagrams). Thus, in the chosen approximation, the $\widehat{M}$-transformation reduces to the nonlinear transformations of the constants K_n and Q_n.

Equations (4.4.126) show that $d = 4$ is the specific (critical) value of the space dimensionality for the following reasons.

- At $d > 4$ and for small positive $u_{\rm st}$, formulae (4.4.126) give the inequalities:

$$0 < Q_{n+1} < 2^{4-d}Q_n \quad (4.4.127)$$

if $0 < Q_0 \ll 1$. Therefore, $Q_n \to 0$ and hence $K_n \to 0$. As a result, the action $S^{(n)}$ in the limit $n \to \infty$ becomes the free-field action:

$$S^{(n)} \xrightarrow[n\to\infty]{} -\tfrac{1}{2}\int d^d k\, k^2 \varphi(\boldsymbol{k})\varphi(-\boldsymbol{k}) \quad (4.4.128)$$

and all correlation functions, thermodynamical functions and critical exponents coincide with those for the free-field theory. In particular, for the critical exponents we have:

$$\eta = 0 \qquad \nu = \tfrac{1}{2} \qquad \gamma = 1. \quad (4.4.129)$$

- For $d = 4$, the $\widehat{M}$-transformation (4.4.126) also leads to zero values of K and Q and the critical exponents take free-field theory values.
- For $d < 4$, a non-trivial stationarity point defined by the equations appears:

$$Q_{\rm st} = (1 - 2^{d-4})\left[\frac{3}{2(2\pi)^d}\int_{(k_0/2)<k<k_0}\frac{dk}{(k^2 + K_n)^2}\right]^{-1}$$
$$K = \frac{4}{9}(2^{d-4} - 1)\int_{(k_0/2)<k<k_0}\frac{dk}{k^2 + K_n}\left[\int_{(k_0/2)<k<k_0}\frac{dk}{(k^2 + K_n)^2}\right]^{-1}. \quad (4.4.130)$$

The linearized operator $\widehat{A}$ has the form of a 2×2 matrix:

$$\begin{pmatrix} 4 - \frac{2Q}{(2\pi)^d}\int_{(k_0/2)<k<k_0}\frac{dk}{(k^2+K_n)^2} & \frac{2}{(2\pi)^d}\int_{(k_0/2)<k<k_0}\frac{dk}{k^2+K_n} \\ \frac{2^{4-d}3Q^2}{(2\pi)^d}\int_{(k_0/2)<k<k_0}\frac{dk}{(k^2+K_n)^3} & 2^{4-d}\left(1 - \frac{3Q}{(2\pi)^d}\int_{(k_0/2)<k<k_0}\frac{dk}{(k^2+K_n)^2}\right) \end{pmatrix}. \quad (4.4.131)$$

At $d < 4$, the largest eigenvalue λ_1 of this matrix differs from the free-field value (i.e. $\lambda_1 = 4$) and, according to (4.4.121), gives the critical exponent ν which also differs from the free-field value

(4.4.129). Note that the exponent η is still zero in this approximation as in free-field theory (cf (4.4.125) and (4.4.118)).

The distinguishing role of the dimensionality $d = 4$ inspired Wilson and Fisher (see Wilson and Fisher (1972)) to develop a series perturbation expansion in the parameter $\varepsilon \stackrel{\text{def}}{\equiv} 4 - d$. This expansion implies the extension of the usual diagram techniques to spaces of non-integer dimensions. In fact, this is not completely new: we have already mentioned such a possibility in chapter 3, in the context of the dimensional regularization of quantum field theories (see section 3.2.7). We shall not present the ε-expansion in this book and refer the reader to the previously cited original paper and to special reviews and monographs, e.g., to Ma (1976).

4.4.5 Quantum field theory at finite energy

In this section, we are going to discuss the path-integral formulation of relativistic quantum field theory at *finite energy*, using the *microcanonical distribution*. This method is a natural extension of the real-time formalism, which we have introduced in section 4.4.3.

In a canonical ensemble, the system of interest is one which is surrounded by a large thermal reservoir such that the system is kept at a certain temperature T. This is the situation in which the quantum field theory at finite temperature is formulated. The microcanonical ensemble, on the other hand, is appropriate for investigating an isolated system with a finite energy E and a volume L^3. In the microcanonical distribution, all microscopic states have equal probabilities; this is based on the ergodic hypothesis. This microcanonical ensemble is considered to be more fundamental than the canonical one and it is well known that in the thermodynamical limit ($E, L^3 \to \infty$ with E/L^3 finite), the canonical and microcanonical ensembles are equivalent. However, for a system at finite E and L^3, there are finite differences between these two ensembles and we should use the microcanonical ensemble to investigate such a system.

One of the advantages of the path-integral approach is that the relation between the microcanonical and canonical cases becomes clear and so is the procedure to take the thermodynamical limit. In this subsection, the field theory is formulated in Minkowski spacetime (i.e. with real time) so that it can also be used to investigate the time evolution of the system (see Chaichian and Senda (1993) and Chaichian *et al* (1994)). Thus, this is indeed an analog of the real-time field theory at *finite temperature*. In order to illustrate how this formulation works, we shall use real scalar field theory with φ^4 interaction.

An example of a physical situation where these results can be applied is the early universe: the universe does not have a thermal reservoir. At a late stage in the evolution of the universe, the use of the canonical ensemble is justified, but at an early stage the microcanonical investigation would be preferred. Another interesting problem would be the finite-energy effect on the phase transition in the Landau–Ginzburg model with ϕ^3 and ϕ^4 potentials and in the so-called *quark bag models*.

◇ **The path integral and the microcanonical distribution**

For the path-integral formulation of quantum field theory at finite energy we shall use the 'time-path' method, which was discussed in section 4.4.3. We present the calculations only for the real scalar boson. The extension to the other fields is straightforward.

Let us start by defining the partition function in the presence of external source $j(x)$ at fixed energy E:

$$\mathcal{Z}_E^{(\text{mc})}[j] = \int d\varphi \, \langle \varphi, t | \delta(\hat{H} - E) \mathbf{T}\left(\exp\left\{ i \int d^4x \, j(x)\hat{\varphi}(x) \right\} \right) | \varphi, t \rangle \qquad (4.4.132)$$

where **T** means the time-ordered product. The sum over the complete set of field configurations (trace) is taken at a certain time t. Using the integral representation of the delta function, $\delta(\hat{H} - E) = \int \frac{d\alpha}{2\pi} \exp(-i\alpha(\hat{H} - E))$, (4.4.132) is written as

$$\mathcal{Z}_E^{(\text{mc})}[j] = \int \frac{d\alpha}{2\pi} e^{i\alpha E} \int d\varphi \, \langle \varphi, t + \alpha | \mathbf{T}\left(\exp\left\{i \int d^4x \, j(x)\hat{\varphi}(x)\right\}\right) | \varphi, t \rangle$$

$$= \int \frac{d\alpha}{2\pi} e^{i\alpha E} \mathsf{N}_\alpha \int \mathcal{D}\varphi \, \exp\left\{i \int_C d^4x \, (\mathcal{L} + j\varphi)\right\}$$

where we have used $e^{i\alpha \hat{H}} |\varphi, t\rangle = |\varphi, t + \alpha\rangle$. The second line is the path-integral representation of the first line. The symbol 'C' in the exponent in the second line means that the time path in the integration over x_0 should be chosen such that it connects t to $t + \alpha$ (see later). The α-dependent normalization factor is represented by N_α. Following the usual procedure, the partition function becomes

$$\mathcal{Z}_E^{(\text{mc})}[j] = \int \frac{d\alpha}{2\pi} e^{i\alpha E} \mathsf{N}'_\alpha \exp\left\{i \int_C d^4x \mathcal{L}_I\left(\frac{1}{i} \frac{\delta}{\delta j(x)}\right)\right\} \exp\left\{-\frac{i}{2} \int_C d^4x \, d^4x' \, j(x) D_\alpha(x - x') j(x')\right\}$$
(4.4.133)

where, as usual, $\mathcal{L}_I$ is the interaction Lagrangian and $D_\alpha(x - x')$ is a α-dependent propagator. Then the properly normalized generating functional of Green functions, $\mathcal{Z}_E[j]$, is given by

$$\mathcal{Z}_E[j] = \mathcal{Z}_E^{(\text{mc})}[j]/\mathcal{Z}_E^{(\text{mc})}[0]$$

$$= \int \frac{d\alpha}{2\pi} e^{i\alpha E} \left(\frac{\mathsf{N}'_\alpha}{\mathcal{Z}_E[0]}\right) \exp\left\{i \int_C d^4x \, \mathcal{L}_I\left(\frac{1}{i}\frac{\delta}{\delta j(x)}\right)\right\}$$

$$\times \exp\left\{-\frac{i}{2} \int_C d^4x \, d^4x' \, j(x) D_\alpha(x - x') j(x')\right\}.$$
(4.4.134)

Since the diagrammatic structure of the last two factors with exponentials in (4.4.134) is the same as the ordinary generating functional in a quantum field theory, we define

$$\mathcal{Z}_\alpha[j] \equiv \exp\left\{i \int_C d^4x \, \mathcal{L}_I\left(\frac{1}{i}\frac{\delta}{\delta j(x)}\right)\right\} \exp\left\{-\frac{i}{2} \int_C d^4x \, d^4x' \, j(x) D_\alpha(x - x') j(x')\right\}$$

$$\equiv B_\alpha \mathcal{Z}_\alpha^c[j]$$
(4.4.135)

where B_α stands for vacuum diagrams and $\mathcal{Z}_\alpha^c[j]$ represents the diagrams connected to the external source. Then (4.4.134) can be written as

$$\mathcal{Z}_E[j] = \int \frac{d\alpha}{2\pi} e^{i\alpha E} \rho_{E,\alpha} \mathcal{Z}_\alpha^c[j]$$
(4.4.136)

where we have defined

$$\rho_{E,\alpha} = \mathsf{N}'_\alpha B_\alpha / \mathcal{Z}_E[0].$$
(4.4.137)

Defining the α-integrations of $\rho_{E,\alpha}$ and $\mathcal{Z}_\alpha^c[j]$ by

$$\rho_{E,E'} \equiv \int \frac{d\alpha}{2\pi} e^{i\alpha E'} \rho_{E,\alpha} \qquad \mathcal{Z}_E^c[j] \equiv \int \frac{d\alpha}{2\pi} e^{i\alpha E} \mathcal{Z}_\alpha^c[j]$$
(4.4.138)

the generating functional in (4.4.134) can be written as

$$\mathcal{Z}_E[j] = \int_0^E dE' \, \rho_{E,E-E'} \mathcal{Z}_{E'}^c[j].$$
(4.4.139)

The N-point Green function in the medium of energy E is given by the derivatives of the generating functional with respect to external sources:

$$G_E(x_1,\ldots,x_N) = \int_0^E dE' \, \rho_{E,E-E'} \left(\prod_{i=1}^N \frac{\delta}{i\delta j(x_i)} \right) \mathcal{Z}_{E'}^c[j]|_{j=0}. \tag{4.4.140}$$

◇ **Transition to the canonical partition function in the thermodynamic limit**

In the thermodynamical limit, $E, V \gg 1$ keeping E/V finite, the statistical weight $\rho_{E,E-E'}$ has a sharp peak at $E' \sim 0$. As we shall discuss later, the partition function (density of the field configuration) $\mathcal{Z}_E^{(\text{mc})}$ is related to the entropy S_E by $\mathcal{Z}_E^{(\text{mc})} \propto \exp(S_E)$. Thus the statistical weight $\rho_{E,E-E'}$ is approximated in the thermodynamical limit by

$$\rho_{E,E-E'} = \exp(S_{E-E'} - S_E) \sim e^{-\beta_E E'} \tag{4.4.141}$$

where $\beta_E \equiv \partial S_E/\partial E$ is the inverse of the temperature of the system with energy E. Therefore, in the thermodynamical limit, the generating functional in (4.4.139) can be approximated as

$$\mathcal{Z}_E[j] \sim \int_0^\infty dE' \, e^{-\beta_E E'} \mathcal{Z}_{E'}^c[j]. \tag{4.4.142}$$

This is nothing other than a *Laplace transformation* of the generating functional for Green functions, $\mathcal{Z}_{E'}^c[j]$. Therefore, the right-hand side of (4.4.142) is identified with the generating functional at the temperature $1/\beta_E$ in the canonical ensemble. This shows the equivalence between the microcanonical method and the canonical method in the thermodynamical limit.

◇ **Evaluation of the generating functional in the case of fixed energy**

Although we are going to discuss the thermodynamical limit in some cases, our main interest is in the evaluation of (4.4.139).

In the case of the real scalar boson, the propagator in (4.4.133) is defined by the differential equation:

$$-(\partial^2 + m^2) D_\alpha(x - x') = \delta^4(x - x').$$

In the derivation of (4.4.133), the requirement that the boundary terms vanish gives a periodic boundary condition:

$$\varphi(t + \alpha, \mathbf{x}) = \varphi(t, \mathbf{x}), \quad D_\alpha(t + \alpha, \mathbf{x}) = D_\alpha(t, \mathbf{x}). \tag{4.4.143}$$

The explicit derivation of this periodic boundary condition pertains to problem 4.4.4, page 254.

In the real-time formulation of finite-temperature quantum field theory, different choices of the time paths define field theories which are *physically* equivalent, but still different, in the sense that quantities such as propagators are different. We find the same situation in the finite-energy quantum field theory. The only restriction in choosing the time path is that it should connect a certain time t and $t+\alpha$. The choice of the time path which proves to be most convenient for real-time formulation is $C = C_1 + C_2 + C_3$, shown in figure 4.4. This consists of three parts: C_1 connecting $-t$ to t, C_2 connecting t to $-t$ backward in time and C_3 connecting $-t$ to $-t + \alpha$. Because of the requirement of causality and the well-definedness of the path integral, the path monotonically decreases in the imaginary time direction by the infinitesimal amount ε. We will take the limit t going to infinity at the end of the calculations. As in the finite-temperature quantum field theory, physical operators, which appear as *external lines* in Green functions, are assumed to have support on the path C_1 and fields having support on C_2 and C_3 are considered to be ghosts.

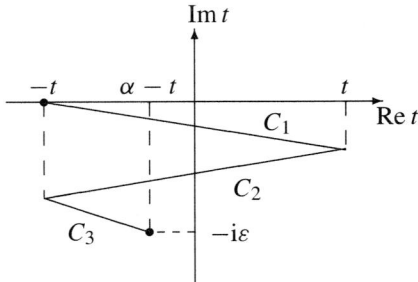

Figure 4.4. The time-path C used to formulate finite-energy quantum field theory.

The great simplification resulting from the choice of this path is that the operators on the C_3 path decouple from those on the C_1 and C_2 paths. This is because, in the limit $t \to \infty$, the propagator connecting the paths C_a ($a = 1, 2$) and C_3 vanishes due to the Riemann–Lebesgue theorem and the adiabatic switching-off of the external source $j(x)$ (see problem 4.4.5, page 255). Therefore, the real-time part $C_1 + C_2$, which is used to evaluate the ensemble average of the time-dependent operators, and the thermodynamical part C_3 are separated. Thus we can write the generating functional of connected diagrams in (4.4.135) as follows:

$$\mathcal{Z}_\alpha^c[j] = \mathcal{Z}_\alpha^{c(1+2)}[j] \cdot \mathcal{Z}_\alpha^{c(3)}[j] \tag{4.4.144}$$

where $\mathcal{Z}^{c(a)}$ is the one for fields on the path C_a. In general, we can choose any path connecting t and $t + \alpha$ within the requirement of consistency of the theory and they give physically equivalent results when the system is in equilibrium. The equivalence of the different choices of path is a simple consequence of the completeness of the set of states inserted at each moment in the path integration, which is the same as the situation in finite-temperature quantum field theory. An example of a calculation showing the equivalence of physical quantities obtained by choosing different paths is suggested in problem 4.4.6, page 255.

◇ The Green function and Feynman rules

Since the $C_1 + C_2$ part decouples from the C_3 part in the limit $t \to \infty$, let us concentrate, at first, on the former. The $C_1 + C_2$ part of the quantity $\mathcal{Z}_\alpha$ defined in (4.4.135) is

$$\mathcal{Z}_\alpha^{(1+2)}[j] = \exp\left\{ i \int_c d^4x \, \mathcal{L}_I\left(\frac{1}{i}\frac{\delta}{\delta j(x)}\right) \right\}$$
$$\times \exp\left\{ -\frac{i}{2} \sum_{a,b} \int_{C_a} d^4x \int_{C_b} d^4x' \, j_a(x) D_\alpha^{ab}(x - x') j_b(x') \right\} \tag{4.4.145}$$

where j_a is an external source having support on C_a and the summation is taken over $a, b = 1, 2$. In (4.4.145), D^{ab} represents a propagator between the fields on C_a and C_b:

$$D_\alpha(x - x') \equiv D_\alpha^{ab}(x - x') \qquad x \in C_a, \ x' \in C_b, \ a = 1, 2.$$

The integration over the path C_a is given by

$$\int_{C_a} d^4x = \epsilon_a \int_{-t}^{t} d^4x \bigg|_{t \to \infty} \tag{4.4.146}$$

where ϵ_a equals 1 for $a = 1$ and -1 for $a = 2$. Therefore, it is convenient to introduce the notation

$$\bar{D}_\alpha^{ab}(x - x') \equiv \epsilon_a \epsilon_b D_\alpha^{ab}(x - x'). \tag{4.4.147}$$

Then, (4.4.145) becomes

$$\mathcal{Z}_\alpha^{(1+2)}[j_1, j_2] = \exp\left\{ i \sum_a \epsilon_a \int d^4x\, \mathcal{L}_1(\delta/i\delta j_a) \right\}$$

$$\times \exp\left\{ -\frac{i}{2} \sum_{a,b} \int d^4x\, d^4x'\, j_a(x) \bar{D}_\alpha^{ab}(x - x') j_b(x') \right\}. \tag{4.4.148}$$

The Feynman rules of the $C_1 + C_2$ part in the ϕ^4-theory become

$$\overset{a\quad\quad b}{\bullet\!\!-\!\!\!-\!\!\bullet} = i\bar{D}_\alpha^{ab}(k) \qquad \overset{a\quad\quad a}{\underset{a\quad\quad a}{\times}} = -i\lambda\epsilon_a$$

where $\bar{D}(k)$ is the Fourier transform of $\bar{D}(x)$. In problem 4.4.4, page 254, we suggest explicitly calculating the propagators and their periodicity properties. The 2×2-matrix propagator $\bar{D}^{ab}$ is given by

$$i\bar{D}_\alpha^{ab}(k) = \begin{pmatrix} \frac{i}{k^2 - m^2 + i\epsilon} + 2\pi\delta(k^2 - m^2) f_\alpha^{(-)}(\omega_k) & -2\pi\epsilon(k_0)\delta(k^2 - m^2) f_\alpha^{(-)}(k_0) \\ -2\pi\epsilon(k_0)\delta(k^2 - m^2) f_\alpha^{(+)}(k_0) & \frac{-i}{k^2 - m^2 - i\epsilon} + 2\pi\delta(k^2 - m^2) f_\alpha^{(-)}(\omega_k) \end{pmatrix}$$

$$f_\alpha^{(-)}(u) = \sum_{n=1}^\infty e^{-in\alpha u} \qquad f_\alpha^{(+)}(u) = 1 + f_\alpha^{(-)}(u) \tag{4.4.149}$$

where $\omega_k = \sqrt{k^2 + m^2}$. Note that α has a small negative imaginary part, $\alpha = \operatorname{Re}\alpha - i\epsilon$, due to our choice of the path C. The propagator given in (4.4.149) has the property

$$\bar{D}_\alpha^{ab}(x) = \bar{D}_\alpha^{ba}(-x) \quad \text{and} \quad \bar{D}_\alpha^{ab}(k) = \bar{D}_\alpha^{ba}(-k).$$

For example, the two-point Green function of fields on C_1 at the tree level is obtained using (4.4.140) and (4.4.149):

$$G_E^{11}(k) = \int \frac{d\alpha}{2\pi} e^{i\alpha E} \rho_{E,\alpha}(iD_\alpha^{11}(k)) \tag{4.4.150}$$

$$= \int_0^E dE'\, \rho_{E,E-E'} \left[\frac{i\delta(E')}{k^2 - m^2 + i\epsilon} + 2\pi\delta(k^2 - m^2) \sum_{n=1}^\infty \delta(E' - n\omega_k) \right].$$

In the thermodynamical limit, using (4.4.141) and (4.4.142), (4.4.150) is approximated as

$$G_E^{11}(k) \sim \int_0^E dE'\, e^{-\beta_E E'} \left[\frac{i\delta(E')}{k^2 - m^2 + i\epsilon} + 2\pi\delta(k^2 - m^2) \sum_{n=1}^\infty \delta(E' - n\omega_k) \right]$$

$$= \frac{i}{k^2 - m^2 + i\epsilon} + 2\pi\delta(k^2 - m^2) \frac{1}{e^{\beta_E \omega_k} - 1}. \tag{4.4.151}$$

Thus, we have recovered the two-point Green function $D_\beta^{11}(k)$ in the finite-temperature quantum field theory, cf (4.4.79). We should note that the temperature in (4.4.151), $1/\beta_E$, is given by the microcanonical ensemble with the energy E, namely $\beta_E = \partial S_E / \partial E$.

Let us turn to the C_3 part. The propagator of the fields on the C_3 path depends on the sign of α. Using the same notation as before, we have:

$$i\bar{D}^{33}_\alpha(x) = \theta(\alpha)\left\{\theta(x_0)\int\frac{d^4k}{(2\pi)^4}e^{-ikx}2\pi\delta(k^2-m^2)f^{(+)}_\alpha(k_0)\right.$$
$$\left.+\theta(-x_0)\int\frac{d^4k}{(2\pi)^4}e^{-ikx}2\pi\delta(k^2-m^2)f^{(-)}_\alpha(k_0)\right\}$$
$$+\theta(-\alpha)\left\{\theta(x_0)\int\frac{d^4k}{(2\pi)^4}e^{-ikx}2\pi\delta(k^2-m^2)f^{(-)}_\alpha(k_0)\right.$$
$$\left.+\theta(-x_0)\int\frac{d^4k}{(2\pi)^4}e^{-ikx}2\pi\delta(k^2-m^2)f^{(+)}_\alpha(k_0)\right\}. \qquad (4.4.152)$$

Because of the finiteness of the time interval on the C_3 path, the propagator in (4.4.152) is not Fourier transformed like those in (4.4.149).

◇ Examples of calculations in ϕ^4 theory

Let us consider the full propagator as a sum of self-energies (problem 4.4.7, page 256 suggests calculating the expectation value of the number operator for a system with fixed energy). In the one-loop approximation, the self-energy Σ^{ab}_α of the scalar boson is given by the diagram

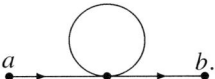

After a straightforward calculation, taking care with the positions of the poles of the propagators, we obtain

$$\Sigma^{ab}_\alpha = (\tau_3)^{ab}(-i\Sigma^{(0)} - i\Sigma^{(1)}_\alpha) \equiv (\tau_3)^{ab}\Sigma_\alpha$$

$$\Sigma^{(0)} = \frac{\lambda}{2}\frac{2\pi^{n/2}\Gamma(1-d/2)}{32\pi^4(m^2)^{1-d/2}} \qquad (4.4.153)$$

$$\Sigma^{(1)}_\alpha = \frac{\lambda}{2}\int\frac{d^3k}{(2\pi)^3\omega_k}f^{(-)}_\alpha(\omega_k)$$

where $\tau_3 = \begin{pmatrix} 1 & 0 \\ 0 & -1 \end{pmatrix}$ and we have used the dimensional regularization, $d = 4 - \delta$ (cf section 3.2.7).

It is useful to rewrite the matrix propagator in the following form:

$$i\bar{D}^{ab}_\alpha(k) = [\theta(k_0)\Phi^{(+)}_{R,\alpha}(\omega_k) + \theta(-k_0)\Phi^{(-)}_{A,\alpha}(\omega_k)]\frac{i}{k^2-m^2+i\varepsilon}$$
$$+ [\theta(k_0)\Phi^{(+)}_{A,\alpha}(\omega_k) + \theta(-k_0)\Phi^{(-)}_{R,\alpha}(\omega_k)]\frac{i}{k^2-m^2-i\varepsilon} \qquad (4.4.154)$$

where A and R stand for the advanced and retarded parts, respectively, and $(\pm)$ for the positive and negative k_0 parts. The 2×2-matrices Φ are defined by

$$\Phi^{(+)}_{R,\alpha}(\omega_k) = \begin{pmatrix} f^{(+)}_\alpha(\omega_k) & -f^{(-)}_\alpha(\omega_k) \\ -f^{(+)}_\alpha(\omega_k) & f^{(-)}_\alpha(\omega_k) \end{pmatrix} \qquad \Phi^{(-)}_{R,\alpha}(\omega_k) = \begin{pmatrix} -f^{(-)}_\alpha(\omega_k) & f^{(+)}_\alpha(\omega_k) \\ f^{(-)}_\alpha(\omega_k) & -f^{(+)}_\alpha(\omega_k) \end{pmatrix}$$

$$\Phi^{(+)}_{A,\alpha}(\omega_k) = \begin{pmatrix} -f^{(-)}_\alpha(\omega_k) & f^{(-)}_\alpha(\omega_k) \\ f^{(+)}_\alpha(\omega_k) & -f^{(+)}_\alpha(\omega_k) \end{pmatrix} \qquad \Phi^{(-)}_{A,\alpha}(\omega_k) = \begin{pmatrix} f^{(+)}_\alpha(\omega_k) & -f^{(+)}_\alpha(\omega_k) \\ -f^{(-)}_\alpha(\omega_k) & f^{(-)}_\alpha(\omega_k) \end{pmatrix}.$$

The matrices Φ have simple product relations

$$\Phi \cdot \Phi \equiv \Phi \tau_3 \Phi \qquad \Phi_R^{(\pm)} \cdot \Phi_R^{(\pm)} = \Phi_R^{(\pm)} \qquad \Phi_R^{(\pm)} \cdot \Phi_A^{(\pm)} = 0$$
$$\Phi_A^{(\pm)} \cdot \Phi_R^{(\pm)} = 0 \qquad \Phi_A^{(\pm)} \cdot \Phi_A^{(\pm)} = \Phi_A^{(\pm)}.$$
(4.4.155)

Using the propagator in (4.4.154) and the product property between $\Phi_{A,R}^{(\pm)}$ in (4.4.155), the sum of self-energies

—●— = —— + —(Σ)— + —(Σ)—(Σ)— + ···

is performed. Let us denote this sum, that is, the two-point Green function at finite energy taking into account the interaction, by $i\Delta_\alpha^{ab}(k)$. Then we find

$$i\Delta_\alpha(k) = \left\{ \theta(k_0) \left[\frac{i}{k^2 - m^2 - i\Sigma_\alpha + i\varepsilon} \Phi_{R,\alpha}^{(+)} + \frac{i}{k^2 - m^2 - i\Sigma_\alpha - i\varepsilon} \Phi_{A,\alpha}^{(+)} \right] \right.$$
$$\left. + \theta(-k_0) \left[\frac{i}{k^2 - m^2 - i\Sigma_\alpha + i\varepsilon} \Phi_{A,\alpha}^{(-)} + \frac{i}{k^2 - m^2 - i\Sigma_\alpha - i\varepsilon} \Phi_{R,\alpha}^{(-)} \right] \right\}. \quad (4.4.156)$$

Here, we impose the renormalization condition that $m^2 + \Sigma^{(0)}$ be the physical mass.

◇ The entropy, effective action and the behaviour of symmetry at high energy

It is an interesting problem to investigate the symmetry behaviour in finite-energy quantum field theory. For this purpose, we can use the general formulation based on the entropy and the effective action adapted to finite-energy quantum field theory. The definition of the effective action in finite-energy quantum field theory has some peculiarities, and we refer the reader for their discussion to Chaichian and Senda (1993) and Chaichian *et al* (1994). The entropy and effective action can be calculated using the perturbation theory with respect to the coupling constants or by the stationary-phase approximation.

Since we are treating the system quantum mechanically, there is a certain width of energy, δE, such that we cannot distinguish between the energy eigenvalues E and $E + \delta E$ because of the uncertainty principle. From the physical point of view, the uncertainty of the energy is of the order of the momentum uncertainty, namely $\delta E \sim L^{-1}$. In the following, we will not specify δE and we will consider it as a constant. The appropriate definition of the effective action $S_{\text{eff},E}$ in the finite-energy field theory reads

$$S_{\text{eff},E}[\phi] \stackrel{\text{def}}{\equiv} \ln \left\{ \delta E \int \frac{d\alpha}{2\pi} e^{i\alpha E} N_\alpha' B_\alpha e^{i\Gamma_\alpha[\phi]} \right\} \quad (4.4.157)$$

where $\Gamma_\alpha[\phi]$ is, as usual, the generating functional of the one-particle irreducible (OPI) Green functions (depending on the parameter α),

$$\phi_\alpha^a(x) \equiv \epsilon_a \frac{\partial W_\alpha[j]}{\partial j_a(x)} \qquad \epsilon_a = \begin{cases} 1 & a = 1 \\ -1 & a = 2 \\ \epsilon(\alpha) & a = 3 \end{cases}$$
$$\Gamma_\alpha[\phi] = W_\alpha[j] - \int_C d^4x \, j(x)\phi_\alpha(x). \quad (4.4.158)$$

Here $\epsilon(\alpha)$ is a step-function: $\epsilon(\alpha) = \alpha/|\alpha|$ for $\alpha \neq 0$ and $\epsilon(0) = 0$. Note that the time integration in the definition of Γ_α is taken over the path C because of the definition of ϕ^a; W is the generating functional of connected Green functions.

It is interesting to see the symmetry behaviour at *high-energy density*. For this purpose, let us examine the case $m^2 < 0$ for φ^4-theory, where spontaneous symmetry-breaking occurs at zero energy. Since we are considering a classical field which is independent of the coordinate, we define the density of the effective potential as in the finite-temperature case (cf (4.4.47)):

$$w_E[\phi] = -\frac{1}{L^3} S_{\text{eff},E}[\phi].$$

The condition that the symmetry is restored is

$$\frac{d^2 w_E[\phi]}{d\phi^2}\bigg|_{\phi=0} > 0.$$

This condition gives, in the stationary-phase approximation, for the α-integration in (4.4.157):

$$\frac{\lambda}{24}\left(\frac{30E}{\pi^2 L^3}\right)^{1/2} > |m^2|.$$

Therefore, the critical energy density $\varepsilon_c = E_c/L^3$, above which the symmetry is restored, is

$$\varepsilon_c \sim \frac{\pi^2}{30}\left(\frac{24|m^2|}{\lambda}\right)^2.$$

In a similar way as that presented for ϕ^4-theory, we can formulate the quantum field theory at finite energy for gauge theories such as QCD. In this case, the formulation would be useful in describing the dynamical evolution of a strongly interacting system, e.g., ion–ion collisions. We could also study the behaviour of the QCD running coupling constant as a function of energy in analogy with temperature behaviour.

4.4.6 Problems

Problem 4.4.1. Calculate the effective potential for φ^4-theory in the one-loop approximation using the stationary-phase method, cf (4.4.36).

Hint. Let us consider the general action for a real scalar field:

$$S = \int d^4x \left[\tfrac{1}{2}(\partial\varphi)^2 + \tfrac{1}{2}m^2\varphi^2 + V(\varphi)\right]$$

and expand it (up to quadratic terms) about the solution φ_c of the stationary equation,

$$\frac{\delta S}{\delta\varphi} + J(x) = 0$$

for the action in the presence of the external source. The Gaussian integration yields

$$\mathcal{Z}[J] \approx \left[\frac{\det(\Box + m^2)}{\det(\Box + m^2 + U''(\varphi_c))}\right]^{1/2} \exp\left\{\frac{i}{\hbar}\left[S[\varphi_c] + \int d^4x \, J\varphi_c\right]\right\}.$$

The approximation for the generating functional for connected Green functions proves to be

$$\frac{\hbar}{i} W[J] = \frac{\hbar}{i} \ln \mathcal{Z} = S[\varphi_c] + \int d^4x \, J\varphi_c$$
$$+ \tfrac{1}{2} i\hbar [\operatorname{Tr} \ln(\Box + m^2 + U''(\varphi_c)) - \operatorname{Tr} \ln(\Box + m^2)] + \mathcal{O}(\hbar^2). \quad (4.4.159)$$

The mean value of the field is defined by

$$\phi(x) = \frac{\hbar}{i} \frac{\delta W[J]}{\delta J(x)}$$

and satisfies $\phi = \varphi_c + \mathcal{O}(\hbar)$. Since φ_c is a solution of the equation

$$\frac{\delta S}{\delta \varphi} + J(x) = 0$$

the relation (4.4.159) can be written as

$$\frac{\hbar}{i} W[J] = S[\phi] + \int d^4x \, J\phi + \tfrac{1}{2} i\hbar [\operatorname{Tr} \ln(\Box + m^2 + U''(\phi)) - \operatorname{Tr} \ln(\Box + m^2) + \mathcal{O}(\hbar^2). \quad (4.4.160)$$

The effective action $\Gamma[\phi]$ is obtained now by a Legendre transform and reads as

$$\Gamma[\phi] = \Gamma^{(0)}[\phi] + \Gamma^{(1)}[\phi] + \mathcal{O}(\hbar^2) \quad (4.4.161)$$

where

$$\Gamma^{(0)}[\phi] = S[\phi]$$

(cf (4.4.30)) and

$$\Gamma^{(1)}[\phi] = \tfrac{1}{2} i\hbar [\operatorname{Tr} \ln(\Box + m^2 + U''(\phi)) - \operatorname{Tr} \ln(\Box + m^2).$$

The transition to the momentum representation (i.e. the Fourier transform) and the addition of the appropriate *counter-terms* (i.e. renormalization) give the required result (4.4.36).

Problem 4.4.2. Give qualitative arguments supporting formula (4.4.120) for the correlation length $\xi(T)$, with the help of the linearized form of the $\widehat{M}$-transformation (4.4.109) and the definition (4.4.89), (4.4.90).

Hint. Since at the critical temperature T_c the set u_{st} is the stationary point of the $\widehat{M}$-transformation, we may assume that in the vicinity of the critical temperature, the difference $u^{(n)} - u_{st}$ for large values of n is proportional to the difference $T - T_c$. On the other hand, if the system is outside the phase transition curve, the iterated application of the $\widehat{M}$-transformation does not result in u_{st}. The closer the system becomes to the phase transition curve, the more iterations are required to remove it from the vicinity of u_{st}. To estimate the number of such iterations, let us linearize the $\widehat{M}$-transformation as in (4.4.119). Then the difference $u^{(n+m)} - u_{st}$ is of the order

$$u^{(n+m)} - u_{st} \sim \lambda_1^m (T - T_c) \quad (4.4.162)$$

where λ_1 is the maximal eigenvalue of the linearized transformation (4.4.119). If the difference (4.4.162) is large enough, e.g., if it is of the order of the critical temperature (which is a characteristic quantity for a system near a phase transition), we can say that the system is not in the vicinity of the phase transition. Thus, supposing $T_c \sim \lambda_1^m (T - T_c)$, we have the following estimation for the necessary number m of $\widehat{M}$-transformations:

$$m \sim \frac{\ln(T_c/(T - T_c))}{\ln \lambda_1}. \quad (4.4.163)$$

254   *Path integrals in statistical physics*

The correlation function of the new system (i.e. after the $\widehat{M}$-transformations) has the exponential asymptotics: $G(r) \xrightarrow[r\to\infty]{} \exp\{-r/\xi_0\}$ (because it is outside the vicinity of the phase transition). But this new system is nothing but the initial one after the scale increasing by 2^m times. Therefore the correlator of the initial system has the asymptotics

$$\exp\left\{-\frac{r}{2^m \xi_0}\right\}. \tag{4.4.164}$$

From (4.4.163) and (4.4.164), we find an expression for the correlation length, $\xi(T) \sim 2^m$, which is equivalent to (4.4.120).

Problem 4.4.3. Show that the calculation of the trace of the density operator for *fermionic* systems amounts to path integration with *antiperiodic* conditions (4.4.13).

Hint. Use the explicit derivation of the path integrals for fermionic systems in problem 2.6.10, page 316, volume I (its generalization to field systems is quite straightforward) and derive the trace of an operator.

Problem 4.4.4. Show explicitly that the propagator of a scalar field at finite energy obeys the periodic boundary conditions (4.4.143).

Hint. Assuming that the volume of the system is sufficiently large, the Fourier components of the free field and their commutation relations are given by

$$\varphi(x) = \int \frac{d^3k}{\sqrt{(2\pi)^3 2\omega_k}} (a_k e^{i(kx-\omega_k x_0)} + a_k^\dagger e^{-i(kx-\omega_k x_0)})$$

$$[a_k, a_{k'}^\dagger] = \delta^3(k-k') \qquad \text{others} = 0. \tag{4.4.165}$$

The two-point Green function in a medium of finite energy E in the time-path formalism is defined as

$$G_E(x-y) = \int \frac{d\alpha}{2\pi} e^{i\alpha E} \rho_{E,\alpha} \langle \mathbf{T}_c(\phi(x)\phi(y))\rangle_\alpha$$

where $\langle \cdot \rangle_\alpha$ on the right-hand side means

$$\langle \widehat{A}\widehat{B}\rangle_\alpha \equiv \mathrm{Tr}(e^{-i\alpha\hat{H}}\widehat{A}\widehat{B})/\mathrm{Tr}(e^{-i\alpha\hat{H}})$$

and $\rho_{E,\alpha}$ is defined in (4.4.137). Because of the contour ordering, we introduce the contour step-function $\theta_c(x_0 - y_0)$, which is one if x_0 is closer to the end of the contour than y_0, $\frac{1}{2}$ when $x_0 = y_0$ and otherwise 0. Then,

$$\langle \mathbf{T}_c(\varphi(x)\varphi(y))\rangle_\alpha = \theta_c(x_0 - y_0)\langle \varphi(x)\varphi(y)\rangle_\alpha + \theta_c(-x_0 + y_0)\langle \varphi(y)\varphi(x)\rangle_\alpha.$$

Using the standard procedure, we obtain

$$\langle \varphi(x)\varphi(y)\rangle_\alpha = \int \frac{d^4k}{(2\pi)^3} e^{-ik(x-y)} \epsilon(k_0)\delta(k^2 - m^2)(1 + f_\alpha^{(-)}(k_0))$$

$$\langle \varphi(y)\varphi(x)\rangle_\alpha = \int \frac{d^4k}{(2\pi)^3} e^{-ik(x-y)} \epsilon(k_0)\delta(k^2 - m^2) f_\alpha^{(-)}(k_0)$$

where we are using the metric $(+, -, -, -)$ and $\epsilon(k)$ equals $k/|k|$ for $k \neq 0$ and 0 for $k = 0$. The quantity $f_\alpha^{(-)}$ is defined in (4.4.149). Let us define $D^>$ and $D^<$ by

$$iD_\alpha^>(k) = 2\pi\epsilon(k_0)\delta(k^2 - m^2) f_\alpha^{(+)}(k_0)$$
$$iD_\alpha^<(k) = 2\pi\epsilon(k_0)\delta(k^2 - m^2) f_\alpha^{(-)}(k_0) \tag{4.4.166}$$

and their Fourier transformation by $D_\alpha(x) = \int \frac{d^4k}{(2\pi)^4} e^{-ik(x-y)} D_\alpha(k)$. Since the two-point Green function and the propagator are related by

$$G_E(x-y) = \int \frac{d\alpha}{2\pi} e^{-i\alpha E} \rho_{E,\alpha} i D_\alpha(x-y)$$

the propagator is obtained as

$$D_\alpha(x-y) = \theta_c(x_0 - y_0) D_\alpha^>(x-y) + \theta_c(-x_0 + y_0) D_\alpha^<(x-y). \quad (4.4.167)$$

For the propagators of the fields on the path $C_1 + C_2$, the result in (4.4.149) is found by Fourier-transforming this expression and using the definition of $\bar{D}$ in (4.4.147). The propagator of the fields on C_3 in (4.4.152) is obtained directly from (4.4.167).

In order to check the periodicity (4.4.143), let us discuss keeping t finite in the definition of the path C. For $x_1 = (-t, \mathbf{x}_1) \in C_1$ and $\forall x_2 \in C$, we find that

$$D_\alpha(x_1 - x_2) = D_\alpha^<(-t - x_{2,0}, \mathbf{x}_1 - \mathbf{x}_2) \quad (4.4.168)$$

where we have used the fact that x_1 is placed at the starting point of the path C. For $x_1 = (\alpha - t, \mathbf{x}_1) \in C_3$, at the endpoint of the path C, and $\forall x_2 \in C$, the propagator becomes

$$D_\alpha(x_1 - x_2) = D_\alpha^>(-t + \alpha - x_{2,0}, \mathbf{x}_1 - \mathbf{x}_2). \quad (4.4.169)$$

Using the expressions of $D^>$ and $D^<$ from (4.4.166) and the property $f_\alpha^{(-)}(-u) = -f_\alpha^{(+)}(u)$ for $u > 0$, we find that

$$D_\alpha^>(-t + \alpha - x_{2,0}, \mathbf{x}_1 - \mathbf{x}_2) = D_\alpha^<(-t - x_{2,0}, \mathbf{x}_1 - \mathbf{x}_2).$$

Therefore, the right-hand side of (4.4.168) is equal to the right-hand side of (4.4.169). Thus we have found the desired periodicity.

Problem 4.4.5. Show the decoupling of the C_3 part from $C_1 + C_2$.

Hint. The decoupling of the C_3 part from $C_1 + C_2$ is equivalent to the vanishing of the propagator connecting C_a, $a = 1, 2$ and C_3. After integrating out the field, the part connecting the sources on C_a and C_3 is given by

$$i \int_{C_a} d^4x \int_{C_3} d^4x' \, j_a(x) D_\alpha(x - x') j_3(x')$$

$$= i \int \frac{d^4k}{(2\pi)^4} e^{-ik_0 t} \int_{C_a} d^4x \int_0^\alpha dt' \int d^3\mathbf{x}' \, e^{-ik(x-x')} j_a(x) D_\alpha^<(k) j_3(-t + t', \mathbf{x}') \quad (4.4.170)$$

where we have changed the integration variable over time on C_3. Since j_3 is chosen to be a smooth function of $x' \in (C_3, r^3)$, (4.4.170) has a form in which we can use the Riemann–Lebesgue theorem with respect to the k_0 integration. Thus, implemented with a standard adiabatic switching on and off mechanism for the external source, the right-hand side of (4.4.170) vanishes in the limit $t \to \infty$.

Problem 4.4.6. Show the equivalence of the different choices of time paths: the one depicted in figure 4.4 and the path presented in figure 4.5

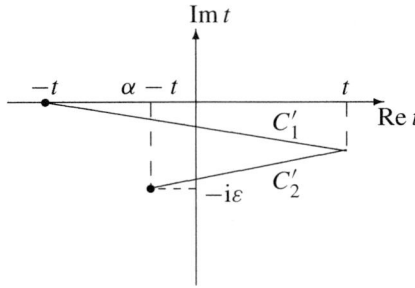

Figure 4.5. The alternative time path C' for the formulation of finite-energy quantum field theory.

Hint. The merit of the path C in figure 4.4 is that the real-time part $C_1 + C_2$ and the equilibrium part C_3 decouple, making the theory simple. However, there is always the possibility of choosing the other path, in particular, the path $C' = C'_1 + C'_2$ depicted in figure 4.5. The general reason for the independence of the results of the choice of paths is the very construction of the path integral based on the completeness of the states at each time. To verify this statement by practical calculation, we can find the one-loop effective action using C'. For further hints, see Chaichian and Senda (1993). Here, we note only that similarly to the discussion in the preceding problem, the paths C'_1 and C'_2 decouple.

Problem 4.4.7. Calculate the expectation value of the number operator at tree level for a system with a microcanonical density operator (i.e. with fixed finite energy).

Hint. The number operator of the state with momentum k is $\hat{n}_k^{(1)} = a_k^{(1)\dagger} a_k^{(1)}$, where $a_k^{(1)}$ is a Fourier component of the field on C_1 (see problem 4.4.4). The expectation value of $\hat{n}_k^{(1)}$ in a medium of energy E is

$$\langle \hat{n}_k^{(1)} \rangle_E = \lim_{\epsilon \to 0} \int \frac{d^3x \, d^3y}{(2\pi)^3 2\omega_k} e^{-ik(x-y)} (\partial_{x_0} + i\omega_k)(\partial_{y_0} - i\omega_k) \langle \mathbf{T}(\varphi_1(x)\varphi_1(y)) \rangle_E \big|_{x_0 = t + \epsilon, \, y_0 = t}$$

$$= \frac{L^3}{(2\pi)^3} \int_0^E dE' \, \rho_{E,E-E'} \sum_{n=1}^{\infty} \delta(E' - n\omega_k)$$

where φ_1 represents a field on the path C_1. To obtain the last line, we have used the explicit form of the two-point Green function $\langle \mathbf{T}(\varphi_1(x)\varphi_1(y)) \rangle_E = G_E^{11}$, where G_E^{11} is given in (4.4.150). The total number operator is given by

$$\hat{N} = \int \frac{d^3k}{(2\pi)^3 2\omega_k} \hat{n}_k^{(1)}.$$

Thus, its expectation value becomes

$$\langle \hat{N} \rangle_E = \frac{L^3}{(2\pi)^3} \int_0^E dE' \, \rho_{E,E-E'} \bar{N}(E')$$

$$\bar{N}(E') \equiv \frac{L^3}{(2\pi)^3} \int \frac{d^3k}{(2\pi)^3 2\omega_k} \sum_{n=1}^{\infty} \delta(E' - n\omega_k) \qquad (4.4.171)$$

where $\bar{N}(E')$ means the number of particles in the subsystem of energy E'.

4.5 Superfluidity, superconductivity, non-equilibrium quantum statistics and the path-integral technique

At the beginning of the last century, it was shown on statistical grounds only (Bose 1924, Einstein 1925, Fermi 1926) that, under extreme conditions, ideal gases of indistinguishable particles have remarkable features on a macroscopic scale. Furthermore, in the case of an interacting gas even more stunning phenomena may occur. A well-known example in this context is *superconductivity*. In a superconducting metal, an attractive interaction between two electrons with opposite momenta causes an instability in the Fermi surface together with the formation of Cooper pairs (Cooper 1956) (see also, e.g., de Gennes (1989) and references therein). The latter are allowed to move freely through the lattice, resulting in a superconducting current and a vanishing resistance. According to the successful BCS theory (Bardeen *et al* 1957) describing this, the attractive interaction is the result of a phonon exchange process, and BCS theory shows that superconductivity can, in a certain sense, be regarded as a result of a *Bose–Einstein condensation* of the Cooper pairs. Another striking example of this condensation process is associated with the superfluid phase of liquid ^{4}He (see, e.g., Griffin (1993) and Griffin *et al* (1995)).

Phenomenologically, the characteristics of *superfluidity* can be explained if the dispersion relation of the elementary excitations differs from the particle-like dispersion $\epsilon(\boldsymbol{k}) = \hbar^2 \boldsymbol{k}^2/2m$ and is linear for small momenta. To see that this feature indeed implies superfluidity, let us consider liquid ^{4}He in a very long cylindrical pipe moving with velocity $\boldsymbol{v}$ along its symmetry axis. Describing the strongly interacting Bose liquid in the pipe as an assembly of non-interacting quasiparticles in a state of thermal equilibrium, we find by the usual methods of statistical mechanics that, in the laboratory frame, the number of quasiparticles with momentum $\hbar\boldsymbol{k}$ and energy $\hbar\omega(\boldsymbol{k})$ is given by

$$N(\boldsymbol{k}) = \frac{1}{\mathrm{e}^{(\hbar\omega(\boldsymbol{k}) - \hbar\boldsymbol{k}\cdot\boldsymbol{v})/k_\mathrm{B}T} - 1}. \tag{4.5.1}$$

The velocity $\boldsymbol{v}$ is still arbitrary at this point, but its magnitude has an upper bound because the occupation numbers $N(\boldsymbol{k})$ must be positive. Therefore, we require for all $\boldsymbol{k} \neq 0$ that $\hbar\omega(\boldsymbol{k}) > \hbar\boldsymbol{k}\cdot\boldsymbol{v}$. In this case, the total momentum $\boldsymbol{P} = \sum_{\boldsymbol{k}} \hbar\boldsymbol{k} N(\boldsymbol{k})$ carried along the walls of the pipe, is, at low temperatures T, clearly much smaller than the momentum $Nm\boldsymbol{v}$ that we obtain when the whole liquid containing N atoms moves rigidly with the walls. Hence, if the dispersion relation is linear, i.e. $\hbar\omega(\boldsymbol{k}) = \hbar c|\boldsymbol{k}|$, we conclude by a Galilean transformation that the fluid can have a stationary (frictionless) flow if the speed is small enough and obeys $v < c$. Herewith, both the existence of superfluidity as well as a critical speed above which the phenomenon cannot take place is explained (of course, only at a heuristic level). The linear dispersion at long wavelengths in a Bose system with short-range interactions is well established by now. However, to calculate its speed c from first principles is not feasible in general and in particular not for the strongly interacting ^{4}He liquid. To achieve that, we have to consider a dilute Bose gas, for which we can rely on the weakness of the interactions or, more precisely, on the smallness of the gas parameter $(na^3)^{1/2}$, where n is the density and a is the radius of interaction of the particles in the gas.

An approach to describing Bose–Einstein condensation (and hence to the phenomena of superfluidity and superconductivity) depends on what we are going to find out about this phenomenon. If we are interested in calculating the magnitude of the Bose–Einstein condensate, the spectrum of (quasi)particles, *spatial* correlators and other *static* characteristics, we can use the methods of the equilibrium quantum statistics based on a path integral with imaginary time (see section 4.4.1). Starting from the non-relativistic secondary quantized action (4.4.1) together with an interaction of the form (4.4.5) and using the standard path-integral technique, we can develop an appropriate approximation method for calculating the static physical characteristics of the condensation. As usual in phase transitions, a finite-order perturbation theory approximation is not suitable for this aim, and we have to sum some *infinite* series of Feynman diagrams. An advantage of the path-integral formalism is that it allows us to develop an improved perturbation expansion in the domain of small momenta of the particles (see section 3.3.1), which is especially important in the case of the condensation. We shall present some essential points of this approach in sections 4.5.1 and 4.5.2.

If we are interested in describing the formation *process* of the Bose–Einstein condensate, we have to use the methods of non-equilibrium quantum statistics. The time-dependent phenomena and the characteristics of a quantum system which is in contact with a thermal reservoir and which undergoes Bose–Einstein condensation are discussed in section 4.5.3.

4.5.1 Perturbation theory for superfluid Bose systems

The perturbation theory for statistical systems (non-relativistic field theory at imaginary time) has been developed in section 4.4.1. This theory is applicable at high temperature (above the critical temperature of the phase transition to the superfluid state) and it below the phase transition should be modified. As discussed in section 4.4.4, the characteristic property of the phase transition is the existence of long-range correlations, so that the correlator $\langle \varphi(\tau, \boldsymbol{x})\varphi(\tau, \boldsymbol{y})\rangle_\beta$ decreases as $r = |\boldsymbol{x} - \boldsymbol{y}| \to \infty$ as a power and not as an exponential. In a three-dimensional space this correlator as $r \to \infty$ tends to a constant ρ_0 which is just the condensate density.

◇ Simplest perturbation theory taking into account a condensate

In order to develop a perturbation theory for the action (4.4.1), (4.4.5) taking into account a possible non-zero condensate, we make a field variable shift:

$$\varphi(\tau, \boldsymbol{x}) \to \varphi'(\tau, \boldsymbol{x}) + \alpha \qquad \varphi^*(\tau, \boldsymbol{x}) \to (\varphi')^*(\tau, \boldsymbol{x}) + \alpha \qquad (4.5.2)$$

where $|\alpha| = \rho_0$. After the transition to the Fourier transform (4.4.7) for the fields $\varphi(\tau, \boldsymbol{x}) \to a(\omega, \boldsymbol{k}) \equiv a(k)$, the shift (4.5.2) reads as

$$a(k) \to b(k) + \alpha\sqrt{\beta L^3}\delta_{k0} \qquad a^*(k) \to b^*(k) + \alpha^*\sqrt{\beta L^3}\delta_{k0} \qquad (4.5.3)$$

and the action takes the form

$$S = \sum_k \left\{ \left(\frac{k^2}{2m} - i\omega - \mu\right)b^*(k)b(k) + |\alpha|^2(\widetilde{u}(0) + \widetilde{u}(k))b^*(k)b(k) \right.$$
$$+ \frac{\widetilde{u}(k)}{2}(\alpha^2 b^*(k)b^*(-k) + (\alpha^*)^2 b(k)b(-k))$$
$$+ \frac{1}{2\sqrt{\beta L^3}} \sum_{k_1+k_2=k_3} [\widetilde{u}(\boldsymbol{k}_1) + \widetilde{u}(\boldsymbol{k}_2)](\alpha b^*(k_1)b^*(k_2)b(k_3) + \alpha^* b(k_1)b(k_2)b^*(k_3))$$
$$+ \frac{1}{4\beta L^3} \sum_{k_1+k_2=k_3+k_4} [\widetilde{u}(\boldsymbol{k}_1 - \boldsymbol{k}_3) + \widetilde{u}(\boldsymbol{k}_1 - \boldsymbol{k}_4)]b^*(k_1)b^*(k_2)b(k_3)b(k_4)$$
$$\left. + \sqrt{\beta L^3}[\gamma^* b(0) + \gamma b^*(0)] - \beta L^3[\mu|\alpha|^2 - \tfrac{1}{2}\widetilde{u}(0)|\alpha|^2] \right\}. \qquad (4.5.4)$$

Here $\gamma = \widetilde{u}(0)|\alpha|^2 - \alpha\mu$. All the terms in this action except the first one (which corresponds to an ideal gas) are considered as perturbations. In addition to the elements mentioned in section 4.4.1, the diagram technique for the perturbation theory in the case of a possible condensate contains the following elements:

Superfluidity, superconductivity, non-equilibrium statistics

Diagrams:
- $k=0$, γ
- $k=0$, γ^*
- $k \to \bullet \leftarrow k$, $|\alpha|^2(\widetilde{u}(0) + \widetilde{u}(k))$
- $k \to \bullet \leftarrow -k$, $\alpha^2 \widetilde{u}(k)$
- $k \leftarrow \bullet \to -k$, $(\alpha^*)^2 \widetilde{u}(k)$
- Three-line vertex with k_1, k_2, k_3: $\alpha[\widetilde{u}(k_1) + \widetilde{u}(k_2)]$
- Three-line vertex with k_1, k_2, k_3: $\alpha^*[\widetilde{u}(k_1) + \widetilde{u}(k_2)]$.

The constant α is chosen so that the sum of all diagrams with one external line are cancelled out; graphically this condition looks as follows:

[diagram: blackened blob with one external line ≡ single line + loop diagram + ⋯ = 0]

This requirement is equivalent to the condition

$$\langle \widehat{b}(0) \rangle_\beta = \langle \widehat{b}^\dagger(0) \rangle_\beta = 0$$

or

$$\langle \widehat{a}(0) \rangle_\beta = \alpha\sqrt{\beta L^3} \qquad \langle \widehat{a}^\dagger(0) \rangle_\beta = \alpha^*\sqrt{\beta L^3}$$

and

$$\langle \widehat{a}(k)\widehat{a}^\dagger(k) \rangle_\beta = \langle \widehat{b}(k)\widehat{b}^\dagger(k) \rangle_\beta + \beta L^3 |\alpha|^2 \delta_{k0}. \tag{4.5.5}$$

The last relation shows that $|\alpha|^2$ has the meaning of a condensate density.

◇ Normal and anomalous Green functions and the spectrum of particles

Later in this section we shall put, for simplicity, $2m = 1$ (together with $k_B = \hbar = 1$).

In a superfluid system, along with the usual ('normal') Green function

$$G(k) = \langle \widehat{b}(k)\widehat{b}^\dagger(k) \rangle_\beta \tag{4.5.6}$$

two *anomalous* functions appear:

$$G^{(\text{an})}(k) = -\langle \widehat{b}(k)\widehat{b}(-k) \rangle_\beta \qquad \bar{G}^{(\text{an})}(k) = \langle \widehat{b}^\dagger(k)\widehat{b}^\dagger(-k) \rangle_\beta. \tag{4.5.7}$$

These Green functions can be expressed by normal and anomalous self-energy parts Σ_1 and Σ_2 (which are sums of the OPI diagrams), according to the following Dyson–Schwinger equations (cf section 3.1.5):

$$\begin{aligned} G(k) &= G_0(k) + G_0(k)\Sigma_1(k)G(k) + G_0(k)\Sigma_2(k)G^{(\text{an})}(-k) \\ G^{(\text{an})}(k) &= G_0(-k)\Sigma_1^*(k)G(k) + G_0(-k)\Sigma_1(-k)G^{(\text{an})}(-k). \end{aligned} \tag{4.5.8}$$

In a dilute gas with a small parameter

$$n^{1/3}a \ll 1 \tag{4.5.9}$$

(this means that the radius of particle interactions a is much smaller than the mean distance between the particles; the latter is obviously related to the density of particles n) we can develop (see Popov (1983)) an appropriate approximation for the calculation of the self-energy parts Σ_1, Σ_2 and solve equations (4.5.7) with the result

$$G(k) = \frac{i\omega + k^2 + \Lambda}{\omega^2 + k^4 + 2\Lambda k^2}$$

$$G^{(\text{an})}(k) = \frac{\Lambda}{\omega^2 + k^4 + 2\Lambda k^2} \tag{4.5.10}$$

where

$$\Lambda \stackrel{\text{def}}{\equiv} t_0\rho_0 = \mu - \frac{2}{(4\pi)^{3/2}}\zeta(3/2)t_0 T^{3/2} \tag{4.5.11}$$

($\zeta(x)$ is the Riemann $\zeta(x)$-function) and the constant t_0 is the characteristic of the interaction between particles (namely, their scattering amplitude in the limit of zero momenta, $k_i \to 0$, $i = 1$–4). Recall that ρ_0 is the condensate density. This solution is valid only for $\Lambda > 0$, that is, for temperatures T below the critical value T_c defined by the condition $\Lambda = 0$. If $\Lambda < 0$ (that is, for a high temperature T, only a non-anomalous Green function exists:

$$G(k) = \frac{1}{i\omega - k^2 + \Lambda}. \tag{4.5.12}$$

After the analytical continuation $i\omega \to E$, we can extract the spectrum of the system (defined by the poles of the Green functions), above and below the critical temperature:

$$E = \sqrt{k^4 + 2\Lambda k^2} \qquad \Lambda > 0 \qquad \text{(superfluid phase)} \tag{4.5.13}$$

$$E = k^2 + |\Lambda| \qquad \Lambda < 0 \qquad \text{(normal phase)}. \tag{4.5.14}$$

It is seen that in the first case for $|k| \ll \sqrt{\Lambda}$, the spectrum of particles becomes *linear*. As we have explained in the introduction to this section, such a spectrum implies *superfluidity*.

◇ Comment on the improved perturbation expansion for superfluid systems by separate path integration over higher and lower modes

The calculation of higher orders of the perturbation theory considered earlier shows that it becomes slowly convergent in the vicinity of both the critical temperature (i.e. $\Lambda \approx 0$) and small momenta of the particles ($k \approx 0$). This situation can be improved (Popov 1983) by separate consecutive path integrations over the higher and lower modes of the fields, as in section 3.3.1. The corresponding calculation in this approach goes through the following steps:

(i) The fields $\varphi(\tau, k)$, $\varphi^*(\tau, k)$ are separated into lower-mode $\varphi_0(\tau, k)$, $\varphi_0^*(\tau, k)$ and higher-mode $\varphi_1(\tau, k)$, $\varphi_1^*(\tau, k)$ parts: φ_0, φ_0^* correspond to Fourier modes with momenta $|k| < \kappa$ in the expansion (4.4.7), while φ_1, φ_1^* have Fourier modes with $|k| \geq \kappa$. Here κ is some small parameter which is adjusted for an optimal convergence of the perturbation expansion.
(ii) Using the standard perturbation theory as before, we calculate the effective action $S^{(\text{eff})}[\varphi_0, \varphi_0^*]$ for the lower mode fields φ_0, φ_0^*:

$$\exp\{-S^{(\text{eff})}[\varphi_0, \varphi_0^*]\} = \int \mathcal{D}\varphi_1^* \mathcal{D}\varphi_1 \exp\{-S\}.$$

The only specific feature of this calculation is that all sums over momenta are cut off at the lower limit κ. Thus, this perturbation expansion has no divergences at small values of momenta.

(iii) The effective action $S^{(\text{eff})}[\varphi_0, \varphi_0^*]$ can be expanded into power series in the field variables and for the small parameter κ, the expansion can be restricted up to quadratic terms. Thus, the remaining path integration over the lower mode fields φ_0, φ_0^* becomes Gaussian and can be carried out exactly.

The improved perturbation expansion allows us to obtain more detailed information about superfluid systems. In particular, we can find (Popov 1983) the $1/k^2$-asymptotics of the Green function

$$G(0, \mathbf{k}) \xrightarrow[k \to 0]{} \text{constant} \frac{1}{\mathbf{k}^2}.$$

Among other things, this asymptotic shows that it is impossible for the condensate to form in one- and two-dimensional spaces since the singularity $1/\mathbf{k}^2$ is non-integrable in the low-dimensional spaces.

4.5.2 Perturbation theory for superconducting Fermi systems

The phenomenon of superconductivity in Fermi systems is quite close to the superfluidity of Bose systems. In the path-integral formalism, the analog of the Bose field φ turns out to be the product of two Fermi fields, $\psi\psi$, and the analog of the one-particle correlator $\langle \varphi(\tau', \mathbf{x}')\varphi(\tau, \mathbf{x})\rangle$ turns out to be the two-particle correlator (mean value of four Fermi fields):

$$\langle \psi(\tau_1, \mathbf{x}_1)\psi(\tau_2, \mathbf{x}_2)\bar{\psi}(\tau_3, \mathbf{x}_3)\bar{\psi}(\tau_4, \mathbf{x}_4)\rangle. \tag{4.5.15}$$

A system of Fermi particles has long-distance correlations if, at fixed differences $\mathbf{x}_1 - \mathbf{x}_2$ and $\mathbf{x}_3 - \mathbf{x}_4$, the correlator (4.5.15) decreases in the limit $\mathbf{x}_1 - \mathbf{x}_3$ slower than an exponential or even has a finite non-zero limiting value. In the latter case, we can expect anomalous Green functions to exist:

$$\langle \psi(\tau_1, \mathbf{x}_1)\psi(\tau_2, \mathbf{x}_2)\rangle \quad \text{and} \quad \langle \bar{\psi}(\tau_3, \mathbf{x}_3)\bar{\psi}(\tau_4, \mathbf{x}_4)\rangle. \tag{4.5.16}$$

In ordinary perturbation theory, such mean values vanish. In order to modify the perturbation expansion taking possible anomalous non-zero mean values into account, let us add to the action the terms with the external sources $\eta(p)$, $\bar{\eta}(p)$ of the form

$$S^{(\text{an})} \sim \sum_p (\bar{\eta}(p)a(p)a(-p) + \eta(p)a^*(p)a^*(-p)) \tag{4.5.17}$$

where $a(p)$, $a^*(p)$ are the amplitudes in the Fourier transform of the fields $\psi, \bar{\psi}$. The peculiarity of this action functional is that now the number of particles is not a conserved quantity and for non-zero $\eta(p)$, $\bar{\eta}(p)$, the anomalous mean values (4.5.16) do not vanish. In order to find out whether the system undergoes a phase transition into the superconducting state, we have to study the limit $\eta(p) \to 0$, $\bar{\eta}(p) \to 0$: if the anomalous mean values (4.5.16) have a non-zero limit, the system is in the superconducting state, otherwise the system is in the normal state. Such a study can be carried out using the perturbation theory and the Dyson–Schwinger equations (cf section 3.1.5).

◇ Superconductivity and perturbation theory for Fermi systems

To introduce the diagram technique we must take into account the spin degrees of freedom for the Fermi particles and describe them by the fields $\psi_s(\tau, \mathbf{x})$, s being the spin projection onto some space direction. Let us restrict ourselves to the following non-zero anomalous mean values:

$$S^{(\text{an})} \stackrel{\text{def}}{\equiv} \langle \psi_s(\tau_1, \mathbf{x}_1)\psi_{-s}(\tau_2, \mathbf{x}_2)\rangle_\beta$$

$$\bar{S}^{(\text{an})} \stackrel{\text{def}}{\equiv} \langle \bar{\psi}_s(\tau_3, \mathbf{x}_3)\bar{\psi}_{-s}(\tau_4, \mathbf{x}_4)\rangle_\beta \tag{4.5.18}$$

while $\langle\psi_s\psi_s\rangle_\beta$ and $\langle\bar\psi_s\bar\psi_s\rangle_\beta$ vanish. Then, the path-integral technique (see section 3.1.5) allows us to derive the following Dyson–Schwinger equations for the normal $S(p) \stackrel{\text{def}}{\equiv} \langle\bar\psi_s(\tau_1,\boldsymbol{x}_1)\psi_s(\tau_2,\boldsymbol{x}_2)\rangle_\beta$ and the anomalous Green functions:

$$S(p) = S_0(p) + S_0(p)\Sigma_1(p)S(p) + S_0(p)\Sigma_2(p)S^{(\text{an})}(p)$$
$$S^{(\text{an})}(p) = -S_0(p)\bar\Sigma_2(p)S(p) + S_0(-p)\Sigma_1(-p)S^{(\text{an})}(p). \qquad (4.5.19)$$

Here Σ_1 and Σ_2 are the self-energy parts, i.e. the sums of one-particle irreducible (cf section 3.1.5) diagrams

[diagrams for Σ_1 and Σ_2]

In these diagrams, the line $\longrightarrow\!\!\!\!\!\longrightarrow$ denotes the normal Green function $S(p)$, while $\longleftrightarrow\!\!\!\!\!\longrightarrow$ denotes the anomalous Green function $S^{(\text{an})}(p)$.

Since the free Green function $S_0(p)$ is given by (see section 4.4.1)

$$S_0(\omega,\boldsymbol{k}) = \frac{1}{-\boldsymbol{k}^2 + i\omega + \lambda} \qquad (4.5.20)$$

(cf (4.4.1) and (4.4.2)), the solution of (4.5.19) can be written as

$$S(p) = \frac{i\omega + \boldsymbol{k}^2 - \mu + \Sigma_1(-p)}{(i\omega + \boldsymbol{k}^2 - \mu + \Sigma_1(-p))(i\omega - \boldsymbol{k}^2 - \mu - \Sigma_1(p)) - |\Sigma_2(p)|^2}$$
$$S(p) = \frac{i\omega + \boldsymbol{k}^2 - \mu + \Sigma_1(-p)}{(i\omega + \boldsymbol{k}^2 - \mu + \Sigma_1(-p))(i\omega - \boldsymbol{k}^2 - \mu - \Sigma_1(p)) - |\Sigma_2(p)|^2}. \qquad (4.5.21)$$

In order to obtain an explicit expression for the anomalous Green function, as in the Bose gas, the expansion for the self-energy parts has to be cut off in a self-consistent way. This can be done, in particular, in the case of the small gas parameter $n^{1/3}a \ll 1$. Then we can show (Popov 1983) that the main contribution to the self-energy parts is given by the diagrams:

[approximate diagrams for Σ_1 and Σ_2]

The four-point Green function entering the diagram for Σ_1 is given by the following sequence of diagrams:

[ladder/bubble diagram series]

Superfluidity, superconductivity, non-equilibrium statistics 263

(which can be summed by the so-called T-matrix method).

As a result of the calculation of the anomalous Green functions, we can find physically important characteristics of superconducting systems, in particular, the energy gap $\Delta(0)$ in the spectrum of a system in a superconducting state (we recover the particle mass m explicitly):

$$\Delta(0) = \frac{4p_F^2}{m} \exp\left\{-\frac{2\pi^2}{m|t_0|p_F} - 2\right\}$$

where $p_F = \sqrt{2m\mu}$ is the Fermi momentum (radius of the Fermi sphere) and t_0 is the value of the T-matrix in the domain $p \leq p_F$, where it can be approximated by the constant $t_0 = T(p_1 = p_2 = p_3 = p_4 = 0)$ (cf (4.5.11) in a Bose gas). The critical temperature of the transition to the superconducting state is expressed in terms of the energy gap as follows

$$T_c = \frac{\exp\{\mathfrak{C}\}}{\pi}\Delta(0) \qquad (4.5.22)$$

($\mathfrak{C} = 0.5772\ldots$ is the Euler constant).

4.5.3 Non-equilibrium quantum statistics and the process of condensation of an ideal Bose gas

Our main goal in this section is to introduce the reader to the problems of formation and evolution of the Bose–Einstein condensation and their description in terms of path integrals. We shall start from the relatively simple case of an ideal Bose gas: at first, from an isolated one and then coupled to a thermal reservoir. In fact, considering the model for an ideal gas provides quite a comprehensive presentation of the path-integral techniques in the theory of superfluidity. The introduction of an interaction, which physically is extremely important, does not require essentially new *path-integral* methods. Therefore, we shall not discuss the Bose–Einstein condensation of a realistic interacting gas in full detail (for a complete consideration of this complicated topic, we refer the reader, e.g., to Griffin (1993), Griffin *et al* (1995), Stoof (1999) and references therein).

◇ **Ideal gas of bosonic quantum point particles**

In the textbooks, an ideal Bose gas of quantum particles is generally discussed in terms of the average occupation numbers of the one-particle states $\chi_\alpha(x)$. Given the canonical density matrix $\hat{\rho}_C(t_0)$ of the gas at an initial time t_0, these occupation numbers obey

$$N_\alpha(t) = \text{Tr}[\hat{\rho}_C(t_0)\hat{\varphi}_\alpha^\dagger(t)\hat{\varphi}_\alpha(t)] \qquad (4.5.23)$$

with $\hat{\varphi}_\alpha^\dagger(t)$ and $\hat{\varphi}_\alpha(t)$ the usual creation and annihilation operators of second quantization in the Heisenberg picture, respectively. Because the Hamiltonian of the gas

$$\hat{H} = \sum_\alpha \epsilon_\alpha \hat{\varphi}_\alpha^\dagger(t)\hat{\varphi}_\alpha(t) \qquad (4.5.24)$$

commutes with the number operators $\hat{N}_\alpha(t) = \hat{\varphi}_\alpha^\dagger(t)\hat{\varphi}_\alpha(t)$, the *non-equilibrium dynamics* of the system is trivial and the average occupation numbers are, at all times, equal to their value at the initial time t_0. If we are also interested in fluctuations, it is convenient to introduce the eigenstates of the number operators, i.e.

$$|\{N_\alpha\}; t\rangle = \prod_\alpha \frac{(\hat{\varphi}_\alpha^\dagger(t))^{N_\alpha}}{\sqrt{N_\alpha!}}|0\rangle \qquad (4.5.25)$$

and to consider the full probability distribution

$$P(\{N_\alpha\}; t) = \text{Tr}[\hat{\rho}(t_0)|\{N_\alpha\}; t\rangle\langle\{N_\alpha\}; t|] \qquad (4.5.26)$$

which is again independent of time for an ideal Bose gas. The average occupation numbers are then determined by

$$N_\alpha(t) = \sum_{\{N_\alpha\}} N_\alpha P(\{N_\alpha\}; t) \qquad (4.5.27)$$

and the characteristics of the fluctuations (dispersions of observables) can be obtained from similar expressions.

It is even more convenient to consider the eigenstates of the field operator

$$\hat{\varphi}(\boldsymbol{x}, t) = \sum_\alpha \hat{\varphi}_\alpha(t) \chi_\alpha(\boldsymbol{x}) \qquad (4.5.28)$$

that is, coherent states for the creation and annihilation operators $\hat{\varphi}_\alpha(t)$, $\hat{\varphi}_\alpha^\dagger(t)$ with the equal-time commutation relation $[\hat{\varphi}_\alpha(t), \hat{\varphi}_\alpha^\dagger(t)] = 1$. As we know from section 2.3.3, an eigenstate of $\hat{\varphi}(\boldsymbol{x}, t)$ with the eigenvalue $\phi(\boldsymbol{x}) = \sum_\alpha \phi_\alpha \chi_\alpha(\boldsymbol{x})$ is given by

$$|\phi; t\rangle = \exp\left\{\int d\boldsymbol{x}\, \phi(\boldsymbol{x}) \hat{\varphi}^\dagger(\boldsymbol{x}, t)\right\} |0\rangle = \exp\left\{\sum_\alpha \phi_\alpha \hat{\varphi}_\alpha^\dagger(t)\right\} |0\rangle \qquad (4.5.29)$$

and is also clearly an eigenstate of $\hat{\varphi}_\alpha(t)$, with the eigenvalue ϕ_α. A straightforward generalization of the coherent state properties discussed in chapter 2 to the case of an infinite number of creation–annihilation operators shows that these eigenstates obey the inner product

$$\langle\phi; t|\phi'; t\rangle = \exp\left\{\int d\boldsymbol{x}\, \phi^*(\boldsymbol{x})\phi'(\boldsymbol{x})\right\} = \exp\left\{\sum_\alpha \phi_\alpha^* \phi'_\alpha\right\} \qquad (4.5.30)$$

and the completeness relation

$$\int \prod_\alpha \frac{d\phi_\alpha^* d\phi_\alpha}{2\pi i} \frac{|\phi; t\rangle\langle\phi; t|}{\langle\phi; t|\phi; t\rangle} = \mathbb{I}. \qquad (4.5.31)$$

Thus, in analogy with the occupation number representation in (4.5.26), we can now develop another description of the Bose gas, by making use of these coherent states and considering the probability distribution

$$P[\phi^*, \phi; t] = \text{Tr}\left[\hat{\rho}_C(t_0) \frac{|\phi; t\rangle\langle\phi; t|}{\langle\phi; t|\phi; t\rangle}\right]. \qquad (4.5.32)$$

Although we expect this probability distribution to be once again independent of time, let us nevertheless proceed to deriving its equation of motion in a way that can be generalized when we consider an interacting Bose gas. First, we need to expand the density matrix $\hat{\rho}(t_0)$ in terms of these coherent states. For an isolated Bose gas, it is appropriate to take an initial density matrix that commutes with the total number operator $\hat{N} = \sum_\alpha \hat{N}_\alpha(t)$ and then we find the expression

$$\hat{\rho}_C(t_0) = \int \prod_\alpha \frac{d(\phi_0^*)_\alpha \, d(\phi_0)_\alpha}{2\pi i} \rho[|\phi_0|^2; t_0] \frac{|\phi_0; t_0\rangle\langle\phi_0; t_0|}{\langle\phi_0; t_0|\phi_0; t_0\rangle} \qquad (4.5.33)$$

in which the expansion coefficients $\rho[|\phi_0|^2; t_0]$ only depend on the amplitude of the field $\phi_0(\boldsymbol{x})$, but not on its phase. This is equivalent to saying that the initial state of the gas does not have a spontaneously broken

$U(1)$ symmetry. Since we are ultimately interested in the dynamics of the Bose–Einstein condensation which implies $U(1)$-symmetry-breaking, it is crucial not to consider an initial state in which this symmetry is already broken.

Next, we substitute this expansion into equation (4.5.32) to obtain

$$P[\phi^*,\phi;t] = \int \prod_\alpha \frac{d(\phi_0^*)_\alpha \, d(\phi_0)_\alpha}{2\pi i} \rho[|\phi_0|^2;t_0] \frac{|\langle\phi;t|\phi_0;t_0\rangle|^2}{\langle\phi;t|\phi;t\rangle\langle\phi_0;t_0|\phi_0;t_0\rangle}. \quad (4.5.34)$$

This is a particularly useful result, because the time dependence is now completely determined by the matrix element $\langle\phi;t|\phi_0;t_0\rangle$, for which we can immediately write the path-integral representation.

◇ Schwinger–Keldysh closed time-path formalism

The next step is again, as in the foregoing sections, based on modification of the integration over the time variable. Recall that $\langle\phi,t|\phi,t_0\rangle$ is given by the path integral in the holomorphic representation (see section 3.1.1) over all complex fields $\varphi(\mathbf{x},t_+) = \sum_\alpha \varphi_\alpha(t_+)\chi_\alpha(\mathbf{x})$, with the asymmetrical boundary conditions (3.1.82):

$$\langle\phi;t|\phi_0;t_0\rangle = \int_{\mathcal{C}\{\varphi^*(\mathbf{x},t)=\phi^*(\mathbf{x});\varphi(\mathbf{x},t_0)=\phi_0(\mathbf{x})\}} \mathcal{D}\varphi^* \mathcal{D}\varphi \, \exp\left\{\frac{i}{\hbar} S_+[\varphi^*,\varphi]\right\} \quad (4.5.35)$$

with the 'forward' action $S_+[\varphi^*,\varphi]$ given by (cf (4.4.1) and (3.1.83))

$$S_+[\varphi^*,\varphi] = \sum_\alpha \left\{ -i\hbar\varphi_\alpha^*(t)\varphi_\alpha(t) + \int_{t_0}^t dt_+ \, \varphi_\alpha^*(t_+)\left(i\hbar\frac{\partial}{\partial t_+} - \epsilon_\alpha\right)\varphi_\alpha(t_+) \right\}. \quad (4.5.36)$$

In the same manner, the matrix element $\langle\phi;t|\phi_0;t_0\rangle^* = \langle\phi_0;t_0|\phi;t\rangle$ can be written as a path integral over all field configurations $\varphi(\mathbf{x},t_-) = \sum_\alpha \varphi_\alpha(t_-)\chi_\alpha(\mathbf{x})$ evolving 'backward' in time from t to t_0, i.e.

$$\langle\phi;t|\phi_0;t_0\rangle^* = \int_{\mathcal{C}\{\varphi(\mathbf{x},t)=\phi(\mathbf{x});\varphi^*(\mathbf{x},t_0)=\phi_0^*(\mathbf{x})\}} \mathcal{D}\varphi^* \mathcal{D}\varphi \, \exp\left\{\frac{i}{\hbar} S_-[\varphi^*,\varphi]\right\} \quad (4.5.37)$$

with the 'backward' action

$$S_-[\varphi^*,\varphi] = \sum_\alpha \left\{ -i\hbar\varphi_\alpha^*(t_0)\varphi_\alpha(t_0) + \int_t^{t_0} dt_- \, \varphi_\alpha^*(t_-)\left(i\hbar\frac{\partial}{\partial t_-} - \epsilon_\alpha\right)\varphi_\alpha(t_-) \right\}$$

$$= \sum_\alpha \left\{ -i\hbar\varphi_\alpha^*(t)\varphi_\alpha(t) + \int_t^{t_0} dt_- \, \varphi_\alpha(t_-)\left(-i\hbar\frac{\partial}{\partial t_-} - \epsilon_\alpha\right)\varphi_\alpha^*(t_-) \right\}. \quad (4.5.38)$$

Putting all these results together, we see that the probability distribution $P[\phi^*,\phi;t]$ can in fact be represented by a path integral over all fields $\varphi(\mathbf{x},t)$ that evolve backwards from t to t_0 and then forward in time from t_0 to t. Absorbing, for brevity, the factor $\rho[|\phi_0|^2;t_0]$ into the normalization factor of the functional integral, we thus arrive at the desired result:

$$P[\phi^*,\phi;t] = \mathfrak{N}^{-1} \int_{\mathcal{C}^t\{\varphi(\mathbf{x},t)=\phi(\mathbf{x});\varphi^*(\mathbf{x},t)=\phi^*(\mathbf{x})\}} \mathcal{D}\varphi^* \mathcal{D}\varphi \, \exp\left\{\frac{i}{\hbar} S[\varphi^*,\varphi]\right\} \quad (4.5.39)$$

where the total (backward–forward) action obeys

$$S[\varphi^*,\varphi] = S_-[\varphi^*,\varphi] + S_+[\varphi^*,\varphi] = -i\hbar \sum_\alpha (\varphi_\alpha^*(t)\varphi_\alpha(t) - |\phi_\alpha|^2)$$

$$+ \sum_\alpha \int_{\mathcal{C}^t} dt' \left\{ \frac{1}{2}\left(\varphi_\alpha^*(t')i\hbar\frac{\partial}{\partial t'}\varphi_\alpha(t') - \varphi_\alpha(t')i\hbar\frac{\partial}{\partial t'}\varphi_\alpha^*(t')\right) - \epsilon_\alpha \varphi_\alpha^*(t')\varphi_\alpha(t') \right\} \quad (4.5.40)$$

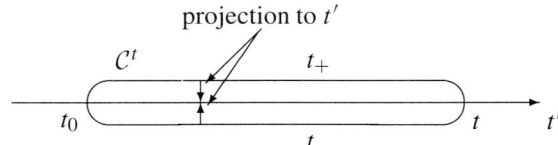

Figure 4.6. The closed time Schwinger–Keldysh contour defining the path integral for the probability distribution.

and the integration along the closed *Schwinger–Keldysh contour* $\mathcal{C}^t$ is defined by $\int_{\mathcal{C}^t} dt' = \int_t^{t_0} dt_- + \int_{t_0}^{t} dt_+$ as depicted in figure 4.6 (this is the so-called *closed time-path Schwinger–Keldysh formalism* (Schwinger 1961, Keldysh 1965)).

◇ The Fokker–Planck equation for the probability distribution of an ideal gas

We are now in a position to derive the equation of motion, i.e. the Fokker–Planck equation, for the probability distribution $P[\phi^*, \phi; t]$. This is most easily achieved by performing the variable transformation

$$\varphi(\boldsymbol{x}, t_\pm) = \phi(\boldsymbol{x}, t') \pm \xi(\boldsymbol{x}, t')/2$$

in (4.5.39), where t' is the projection of the t_+, t_- times, as shown in figure 4.6. In this manner, the fields $\varphi(\boldsymbol{x}, t_-)$ and $\varphi(\boldsymbol{x}, t_+)$ which exist on the backward and forward branch of the Schwinger–Keldysh contour, respectively, are 'projected' onto the real-time axis. Moreover, at the same time, we perform a separation between the (semi)classical dynamics described by $\phi(\boldsymbol{x}, t')$ and the quantum fluctuations determined by $\xi(\boldsymbol{x}, t')$. After the transformation, we have

$$P[\phi^*, \phi; t] = \int_{\mathcal{C}\{\phi(\boldsymbol{x},t)=\phi(\boldsymbol{x}); \phi^*(\boldsymbol{x},t)=\phi^*(\boldsymbol{x})\}} \mathcal{D}\phi^* \, \mathcal{D}\phi \int \mathcal{D}\xi^* \, \mathcal{D}\xi \, \exp\left\{\frac{i}{\hbar} S[\phi^*, \phi; \xi^*, \xi]\right\} \qquad (4.5.41)$$

with

$$S[\phi^*, \phi; \xi^*, \xi] = \sum_\alpha \int_{t_0}^t dt' \left\{ \phi_\alpha^*(t') \left(i\hbar \frac{\partial}{\partial t'} - \epsilon_\alpha \right) \xi_\alpha(t') + \xi_\alpha^*(t') \left(i\hbar \frac{\partial}{\partial t'} - \epsilon_\alpha \right) \phi_\alpha(t') \right\}. \qquad (4.5.42)$$

Since this action is linear in $\xi_\alpha(t')$ and $\xi_\alpha^*(t')$, the integration over the quantum fluctuations leads only to a constraint and we find that

$$P[\phi^*, \phi; t] = \mathfrak{N}^{-1} \int_{\mathcal{C}\{\phi(\boldsymbol{x},t)=\phi(\boldsymbol{x}); \phi^*(\boldsymbol{x},t)=\phi^*(\boldsymbol{x})\}} \mathcal{D}\phi^* \, \mathcal{D}\phi$$

$$\times \prod_\alpha \delta\left[\left(-i\frac{\partial}{\partial t'} - \frac{\epsilon_\alpha}{\hbar} \right) \phi_\alpha^*(t') \cdot \left(i\frac{\partial}{\partial t'} - \frac{\epsilon_\alpha}{\hbar} \right) \phi_\alpha(t') \right] \qquad (4.5.43)$$

or equivalently that (problem 4.5.2, page 278)

$$P[\phi^*, \phi; t] = \int \left(\prod_\alpha \frac{d(\phi_0^*)_\alpha d(\phi_0)_\alpha}{2\pi i} \right) P[|\phi_0|^2; t_0] \prod_\alpha \delta(|\phi_\alpha - \phi_\alpha^{\mathrm{cl}}(t)|^2) \qquad (4.5.44)$$

where we have used the fact that $P[\phi^*, \phi; t_0]$ is only a function of the amplitude $|\phi|^2$ and also introduced the quantity $\phi_\alpha^{\mathrm{cl}}(t)$ obeying the semiclassical equation of motion

$$i\hbar \frac{\partial}{\partial t} \phi_\alpha^{\mathrm{cl}}(t) = \epsilon_\alpha \phi_\alpha^{\mathrm{cl}}(t) \qquad (4.5.45)$$

and the initial condition $\phi_\alpha^{\text{cl}}(t_0) = \phi_{0,\alpha}$.

The latter equation is thus solved by $\phi_\alpha^{\text{cl}}(t) = \phi_{0,\alpha} e^{-i\epsilon_\alpha(t-t_0)/\hbar}$ and we conclude from a simple change of variables in (4.5.44) that for an ideal Bose gas, $P[\phi^*, \phi; t] = P[|\phi|^2; t_0]$, as expected (there is no $U(1)$-symmetry-breaking). We also see from (4.5.44) that the desired equation of motion for $P[\phi^*, \phi; t]$ reads as

$$i\hbar \frac{\partial}{\partial t} P[\phi^*, \phi; t] = -\left(\sum_\alpha \frac{\partial}{\partial \phi_\alpha} \epsilon_\alpha \phi_\alpha\right) P[\phi^*, \phi; t] + \left(\sum_\alpha \frac{\partial}{\partial \phi_\alpha^*} \epsilon_\alpha \phi_\alpha^*\right) P[\phi^*, \phi; t]. \quad (4.5.46)$$

This is the Fokker–Planck equation for an ideal Bose gas. We expect this result to be related to the fact that in the operator formalism the occupation numbers $N_\alpha(t)$ are independent of time.

It is not difficult to show that any functional that only depends on the amplitudes $|\phi_\alpha|^2$ is a solution of the stationary solutions of the Fokker–Planck equation. As it stands, the Fokker–Planck equation, therefore, does not lead to a unique equilibrium distribution. This is not surprising because for an isolated, ideal Bose gas there is no mechanism for redistributing the particles over the various energy levels and thus for relaxation towards equilibrium. However, the situation changes when we allow the bosons in the trap to tunnel back and forth to a reservoir at a temperature T. The corrections to the Fokker–Planck equation that are required to describe the physics in this case are considered next. However, to determine these corrections in the most convenient way, we have to generalize the theory slightly because with the probability distribution $P[\phi^*, \phi; t]$ we are only able to study *spatial*, but not temporal correlations in the Bose gas.

To study these as well, we construct, as usual, a generating functional $\mathcal{Z}[J, J^*]$ for the time-ordered correlation functions. To this aim, we introduce the probability distribution $P_J[\phi^*, \phi; t]$ for a Bose gas in the presence of the external currents $J(\mathbf{x}, t)$ and $J^*(\mathbf{x}, t)$, by adding to the Hamiltonian the terms

$$-\hbar \int d\mathbf{x} \, (\hat{\varphi}(\mathbf{x}, t) J^*(\mathbf{x}, t) + J(\mathbf{x}, t) \hat{\varphi}^\dagger(\mathbf{x}, t)) = -\hbar \sum_\alpha (\hat{\varphi}_\alpha(t) J_\alpha^*(t) + J_\alpha(t) \hat{\varphi}_\alpha^\dagger(t))$$

and integrate this expression over $\phi(\mathbf{x})$ to obtain the desired generating functional:

$$\mathcal{Z}_\beta[J, J^*] = \int \prod_\alpha \frac{d\phi_\alpha^* d\phi_\alpha}{2\pi i} P_J[\phi^*, \phi; t] = \int \mathcal{D}\varphi^* \mathcal{D}\varphi \, \exp\left\{\frac{i}{\hbar} S[\varphi^*, \varphi]\right\}$$
$$\times \exp\left\{i \int_{\mathcal{C}^\infty} dt \int d\mathbf{x} \, (\varphi(\mathbf{x}, t) J^*(\mathbf{x}, t) + J(\mathbf{x}, t) \varphi^*(\mathbf{x}, t))\right\}. \quad (4.5.47)$$

Note that $\mathcal{Z}_\beta[J, J^*]$ is indeed independent of the time t because of the fact that $P_J[\phi^*, \phi; t]$ is a probability distribution (cf (4.5.31) and (4.5.32)) and thus properly normalized. We are therefore allowed to deform the contour $\mathcal{C}^t$ to any closed contour that runs through t_0. Since we are, in principle, interested in all times $t \geq t_0$, the most convenient choice is the contour that runs backward from infinity to t_0 and then forward from t_0 to infinity. This contour is denoted by $\mathcal{C}^\infty$ and also called the Schwinger–Keldysh contour in the following, because in practice there is never any confusion with the more restricted contour $\mathcal{C}^t$ that is required when we consider a probability distribution. Now all time-ordered correlation functions can be obtained by functional differentiation with respect to the currents $J(\mathbf{x}, t)$ and $J^*(\mathbf{x}, t)$. We have, for instance, that

$$\text{Tr}[\hat{\rho}_C(t_0) \hat{\varphi}(\mathbf{x}, t)] = \frac{1}{i} \frac{\delta}{\delta J^*(\mathbf{x}, t)} \mathcal{Z}_\beta[J, J^*]\bigg|_{J, J^*=0} \quad (4.5.48)$$

and similarly that

$$\text{Tr}[\hat{\rho}_C(t_0) T_{\mathcal{C}^\infty}(\hat{\varphi}(\mathbf{x}, t) \hat{\varphi}^\dagger(\mathbf{x}', t'))] = \frac{1}{i^2} \frac{\delta^2}{\delta J^*(\mathbf{x}, t) \delta J(\mathbf{x}', t')} \mathcal{Z}_\beta[J, J^*]\bigg|_{J, J^*=0}. \quad (4.5.49)$$

268 *Path integrals in statistical physics*

Note that the times t and t' always have to be larger than or equal to t_0 for these identities to be valid.

◇ **Bosonic quantum particles coupled to a reservoir**

As a reservoir, we take an ideal gas of N bosons in a box with volume L^3. The states in this box are labeled by the momentum $\hbar \boldsymbol{k}$ and equal to $\chi_{\boldsymbol{k}}(\boldsymbol{x}) = \exp\{i\boldsymbol{k}\cdot\boldsymbol{x}\}/L^{3/2}$. They are created and annihilated by second-quantized fields $\hat{\Psi}_{\boldsymbol{k}}^{\dagger}, \hat{\Psi}_{\boldsymbol{k}}$ and have an energy $\epsilon(\boldsymbol{k}) = \hbar^2 \boldsymbol{k}^2/2m + \Delta V^{\mathrm{ex}}$, where ΔV^{ex} accounts for a possible bias between the potential energies of a particle in the centre of the trap and a particle in the reservoir. The reservoir is also taken to be sufficiently large that it can be treated in the thermodynamic limit and is in equilibrium, with the temperature T and the chemical potential μ, for the times $t < t_0$. At t_0 it is brought into contact with the Bose gas under consideration by means of the tunnel Hamiltonian

$$\hat{H}^{\mathrm{int}} = \frac{1}{L^{3/2}} \sum_{\alpha} \sum_{\boldsymbol{k}} (t_{\alpha}(\boldsymbol{k})\hat{\varphi}_{\alpha}(t)\hat{\Psi}_{\boldsymbol{k}}^{\dagger}(t) + t_{\alpha}^{*}(\boldsymbol{k})\hat{\Psi}_{\boldsymbol{k}}(t)\hat{\varphi}_{\alpha}^{\dagger}(t)) \tag{4.5.50}$$

with complex tunneling matrix elements $t_{\alpha}(\boldsymbol{k})$ that, for simplicity, are assumed to be almost constant for momenta $\hbar \boldsymbol{k}$ smaller than some fixed momentum $\hbar k_c$, but to vanish rapidly for momenta larger than this cutoff.

To study the evolution of the combined system for times $t \geq t_0$, we thus have to deal with the action

$$\begin{aligned}
S[\varphi^{*}, \varphi; \Psi^{*}, \Psi] = & -\frac{1}{L^{3/2}} \sum_{\alpha} \sum_{\boldsymbol{k}} \int_{C^{\infty}} dt\, (t_{\alpha}(\boldsymbol{k})\varphi_{\alpha}(t)\Psi_{\boldsymbol{k}}^{*}(t) + t_{\alpha}^{*}(\boldsymbol{k})\Psi_{\boldsymbol{k}}(t)\varphi_{\alpha}^{*}(t)) \\
& + \sum_{\alpha} \int_{C^{\infty}} dt\, \varphi_{\alpha}^{*}(t) \left(i\hbar \frac{\partial}{\partial t} - \epsilon_{\alpha} + \mu \right) \varphi_{\alpha}(t) \\
& + \sum_{\boldsymbol{k}} \int_{C^{\infty}} dt\, \Psi_{\boldsymbol{k}}^{*}(t) \left(i\hbar \frac{\partial}{\partial t} - \epsilon(\boldsymbol{k}) + \mu \right) \Psi_{\boldsymbol{k}}(t)
\end{aligned} \tag{4.5.51}$$

if we measure all energies relative to the chemical potential. Let us also introduce the complex field $\Psi(\boldsymbol{x}, t) = \sum_{\boldsymbol{k}} \Psi_{\boldsymbol{k}}(t)\chi_{\boldsymbol{k}}(\boldsymbol{x})$ for the degrees of freedom of the reservoir. However, we are only interested in the evolution of the Bose gas in the trap and therefore only in the time-ordered correlation functions of this part of the system. The corresponding generating functional

$$\begin{aligned}
\mathcal{Z}_{\beta}[J, J^{*}] = & \int \mathcal{D}\varphi^{*}\, \mathcal{D}\varphi \int \mathcal{D}\Psi^{*}\, \mathcal{D}\Psi\, \exp\left\{\frac{i}{\hbar} S[\varphi^{*}, \varphi; \Psi^{*}, \Psi]\right\} \\
& \times \exp\left\{i \int_{C^{\infty}} dt \int d\boldsymbol{x}\, (\varphi(\boldsymbol{x}, t)J^{*}(\boldsymbol{x}, t) + J(\boldsymbol{x}, t)\varphi^{*}(\boldsymbol{x}, t))\right\}
\end{aligned} \tag{4.5.52}$$

is of the same form as the functional integral in (4.5.47), but now with an effective action, that is defined by

$$\exp\left\{\frac{i}{\hbar} S^{(\mathrm{eff})}[\varphi^{*}, \varphi]\right\} \equiv \int \mathcal{D}\Psi^{*}\, \mathcal{D}\Psi\, \exp\left\{\frac{i}{\hbar} S[\varphi^{*}, \varphi; \Psi^{*}, \Psi]\right\}. \tag{4.5.53}$$

Hence, our next task is to integrate out the field $\Psi(\boldsymbol{x}, t)$, which can be done exactly because it only requires a Gaussian integration (see problem 4.5.5, page 279):

$$\begin{aligned}
S^{(\mathrm{eff})}[\varphi^{*}, \varphi] = & \sum_{\alpha, \alpha'} \int_{C^{\infty}} dt \int_{C^{\infty}} dt'\, \varphi_{\alpha}^{*}(t) \\
& \times \left\{ \left(i\hbar \frac{\partial}{\partial t} - \epsilon_{\alpha} + \mu \right) \delta_{\alpha, \alpha'} \delta(t, t') - \hbar \Sigma_{\alpha, \alpha'}(t, t') \right\} \varphi_{\alpha'}(t')
\end{aligned} \tag{4.5.54}$$

with the self-energy $\Sigma_{\alpha,\alpha'}(t,t')$ of the form

$$\hbar\Sigma_{\alpha,\alpha'}(t,t') = \frac{1}{\hbar L^3} \sum_{k} t_\alpha^*(k) G(k;t,t') t_{\alpha'}(k) \tag{4.5.55}$$

where $G(k;t,t')$ is the Green function obeying

$$\left(i\hbar\frac{\partial}{\partial t} - \epsilon(k) + \mu\right) G(k;t,t') = \hbar\delta_C(t,t'). \tag{4.5.56}$$

Here we have introduced the δ_C function on the Schwinger–Keldysh contour defined by

$$\int_{C^\infty} dt' \, \delta_C(t,t') = 1.$$

In order to solve equation (4.5.56), we need to know the appropriate boundary conditions at $t = t'$. To derive them, we note that $G(k;t,t')$ describes the properties of the reservoir and, similarly to (4.5.49), it can be expressed via the density matrix $\hat{\rho}_{C,R}$ of the reservoir as follows:

$$iG(k;t,t') = \mathrm{Tr}[\hat{\rho}_{C,R}(t_0) T_{C^\infty}(\hat{\Psi}_k(t) \hat{\Psi}_k^\dagger(t'))]. \tag{4.5.57}$$

From this identification, we see that the desired solution fulfilling the appropriate boundary conditions

$$G(k;t,t')|_{t=t'} = -i(N^{(B)}(k) + \tfrac{1}{2})$$

is apparently

$$G(k;t,t') = -ie^{-i(\epsilon(k)-\mu)(t-t')/\hbar}\{\tilde{\theta}(t-t')(1 + N^{(B)}(k)) + \tilde{\theta}(t'-t)N^{(B)}(k)\} \tag{4.5.58}$$

with $N^{(B)}(k) = 1/(e^{\beta(\epsilon(k)-\mu)} - 1)$ being the appropriate Bose distribution function and $\beta = 1/(k_B T)$. It is convenient to decompose the Green function into its pieces $G^{>}(k;t-t')$ and $G^{<}(k;t-t')$ by means of

$$G(k;t,t') = \tilde{\theta}(t-t') G^{>}(k;t-t') + \tilde{\theta}(t'-t) G^{<}(k;t-t'). \tag{4.5.59}$$

Due to the fact that we are always dealing with time-ordered correlation functions, such a decomposition turns out to be a generic feature of all the functions on the Schwinger–Keldysh contour that we will encounter in the following.

◇ **Properties of the effective action for an ideal gas in a reservoir**

Having obtained the Green function of the reservoir, we can now return to our discussion of the effective action $S^{(\mathrm{eff})}[\varphi^*, \varphi]$ for the Bose gas in the trap. We again perform the transformation $\varphi_\alpha(t_\pm) = \phi_\alpha(t) \pm \xi_\alpha(t)/2$ to explicitly separate the (semi)classical dynamics from the effect of fluctuations. This leads to the following expression for the effective action

$$S^{(\mathrm{eff})}[\phi^*, \phi; \xi^*, \xi] = \sum_\alpha \int dt \left\{\phi_\alpha^*(t)\left(i\hbar\frac{\partial}{\partial t} - \epsilon_\alpha + \mu\right)\xi_\alpha(t) + \xi_\alpha^*(t)\left(i\hbar\frac{\partial}{\partial t} - \epsilon_\alpha + \mu\right)\phi_\alpha(t)\right\}$$

$$- \sum_{\alpha,\alpha'} \int dt \int dt' (\phi_\alpha^*(t)\hbar\Sigma_{\alpha,\alpha'}^{(-)}(t-t')\xi_{\alpha'}(t') + \xi_\alpha^*(t)\hbar\Sigma_{\alpha,\alpha'}^{(+)}(t-t')\phi_{\alpha'}(t'))$$

$$- \tfrac{1}{2}\sum_{\alpha,\alpha'} \int dt \int dt' \, \xi_\alpha^*(t)\hbar\Sigma_{\alpha,\alpha'}^{K}(t-t')\xi_{\alpha'}(t') \tag{4.5.60}$$

where we have introduced the retarded and advanced components of the self-energy (4.5.55) defined by

$$\Sigma^{(\pm)}_{\alpha,\alpha'}(t-t') = \pm\tilde{\theta}(\pm(t-t'))(\Sigma^{>}_{\alpha,\alpha'}(t-t') - \Sigma^{<}_{\alpha,\alpha'}(t-t')) \tag{4.5.61}$$

and the so-called *Keldysh component*

$$\Sigma^{K}_{\alpha,\alpha'}(t-t') = \Sigma^{>}_{\alpha,\alpha'}(t-t') + \Sigma^{<}_{\alpha,\alpha'}(t-t') \tag{4.5.62}$$

which is associated with the part which is quadratic in the fluctuations.

In order to clarify the physical content of these various components of the self-energy, it is useful to write the factor

$$\exp\left\{-\frac{i}{2}\sum_{\alpha,\alpha'}\int_{t_0}dt\int_{t_0}dt'\,\xi^*_\alpha(t)\Sigma^{K}_{\alpha,\alpha'}(t-t')\xi_{\alpha'}(t')\right\}$$

in the integrand of the functional integral $\int \mathcal{D}\phi^* \mathcal{D}\phi \mathcal{D}\xi^* \mathcal{D}\xi \exp\{iS^{(\text{eff})}[\phi^*,\phi;\xi^*,\xi]/\hbar\}$ as a Gaussian integral over a complex field $\eta(\boldsymbol{x},t)$. Then the total effective action becomes:

$$\begin{aligned}
&S^{(\text{eff})}[\phi^*,\phi;\xi^*,\xi;\eta^*,\eta] \\
&= \sum_{\alpha,\alpha'}\int_{t_0}dt\int_{t_0}dt'\,\phi^*_\alpha(t)\left\{\left(i\hbar\frac{\partial}{\partial t}-\epsilon_\alpha+\mu-\eta^*_\alpha(t)\right)\delta_{\alpha,\alpha'}\delta(t-t')-\hbar\Sigma^{(-)}_{\alpha,\alpha'}(t-t')\right\}\xi_{\alpha'}(t') \\
&+ \sum_{\alpha,\alpha'}\int_{t_0}dt\int_{t_0}dt'\,\xi^*_\alpha(t)\left\{\left(i\hbar\frac{\partial}{\partial t}-\epsilon_\alpha+\mu-\eta_\alpha(t)\right)\delta_{\alpha,\alpha'}\delta(t-t')-\hbar\Sigma^{(+)}_{\alpha,\alpha'}(t-t')\right\}\phi_{\alpha'}(t') \\
&+ 2\sum_{\alpha,\alpha'}\int_{t_0}dt\int_{t_0}dt'\,\eta^*_\alpha(t)(\hbar\Sigma^K)^{-1}_{\alpha,\alpha'}(t-t')\eta_{\alpha'}(t')
\end{aligned} \tag{4.5.63}$$

and is thus linear in $\xi_\alpha(t)$ and $\xi^*_\alpha(t)$. Integrating over these fluctuations, we conclude from this action that the field $\phi(\boldsymbol{x},t)$ is constrained to obey the *Langevin equations*

$$i\hbar\frac{\partial}{\partial t}\phi_\alpha(t) = (\epsilon_\alpha-\mu)\phi_\alpha(t) + \sum_{\alpha'}\int_{t_0}^{\infty}dt'\,\hbar\Sigma^{(+)}_{\alpha,\alpha'}(t-t')\phi_{\alpha'}(t') + \eta_\alpha(t) \tag{4.5.64}$$

and

$$-i\hbar\frac{\partial}{\partial t}\phi^*_\alpha(t) = (\epsilon_\alpha-\mu)\phi^*_\alpha(t) + \sum_{\alpha'}\int_{t_0}^{\infty}dt'\,\phi^*_{\alpha'}(t')\hbar\Sigma^{(-)}_{\alpha',\alpha}(t'-t) + \eta^*_\alpha(t) \tag{4.5.65}$$

with the Gaussian noise terms $\eta_\alpha(t)$ and $\eta^*_\alpha(t)$ which, from the last term in the right-hand side of (4.5.63), are seen to have the time correlations

$$\langle \eta^*_\alpha(t)\eta_{\alpha'}(t')\rangle = \frac{i\hbar^2}{2}\Sigma^{K}_{\alpha,\alpha'}(t-t') = \frac{1}{2}\int\frac{d\boldsymbol{k}}{(2\pi)^3}(1+2N^{(\text{B})}(\boldsymbol{k}))t^*_\alpha(\boldsymbol{k})e^{-i(\epsilon(\boldsymbol{k})-\mu)(t-t')/\hbar}t_{\alpha'}(\boldsymbol{k}) \tag{4.5.66}$$

in the thermodynamic (infinite-volume) limit.

◇ Long-time behaviour of the ideal gas coupled to a reservoir

Let us consider the limit $t_0 \to -\infty$, which physically means that we neglect the initial transients that are due to the precise way in which the contact between the trap and the reservoir is made, and focus on the 'universal' dynamics which is independent of these details. In addition, at long times, the dynamics of the gas is expected to be sufficiently slow and we can neglect the memory effects altogether. Also, we consider the case of a reservoir that is so weakly coupled to the gas in the trap that we can treat the coupling in the second order of perturbation theory. As a result, we can also neglect the non-diagonal elements of the self-energies. Eventually, we are then allowed to put

$$\Sigma_{\alpha,\alpha'}^{(\pm),K}(t-t') = \Sigma_{\alpha}^{(\pm),K}\delta_{\alpha,\alpha'}\delta(t-t'). \tag{4.5.67}$$

With these simplifications, the Langevin equations (4.5.64) and (4.5.65) become

$$i\hbar\frac{\partial}{\partial t}\phi_\alpha(t) = (\epsilon_\alpha + \hbar\Sigma_\alpha^{(+)} - \mu)\phi_\alpha(t) + \eta_\alpha(t) \tag{4.5.68}$$

and the complex conjugate equation for $\phi_\alpha^*(t)$. The retarded self-energy in this equation is given by

$$\hbar\Sigma_\alpha^{(+)} = \int \frac{d\mathbf{k}}{(2\pi)^3} t_\alpha^*(\mathbf{k}) \frac{1}{\epsilon_\alpha + i0 - \epsilon(\mathbf{k})} t_\alpha(\mathbf{k}) \tag{4.5.69}$$

and using the well-known formula for the distributions (generalized functions), i.e.

$$\frac{1}{\omega - \omega' \pm i0} = \frac{\mathcal{P}}{\omega - \omega'} \mp i\pi\delta(\omega - \omega')$$

($\mathcal{P}$ denotes the principal value part of an integral) the result can be decomposed into real and imaginary parts:

$$S_\alpha \equiv \mathrm{Re}\,\Sigma_\alpha^+ = \int \frac{d\mathbf{k}}{(2\pi)^3} t_\alpha^*(\mathbf{k}) \frac{\mathcal{P}}{\epsilon_\alpha - \epsilon(\mathbf{k})} t_\alpha(\mathbf{k}) \tag{4.5.70}$$

and

$$R_\alpha \equiv -\mathrm{Im}\,\Sigma_\alpha^+ = \pi \int \frac{d\mathbf{k}}{(2\pi)^3} \delta(\epsilon_\alpha - \epsilon(\mathbf{k}))|t_\alpha(\mathbf{k})|^2. \tag{4.5.71}$$

The interpretation of these results is quite obvious if we consider the average of the Langevin equation, i.e.

$$i\hbar\frac{\partial}{\partial t}\langle\phi_\alpha\rangle(t) = (\epsilon_\alpha + S_\alpha - iR_\alpha - \mu)\langle\phi_\alpha\rangle(t) \tag{4.5.72}$$

which is solved by

$$\langle\phi_\alpha\rangle(t) = \langle\phi_\alpha\rangle(0)e^{-i(\epsilon_\alpha+S_\alpha-\mu)t/\hbar}e^{-R_\alpha t/\hbar}. \tag{4.5.73}$$

Hence, the real part of the retarded self-energy S_α represents the shift in the energy of state $\chi_\alpha(\mathbf{x})$, due to the coupling with the reservoir, while the fact that $|\langle\phi_\alpha\rangle(t)|^2 = |\langle\phi_\alpha\rangle(0)|^2 e^{-2R_\alpha t/\hbar} \equiv |\langle\phi_\alpha\rangle(0)|^2 e^{-\Gamma_\alpha t}$ shows that the average rate of decay Γ_α of the state $\chi_\alpha(\mathbf{x})$ is equal to $2R_\alpha/\hbar$.

Next, we are going to determine $i\hbar\partial\langle|\phi_\alpha|^2\rangle(t)/\partial t$. To do so we first formally solve the Langevin equation by

$$\phi_\alpha(t) = e^{-i(\epsilon_\alpha+\hbar\Sigma_\alpha^{(+)}-\mu)t/\hbar}\left\{\phi_\alpha(0) - \frac{i}{\hbar}\int_0^t dt'\,\eta(t')e^{i(\epsilon_\alpha+\hbar\Sigma_\alpha^{(+)}-\mu)t'/\hbar}\right\}. \tag{4.5.74}$$

Multiplying this with the complex conjugate expression and taking the average, we obtain:

$$\langle|\phi_\alpha|^2\rangle(t) = e^{-2R_\alpha t/\hbar}\left\{\langle|\phi_\alpha|^2\rangle(0) + \frac{i}{2}\Sigma_\alpha^K \int_0^t dt'\,e^{2R_\alpha t'/\hbar}\right\} \tag{4.5.75}$$

which shows that

$$i\hbar\frac{\partial}{\partial t}\langle|\phi_\alpha|^2\rangle(t) = -2iR_\alpha\langle|\phi_\alpha|^2\rangle(t) - \frac{1}{2}\hbar\Sigma_\alpha^K. \tag{4.5.76}$$

On the other hand, the Keldysh component of the self-energy is given by

$$\hbar\Sigma_\alpha^K = -2\pi i \int \frac{d\mathbf{k}}{(2\pi)^3}(1 + 2N^{(B)}(\mathbf{k}))\delta(\epsilon_\alpha - \epsilon(\mathbf{k}))|t_\alpha(\mathbf{k})|^2 \qquad (4.5.77)$$

and therefore obeys

$$\hbar\Sigma_\alpha^K = -2i(1 + 2N_\alpha^{(B)})R_\alpha \qquad (4.5.78)$$

with $N_\alpha^{(B)} = (e^{\beta(\epsilon_\alpha - \mu)} - 1)^{-1}$ being the Bose distribution function. Relation (4.5.78) is a particular case of the important *fluctuation–dissipation theorem* in quantum statistics, because it relates the strength of the fluctuations determined by $\hbar\Sigma_\alpha^K$, to the amount of dissipation that is given by R_α. The fluctuation–dissipation theorem ensures that the gas relaxes to thermal equilibrium. This can be seen from (4.5.76), because substituting the fluctuation–dissipation theorem leads to

$$i\hbar\frac{\partial}{\partial t}\langle|\phi_\alpha|^2\rangle(t) = -2iR_\alpha\langle|\phi_\alpha|^2\rangle(t) + iR_\alpha(1 + 2N_\alpha^{(B)}) \qquad (4.5.79)$$

which tells us that at equilibrium (that is, if $\partial\langle|\phi_\alpha|^2\rangle(t)/\partial t = 0$) we have $\langle|\phi_\alpha|^2\rangle = N_\alpha^{(B)} + \frac{1}{2}$, as it should be. Substituting this identity in (4.5.79), we indeed obtain the correct rate equation for the average occupation numbers

$$\frac{\partial}{\partial t}N_\alpha(t) = -\Gamma_\alpha N_\alpha(t) + \Gamma_\alpha N_\alpha^{(B)} = -\Gamma_\alpha N_\alpha(t)(1 + N^{(B)}(\mathbf{k})) + \Gamma_\alpha(1 + N_\alpha(t))N^{(B)}(\mathbf{k}) \qquad (4.5.80)$$

that might justly be called the *quantum Boltzmann equation* for the gas, because the right-hand side contains precisely the rates for scattering into and out of the reservoir.

From the time-evolution equations for $\langle\phi_\alpha\rangle(t)$ and $\langle|\phi_\alpha|^2\rangle(t)$ we can read off the corresponding Fokker–Planck equation for the ideal gas coupled to the reservoir (cf also problem 4.5.1, page 277 for a direct derivation of the Fokker–Planck equation):

$$i\hbar\frac{\partial}{\partial t}P[\phi^*,\phi;t] = -\left(\sum_\alpha \frac{\partial}{\partial \phi_\alpha}(\epsilon_\alpha + \hbar\Sigma_\alpha^{(+)} - \mu)\phi_\alpha\right)P[\phi^*,\phi;t]$$
$$+ \left(\sum_\alpha \frac{\partial}{\partial \phi_\alpha^*}(\epsilon_\alpha + \hbar\Sigma_\alpha^{(-)} - \mu)\phi_\alpha^*\right)P[\phi^*,\phi;t]$$
$$- \left(\frac{1}{2}\sum_\alpha \frac{\partial^2}{\partial\phi_\alpha^*\partial\phi_\alpha}\hbar\Sigma_\alpha^K\right)P[\phi^*,\phi;t]. \qquad (4.5.81)$$

Using again the fluctuation–dissipation theorem, it is not difficult to show that the stationary solution of this Fokker–Planck equation is

$$P[\phi^*,\phi;\infty] = \prod_\alpha \frac{1}{N_\alpha^{(B)} + \frac{1}{2}}\exp\left\{-\frac{1}{N_\alpha^{(B)} + \frac{1}{2}}|\phi_\alpha|^2\right\}. \qquad (4.5.82)$$

Summarizing, the dynamics of a gas coupled to a reservoir is solved by

$$\langle\phi_\alpha\rangle(t) = \langle\phi_\alpha\rangle(0)e^{-i(\epsilon_\alpha' - \mu)t/\hbar}e^{-\Gamma_\alpha t/2} \qquad (4.5.83)$$

and

$$N_\alpha(t) = N_\alpha(0)e^{-\Gamma_\alpha t} + N_\alpha^{(B)}(1 - e^{-\Gamma_\alpha t}). \qquad (4.5.84)$$

In the limit $t \to \infty$, the average of the annihilation operators $\langle \phi_\alpha \rangle(t)$ thus always vanishes, but the average occupation numbers $N_\alpha(t)$ relax to the equilibrium distribution $N_\alpha = (e^{\beta(\epsilon_\alpha - \mu)} - 1)^{-1}$. Although this appears to be an immediately obvious result, its importance stems from the fact that it is also true if we tune the potential energy bias ΔV^{ex}, such that at low temperatures the ground state $\chi'_0(x)$ acquires a macroscopic occupation, i.e. $N_0 \gg 1$. The gas therefore never shows a spontaneous breaking of the $U(1)$ symmetry, in agreement with the notion that we are essentially dealing with an ideal Bose gas in the grand canonical ensemble. The reason for the absence of spontaneous symmetry-breaking can also be understood from our stationary solution of the Fokker–Planck equation in (4.5.82), which shows that the probability distribution for $|\phi_0|$ is proportional to the Boltzmann factor $e^{-\beta(\epsilon_0 - \mu)|\phi_0|^2}$ in the degenerate regime of interest and the corresponding free energy $F(|\phi_0|) = (\epsilon_0 - \mu)|\phi_0|^2$ never shows an instability towards the formation of a non-zero average of $|\phi_0|$, due to the fact that $\epsilon_0 - \mu$ can never become less then zero. Once we introduce interactions between the atoms in the gas, this picture changes completely.

◇ A comment on the non-diagonal parts of the self-energies

Before we start a short discussion of a weakly interacting Bose gas, it is necessary to make a final remark about the effect of the non-diagonal elements of the self-energies (recall that we used the simplified diagonal ansatz (4.5.67)). Physically, including these non-diagonal elements accounts for the change in the wavefunctions $\chi_\alpha(x)$, due to the interaction with the reservoir. This can clearly be neglected if the coupling with the reservoir is sufficiently weak or, more precisely, if $|\hbar \Sigma^{(+)}_{\alpha', \alpha}(\epsilon_\alpha + S_\alpha - \mu)|$ is much smaller than the energy splitting $|\epsilon_{\alpha'} - \epsilon_\alpha|$ between the states of the gas. A strong-coupling situation can also be studied and the main difference is that we need to expand our various fields not in terms of the eigenstates $\chi_\alpha(x)$ but in the eigenstates $\chi'_\alpha(x)$ of the non-local Schrödinger equation

$$\epsilon'_\alpha \chi'_\alpha(x) = \left(-\frac{\hbar^2 \nabla^2}{2m} + V^{\text{ex}}(x)\right)\chi'_\alpha(x) + \int dx' \, \text{Re}[\hbar \Sigma^{(+)}(x, x'; \epsilon'_\alpha - \mu)]\chi'_\alpha(x') \tag{4.5.85}$$

where ϵ'_α are the new eigenvalues and $\hbar \Sigma^{(+)}(x, x'; \epsilon) = \sum_{\alpha, \alpha'} \chi_\alpha(x) \hbar \Sigma^{(+)}_{\alpha, \alpha'}(\epsilon) \chi^*_{\alpha'}(x')$. In this new basis, the non-diagonal elements of the self-energies can now be neglected and we find essentially the same results as before. We only need to replace $\epsilon_\alpha + S_\alpha$ by ϵ'_α and $t_\alpha(k)$ by

$$\sum_{\alpha'} \left(\int dx \, \chi'_\alpha(x') \chi^*_{\alpha'}(x)\right) t_{\alpha'}(k).$$

Neglecting the non-diagonal elements in this basis only requires that the real part of the retarded self-energy is much larger than its imaginary part, which is always fulfilled in the low-energy regime.

◇ Bose–Einstein condensation of an interacting gas

As well as the majority of non-trivial realistic physical models, the Bose gas with interaction cannot be solved exactly and we should find appropriate approximation methods. The usual perturbation theory is not suitable for the consideration of phase transitions. Thus, further investigations amount essentially to the summation of an appropriate infinite series of Feynman diagrams of the perturbation theory, or, equivalently, to approximate solution of integro-differential equations of the type (4.5.85). We shall present, for completeness, a very brief sketch of such an analysis in the rest of this section. We shall consider only a homogeneous Bose gas: in this case, we are allowed to take the thermodynamic limit in which the Bose–Einstein condensation becomes a true second-order phase transition. We are then, in fact, studying the dynamics of a spontaneous symmetry-breaking under the most ideal circumstances.

As we have mentioned earlier, the energies ϵ'_α and the corresponding eigenstates $\chi'_\alpha(x)$ for an ideal gas can be determined from a non-local Schrödinger equation of the type (4.5.85), once we know the retarded self-energy $\hbar\Sigma^{(+)}(x, x'; t - t')$. The same is true for a homogeneous atomic Bose gas with interactions, only with the exception that the non-zero self-energy is now due to the interatomic interactions and not to the presence of a reservoir. In a sense, an interacting gas also plays the role of its own reservoir. Moreover, the homogeneity of the gas leads to an important simplification, because translational invariance requires that the retarded self-energy is only a function of the relative distance $x - x'$. Therefore, the Schrödinger equation in (4.5.85) is solved by $\chi'_k(x) = \chi_k(x)$ and

$$\epsilon'(k) = \epsilon(k) + \text{Re}[\hbar\Sigma^{(+)}(k; \epsilon'(k) - \mu)] \tag{4.5.86}$$

where $\epsilon(k) = \hbar^2 k^2/2m$ again and

$$\Sigma^{(+)}(x - x'; t - t') = \int \frac{dk}{(2\pi)^3} \int \frac{d\epsilon}{2\pi\hbar} \Sigma^{(+)}(k; \epsilon) e^{i(k \cdot (x-x') - \epsilon(t-t')/\hbar)}. \tag{4.5.87}$$

To solve (4.5.86), we need of course an expression for the retarded self-energy of a weakly interacting Bose gas, which follows once we know the full self-energy $\hbar\Sigma(k; t, t')$ defined on the Schwinger–Keldysh contour C^∞. Unfortunately, even for a dilute system this quantity cannot be calculated exactly and some approximation is called for. The approximation that we will make here is the so-called many-body T-matrix approximation. The main motivation for this approximation is that due to the smallness of the gas parameter $(na^3)^{1/2}$ (cf (4.5.9)), it is very unlikely for three or more particles to be within the range of the interaction and we only need to account for all possible two-body processes taking place in the gas.

Given the effective interaction $V(k, k', K; t, t')$ for the scattering of two atoms which at the time t' have the momenta $\hbar(K/2 \pm k')$ and at the time t, the momenta $\hbar(K/2 \pm k)$, respectively, the exact self-energy obeys a Hartree–Fock-like relation

$$\hbar\Sigma(k; t, t') = 2i \int \frac{dk'}{(2\pi)^3} V(k - k', k - k', k + k'; t, t') G(k'; t', t) \tag{4.5.88}$$

where the Green function equals again (4.5.58), but with $\epsilon(k)$, replaced by $\epsilon'(k)$ to make the theory self-consistent. Since we are dealing with bosons, the effective interaction is a sum of a direct and an exchange term and can be written as $V(k, k', K; t, t') = (T(k, k', K; t, t') + T(-k, k', K; t, t'))/2$ in terms of the many-body T-matrix that obeys the Lippmann–Schwinger equation (see, e.g., Taylor (1972))

$$T(k, k', K; t, t') = V(k - k')\delta(t, t') + \frac{i}{\hbar} \int_{C^\infty} dt'' \int \frac{dk''}{(2\pi)^3} V(k - k'') \\ \times G(K/2 + k''; t, t'') G(K/2 - k''; t, t'') T(k'', k', K; t'', t') \tag{4.5.89}$$

with $V(k - k')$ being the Fourier transform of the interatomic interaction potential. By iterating this equation, we see that the many-body T-matrix indeed sums all possible collisions between two particles. Moreover, the Green functions $G(K/2 \pm k''; t, t'')$ describe the propagation of an atom with momentum $\hbar(K/2 \pm k'')$ from the time t'' to the time t in the gas. Therefore, we also see that the many-body T-matrix incorporates the effect of the surrounding gaseous medium on the propagation of the atoms between two collisions. A detailed consideration of this equation (Stoof 1999) allows us to conclude that for the thermal momenta, i.e. for momenta $\hbar k$ which are of the order of $\hbar/\lambda_B$ (λ_B is defined in (4.1.25)), we have

$$\epsilon'(k; t) = \epsilon(k) + \frac{8\pi n(t) a \hbar^2}{m} \tag{4.5.90}$$

(a is the radius of particle interactions). However, this conclusion is not valid for momenta $\hbar\boldsymbol{k}$ that are much smaller than the thermal momenta, because in that case the energy denominator in the integrant favors the small momenta where the occupation numbers are especially large in the degenerate regime. For such momenta and, in particular, for the momenta obeying $\hbar k < \hbar\sqrt{8\pi na} \ll \hbar/\lambda_B$, it has been found in a good approximation that

$$\epsilon'(\boldsymbol{k};t) = \epsilon(\boldsymbol{k}) + \frac{8\pi n(t)a\hbar^2}{m} - 2\int \frac{d^3k'}{(2\pi)^3} \int \frac{d^3k''}{(2\pi)^3} |V^{(+)}(\boldsymbol{0},\boldsymbol{k}',\boldsymbol{k}'')|^2 N(\boldsymbol{k}';t) N(\boldsymbol{k}'';t) \frac{\mathcal{P}}{\hbar^2 \boldsymbol{k}'\cdot\boldsymbol{k}''/m} \quad (4.5.91)$$

(the $V^{(+)}$ component of V is defined similarly to (4.5.62)).

In principle, equations (4.5.90) and (4.5.91) already show clearly the tendency of the gas to become unstable towards Bose–Einstein condensation, because the energy of the one-particle *ground* state is shifted less upwards compared to the one-particle states with *thermal* energies. To show when the gas is actually unstable, we need to compare the energy of the zero-momentum state with the instantaneous chemical potential which is found to be

$$\mu(t) \simeq \frac{8\pi n(t)a\hbar^2}{m} + \mu_0(t) \quad (4.5.92)$$

where the time dependence of the ideal gas chemical potential $\mu_0(t) \equiv \mu_0(n(t)$ and $T(t))$ is related to the precise path in the density–temperature plane that is followed during the cooling process. An instability therefore occurs once the quantity

$$\epsilon'(\boldsymbol{0};t) - \mu(t) \simeq -\mu_0(t) - 2\int \frac{d\boldsymbol{k}'}{(2\pi)^3} \int \frac{d\boldsymbol{k}''}{(2\pi)^3} |V^{(+)}(\boldsymbol{0},\boldsymbol{k}',\boldsymbol{k}'')|^2 N(\boldsymbol{k}';t) N(\boldsymbol{k}'';t) \frac{\mathcal{P}}{\hbar^2 \boldsymbol{k}'\cdot\boldsymbol{k}''/m}$$

becomes less than zero (cf the comment after (4.5.84)). It can be shown that the gas indeed develops the required instability for Bose–Einstein condensation if $a > 0$ and the temperature is less than a critical temperature T_c.

◇ **Dynamics of the zero-momentum part of the order parameter**

At the semiclassical level, the results obtained earlier show that the effective action for the long-wavelength dynamics of the gas, i.e. for states with momenta $\hbar k < \hbar\sqrt{8\pi na}$, is given by

$$S^{cl}[\phi^*,\phi] = \int dt \Bigg\{ \sum_{\boldsymbol{k}} \phi_{\boldsymbol{k}}^*(t)\left(i\hbar\frac{\partial}{\partial t} - \epsilon'(\boldsymbol{k};t) + \mu(t)\right)\phi_{\boldsymbol{k}}(t)$$
$$- \frac{1}{2V}\sum_{\boldsymbol{k},\boldsymbol{k}',\boldsymbol{K}} T^{(+)}(\boldsymbol{0},\boldsymbol{0},\boldsymbol{0};0) \phi_{\boldsymbol{K}/2+\boldsymbol{k}}^*(t) \phi_{\boldsymbol{K}/2-\boldsymbol{k}}^*(t) \phi_{\boldsymbol{K}/2-\boldsymbol{k}'}(t) \phi_{\boldsymbol{K}/2+\boldsymbol{k}'}(t) \Bigg\} \quad (4.5.93)$$

(the $T^{(+)}$ component of T is defined similarly to (4.5.62)). The field $\phi(\boldsymbol{x},t)$ can be considered as the order parameter of the Bose gas. The effective action $S^{cl}[\phi^*,\phi]$ defines the *Landau–Ginzburg theory* for this order parameter. In particular, we thus find that the dynamics of the zero-momentum part of the order parameter, i.e. the condensate, is determined by

$$S_0^{cl}[\phi_0^*,\phi_0] = \int dt \left\{ \phi_0^*(t)\left(i\hbar\frac{\partial}{\partial t} - \epsilon'(\boldsymbol{0};t) + \mu(t)\right)\phi_0(t) - \frac{T^{(+)}(\boldsymbol{0},\boldsymbol{0},\boldsymbol{0};0)}{2V}|\phi_0(t)|^4 \right\}. \quad (4.5.94)$$

Introducing the density $\rho_0(t)$ of the condensate and its phase θ_0 by means of the relation $\phi_{\mathbf{0}}(t) \equiv \sqrt{\rho_0(t)V}e^{i\theta_0}$, this simply leads to

$$S_0^{\mathrm{cl}}[\rho_0, \mu] = V \int dt \left(-\epsilon'(\mathbf{0}; t) + \mu(t) - \frac{T^{(+)}(\mathbf{0}, \mathbf{0}, \mathbf{0}; 0)}{2}\rho_0(t) \right) \rho_0(t). \tag{4.5.95}$$

Note that in the process of deriving the last equation, we have omitted the topological term $\int dt\, (i\hbar \partial \rho_0(t)/\partial t)$ as it does not affect the equations of motion and is therefore irrelevant at the semiclassical level. This is, in principle, important only when we also want to consider the quantum fluctuations of the condensate. Clearly, this action is minimized by $\rho_0(t) = 0$, if $\epsilon'(\mathbf{0}; t) - \mu(t) > 0$. However, we have a *non-trivial* minimum at

$$\rho_0(t) = -\frac{\epsilon'(\mathbf{0}; t) - \mu(t)}{T^{(+)}(\mathbf{0}, \mathbf{0}, \mathbf{0}; 0)} \tag{4.5.96}$$

if $\epsilon'(\mathbf{0}; t) - \mu(t) < 0$. This result gives the desired evolution of the condensate density after the gas has become unstable, in terms of the time-dependent chemical potential $\mu(t)$. The next task is therefore to determine an equation of motion for the chemical potential.

To achieve this, we have to consider the interactions between the condensed and non-condensed parts of the gas, which have been neglected so far. Substituting $\phi_{\mathbf{k}}(t) = \phi_{\mathbf{0}}(t)\delta_{\mathbf{k},\mathbf{0}} + \phi'_{\mathbf{k}}(t)(1 - \delta_{\mathbf{k},\mathbf{0}})$ into the semiclassical action and integrating over the fluctuations $\phi'_{\mathbf{k}}(t)$ describing the non-condensed part of the gas, we find that the correct semiclassical action for the condensate reads as

$$S^{\mathrm{cl}}[\rho_0, \mu] = S_0^{\mathrm{cl}}[\rho_0, \mu] - i\hbar \ln(Z^{\mathrm{cl}}[\rho_0, \mu]) \tag{4.5.97}$$

where $Z^{\mathrm{cl}}[\rho_0, \mu]$ represents the functional integral over the fluctuations for given evolutions of the condensate density and the chemical potential. Writing $S^{\mathrm{cl}}[\phi_{\mathbf{0}}^* + \phi'^*, \phi_{\mathbf{0}} + \phi']$ as $S_0^{\mathrm{cl}}[\phi_{\mathbf{0}}^*, \phi_{\mathbf{0}}] + S_1^{\mathrm{cl}}[\phi'^*, \phi'; \rho_0, \mu]$, we obtain

$$Z^{\mathrm{cl}}[\rho_0, \mu] = \int \mathcal{D}\phi'^* \mathcal{D}\phi' \, \exp\left\{ \frac{i}{\hbar} S_1^{\mathrm{cl}}[\phi'^*, \phi'; \rho_0, \mu] \right\}. \tag{4.5.98}$$

With this action, the total density of the gas is calculated in the thermodynamic limit as

$$\rho(t) = \frac{1}{V} \frac{\delta S^{\mathrm{cl}}[\rho_0, \mu]}{\delta \mu(t)} = \rho_0(t) + \int \frac{d^3k}{(2\pi)^3} N(\mathbf{k}; t) \tag{4.5.99}$$

where the occupation numbers are found from

$$N(\mathbf{k}; t) = \frac{\int \mathcal{D}\phi'^* \mathcal{D}\phi' \, \phi_{\mathbf{k}}'^*(t)\phi_{\mathbf{k}}'(t) \exp\{iS_1^{\mathrm{cl}}[\phi'^*, \phi'; \rho_0, \mu]/\hbar\}}{Z^{\mathrm{cl}}[\rho_0, \mu]}. \tag{4.5.100}$$

The latter two equations, together with (4.5.96), both give the condensate density $\rho_0(t)$ and the chemical potential $\mu(t)$ as a function of the total density $\rho(t)$ and formally thus completely solve the semiclassical dynamics of the gas.

Determining the occupation numbers $N(\mathbf{k}, t)$ requires solving an interacting quantum field theory, which cannot be done exactly. An approximation is thus called for. Taking only the quadratic terms in $S_1^{\mathrm{cl}}[\phi'^*, \phi'; \rho_0, \mu]$ into account amounts to the *Bogoliubov approximation*. Indeed, the action for the fluctuations then becomes equal to

$$S_{\mathrm{B}}[\phi'^*, \phi'] = \int dt \left\{ \sum_{\mathbf{k} \neq \mathbf{0}} \phi_{\mathbf{k}}'^*(t) \left(i\hbar \frac{\partial}{\partial t} - \epsilon(\mathbf{k}) - \rho_0(t)T^{(+)}(\mathbf{0}, \mathbf{0}, \mathbf{0}; 0) \right) \phi_{\mathbf{k}}'(t) \right.$$
$$\left. - \frac{1}{2} T^{(+)}(\mathbf{0}, \mathbf{0}, \mathbf{0}; 0)\rho_0(t) \sum_{\mathbf{k} \neq \mathbf{0}} (\phi_{\mathbf{k}}'^*(t)\phi_{-\mathbf{k}}'^*(t) + \phi_{-\mathbf{k}}'(t)\phi_{\mathbf{k}}'(t)) \right\} \tag{4.5.101}$$

if we use (4.5.91) to evaluate the energy difference $\epsilon'(\mathbf{k}; t) - \epsilon'(\mathbf{0}; t) = \epsilon(\mathbf{k})$ at the long wavelengths of interest here. The energies of the Bogoliubov quasiparticles, at this level of approximation, thus obey

$$\hbar\omega(\mathbf{k}; t) = \sqrt{\epsilon^2(\mathbf{k}) + 2\rho_0(t)T^{(+)}(\mathbf{0}, \mathbf{0}, \mathbf{0}; 0)\epsilon(\mathbf{k})}. \quad (4.5.102)$$

They are purely real and correspond to the *linear* dispersion relation for small momenta. However, in the Bogoliubov approximation, the quasiparticles are non-interacting. This is reasonable sufficiently far below the critical temperature when the condensate density is large, but not very close to the critical temperature. In that case, the interactions between the quasiparticles are very important and cannot be neglected.

We refer the reader for details and for further consideration (in particular, for the study of fluctuations and determination of the full probability distribution $P[\phi^*, \phi; t]$) to van Kampen (1981), Popov (1983), Zinn-Justin (1989), Griffin (1993), Griffin *et al* (1995) and Stoof (1999).

4.5.4 Problems

Problem 4.5.1. Derive the Fokker–Planck equation (4.5.81) for an ideal Bose gas coupled to a reservoir, *directly* from the effective action $S^{(\mathrm{eff})}[\phi^*, \phi; \xi^*, \xi]$ defined in (4.5.60), making the same approximations on the self-energies as in (4.5.67) (and without making use of the Langevin equations).

Hint. The effective action for the probability distribution $P[\phi^*, \phi; t]$ in the required approximation reads

$$S^{(\mathrm{eff})}[\phi^*, \phi; \xi^*, \xi] = \sum_\alpha \int_{t_0}^t dt'\, \phi_\alpha^*(t') \left(i\hbar \frac{\partial}{\partial t'} - \epsilon_\alpha - \hbar\Sigma_\alpha^{(-)} + \mu \right) \xi_\alpha(t')$$

$$+ \sum_\alpha \int_{t_0}^t dt'\, \xi_\alpha^*(t') \left(i\hbar \frac{\partial}{\partial t'} - \epsilon_\alpha - \hbar\Sigma_\alpha^{(+)} + \mu \right) \phi_\alpha(t')$$

$$- \tfrac{1}{2} \sum_\alpha \int_{t_0}^t dt'\, \xi_\alpha^*(t') \hbar \Sigma_\alpha^K \xi_\alpha(t') \quad (4.5.103)$$

and is quadratic in the fluctuation field $\xi(\mathbf{x}, t)$. We can thus again perform the integration over this field exactly. The result is

$$S^{(\mathrm{eff})}[\phi^*, \phi] = \sum_\alpha \int_{t_0}^t dt'\, \frac{2}{\hbar \Sigma_\alpha^K} \left| \left(i\hbar \frac{\partial}{\partial t'} - \epsilon_\alpha - \hbar\Sigma_\alpha^{(+)} + \mu \right) \phi_\alpha(t') \right|^2 \equiv \int_{t_0}^t dt\, L(t'). \quad (4.5.104)$$

Since the probability distribution $P[\phi^*, \phi; t]$ is equal to the functional integral

$$P[\phi^*, \phi; t] = \int_{\mathcal{C}\{\phi(\mathbf{x},t)=\phi(\mathbf{x}); \phi^*(\mathbf{x},t)=\phi^*(\mathbf{x})\}} \mathcal{D}\phi^*\, \mathcal{D}\phi\, \exp\left\{ \frac{i}{\hbar} S^{(\mathrm{eff})}[\phi^*, \phi] \right\} \quad (4.5.105)$$

we know that $P[\phi^*, \phi; t]$ must obey the 'Schrödinger equation' that results from quantizing the classical theory with the Lagrangian $L(t)$.

The quantization of this theory is straightforward. The momentum conjugate to $\phi_\alpha(t)$ is

$$\pi_\alpha(t) = \frac{\partial L(t)}{\partial(\partial \phi_\alpha(t)/\partial t)} = \frac{2i}{\Sigma_\alpha^K} \left(-i\hbar \frac{\partial}{\partial t} - \epsilon_\alpha - \hbar\Sigma_\alpha^{(-)} + \mu \right) \phi_\alpha^*(t) \quad (4.5.106)$$

whereas the momentum conjugate to $\phi_\alpha^*(t)$, i.e. $\pi_\alpha^*(t)$, is given by the complex conjugate expression. The corresponding Hamiltonian is therefore

$$H = \sum_\alpha \left\{ \pi_\alpha(t) \frac{\partial \phi_\alpha(t)}{\partial t} + \pi_\alpha^*(t) \frac{\partial \phi_\alpha^*(t)}{\partial t} \right\} - L(t)$$

$$= \sum_\alpha \left\{ -\frac{i}{\hbar} \pi_\alpha(t)(\epsilon_\alpha + \hbar \Sigma_\alpha^{(+)} - \mu)\phi_\alpha(t) + \frac{i}{\hbar} \pi_\alpha^*(t)(\epsilon_\alpha + \hbar \Sigma_\alpha^{(-)} - \mu)\phi_\alpha^*(t) \right\}$$

$$+ \sum_\alpha \frac{\Sigma_\alpha^K}{2\hbar} |\pi_\alpha(t)|^2. \tag{4.5.107}$$

Applying now the usual quantum-mechanical recipe of demanding non-vanishing commutation relations between the coordinates and their conjugate momenta, we can put in this case $\hat{\pi}_\alpha = (\hbar/i)\partial/\partial\phi_\alpha$ and similarly $\hat{\pi}_\alpha^* = (\hbar/i)\partial/\partial\phi_\alpha^*$. The 'Schrödinger equation'

$$i\hbar \frac{\partial}{\partial t} P[\phi^*, \phi; t] = \widehat{H} P[\phi^*, \phi; t] \tag{4.5.108}$$

then indeed reproduces the Fokker–Planck equation in (4.5.81) exactly.

Problem 4.5.2. Prove the equivalence of expressions (4.5.43) and (4.5.44), (4.5.45) for the probability distribution $P[\phi^*, \phi; t]$ of an ideal Bose gas.

Hint. Use the result of problem (1.2.6) and the fact that $\rho[|\phi_0|^2; t_0]$ enters the normalization constant $\mathfrak{N}^{-1}$ (see the comment before (4.5.39)).

Problem 4.5.3. Derive the time-evolution equation for the mean values $\langle \phi_\alpha \rangle(t)$ and $\langle |\phi_\alpha|^2 \rangle(t)$ for an ideal Bose gas with the probability distribution $P[\phi^*, \phi; t]$ satisfying the Fokker–Planck equation (4.5.46).

Hint. We first consider the average $\langle \phi_\alpha \rangle(t) = \int \mathcal{D}\phi^* \mathcal{D}\phi \, \phi_\alpha P[\phi^*, \phi; t]$. Multiplying (4.5.46) with ϕ_α and integrating over $\phi(x)$, we find after a partial integration that

$$i\hbar \frac{\partial}{\partial t} \langle \phi_\alpha \rangle(t) = \epsilon_\alpha \langle \phi_\alpha \rangle(t) \tag{4.5.109}$$

which precisely corresponds to the equation of motion of $\langle \hat{\varphi}_\alpha(t) \rangle = \text{Tr}[\hat{\rho}_C(t_0)\hat{\varphi}_\alpha(t)]$ in the operator formalism. Similarly, we find that

$$i\hbar \frac{\partial}{\partial t} \langle \phi_\alpha^* \rangle(t) = -\epsilon_\alpha \langle \phi_\alpha^* \rangle(t) \tag{4.5.110}$$

in agreement with the result for $\langle \hat{\varphi}_\alpha^\dagger(t) \rangle = \text{Tr}[\hat{\rho}_C(t_0)\hat{\varphi}_\alpha^\dagger(t)]$.

Next, we consider the average of $|\phi_\alpha|^2$, for which we immediately obtain

$$i\hbar \frac{\partial}{\partial t} \langle |\phi_\alpha|^2 \rangle(t) = 0. \tag{4.5.111}$$

Problem 4.5.4. Derive a relation between the mean value $\langle |\phi_\alpha|^2 \rangle(t)$ of an ideal Bose gas and the occupation numbers N_a of the one-particle states.

Hint. To give the relation between $\langle |\phi_\alpha|^2 \rangle(t)$ and $N_\alpha(t)$ is complicated by the fact that at equal times the operators $\hat{\varphi}_\alpha(t)$ and $\hat{\varphi}_\alpha^\dagger(t)$ do not commute. However, the path integral produces time-ordered operator products. This implies that $\langle |\phi_\alpha|^2 \rangle(t)$ is the value at $t' = t$ of

$$\text{Tr}[\hat{\rho}_C(t_0)\mathbf{T}_{C'}(\hat{\varphi}_\alpha(t)\hat{\varphi}_\alpha^\dagger(t'))] = \widetilde{\theta}(t - t') \text{Tr}[\hat{\rho}_C(t_0)\hat{\varphi}_\alpha(t)\hat{\varphi}_\alpha^\dagger(t')] + \widetilde{\theta}(t' - t) \text{Tr}[\hat{\rho}_C(t_0)\hat{\varphi}_\alpha^\dagger(t')\hat{\varphi}_\alpha(t)]$$

with $\mathbf{T}_{\mathcal{C}'}$ being the time-ordering operator on the Schwinger–Keldysh contour and $\widetilde{\theta}(t, t')$ the symmetrical Heaviside function. Since the latter is equal to $\frac{1}{2}$ at equal times, we conclude that

$$\langle |\phi_\alpha|^2 \rangle(t) = N_\alpha(t) + \tfrac{1}{2}. \tag{4.5.112}$$

Thus, the stationarity of $\langle |\phi_\alpha|^2 \rangle$ (proved in the preceding problem) is related to the fact that the occupation numbers $N_\alpha(t)$ for an ideal gas are independent of time (see the beginning of section 4.5.3).

Problem 4.5.5. Calculate the effective action $S^{(\text{eff})}$ for an ideal Bose gas in a thermal reservoir defined by the Gaussian path integral (4.5.53).

Hint. Let us introduce the $\delta_{\mathcal{C}}$ function on the Schwinger–Keldysh contour defined by

$$\int_{\mathcal{C}^\infty} dt' \, \delta_{\mathcal{C}}(t, t') = 1$$

and the Green function $G(\mathbf{k}; t, t')$ obeying (4.5.56), i.e.

$$\left(i\hbar \frac{\partial}{\partial t} - \epsilon(\mathbf{k}) + \mu \right) G(\mathbf{k}; t, t') = \hbar \delta_{\mathcal{C}}(t, t'). \tag{4.5.113}$$

The action $S[\varphi^*, \varphi; \Psi^*, \Psi]$ can be written as a complete square or, more precisely, as the sum of two squares $S_1[\varphi^*, \varphi]$ and $S_2[\varphi^*, \varphi; \Psi^*, \Psi]$ that are given by

$$S_1[\varphi^*, \varphi] = \sum_\alpha \int_{\mathcal{C}^\infty} dt \, \varphi_\alpha^*(t) \left(i\hbar \frac{\partial}{\partial t} - \epsilon_\alpha + \mu \right) \varphi_\alpha(t)$$
$$- \frac{1}{\hbar L^3} \sum_{\alpha,\alpha'} \sum_{\mathbf{k}} \int_{\mathcal{C}^\infty} dt \int_{\mathcal{C}^\infty} dt' \, \varphi_\alpha^*(t) t_\alpha^*(\mathbf{k}) G(\mathbf{k}; t, t') t_{\alpha'}(\mathbf{k}) \varphi_{\alpha'}(t') \tag{4.5.114}$$

and

$$S_2[\varphi^*, \varphi; \Psi^*, \Psi] = \sum_{\mathbf{k}} \int_{\mathcal{C}^\infty} dt \left(\Psi_{\mathbf{k}}^*(t) - \frac{1}{\hbar L^{3/2}} \sum_\alpha \int_{\mathcal{C}^\infty} dt' \, t_\alpha^*(\mathbf{k}) \varphi_\alpha^*(t') G(\mathbf{k}; t', t) \right)$$
$$\times \left(i\hbar \frac{\partial}{\partial t} - \epsilon(\mathbf{k}) + \mu \right) \left(\Psi_{\mathbf{k}}(t) - \frac{1}{\hbar L^{3/2}} \sum_\alpha \int_{\mathcal{C}^\infty} dt' \, G(\mathbf{k}; t, t') \varphi_\alpha(t') t_\alpha(\mathbf{k}) \right)$$
$$\tag{4.5.115}$$

respectively. Since the first term is independent of the field $\Psi(\mathbf{x}, t)$, we only need to evaluate the functional integral $\int \mathcal{D}\Psi^* \mathcal{D}\Psi \exp(i S_2[\varphi^*, \varphi; \Psi^*, \Psi]/\hbar)$. Performing a shift in the integration variables, we see, however, that this functional integral is equal to

$$\int \mathcal{D}\Psi^* \mathcal{D}\Psi \, \exp\left\{ \frac{i}{\hbar} \sum_{\mathbf{k}} \int_{\mathcal{C}^\infty} dt \, \Psi_{\mathbf{k}}^*(t) \left(i\hbar \frac{\partial}{\partial t} - \epsilon(\mathbf{k}) + \mu \right) \Psi_{\mathbf{k}}(t) \right\} = \text{Tr}[\hat{\rho}_{C,R}(t_0)] = 1. \tag{4.5.116}$$

As a result, the effective action $S^{(\text{eff})}[\varphi^*, \varphi]$ is just equal to $S_1[\varphi^*, \varphi]$, which can be rewritten as in (4.5.54).

Problem 4.5.6. Derive the Bose gas effective action $S^{(\text{eff})}[\phi^*, \phi; \xi^*, \xi; \eta^*, \eta]$ with the auxiliary stochastic field η^*, η (the result is given by (4.5.63)), which leads to the Langevin equations (4.5.64) and (4.5.65).

Hint. Multiply the integrand of the path integral defining the effective action $S^{(\text{eff})}[\phi^*, \phi; \xi^*, \xi]$ in (4.5.60) by a factor 1, which is written as the Gaussian integral $\int \mathcal{D}\eta^* \mathcal{D}\eta \, \exp\{iS^{(\text{eff})}_{\text{aux}}[\eta^*, \eta]/\hbar\}$, with

$$S^{(\text{eff})}_{\text{aux}}[\eta^*, \eta] = \frac{1}{2} \sum_{\alpha,\alpha'} \int dt \int dt' \left(2\eta_\alpha^*(t) - \sum_{\alpha''} \int dt'' \, \xi_{\alpha''}^*(t'') \hbar \Sigma^K_{\alpha'',\alpha}(t'' - t) \right)$$
$$\times (\hbar \Sigma^K)^{-1}_{\alpha,\alpha'}(t - t') \left(2\eta_{\alpha'}(t') - \sum_{\alpha''} \int dt'' \, \hbar \Sigma^K_{\alpha',\alpha''}(t' - t'') \xi_{\alpha''}(t'') \right) \quad (4.5.117)$$

and add this action to $S^{(\text{eff})}[\phi^*, \phi; \xi^*, \xi]$.

4.6 Non-equilibrium statistical physics in the path-integral formalism and stochastic quantization

This section is devoted to one of the most striking manifestations of the deep interrelations between stochastic processes, quantum mechanics and statistical physics, which we have stressed throughout the book. Namely, we shall discuss the method of the so-called stochastic quantization.

An alternative scheme for quantization based on *stochastic averages* has been presented by Parisi and Wu (1981) (see also, e.g., Damgaard and Hüffel (1987) and Namiki (1992) for comprehensive reviews and references). The main idea of this *stochastic quantization* approach to quantum mechanics is to view the quantum theory in the Euclidean time as the *equilibrium limit* of a statistical system coupled to a thermal reservoir. This system evolves in a new additional time direction, which is called *stochastic time* until it reaches the equilibrium limit for infinite stochastic time. The coupling to the heat reservoir is simulated by means of a stochastic noise field which forces the original dynamical variables to wander randomly similarly to the Brownian particle (Wiener process) which we discussed in detail in chapter 1. In the equilibrium limit, the stochastic averages become identical to the ordinary Euclidean vacuum expectation values. The Minkowski (real) time variant of the formalism (with the usual $i\varepsilon$-prescription for the regularization of the corresponding path integrals) is also available and we shall discuss it as well as the Euclidean variant.

There are two equivalent formulations of stochastic quantization due to the general properties of the stochastic processes (see chapter 1):

- In one formulation, all fields have an additional dependence on stochastic time. Their stochastic time evolution is determined by a *Langevin equation* which has a drift term constructed from the gradient of the classical action of the system. The expectation values of observables are obtained by ensemble averages over the *Wiener measure*.
- Corresponding to this Langevin equation, there is an equivalent diffusion process which is defined in terms of the *Fokker–Planck equation* for the probability distribution characterizing the stochastic evolution of the system. Now, the expectation values of the observables are defined by functionally integrating them with the stochastic *time-dependent* Fokker–Planck probability distribution.

In several models, we can mathematically rigorously verify that in the infinite stochastic time limit, this Fokker–Planck probability distribution converges to the standard Euclidean configuration-space path-integral density.

Over the past years, stochastic quantization has been successfully applied to different problems, in particular, for the quantization of gauge-field theory without gauge-fixing terms, for studies of quantum anomalies, investigations of supersymmetric models and for the regularization and renormalization program. An important area of application of this approach consists of the numerical (Monte Carlo) simulations for a non-perturbative study of field theories.

In this section, we shall outline the prescription of the Parisi–Wu stochastic quantization method starting from the elementary example of *ordinary integrals* ('zero-dimensional' quantum system), then consider how to calculate the transition amplitudes in ordinary quantum mechanics within the stochastic scheme and, finally, we shall discuss the stochastic quantization of field theories, including gauge-invariant models.

4.6.1 A zero-dimensional model: calculation of usual integrals by the method of 'stochastic quantization'

Let us start by considering the generating functional for the zero-dimensional model, that is, from the following (one-dimensional) integral

$$Z(j) = \mathfrak{N}^{-1} \int dx\, e^{-S(x)} e^{jx} \qquad (4.6.1)$$

where $S(x)$ is some polynomial (e.g., $S(x) = (x^2/2 + gx^4/4)$) and $\mathfrak{N} = Z(0)$. The main idea of the stochastic quantization approach is to express the mean value

$$G_m = \left.\frac{\partial^m}{\partial x^m} Z(j)\right|_{j=0} = \mathfrak{N}^{-1} \int dx\, x^m e^{-S(x)} \qquad (4.6.2)$$

as a large-'time' mean value over a non-equilibrium statistical ensemble, as it relaxes to the equilibrium distribution $\sim \exp\{-\beta S(x)\}$. To this aim, we generalize the integration variable x to a *random variable* $x(s)$ (cf chapter 1), depending on an auxiliary 'time' s.

Let $W(x_1, s_1|x_0, s_0)$ be the (conditional) transition probability that $x(s)$ has the value x_1 at the time s_1, if it had value x_0 at time s_0. The stochastic process $x(s)$ is chosen to be *Markovian* (see chapter 1) and *stationary*, so that W depends only on the time difference $s_1 - s_0$ and, for brevity, we put $s_0 = 0$. Thus, the requirement of relaxation of the fictitious statistical system to the equilibrium state can be formulated as follows:

$$\lim_{s\to\infty} W(x, s|x_0, 0) = \mathfrak{N}^{-1} \exp\{-S(x)\} \qquad (4.6.3)$$

and it is obviously independent of the initial position x_0.

From the computational point of view, the condition (4.6.3) is important because it implies the *ergodic property* of the stochastic process:

$$\lim_{s\to\infty} \int_{-\infty}^{\infty} dx\, x^n W(x, s|x_0, 0) = \mathfrak{N}^{-1} \int_{-\infty}^{\infty} dx\, x^n \exp\{-S(x)\}$$
$$= \lim_{s\to\infty} \frac{1}{s} \int_0^s ds'\, x^n(s') \qquad (4.6.4)$$

(see, e.g., Klauder (1983)), that is, the ensemble average is identical to the temporal average for a sample path. Once we have chosen an evolution equation for $x(s)$ such that (4.6.3) follows, the computer simulation of the time average is straightforward. This way of calculation seems to be too cumbersome for ordinary integrals of the type (4.6.1), but in field theory, it can be quite reasonable.

Next, we proceed to establish the equation of evolution for $x(s)$.

◊ **Langevin equation for the auxiliary stochastic process $x(s)$ and the Fokker–Planck equation for the transition probability**

As we learned from section 1.2, the time evolution of a stochastic process at the microscopic level is governed by the *Langevin equation* (cf (1.2.24)). On the other hand, at the macroscopic level its time behaviour is defined by the corresponding *Fokker–Planck* equation. Let us show that the required relaxation of the probability W (i.e. the limit (4.6.3)) is guaranteed if the stochastic process satisfies the following Langevin equation:

$$\frac{\partial x(s)}{\partial s} = -\frac{\partial S(x)}{\partial x} + \eta(s) \tag{4.6.5}$$

where $\eta(s)$ is the white-noise stochastic process, so that (cf (1.2.13))

$$\langle \eta(s) \rangle_{\text{stoch}} = 0 \qquad \langle \eta(s_2)\eta(s_1) \rangle_{\text{stoch}} = 2\delta(s_2 - s_1) \tag{4.6.6}$$

with all higher connected correlation functions vanishing.

To show that the Langevin equation (4.6.5) gives the correct behaviour for W, we integrate it for a small time interval Δs:

$$\Delta x(s) \approx -\frac{\partial S(x)}{\partial x}\Delta s + \int_s^{s+\Delta s} ds'\, \eta(s'). \tag{4.6.7}$$

From (4.6.6) and (4.6.7), we can now deduce that

$$\lim_{\Delta s \to 0} \left\langle \frac{\Delta x(s)}{\Delta s} \right\rangle_{\text{stoch}} = -\frac{\partial S(x)}{\partial x}$$

$$\lim_{\Delta s \to 0} \left\langle \frac{(\Delta x(s))^2}{\Delta s} \right\rangle_{\text{stoch}} = 2 \tag{4.6.8}$$

$$\lim_{\Delta s \to 0} \left\langle \frac{(\Delta x(s))^n}{\Delta s} \right\rangle_{\text{stoch}} = 0 \qquad n > 2.$$

The last equality follows from the fact that the higher-order correlation functions of $\eta(s)$ vanish. These are exactly the conditions (1.2.15)–(1.2.17) for a transition probability which leads to a Fokker–Planck equation of the form (1.2.21). Thus in the concrete case (4.6.8), the Fokker–Planck equation reads as

$$\frac{\partial}{\partial s} W(x, s | x_0, 0) = \frac{\partial}{\partial x}\left[\frac{\partial S(x)}{\partial x} + \frac{\partial}{\partial x}\right] W(x, s | x_0, 0). \tag{4.6.9}$$

It is easy to see that the equilibrium solution (that is, satisfying $\partial_s W = 0$) of equation (4.6.9) is indeed given by the required Boltzmann-like distribution: $W = \text{constant} \times \exp\{-S(x)\}$.

◇ **The generating functional for Green functions in the approach of stochastic quantization**

We wish to construct a generating functional for the stochastic correlation functions

$$\left.\frac{\delta^m Z(J)}{\delta J(s_1) \cdots \delta J(s_m)}\right|_{J=0} = \langle x(s_1) \cdots x(s_m) \rangle_{\text{stoch}}. \tag{4.6.10}$$

The practical meaning of these correlation functions is that in the limit of the infinite stochastic time s, they are equal to the integrals $\int dx\, x^m \exp\{-\beta S(x)\}$, which we study by the stochastic quantization method. Correspondingly, after the generalization to higher-dimensional models, such correlators in the infinite time limit become the usual Green functions for quantum-mechanical or field operators.

The generating functional $Z[J]$ is constructed in two steps (Gozzi 1983). First we note that the form (4.6.6) of the correlation functions for the white noise $\eta(s)$ shows that the ensemble average of any functional $F[\eta]$ can be written as the following path integral:

$$\langle F[\eta] \rangle_{\text{stoch}} = \int \mathcal{D}\eta(s)\, F[\eta(s)] \exp\left\{-\int_0^\infty ds\, \eta^2(s)\right\}. \tag{4.6.11}$$

The reader may easily verify that correlation functions (4.6.6) are indeed reproduced by this path integral. Thus, the required generating functional can be presented as the integral (4.6.11), with functional $F[J]$ of the form

$$F[J, \eta] = \exp\left\{\int_0^\infty ds\, J(s) x_\eta(s)\right\}$$

where $x_\eta(s)$ satisfies Langevin equation (4.6.5) and, hence, implicitly depends on $\eta(s)$. At the second step, we introduce a new integration variable $y(s)$ and rewrite the functional $F[J, \eta]$ as the path integral over it:

$$F[J, \eta] = \int \mathcal{D}y(s)\, \delta[y(s) - x_\eta(s)] \exp\left\{\int_0^\infty ds\, J(s) y(s)\right\} \qquad (4.6.12)$$

so that the generating functional reads as

$$Z[J] = \int \mathcal{D}\eta\, \mathcal{D}y(s)\, \delta[y(s) - x_\eta(s)] \exp\left\{-\int_0^\infty ds\, \eta^2(s) + \int_0^\infty ds\, J(s) y(s)\right\}. \qquad (4.6.13)$$

Now we want to recover the dependence of x_η on η, rewrite the δ-functional as a δ-functional explicitly depending on η and, using this, integrate over the latter. The method of calculation is quite analogous to the Faddeev–Popov trick which we used in section 3.2.4 for the transition from a gauge field to the gauge parameter integration variables. To achieve this, we change the order of the integrations in (4.6.13): $\int \mathcal{D}\eta\, \mathcal{D}y(s) \to \int \mathcal{D}y(s)\, \mathcal{D}\eta$ and denote by $\widetilde{\eta}$ the solution of the equation

$$y(s) - x_{\widetilde{\eta}(s)} = 0$$

for a fixed function $y(s)$. It is clear (cf the similar consideration after (3.2.132)) that the only contribution to the path integral over η comes from the infinitesimal vicinity of $\widetilde{\eta}$. Then, for small variations $\xi(s)$ around $\eta(s)$ the Langevin equation (4.6.5) which relates x_η and η reads as

$$\frac{\partial}{\partial s}(x_{\widetilde{\eta}} + \delta x) + [S'(x_{\widetilde{\eta}}) + S''(x_{\widetilde{\eta}})\delta x] = \widetilde{\eta} + \xi \qquad (4.6.14)$$

where the prime denotes the derivative in x and where we have kept only the first-order terms in the variation δx corresponding to the infinitesimal variation ξ. This variation δx is related to ξ by the *linear differential operator*:

$$\widehat{M}\delta x \stackrel{\text{def}}{\equiv} \frac{\partial}{\partial s}\delta x + S''(x_{\widetilde{\eta}})\delta x = \xi. \qquad (4.6.15)$$

Substituting (4.6.15) into the path integrals (4.6.13) yields

$$\begin{aligned}Z[J] &= \int \mathcal{D}y(s)\, \mathcal{D}\xi\, \delta[\widehat{M}^{-1}\xi] \exp\left\{-\int_0^\infty ds\, \eta^2(s) + \int_0^\infty ds\, J(s)y(s)\right\} \\ &= \int \mathcal{D}y(s)\, \det\widehat{M} \exp\left\{-\int_0^\infty ds\, \widetilde{\eta}^2(s) + \int_0^\infty ds\, J(s)y(s)\right\} \\ &= \int \mathcal{D}y(s)\, \det\widehat{M} \exp\left\{-\int_0^\infty ds\left[\left(\frac{\partial y}{\partial s'} + S'(x)\right)^2 - J(s)y(s)\right]\right\}. \end{aligned} \qquad (4.6.16)$$

Calculating the determinant of the operator $\widehat{M}$ results in the additional exponential (problem 4.6.1, page 293):

$$\det\widehat{M} = \exp\left\{\int_{-\infty}^\infty ds\, \widetilde{\theta}(0) S''(y(s))\right\} \qquad (4.6.17)$$

where $\widetilde{\theta}(0) = \frac{1}{2}$ is the value of the symmetrical $\widetilde{\theta}$-function (cf (1.2.281)). The choice of the type of step-function (θ, see (1.1.43), or $\widetilde{\theta}$) depends on the way in which the discrete approximation of the path integral is defined. In particular, the *symmetrical* $\widetilde{\theta}$-function corresponds to the *midpoint* prescription (see Gozzi (1983) for details).

Substituting (4.6.17) into (4.6.16) enables us to write the generating functional $Z[J]$ in terms of the path integral

$$Z[J] = \int_{\mathcal{C}\{x_0, 0; x, s\}} \mathcal{D}x \, \exp\left\{-S^{(\text{eff})} + \int_0^\infty ds \, J(s)x(s)\right\} \qquad (4.6.18)$$

with 'effective action'

$$S^{(\text{eff})} \stackrel{\text{def}}{\equiv} \int_0^\infty ds \left\{\left[\frac{\partial x}{\partial s} + S'(x)\right]^2 - \frac{1}{2}S''(x)\right\}. \qquad (4.6.19)$$

Note that in (4.6.19), the functional integration is over all paths with a fixed starting point: $x(s = 0) = x_0$. Sometimes, this proves to be too restrictive and it is convenient to choose x_0 to have a probability distribution $p(x_0)$. This leads to an extra integration in (4.6.18):

$$Z[J] = \int dx_0 \, p(x_0) \int_{\mathcal{C}\{x_0, 0; x, s\}} \mathcal{D}x \, \exp\{-S^{(\text{eff})} + \int_0^\infty ds \, J(s)x(s)\}. \qquad (4.6.20)$$

In some cases, particular choices of $p(x_0)$ make calculations simpler.

A similar approach based on the idea of stochastic quantization is possible for the analysis of integrals with *complex* $S(x)$ in (4.6.1). This allows us to consider the path-integral formulation of the stochastic quantization in *real time*. We shall use this possibility in the next subsection for the stochastic quantization of quantum-mechanical systems (i.e. systems depending on real physical time in addition to the auxiliary stochastic one).

4.6.2 Real-time quantum mechanics within the stochastic quantization scheme

Now let us expand the stochastic quantization method to a non-relativistic quantum-mechanical particle with coordinates $x(t)$, that is, to systems which depend on the physical time variable t.

The generating functional $\mathcal{Z}[J]$ produces vacuum expectations of the dynamical operators under consideration (see chapter 3). Therefore, if we are interested in vacuum expectations, generalizing the formalism presented in the foregoing subsection to the case of dependence on an additional (physical) time variable t is straightforward (with the obvious substitution of the exponents $-S \to iS$ in all the formulae). Given some arbitrary functional f of the coordinates $x(t)$, its vacuum expectation value can be obtained as the equilibrium limit $s \to \infty$ of the stochastic correlation function $\langle f[x; s]\rangle$

$$\langle 0|f(\widehat{x})|0\rangle = \lim_{s \to \infty} \langle f[x; s]\rangle \qquad (4.6.21)$$

which can be defined as

$$\langle f[x; s]\rangle = \int \mathcal{D}x \, f[x] W[x; s]. \qquad (4.6.22)$$

Here $P[x; s]$ is a normalized (generalized) probability distribution, obeying the *Fokker–Planck equation*

$$\frac{\partial}{\partial s} W[x; s] = \int_{-\infty}^\infty dt \, \frac{\delta}{\delta x(t)} \left(\frac{\delta}{\delta x(t)} - i \frac{\delta S}{\delta x(t)}\right) W[x; s]. \qquad (4.6.23)$$

In this subsection, we prefer to work in real-time quantum mechanics to demonstrate the possibility of stochastic quantization directly in real time and because our main interest now concerns the calculation of quantum-mechanical transition *amplitudes*. But we implicitly assume the iε-prescription for the regularization (convergence) of the path integrals which we shall encounter in this formalism.

As we discussed in the preceding subsection, stochastic correlations can be obtained alternatively by assigning an additional stochastic time dependence $x(t) \rightarrow x(t, s)$ to $x(t)$ via the *Langevin equation*

$$\frac{\partial x(t, s)}{\partial s} = \mathrm{i}\frac{\delta S}{\delta x(t, s)} + \eta(t, s). \qquad (4.6.24)$$

Its solution has to be inserted into the given functional f and the average over the white noise η, characterized by

$$\langle \eta(t, s)\eta(t', s')\rangle = 2\delta(t - t')\delta(s - s') \qquad (4.6.25)$$

finally has to be worked out.

However, in quantum mechanics, together with the calculation of vacuum expectation values (mainly used in field theory), we are frequently interested in the normalized matrix elements of time-ordered products of operators between the state vectors $|x, t\rangle$, which are eigenstates of the Heisenberg operators $\widehat{x}(t)$ belonging to the eigenvalues x. We therefore introduce new stochastic correlation functions, taking care of the boundary values x_i at t_i and x_f at t_f. We consider, for example,

$$\frac{\langle x_f, t_f|\mathbf{T}(x(t_1)x(t_2))|x_i, t_i\rangle}{\langle x_f, t_f|x_i, t_i\rangle} = \lim_{s \to \infty} \langle x(t_1, s)x(t_2, s)\rangle_{\text{stoch}} \qquad (4.6.26)$$

and define, in analogy to (4.6.21),

$$\langle x(t_1, s)x(t_2, s)\rangle_{\text{stoch}} = \int_{\mathcal{C}\{x_i, t_i; x_f, t_f\}} \mathcal{D}x\, x(t_1)x(t_2)W[x; s]. \qquad (4.6.27)$$

W now satisfies Fokker–Planck equation (4.6.23) with the time integration restricted to the interval $t \in [t_i, t_f]$. It follows that the stationary solution is given by (constant) $\times\, \mathrm{e}^{\mathrm{i}S}$, so that the standard path-integral representation emerges at the equilibrium limit of (4.6.27).

◇ Transition amplitudes in the formalism of the stochastic quantization

Due to the normalization condition for the probability density W we cannot directly express the transition amplitude $\langle x_f, t_f|x_i, t_i\rangle$ in terms of it. The stochastic expectation values calculated either with Fokker–Planck or Langevin equation techniques are always normalized automatically and it seems difficult to find a way to reproduce such quantities as transition amplitudes within the framework of the stochastic quantization. It is, however, possible to relate $\langle x_f, t_f|x_i, t_i\rangle$ to the normalized stochastic expectation value of the Hamiltonian $\widehat{H}$ (Hüffel and Nakazato 1994).

The key observation for achieving such a relation is that the Schrödinger equation implies the equality

$$\frac{\partial}{\partial t_i}\ln\langle x_f, t_f|x_i, t_i\rangle = \mathrm{i}\frac{\langle x_f, t_f|H(t_i)|x_i, t_i\rangle}{\langle x_f, t_f|x_i, t_i\rangle}. \qquad (4.6.28)$$

The right-hand side is a *normalized* expectation value of the Hamiltonian at t_i, which provides that this quantity is obtainable as a limit of the stochastic average $\langle H(t_i, s)\rangle_{\text{stoch}}$. Note that for conservative systems the stochastic average of the Hamiltonian is t-independent at equilibrium (i.e. $s \to \infty$). Therefore, for calculational simplicity, we evaluate it at the initial time $t = t_i$. Substituting the stochastic average instead of the right-hand side of (4.6.28), we have for its solution:

$$\langle x_f, t_f|x_i, t_i\rangle = \widetilde{c}\exp\left\{\mathrm{i}\int^{t_i} dt'\, \lim_{s \to \infty}\langle H(t', s)\rangle_{\text{stoch}}\right\} \qquad (4.6.29)$$

with $\tilde{c}$ being a constant independent of t_i. For a transparent presentation, we restrict ourselves to Hamiltonians quadratic in the canonical variables. In this case, we can separate the contributions from the classical trajectory and arrive at

$$\langle x_f, t_f | x_i, t_i \rangle = \tilde{c} \exp \left\{ i \int^{t_i} dt' \, H_{\text{cl}}(t') \right\} \exp \left\{ i \int^{t_i} dt' \, \lim_{s \to \infty} \langle H_{\text{fl}}(t', s) \rangle_{\text{stoch}} \right\}$$

$$= c \exp\{i S_{\text{cl}}\} \exp \left\{ i \int^{t_i} dt' \, \lim_{s \to \infty} \langle H_{\text{fl}}(t', s) \rangle_{\text{stoch}} \right\}. \quad (4.6.30)$$

Here the classical action S_{cl} has appeared as a result of the well-known relation $\partial S_{\text{cl}} / \partial t_i = H_{\text{cl}}(t_i)$ (see, e.g., ter Haar (1971)) and is composed of the classical path $x_{\text{c}}(t)$, as usual.

The constant c, in principle, could depend on x_f and x_i which, however, is not the case. This follows from a similar variational principle as before for $\langle x_f, t_f | x_i, t_i \rangle$ with respect to x_f and x_i and the relations $\partial S_{\text{cl}} / \partial x_f = p_{\text{c}}(t_f)$ and $\partial S_{\text{cl}} / \partial x_i = -p_{\text{c}}(t_i)$ (problem 4.6.2, page 293). So the constant c is indeed t_f, t_i, x_f, x_i independent and can be fixed by requiring the transition amplitude to approach a Dirac δ-function $\delta(x_f - x_i)$ in the limit of $t_i = t_f$.

◇ **Phase-space formulation**

Since we have related the transition amplitude to an expectation value of the Hamiltonian, it is natural to rely on a phase-space formulation of the stochastic quantization (Ohba 1987). To this aim, we separate the dynamical variables $x(t)$ and $p(t)$ into classical and fluctuation parts in the usual way:

$$x(t) = x_{\text{c}}(t) + X(t) \qquad p(t) = p_{\text{c}}(t) + P(t) \quad (4.6.31)$$

and implement the boundary conditions for the fluctuations $X(t_i) = X(t_f) = 0$ by the Fourier decompositions

$$X(t) = \sum_{n=1}^{\infty} x_n \sin \frac{n\pi}{T} (t - t_i) \quad (4.6.32)$$

$$P(t) = \sum_{n=1}^{\infty} p_n \cos \frac{n\pi}{T} (t - t_i) + \frac{p_0}{2} \qquad T = t_f - t_i. \quad (4.6.33)$$

Note that no boundary conditions are imposed on the momentum variable $P(t)$ and that any function defined in $[t_i, t_i + T]$ can continuously be extended to $[t_i - T, t_i + T]$ as an even function, like the previous P. Stochastic quantization proceeds by introducing the s-dependence for the Fourier modes $x_n \to x_n(s)$, $p_n \to p_n(s)$, according to the phase-space Langevin equations

$$\frac{d}{ds} x_n = i \frac{\delta S}{\delta x_n} + \xi_n \qquad \frac{d}{ds} p_n = i \frac{\delta S}{\delta p_n} + \eta_n \quad (4.6.34)$$

where the noises fulfill

$$\langle \xi_n(s) \xi_m(s') \rangle_{\text{stoch}} = \langle \eta_n(s) \eta_m(s') \rangle_{\text{stoch}} = 2 \delta_{nm} \delta(s - s')$$
$$\langle \eta_n(s) \xi_m(s') \rangle_{\text{stoch}} = 0. \quad (4.6.35)$$

For the quadratic case, (4.6.34) is explicitly solved by

$$\begin{pmatrix} x_n \\ p_n \end{pmatrix} (s) = \int_0^s d\sigma \, \exp\{i A_n (s - \sigma)\} \begin{pmatrix} \xi_n \\ \eta_n \end{pmatrix} (\sigma) \quad (4.6.36)$$

where A_n is a model-dependent matrix. Using the stochastic calculus, we can derive the following correlators:

$$\langle x_n(s)\xi_m(s)\rangle_{\text{stoch}} = \langle p_n(s)\eta_m(s)\rangle_{\text{stoch}} = \delta_{nm}$$
$$\langle x_n(s)\eta_m(s)\rangle_{\text{stoch}} = \langle p_n(s)\xi_m(s)\rangle_{\text{stoch}} = 0 \qquad (4.6.37)$$

and find that the correlations containing stochastic time derivatives vanish in the equilibrium limit $s \to \infty$

$$\left\langle x_n(s)\frac{dx_m(s)}{ds}\right\rangle_{\text{stoch}} = \left\langle x_n(s)\frac{dp_m(s)}{ds}\right\rangle_{\text{stoch}} = \left\langle p_n(s)\frac{dx_m(s)}{ds}\right\rangle_{\text{stoch}} = \left\langle p_n(s)\frac{dp_m(s)}{ds}\right\rangle_{\text{stoch}} \to 0. \qquad (4.6.38)$$

◇ **Example: a free particle in the phase-space formalism of the stochastic quantization**

The easiest example to discuss is the non-relativistic free particle with the Hamiltonian

$$H = \frac{p^2}{2m}. \qquad (4.6.39)$$

We have now

$$S_{\text{c}} = \frac{m}{2T}(x_f - x_i)^2 \qquad S_{\text{fl}} = \sum_{n=1}^{\infty}\left(\frac{n\pi}{T}x_n p_n - \frac{p_n^2}{2m}\right)\frac{T}{2} - p_0^2\frac{T}{8m} \qquad (4.6.40)$$

and get

$$\frac{dx_n}{ds} = \frac{in\pi}{2}p_n + \xi_n$$
$$\frac{dp_n}{ds} = i\left(\frac{n\pi}{T}x_n - \frac{p_n}{M}\right)\frac{T}{2} + \eta_n$$
$$\frac{dp_0}{ds} = -i\frac{T}{4M}p_0 + \eta_0. \qquad (4.6.41)$$

From the fact (see (4.6.38)) that in the equilibrium limit the correlators

$$\left\langle p_m(s)\frac{dx_n(s)}{ds}\right\rangle_{\text{stoch}} \qquad \left\langle p_m(s)\frac{dp_0(s)}{ds}\right\rangle_{\text{stoch}} \quad \text{and} \quad \left\langle p_0(s)\frac{dp_0(s)}{ds}\right\rangle_{\text{stoch}}$$

vanish, we immediately find

$$\langle p_m p_n\rangle \equiv \lim_{s\to\infty}\langle p_m(s)p_n(s)\rangle_{\text{stoch}} = 0 \qquad \langle p_m p_0\rangle = 0 \qquad \langle p_0^2\rangle = -\frac{4im}{T}. \qquad (4.6.42)$$

Furthermore,

$$\lim_{s\to\infty}\langle H_{\text{fl}}(t_i,s)\rangle_{\text{stoch}} = \lim_{s\to\infty}\frac{\langle P^2(t_i,s)\rangle_{\text{stoch}}}{2m} = \frac{\langle p_0^2\rangle}{8m} = -\frac{i}{2T} \qquad (4.6.43)$$

so that

$$i\int^{t_i}dt'\lim_{s\to\infty}\langle H_{\text{fl}}(t',s)\rangle_{\text{stoch}} = -i\int^{T}dT\left(-\frac{i}{2T}\right) = -\frac{1}{2}\ln T \qquad (4.6.44)$$

and finally

$$\langle x_f,t_f|x_i,t_i\rangle = c\frac{1}{\sqrt{T}}e^{iS_c} \qquad c = \sqrt{\frac{m}{2\pi i}} \qquad (4.6.45)$$

(cf (2.2.41)). The transition amplitudes for another standard example, i.e. for the harmonic oscillator also can be found by the stochastic quantization method; this calculation pertains to problem 4.6.3, page 294.

◇ Configuration-space formulation

Although the phase-space formulation is the natural one, sometimes it is technically more convenient to use the configuration-space formulation. The transition to the Lagrangian (configuration-space) formulation is achieved by stochastic averaging the Legendre transformation:

$$\langle H_{\text{fl}}(t_i, s) \rangle_{\text{stoch}} = \left\langle P(t_i, s) \frac{\partial X}{\partial t}(t_i, s) \right\rangle_{\text{stoch}} - \langle \mathcal{L}_{\text{fl}}(t_i, s) \rangle_{\text{stoch}}. \tag{4.6.46}$$

Now the stochastic average of the fluctuation part of the Hamiltonian can be expressed through the average of the derivatives of the coordinates (Hüffel and Nakazato 1994):

$$\lim_{s \to \infty} \langle H_{\text{fl}}(t_i, s) \rangle_{\text{stoch}} = \frac{m}{2} \lim_{t_1, t_2 \to t_i} \partial_{t_1} \partial_{t_2} \lim_{s \to \infty} \langle X(t_1, s) X(t_2, s) \rangle_{\text{stoch}} \tag{4.6.47}$$

where we have used the time-splitting procedure to define the averaging of the velocities properly.

Substituting this average into (4.6.30) again gives the transition amplitudes. In particular, for a free particle or a harmonic oscillator, we obtain the same results (4.6.45) and (4.6.86).

Having these formulae to hand, we can develop a perturbation theory for systems with non-quadratic Hamiltonians based on an iterative solution of the Langevin equation under given boundary conditions.

4.6.3 Stochastic quantization of field theories

The original aim of Parisi and Wu (1981) was to use stochastic methods to quantize *gauge*-field theories without adding gauge-fixing terms. We shall sketch how this comes about later in this section, but we now start from the more simple example of scalar-field theory.

The formal extension of the stochastic quantization methods presented in the two preceding subsections to scalar-field theory in d dimensions encounters a little difficulty. Since in field theories we usually deal (at least at the first step of an investigation) with Green functions and not *directly* with transition amplitudes, it is more convenient to consider the stochastic quantization in the Euclidean-time formalism.

◇ Stochastic quantization of the scalar field theory

Similar to the one-dimensional case in section 4.6.1, the integrand in the path-integral representation of the generating functional for Green functions of this theory, that is $\mathfrak{N}^{-1} \exp\{-S[\varphi]\}$, is interpreted as the equilibrium distribution of the non-equilibrium statistical system. The scalar field $\varphi(x)$, $x = \{\boldsymbol{x}, t = x^4\}$, is generalized to a *stochastic field* $\varphi(x, s)$, driven by the Langevin equation

$$\frac{\partial \varphi(x, s)}{\partial s} = -\frac{\delta S[\varphi]}{\delta \varphi(x, s)} + \eta(x, s) \tag{4.6.48}$$

which is the direct generalization of (4.6.5). Note that s is the auxiliary 'stochastic' time, while the physical (Euclidean) time x^4 is included in the argument x of the field. The stochastic external field $\eta(x, s)$ have correlator functions which are straightforward generalizations of (4.6.6):

$$\langle \eta(x, s) \rangle_{\text{stoch}} = 0 \qquad \langle \eta(x_2, s_2) \eta(x_1, s_1) \rangle_{\text{stoch}} = 2\delta(s_2 - s_1) \delta(x_2 - x_1) \tag{4.6.49}$$

and with all higher connected correlation functions vanishing.

The Euclidean Green functions appeared as the large stochastic time limits:

$$\lim_{s \to \infty} \langle \varphi_\eta(x_1, s_1) \varphi_\eta(x_2, s_2) \cdots \varphi_\eta(x_m, s_m) \rangle_{\text{stoch}} = \mathfrak{N}^{-1} \int \mathcal{D}\varphi(x) \, \varphi_\eta(x_1) \varphi_\eta(x_2) \cdots \varphi_\eta(x_m) \exp\{-S[\varphi]\} \tag{4.6.50}$$

where the field $\varphi_\eta(x, s)$ satisfies the Langevin equation and hence implicitly depends on the external stochastic field η. A strict proof of this relation requires a suitable regularization of the ultraviolet divergences in the quantum field theory and is a rather complicated problem. However, we can show that it is indeed correct in the framework of the perturbation theory (Floratos and Iliopoulos 1983) (see also Rivers (1987)).

◇ **Perturbation theory based on the Langevin equation**

Let us consider φ^4-theory with a standard interaction Lagrangian $\mathcal{L}_{\text{int}} = g\varphi^4/4!$. In order to develop the perturbation theory, we start from the *free-field* Langevin equation which according to (4.6.48), reads as follows:

$$\frac{\partial \varphi(x, s)}{\partial s} = (\partial_\mu^2 - m^2)\varphi(x, s) + \eta(x, s) \quad (4.6.51)$$

with a solution of the form

$$\varphi_\eta(x, s) = \int_0^s ds' \int d^4x' \, D^{(\text{ret})}(x - x', s - s')\eta(x', s') \quad (4.6.52)$$

expressed in terms of the retarded Green function (i.e. $D^{(\text{ret})}(x, s) = 0$ for $s < 0$) and satisfying

$$\left[\frac{\partial}{\partial s} - (\partial_\mu^2 - m^2)\right] D^{(\text{ret})}(x - x', s - s') = \delta(x - x')\delta(s - s') \quad (4.6.53)$$

(for simplicity, we have chosen the vanishing initial condition: $\varphi_\eta(x, 0) = 0$). It is not difficult to see that the Green function is given by

$$D^{(\text{ret})}(x, s) = \theta(s) \int \frac{d^4k}{(2\pi)^4} \exp\{-s(k^2 + m^2) + ikx\}. \quad (4.6.54)$$

Using (4.6.52), we can now find the *non-equilibrium* correlation function $D^{(\text{st})}(x - x'; s, s') \stackrel{\text{def}}{\equiv} \langle \varphi_\eta(x, s)\varphi_\eta(x', s')\rangle_{\text{stoch}}$

$$\begin{aligned}D^{(\text{st})}(x - x'; s, s') &= \int_0^s d\sigma \int_0^{s'} d\sigma' \int dy \, dy' \, D^{(\text{ret})}(x - y, s - \sigma) \\&\quad \times D^{(\text{ret})}(x' - y', s' - \sigma')\langle \eta(y, \sigma)\eta(y', \sigma')\rangle_{\text{stoch}} \\&= 2\int_0^\infty d\sigma \int dy \, D^{(\text{ret})}(x - y, s - \sigma)D^{(\text{ret})}(x' - y, s' - \sigma). \quad (4.6.55)\end{aligned}$$

Substituting (4.6.54) in (4.6.55) yields

$$D^{(\text{st})}(x - x'; s, s') = \int \frac{d^4k}{(2\pi)^4} \widetilde{D}^{(\text{st})}(k; s, s') e^{ikx} \quad (4.6.56)$$

where

$$\widetilde{D}^{(\text{st})}(k; s, s') = \widetilde{D}_c(k)[\exp\{-(k^2 + m^2)|s - s'|\} - \exp\{-(k^2 + m^2)(s + s')\}] \quad (4.6.57)$$

with $\widetilde{D}_c(k) = (k^2 + m^2)^{-1}$ being the usual Euclidean free-field propagator (cf (3.1.93)). It is seen that at equal large stochastic time ($s = s' \to \infty$), the stochastic Green function converts into the usual field theoretical Green function:

$$\lim_{s \to \infty} \widetilde{D}^{(\text{st})}(k; s, s) = \widetilde{D}_c(k). \quad (4.6.58)$$

We used the simplest initial condition, $\varphi_\eta(x, 0) = 0$, but actually the results at large s do not depend on the choice of the initial condition (see problem 4.6.4, page 295).

If the interaction term $\mathcal{L}_{\text{int}} = \lambda \varphi^4/4!$ is included in the Euclidean action, the Langevin equation can *partially* be integrated out as

$$\varphi_\eta(x, s) = \int_0^s ds' \int d^4x'\, D^{(\text{ret})}(x - x', s - s')[\eta(x', s') + \tfrac{1}{6}\lambda \varphi_\eta^3(x', s')]. \tag{4.6.59}$$

This equation permits an iterative solution in powers of the coupling constant λ:

$$\varphi_\eta(x, s) = \int_0^s ds' \int d^4x'\, D^{(\text{ret})}(x - x', s - s')$$
$$\times \left\{ \eta(x', s') + \tfrac{1}{6}\lambda\left[\int_0^{s'} d\sigma \int d^4y\, D^{(\text{ret})}(x' - y, s' - \sigma)\eta(y, \sigma)\right]^3 + \mathcal{O}(g^2)\right\}. \tag{4.6.60}$$

From this approximate solution, we can again construct $D^{(\text{st})}(x - x'; s, s') \stackrel{\text{def}}{=} \langle \varphi_\eta(x, s)\varphi_\eta(x', s')\rangle_{\text{stoch}}$ or any other correlation function up to the order $\mathcal{O}(g^2)$. Further iterations generate higher powers of λ.

◊ The generating functional for Green functions in the stochastically quantized field theory

The generating functional for Green functions in scalar-field theory is the straightforward generalization of that for the zero-dimensional model (cf (4.6.13)–(4.6.20)). Thus we present only the result:

$$\mathcal{Z}[J] = \int \mathcal{D}\varphi\, p(\varphi(x, 0)) \exp\left\{ -S^{(\text{eff})}[\varphi] + \int d^4x \int_0^\infty ds\, J(x, s)\varphi(x, s)\right\} \tag{4.6.61}$$

where

$$S^{(\text{eff})}[\varphi] = \int dx\, ds \left\{ \frac{1}{4}\left(\frac{\partial \varphi}{\partial s}\right)^2 + \frac{1}{4}\left(\frac{\delta S[\varphi]}{\delta \varphi(x, s)}\right)^2 - \frac{1}{2}\frac{\delta^2 S[\varphi]}{\delta \varphi^2(x, s)}\right\} \tag{4.6.62}$$

is the effective Fokker–Planck action (S is the initial Euclidean action for field theory).

In φ^4-field theory, the effective action takes the form

$$S^{(\text{eff})}[\varphi] = S_0^{(\text{eff})}[\varphi] + S_{\text{int}}^{(\text{eff})}[\varphi] \tag{4.6.63}$$

where the free part $S_0^{(\text{eff})}[\varphi]$ reads as

$$S_0^{(\text{eff})}[\varphi] = \int dx\, ds \left\{ \frac{1}{4}\left(\frac{\partial \varphi}{\partial s}\right)^2 + ((\partial_\mu^2 - m^2)\varphi)^2 - \lambda \varphi^2\right\} \tag{4.6.64}$$

and the interaction part $S_{\text{int}}^{(\text{eff})}[\varphi]$ is

$$S_{\text{int}}^{(\text{eff})}[\varphi] = \frac{1}{4!}\int dx\, ds\left\{ \frac{\lambda^2 \varphi^6}{6} - 2\lambda \varphi^3(\partial_\mu^2 - m^2)\varphi\right\}. \tag{4.6.65}$$

The leading term of the finite-time free-field propagator (4.6.56) can be read immediately from (4.6.64) for $\lambda = 0$. For $\lambda \neq 0$, $S_0^{(\text{eff})}[\varphi]$ and $S_{\text{int}}^{(\text{eff})}[\varphi]$ provide the general rules for a diagrammatic representation of the stochastic Green functions.

The renormalizability of such a $(4 + 1)$-dimensional theory is not obvious, but it has been proven (Klauder and Ezawa 1983) that it is indeed renormalizable similarly to conventional φ^4-field theory.

◇ **Stochastic quantization and gauge fixing**

As we have already mentioned, the original aim of Parisi and Wu (1981) when they suggested the stochastic quantization method was to simplify quantization of *gauge*-field theories avoiding gauge fixing. We shall explain why the stochastic approach allows us to quantize without gauge condition on the example of the simplest Abelian gauge theory without matter (i.e. the theory of free photons).

Recall (see section 3.2) that in the case of the action for the Abelian gauge field A_μ (see (3.2.18) and (3.2.19))

$$S_{\text{YM}} = \tfrac{1}{4} \int dx\, F_{\mu\nu} F^{\mu\nu} = \tfrac{1}{2} \int dx\, A^\mu (-\partial^2 \delta_{\mu\nu} + \partial_\mu \partial_\nu) A^\nu \qquad (4.6.66)$$

(with $F_{\mu\nu} = \partial_\mu A_\nu - \partial_\nu A_\mu$) we are unable to construct a propagator by inserting the action directly into the path integral without modification, because the differential operator in this action is not invertible. However, suppose we extend A_μ to a stochastic field $A_\mu(x, s)$ satisfying the Langevin equation

$$\frac{\partial A_\mu(x, s)}{\partial s} = (\partial^2 \delta_{\mu\nu} - \partial_\mu \partial_\nu) A^\nu(x, s) + \eta_\mu(x, s). \qquad (4.6.67)$$

We take the external stochastic field $\eta_\mu(x, s)$ to have the following correlator functions (cf (4.6.6))

$$\langle \eta_\mu(x, s) \rangle_{\text{stoch}} = 0$$
$$\langle \eta_\mu(x_2, s_2) \eta_\nu(x_1, s_1) \rangle_{\text{stoch}} = 2\delta_{\mu\nu} \delta(x_2 - x_1) \delta(s_2 - s_1).$$

Then equation (4.6.67) with the boundary condition $A_\mu(x, 0) = 0$ is solved, in analogy with (4.6.52), as

$$A_\mu(x, s) = \int_0^s ds' \int d^4 x'\, D^{(\text{ret})}_{\mu\nu}(x - x', s - s') \eta_\mu(x', s') \qquad (4.6.68)$$

where $D^{(\text{ret})}_{\mu\nu}$ is the retarded Green function, satisfying

$$\left[\left(\frac{\partial}{\partial s} - \partial^2\right) \delta^{\mu\nu} + \partial^\mu \partial^\nu \right] D^{(\text{ret})}_{\nu\rho}(x, s) = \delta^\mu{}_\rho \delta(x) \delta(s). \qquad (4.6.69)$$

The advantage of (4.6.69) is that whereas $(-\partial^2 \delta_{\mu\nu} + \partial_\mu \partial_\nu)$ has no inverse, the operator

$$\left[\left(\frac{\partial}{\partial s} - \partial^2\right) \delta_{\mu\nu} + \partial_\mu \partial_\nu \right],$$

is non-singular. This means that the photon propagator in the formalism of the stochastic quantization can be constructed *without gauge fixing*.

To see this, let us pass to the Fourier transformed Langevin equation

$$\frac{\partial \widetilde{A}_\mu(k, s)}{\partial s} = -k^2 \left(\delta_{\mu\nu} - \frac{k_\mu k_\nu}{k^2}\right) \widetilde{A}^\nu(k, s) + \widetilde{\eta}_\mu(k, s). \qquad (4.6.70)$$

The corresponding Fourier transform of the retarded Green function reads as

$$\widetilde{D}^{(\text{ret})}_{\mu\nu}(k, s - s') = \left[\left(\delta_{\mu\nu} - \frac{k_\mu k_\nu}{k^2}\right) e^{-k^2 |s - s'|} + \frac{k_\mu k_\nu}{k^2}\right] \theta(s - s'). \qquad (4.6.71)$$

Now we can find the stochastic photon propagator:

$$\widetilde{D}^{(\text{st})}_{\mu\nu}(k; s, s')\delta(k + k') \stackrel{\text{def}}{\equiv} \langle \widetilde{A}_\mu(k, s)\widetilde{A}_\nu(k, s')\rangle_{\text{stoch}}$$

$$= \int_0^s d\sigma \int_0^{\sigma'} d\sigma' \int dy\, dy'\, \widetilde{D}^{(\text{ret})}_{\mu\lambda}(k, s - \sigma)\widetilde{D}^{(\text{ret})}_{\nu\rho}(k', s' - \sigma')\langle \widetilde{\eta}^\lambda(k, \sigma)\widetilde{\eta}^\rho(k', \sigma')\rangle_{\text{stoch}}$$

$$= 2\delta(k + k') \int_0^\infty ds \int dy\, \widetilde{D}^{(\text{ret})}_{\mu\lambda}(k, s - \sigma)\widetilde{D}^{(\text{ret})}_{\nu\rho}(k', s' - \sigma)\delta^{\lambda\rho}. \qquad (4.6.72)$$

After substituting the solution (4.6.71), we arrive at the final result:

$$\widetilde{D}^{(\text{st})}_{\mu\nu}(k; s, s') = \widetilde{D}^{\text{L}}_{\mu\nu}(k)[\exp\{-k^2|s - s'|\} - \exp\{-k^2|s + s'|\} + 2\min(s, s')\frac{k_\mu k_\nu}{k^2} \qquad (4.6.73)$$

where $\widetilde{D}^{\text{L}}_{\mu\nu}(k)$ coincides with the transverse gauge propagator (i.e. in the Lorentz gauge, cf (3.2.144)):

$$\widetilde{D}^{\text{L}}_{\mu\nu}(k) = \frac{1}{k^2}\left(\delta_{\mu\nu} - \frac{k_\mu k_\nu}{k^2}\right). \qquad (4.6.74)$$

As has been shown by Parisi and Wu (1981), the last term in (4.6.73) does not contribute to the gauge-invariant quantities. The remaining term, for large stochastic times $s = s' \to \infty$, gives the photon with the propagator $\widetilde{D}^{\text{L}}_{\mu\nu}(k)$.

◇ **Gauge freedom in the formalism of the stochastic quantization**

Stochastic quantization seems to lead to a specific choice of the gauge condition (namely, the Lorentz gauge). However, this is not true. To understand this, let us decompose the gauge field into transverse and longitudinal components:

$$A^{\text{T}}_\mu(k, s) = \left(\delta_{\mu\nu} - \frac{k_\mu k_\nu}{k^2}\right) A^\nu(k, s)$$

$$A^{\text{L}}_\mu(k, s) = \frac{k_\mu k_\nu}{k^2} A^\nu(k, s).$$

After a similar decomposition of η, Langevin equation (4.6.68) decomposes as

$$\frac{\partial A^{\text{T}}_\mu}{\partial s} = -k^2 A^{\text{T}}_\mu + \eta^{\text{T}}_\mu \qquad (4.6.75)$$

$$\frac{\partial A^{\text{L}}_\mu}{\partial s} = \eta^{\text{L}}_\mu. \qquad (4.6.76)$$

We can see that the longitudinal field satisfies the *frictionless* Brownian equation. This is a consequence of the gauge invariance of the photon action, which implies $\delta S_{\text{YM}}/\delta A^{\text{L}}_\mu = 0$. The consequence of this fact is the following: if we use, instead of $A_\mu(x, 0) = A_\mu(k, 0) = 0$, some non-zero initial condition $A_\mu(x, 0)$, the solution to (4.6.75) for the transverse part is (cf (4.6.88)):

$$A^{\text{T}}_\mu(k, s) = A^{\text{T(0)}}_\mu(k, s) + A_\mu(k, 0) \exp\{-k^2 s\} \qquad (4.6.77)$$

where $A^{\text{T(0)}}_\mu(k, s)$ is the solution for the vanishing initial condition. As $s \to 0$, $A^{\text{T}}_\mu(k, s) \to A^{\text{T(0)}}_\mu(k, s)$ irrespective of the boundary condition $A^{\text{T}}_\mu(k, 0)$. Thus, the transverse part of the photon field behaves as a Markov variable, losing all memory of the initial condition. This is not the case for the longitudinal part A_μ^L of the field which, with no damping term, always remembers its initial configuration $A_\mu^L(k, 0)$. In particular, if

$$A_\mu^L(k, 0) = k_\mu \alpha(k) \qquad (4.6.78)$$

equation (4.6.76) is solved as

$$A_\mu^L(k, 0) = \int_0^s ds'\, \theta(s - s') \eta_\mu^L(k, s') + k_\mu \alpha(k). \qquad (4.6.79)$$

As a result, the large-time stochastic propagator $\widetilde{D}^{(\text{st})}(k; s, s')$ does depend on the initial condition for the longitudinal field. Specifically, with the choice (4.6.78), the large-time behaviour is modified by the addition of the term $k_\mu k_\nu \alpha(k^2)/k^2$. Choosing an appropriate gauge function $\alpha(k)$, we can reproduce any gauge-equivalent form of the gauge propagator.

Thus, while the stochastic quantization method does not require the introduction of a gauge-fixing term into the Yang–Mills action, it proves to be totally equivalent to the standard path-integral quantization of gauge fields considered in chapter 3.

4.6.4 Problems

Problem 4.6.1. Calculate the determinant of the operator $\widehat{M}$ in (4.6.15), that is

$$\widehat{M} = \frac{\partial}{\partial s} + S''(x_{\widetilde{\eta}})$$

using its representation via path integrals over auxiliary (Grassmann) variables.

Hint. We introduce two Grassmann variables ('ghosts') $c(s)$ and $c^*(s)$ (cf (3.2.168)) and present (det $\widehat{M}$) in terms of the path integral:

$$\det \widehat{M} = \int \mathcal{D}c^*(s)\, \mathcal{D}c(s)\, \exp\left\{ -\int ds\, c^*(s) \widehat{M} c(s) \right\}.$$

We can calculate this path integral by a series expansion in analogy with the usual perturbation theory, in which the role of the 'free action' S_0 is attributed to the part of the exponent defined by the operator $\partial/\partial s$, that is, $S_0 = \int ds\, (c^* \partial c/\partial s)$. This series can be easily summed due to the specific form of the corresponding 'free propagator': $D_0(s - s') = \widetilde{\theta}(s - s')$. This leads to the vanishing of all loop diagrams but one (since $\widetilde{\theta}(s - s')\widetilde{\theta}(s' - s) = 0$ if $s' \neq s$). The result of the calculation is

$$\det \widehat{M} = \exp\left\{ \widetilde{\theta}(0) \int ds\, S''(x(s)) \right\}.$$

Problem 4.6.2. Prove that the constant c in (4.6.30) is indeed independent of x_f and x_i.

Hint. Use the fact that

$$\frac{\partial}{\partial x}|x\rangle = i\widehat{p}|x\rangle$$

(in coordinate representation) and carry out the calculations similarly to (4.6.28)–(4.6.30), but where the time derivatives are substituted with spatial ones and the Hamiltonian with the momentum operator. Then, use the relation from the classical Hamiltonian mechanics:

$$\frac{\partial S_{\text{cl}}}{\partial x_f} = p_{\text{c}}(t_f) \qquad \frac{\partial S_{\text{cl}}}{\partial x_i} = -p_{\text{c}}(t_i).$$

Problem 4.6.3. Using as a guide the example of a free particle considered in section 4.6.2, derive the transition amplitude for a quantum-mechanical *harmonic oscillator* in the framework of the *phase-space stochastic quantization* formalism (by the method of the Langevin equation).

Hint. Using the results of chapter 2, we find that the harmonic oscillator Hamiltonian

$$H = \frac{p^2}{2m} + \frac{1}{2}m\omega^2 x^2 \qquad (4.6.80)$$

corresponds to the following classical and quantum actions:

$$S_{\text{cl}} = \frac{\omega m}{2\sin\omega T}[(x_i^2 + x_f^2)\cos\omega T - 2x_i x_f]$$

$$S = \sum_{n=1}^{\infty}\left(\frac{n\pi}{T}x_n p_n - \frac{p_n^2}{2m} - \frac{1}{2}m\omega^2 x_n^2\right)\frac{T}{2} - p_0^2\frac{T}{8m} \qquad (4.6.81)$$

so that

$$\frac{dx_n}{ds} = \mathrm{i}\left(\frac{n\pi}{T}p_n - m\omega^2 x_n\right)\frac{T}{2} + \xi_n$$

$$\frac{dp_n}{ds} = \mathrm{i}\left(\frac{n\pi}{T}x_n - \frac{p_n}{m}\right)\frac{T}{2} + \eta_n$$

$$\frac{dp_0}{ds} = -\mathrm{i}\frac{T}{4m}p_0 + \eta_0. \qquad (4.6.82)$$

From

$$\left\langle p_n\frac{dp_0}{ds}\right\rangle_{\text{stoch}} \qquad \left\langle p_0\frac{dp_0}{ds}\right\rangle_{\text{stoch}}$$

we find, in the equilibrium limit,

$$\langle p_n p_0\rangle = 0 \qquad \langle p_0^2\rangle = -\mathrm{i}\frac{4m}{T} \qquad (4.6.83)$$

and from

$$\left\langle p_m\frac{dx_n}{ds}\right\rangle_{\text{stoch}}$$

combined with

$$\left\langle p_m\frac{dp_n}{ds}\right\rangle_{\text{stoch}}$$

we find

$$\langle p_n p_m\rangle = \frac{2\mathrm{i}m}{T}\frac{\delta_{nm}}{\left(\frac{n\pi}{T\omega}\right)^2 - 1}. \qquad (4.6.84)$$

Therefore (recalling the boundary condition $X(t_i) = 0$),

$$\lim_{s\to\infty}\langle H_{\text{fl}}(t_i, s)\rangle_{\text{stoch}} = \frac{1}{2m}\left(\sum_{n,m=1}^{\infty}\langle p_n p_m\rangle + \frac{\langle p_0^2\rangle}{4}\right) = -\frac{\mathrm{i}\omega}{2}\cot T\omega \qquad (4.6.85)$$

and the transition amplitude is obtained as

$$\langle x_f, t_f | x_i, t_i\rangle = c\frac{1}{\sqrt{\sin T\omega}}\mathrm{e}^{\mathrm{i}S_c} \qquad c = \sqrt{\frac{m\omega}{2\pi\mathrm{i}}}. \qquad (4.6.86)$$

Thus, the result for the harmonic oscillator does coincide with that obtained by the usual path-integral method (see chapter 2).

Problem 4.6.4. Prove that the stochastic Green function (4.6.55) for large $s = s'$ does not depend on a concrete choice of the initial condition for the Langevin equation (4.6.51).

Hint. The Fourier transformed Langevin equation (4.6.51) reads as

$$\frac{\partial \varphi(k,s)}{\partial s} = -(k^2 - m^2)\varphi(k,s) + \eta(k,s). \qquad (4.6.87)$$

If, instead of choosing $\varphi(k, 0) = 0$, we take an arbitrary initial condition, the solution is

$$\varphi_\eta(k,s) = \varphi_\eta^{(0)}(k,s) + \varphi_\eta(k,0)\exp\{-(k^2 + m^2)s\} \qquad (4.6.88)$$

where $\varphi_\eta^{(0)}(k,s)$ is the solution for the vanishing initial condition. As $s \to \infty$, $\varphi_\eta(k,s)$ relaxes to $\varphi_\eta^{(0)}(k,s)$. Therefore, $D^{(\text{st})}(x - x'; s, s') \stackrel{\text{def}}{\equiv} \langle\varphi_\eta(x,s)\varphi_\eta(x',s')\rangle_{\text{stoch}}$ does not depend on the choice of the initial condition for large $s = s'$.

4.7 Path-integral formalism and lattice systems

In the last section of this book we return essentially to our starting point. Recall that in all chapters of this book, to construct path integrals in stochastic processes, quantum mechanics, field theory and statistical physics, we have started from *discrete*-time or *discrete*-spacetime approximations. Then, our aim was to pass to the corresponding continuum limits which just leads to what is called a 'path integral'. However, in many cases, there are strong reasons for the direct investigation of the *discrete approximations* of the path integrals without going to the continuum limit. These cases can be separated into two groups:

(i) investigations of genuine lattice (physically discrete) systems (for example, spin systems in solid state physics);
(ii) investigations of physically continuous systems by non-perturbative methods based on the discretization of the spacetime (for example, computer simulations in the so-called *lattice gauge theory* of the fundamental interactions).

The objects which appear in the description of physically discrete systems (for example, an expression for the partition function for a lattice system) cannot be truly named 'path integrals'. In fact, they are usual multiple integrals.

To discretize continuous time or spacetime variables to calculate path integrals related to physically *continuous* systems, we have used this approach many times in this book (recall, for example, the calculation of the Wiener integrals in section 1.1.4 or the calculations for the quantum-mechanical oscillator potential in section 2.2.2). In fact, since the very definition of a path integral is heavily based on the discrete approximation (especially, in quantum mechanics), this proves to be a most reliable method of calculation. Such calculations become extremely important and fruitful in situations when there are simply no other suitable exact or approximate ways to reach physical results. This is true, in particular, for the gauge theory of *strong interactions* (QCD), cf section 3.2.8. The problem of self-consistent calculations of physical quantities for field systems in discrete spacetime in such a way that their results are valuable for the continuum limit, is a central issue for lattice gauge theory.

Both the theory of physical lattice systems and lattice gauge theory are very extensive subjects (for a general introduction, see, e.g., Kogut (1979); for a more extensive discussion of physical lattice systems, see, e.g., Feynman (1972a), Baxter (1982) and Izumov and Skryabin (1988); for the lattice gauge theory, see Creutz (1983) and Montvay and Münster (1994)). In this section, we shall discuss and briefly review some elements of these theories and their relation to the path-integral techniques.

4.7.1 Ising model as an example of genuine discrete physical systems

The simplest and most studied models of discrete physical systems are formulated according to the following general rules:

(i) The random variables s are chosen to lie on the intersections of a rectangular lattice with L sites in each dimension ($d = 1, 2, 3$).
(ii) The random variable s at each site takes on a finite number N of values.
(iii) The energy functional $H[s]$ is translationally invariant with periodic boundary conditions.
(iv) The interaction between variables is of finite range.

Some examples of such energy functionals in two dimensions are:

(i) The nearest-neighbour two-dimensional *Ising model* in a magnetic field

$$H[s] = -\sum_{j,k=1}^{L} [J_1 s_{jk} s_{(j+1)k} + J_2 s_{jk} s_{j(k+1)} + h s_{jk}] \qquad (4.7.1)$$

where $s_{jk} = \pm 1$.
(ii) The N-state chiral *Potts model*

$$H[s] = -\sum_{j,k=1}^{L} \sum_{n=1}^{N-1} [E_n^v (\sigma_{jk} \sigma_{(j+1)k}^*)^n + E_n^h (\sigma_{jk} \sigma_{j(k+1)}^*)^n + H_n \sigma_{jk}^n] \qquad (4.7.2)$$

where $\sigma_{jk}^N = 1$. When $N = 2$, we note that (4.7.2) reduces to (4.7.1).

The class of models restricted by conditions (i)–(iv) can be expanded. The variables could lie on the bonds as well as on the lattice sites. Continuous variables could be used and a familiar example of such a system is the n-component classical Heisenberg magnet

$$H[v] = -\sum_{j,k=1}^{L} [\boldsymbol{v}_{jk} \cdot \boldsymbol{v}_{(j+1)k} + \boldsymbol{v}_{jk} \cdot \boldsymbol{v}_{j(k+1)}] \qquad (4.7.3)$$

where $\boldsymbol{v} = \{v_1, \ldots, v_n\}$, $\boldsymbol{v}^2 = 1$. The lattice can also be triangular, hexagonal, etc.

As always in statistical physics, the principal problem is to calculate the partition function for this models:

$$\mathcal{Z}_\beta = \sum_\varphi e^{-\beta H[\varphi]} \qquad (4.7.4)$$

where φ is any of the lattice variables s, σ or $\boldsymbol{v}$.

Eventually, we are usually interested in the thermodynamic limit when

$$L \to \infty \qquad T \text{ fixed and positive} \qquad (4.7.5)$$

and the main quantity to be calculated in this case is the free energy per site in the dimension d:

$$f = -kT \lim_{L \to \infty} \frac{1}{L^d} \ln \mathcal{Z}. \qquad (4.7.6)$$

◇ **One-dimensional Ising model**

We start from the simplest example of the Ising model on a one-dimensional lattice.

The Hamiltonian for spins interacting through nearest neighbours on a one-dimensional lattice (chain) is given by

$$H = -J \sum_{i=1}^{N} s_i s_{i+1} + h \sum_{i=1}^{N} s_i. \tag{4.7.7}$$

This Hamiltonian describes a spin system on a lattice with N sites and subjected to an external magnetic field h, which is constant. The *classical* partition function for this system is

$$\mathcal{Z}_\beta = \sum_{s_i = \pm 1} e^{-\beta H}$$

$$= \sum_{s_i = \pm 1} \exp\left\{-\beta\left[-J \sum_{i=1}^{N} s_i s_{i+1} + h \sum_{i=1}^{N} s_i\right]\right\}. \tag{4.7.8}$$

The spin variables satisfy the periodic boundary condition (also called the *cyclicity condition*)

$$s_{i+N} = s_i. \tag{4.7.9}$$

◇ **Relating the one-dimensional Ising model to a quantum-mechanical system through the path integral**

To study the one-dimensional Ising model, let us use our experience in quantum mechanics. To this aim, consider the quantum-mechanical system described by the Hamiltonian

$$\widehat{H} = -\alpha \sigma_1 + \gamma \sigma_3 \tag{4.7.10}$$

where σ_i are the Pauli matrices and α and γ are two (for the time being) arbitrary constant parameters. Let $|s\rangle$ denote the two-component eigenstates of σ_3, such that

$$\sigma_3 |s\rangle = s|s\rangle \qquad s = \pm 1. \tag{4.7.11}$$

The matrix element

$$\langle s_f | e^{-t\widehat{H}} | s_i \rangle \tag{4.7.12}$$

can be approximated by the discrete version of the path integral

$$\langle s_f | e^{-t\widehat{H}} | s_i \rangle \approx \sum_{s_i = \pm 1} \langle s_f | e^{-\varepsilon \widehat{H}} | s_N \rangle \langle s_N | e^{-\varepsilon \widehat{H}} | s_{N-1} \rangle \cdots \langle s_2 | e^{-\varepsilon \widehat{H}} | s_i \rangle \tag{4.7.13}$$

where $\varepsilon = t/N$. The individual factors in (4.7.13) can be presented (up to $\mathcal{O}(\varepsilon^2)$-terms) in the following form (see problem 4.7.1, page 308)

$$\langle s_{i+1} | e^{-\varepsilon \widehat{H}} | s_i \rangle \approx \langle s_{i+1} | (1 + \varepsilon \alpha \sigma_1 - \varepsilon \gamma \sigma_3) | s_i \rangle + \mathcal{O}(\varepsilon^2)$$
$$= \tfrac{1}{4}(s_i + s_{i+1})^2 + \varepsilon \alpha \tfrac{1}{4}(s_i - s_{i+1})^2 - \varepsilon \gamma \tfrac{1}{2}(s_i + s_{i+1}) + \mathcal{O}(\varepsilon^2)$$
$$= \exp\left\{\frac{\Lambda}{4}(s_i - s_{i+1})^2 + \frac{\lambda}{2}(s_i + s_{i+1})\right\} + \mathcal{O}(\varepsilon^2) \tag{4.7.14}$$

where the constants Λ and λ are defined by the relations:
$$e^\Lambda = \varepsilon\alpha \qquad \sinh(\lambda) = -\varepsilon\gamma.$$

Therefore, we can write

$$\begin{aligned}
\mathrm{Tr}\, e^{-t\widehat{H}} &= \lim_{N\to\infty} \sum_{s_i=\pm 1} \exp\left\{\Lambda \sum_{i=1}^N \left(\frac{1}{2}(s_i - s_{i+1})\right)^2 + \lambda \sum_{i=1}^N \frac{1}{2}(s_i + s_{i+1})\right\} \\
&= \lim_{N\to\infty} \sum_{s_i=\pm 1} \exp\left\{\Lambda \sum_{i=1}^N \frac{1}{2}(1 - s_i s_{i+1}) + \lambda \sum_{i=1}^N s_i\right\} \\
&= \lim_{N\to\infty} e^{N\Lambda/2} \sum_{s_i=\pm 1} \exp\left\{-\frac{\Lambda}{2}\sum_{i=1}^N s_i s_{i+1} + \lambda \sum_{i=1}^N s_i\right\} \\
&= \lim_{N\to\infty} \left(\frac{t}{N}\alpha\right)^{N/2} \sum_{s_i=\pm 1} \exp\left\{-\frac{1}{2}\ln\frac{t\alpha}{N}\sum_{i=1}^N s_i s_{i+1} - \frac{t}{N\gamma}\sum_{i=1}^N s_i\right\}. \quad (4.7.15)
\end{aligned}$$

Thus, if we identify the parameters as follows:

$$-\frac{1}{2}\ln\frac{t\alpha}{N} = \beta \qquad \frac{t}{N\gamma} = h \qquad (4.7.16)$$

we find the following formal relation between the quantum-mechanical trace (with the Hamiltonian (4.7.10)) and the partition function $\mathcal{Z}_\beta$ of the one-dimensional classical Ising model defined by (4.7.8):

$$\mathrm{Tr}\, e^{-t\widehat{H}} = \lim_{N\to\infty} \left(\frac{t\alpha}{N}\right)^{N/2} \mathcal{Z}_\beta$$

where the temperature and the magnetic field are given by (4.7.16).

◇ Calculation of the partition function for the one-dimensional Ising model via the path-integral technique

It is interesting that a very simple way of explicit calculation of the partition function for the Ising model is to express it in a form quite similar to the discrete approximation of the path integral. Namely, manipulations analogous to those we have previously used to establish the relation with the quantum-mechanical problem, allow us to present the partition function as follows (see problem 4.7.2, page 309):

$$\begin{aligned}
\mathcal{Z}_\beta &= \sum_{s_i=\pm 1} \exp\left\{\beta\left[J\sum_{i=1}^N s_i s_{i+1} - h\sum_{i=1}^N s_i\right]\right\} \\
&= \sum_{s_i=\pm 1} \langle s_1|\mathsf{K}|s_N\rangle \langle s_N|\mathsf{K}|s_{N-1}\rangle \cdots \langle s_2|\mathsf{K}|s_1\rangle \\
&= \mathrm{Tr}\,\mathsf{K}^N = \xi_1^N + \xi_2^N \qquad (4.7.17)
\end{aligned}$$

where K is the 2×2 matrix

$$\begin{aligned}
\mathsf{K} &= e^{\beta J}[\cosh\beta h\, \mathbb{I} + e^{-2\beta J}\sigma_1 - \sinh\beta h\sigma_3] \\
&= \begin{pmatrix} e^{\beta(J-h)} & e^{-\beta J} \\ e^{-\beta J} & e^{\beta(J+h)} \end{pmatrix} \qquad (4.7.18)
\end{aligned}$$

and $\xi_{1,2}$ are its eigenvalues:

$$\xi_{1,2} = e^{\beta J}\cosh\beta h \pm (e^{2\beta J}\sinh^2\beta h + e^{-2\beta J})^{1/2}. \tag{4.7.19}$$

Knowledge of the explicit form of the partition function opens the possibility to obtain all the thermodynamical quantities, e.g., the magnetization $M = -(N\beta)^{-1}\partial \ln \mathcal{Z}_\beta/\partial h$ etc.

◇ Two-dimensional Ising model

The Hamiltonian of the Ising model can be straightforwardly generalized to higher space dimensions, but the calculation of the explicit expression for the partition function becomes more complicated. We shall consider only the simplest generalization: the *two*-dimensional Ising model. Once again, we shall consider the variables $s_{i_1 i_2} = \pm 1$, $i_1, i_2 = 1, \ldots, L$ on a *two-dimensional rectangular lattice*, the number L denoting the total number of sites along either of the two axes, while $N = L^2$ is the total number of sites and spins on the lattice. At the final stages, we assume $L^2 \to \infty$. The Hamiltonian describing the interaction of the spins (in the absence, for simplicity, of an external magnetic field) is now given by (we have already mentioned it as an example in (4.7.1)):

$$-\beta H(s) = \sum_{mn}[b_1 s_{mn} s_{(m+1)n} + b_2 s_{mn} s_{m(n+1)}] \qquad \beta = 1/(k_B T) \tag{4.7.20}$$

where $b_\alpha = J_\alpha/k_B T$ are the dimensionless coupling constants and J_α are the magnetic exchange energies. The partition function and the free energy per site are:

$$\mathcal{Z}_\beta = \sum_{s=\pm 1} e^{-\beta H(s)} \tag{4.7.21}$$

$$-\beta f = \lim_{N\to\infty} \frac{1}{N} \ln \mathcal{Z}_\beta \tag{4.7.22}$$

where the sum is taken over 2^N spin configurations provided by $s_{mn} = \pm 1$ at each site.

Making use of the identity for the typical bond weight: $e^{bss'} = \cosh b + \sinh b \cdot ss'$, which readily follows from $(ss')^2 = +1$, we find that

$$\mathcal{Z}_\beta = (2\cosh b_1 \cosh b_2)^N Q \tag{4.7.23}$$

where Q is the *reduced partition function*

$$Q \stackrel{\text{def}}{\equiv} 2^{-N} \sum_{s=\pm 1} \left[\prod_{mn}(1 + t_1 s_{mn} s_{(m+1)n})(1 + t_2 s_{mn} s_{m(n+1)})\right] \tag{4.7.24}$$

where $t_\alpha = \tanh b_\alpha$.

Noting that $s_{mn}^2 = 1$, we reach the following form of the reduced partition function:

$$Q = 2^{-N} \sum_{s=\pm 1} \prod_{mn}(1 + t_1 s_{mn} s_{(m+1)n} + t_2 s_{mn} s_{m(n+1)} + t_1 t_2 s_{mn} s_{(m+1)(n+1)}). \tag{4.7.25}$$

This form or the direct the expression (4.7.24) for the reduced partition function are suitable for high- and low-temperature expansions. We shall not go into the details of this perturbative technique which can be found in, e.g., Kogut (1979). Instead, since we are mainly interested in possible applications of path integrals, we shall show that the partition function can be presented as a discrete version of the path integral over *Grassmann variables*.

◇ Discrete approximation for fermionic path integrals and analytic solutions for two-dimensional Ising models

It was realized that the notion of integration over anticommuting Grassmann variables (see section 2.6.1) is a powerful tool to study the two-dimensional Ising model (Fradkin and Shteingradt (1978), Popov (1983), Plechko (1988), Itzykson and Drouffe (1989) and references therein).

A straightforward calculation shows that the following Gaussian fermionic integral:

$$Q = \int \prod_{mn} dc^*_{mn} dc_{mn} \exp\left\{ \sum_{mn} [(c_{mn}c^*_{mn} - t_1 c_{mn}c^*_{(m-1)n} - t_2 c_{mn}c^*_{mn-1} \right.$$
$$\left. - t_1 t_2 c_{mn}c^*_{(m-1)(n-1)}) - t_1 c_{mn}c_{(m-1)n} - t_2 c^*_{mn}c^*_{m(n-1)}] \right\} \quad (4.7.26)$$

is equal to the reduced partition function up to the boundary effects, which are inessential as $L^2 \to \infty$. Here c_{mn}, c^*_{mn} are totally anticommuting Grassmann variables, two per site. It is seen that this fermionic integral is quite similar to the discrete approximation of the fermionic *path integral* (cf section 2.6.1). The advantage of representation (4.7.26) is that it can be evaluated explicitly. This can be performed by passing to the momentum space by means of the following Fourier substitution:

$$c_{mn} = \frac{1}{L} \sum_{pq} c_{pq} \exp\left\{ i\frac{2\pi}{L}(mp + nq) \right\} \qquad c^*_{mn} = \frac{1}{L} \sum_{pq} c^*_{pq} \exp\left\{ -i\frac{2\pi}{L}(mp + nq) \right\} \quad (4.7.27)$$

where c_{pq}, c^*_{pq} are the new fermionic variables of integration. Integral (4.7.26) now appears in the form

$$Q = \int \prod_{pq} dc^*_{pq} dc_{pq} \exp\left\{ \sum_{pq} \left[c_{pq}c^*_{pq} \left(1 - t_1 \exp\left\{ i\frac{2\pi p}{L} \right\} \right.\right.\right.$$
$$\left. - t_2 \exp\left\{ i\frac{2\pi q}{L} \right\} - t_1 t_2 \exp\left\{ i\frac{2\pi}{L}(p+q) \right\} \right)$$
$$\left.\left. - t_1 c_{pq}c_{(L-p)(L-q)} \exp\left\{ i\frac{2\pi p}{L} \right\} - t_2 c^*_{(L-p)(L-q)} c^*_{pq} \exp\left\{ i\frac{2\pi q}{L} \right\} \right] \right\}. \quad (4.7.28)$$

Note that the fermionic measure transforms in a trivial way by passing from (4.7.26) to (4.7.28). This is because the Jacobian of substitution (4.7.27) is unity, as follows from the orthogonality of the Fourier eigenfunctions.

Integral (4.7.28) decouples into a product of simple low-dimensional integrals. Since only the variables with momenta pq and $(L-p)(L-q)$ interact in (4.7.28), integral (4.7.28) can be expressed as the product of the following independent factors:

$$\int dc^*_{(L-p)(L-q)} dc_{(L-p)(L-q)} \exp\{S_{pq} + S_{(L-p)(L-q)}\} \quad (4.7.29)$$

where

$$S_{pq} = \left[c_{pq}c^*_{pq} \left(1 - t_1 \exp\left\{ i\frac{2\pi p}{L} \right\} - t_2 \exp\left\{ i\frac{2\pi q}{L} \right\} - t_1 t_2 \exp\left\{ i\frac{2\pi}{L}(p+q) \right\} \right) \right.$$
$$\left. - t_1 c_{pq}c_{(L-p)(L-q)} \exp\left\{ i\frac{2\pi p}{L} \right\} - t_2 c^*_{(L-p)(L-q)} c^*_{pq} \exp\left\{ i\frac{2\pi q}{L} \right\} \right].$$

The elementary integral (4.7.29) can be evaluated readily by making use of definitions (2.6.33)–(2.6.36). By comparing (4.7.28) and (4.7.29), it follows that the partition function Q arises if we multiply the factors (4.7.29) over only one-half of the points in the momentum space. That is, we have to multiply factors (4.7.29) in such a way that if the given mode pq is already included into the product, then the conjugated mode $(L-p)(L-q)$ is not to be included, and vice versa. This means, in turn, that the total product of factors (4.7.29) over the complete set of momentum modes, $0 \leq p, q \leq L-1$, yields the *squared* partition function:

$$Q^2 = \prod_{p=0}^{L-1}\prod_{q=0}^{L-1}\left[(1+t_1^2)(1+t_2^2) - 2t_1(1-t_2^2)\cos\frac{2\pi p}{L} - 2t_2(1-t_1^2)\cos\frac{2\pi q}{L}\right]. \quad (4.7.30)$$

Respectively, the reduced free energy per site is:

$$-\beta f_Q = \frac{1}{2}\int_0^{2\pi}\int_0^{2\pi}\frac{dp\,dq}{2\pi\,2\pi}\ln[(1+t_1^2)(1+t_2^2) - 2t_1(1-t_2^2)\cos p - 2t_2(1-t_1^2)\cos q]. \quad (4.7.31)$$

The free energy (4.7.31) is associated with the *reduced* partition function, Q, while the true partition function is given by (4.7.23). Thus, the true free energy per site appears in the form

$$-\beta f = \ln 2 + \frac{1}{2}\int_0^{2\pi}\int_0^{2\pi}\frac{dp\,dq}{2\pi\,2\pi}\ln[\cosh 2b_1 \cosh 2b_2 - \sinh 2b_1 \cos p - \sinh 2b_2 \cos q]. \quad (4.7.32)$$

This is the well-known *Onsager solution* (Onsager 1944) for the free energy of the two-dimensional Ising model on a rectangular lattice. Let us assume that $t_1, t_2 > 0$ (or, equivalently, $J_1, J_2 > 0$, i.e. the ferromagnetic case). It is seen that if

$$1 - t_1 - t_2 - t_1 t_2 = 0 \qquad t_\alpha \equiv \tanh(J_\alpha/k_B T) \quad (4.7.33)$$

the integrand in (4.7.31) has the logarithmic singularity at $p = q = 0$. Thus, the condition (4.7.33) fixes the critical point of the Ising model. Equivalently, this condition can be written in the more usual form:

$$\sinh\left(\frac{2J_1}{k_B T}\right)\sinh\left(\frac{2J_2}{k_B T}\right) = 1. \quad (4.7.34)$$

Since the free energy is known, the specific heat C can be obtained by differentiation with respect to the temperature: $C = k_B\beta^2(\partial^2(-\beta f)/\partial\beta^2)$. The singularity in the specific heat appears to be logarithmic near T_c: $C/k_B \sim |\ln|\tau||$, $\tau = (T-T_c)/T_c \to 0$, as can be deduced from (4.7.31) or (4.7.32). Also, it is not difficult to find the spontaneous magnetization $M \equiv \langle s_{mn}\rangle$ for the model:

$$M^8 = 1 - \frac{1}{8}\frac{(1-t_1^2)^2(1-t_2^2)^2}{t_1^2 t_2^2}. \quad (4.7.35)$$

From (4.7.35), we find $M \sim |T - T_c|^{1/8}$ as $|T - T_c| \to 0$, with the universal value of the critical index $\beta = 1/8$ for the magnetization.

Thus, the integration over the anticommuting Grassmann fields is a powerful tool with which to analyze the two-dimensional Ising models. However, it is worth mentioning that in the presence of an external magnetic field, the Grassmann factorization method discussed earlier requires additional rather complicated combinatorial consideration (see, e.g., Popov (1983)).

4.7.2 Lattice gauge theory

Lattice gauge theory is a primary tool for the study of non-perturbative phenomena in hadronic physics. In addition to giving quantitative information on the confinement of quarks and gluons (i.e. the absence of free quarks and gluons), the approach yields first-principle calculations of hadronic spectra and matrix elements. At fixed lattice spacing, the quantities of interest are expressed as a discrete version of path integrals. These multiple integrals can be approximated by a variety of techniques borrowed from statistical mechanics. Especially promising is a numerical technique, the Monte Carlo method. The lattice provides an ultraviolet cutoff which allows the system to be placed on a computer. Perhaps most importantly, the cutoff is not based on perturbation theory, and thus we can study non-perturbative physics, such as confinement. Also, unlike some other non-perturbative schemes, in this case we have a well-defined system to study.

Lattice gauge theory is rather an old subject, going back to Wilson's work of the early 1970s (for a review, see, e.g., Kogut (1979), Creutz (1983) and Montvay and Münster (1994)). Through the 1980s, it grew into a major industry. The field is currently dominated by computer simulations, although it is, in fact, considerably broader.

It is worth noting that the ultimate goal of lattice gauge theory is to obtain results for the corresponding field theoretical systems in the continuum spacetime. Recall that in chapter 3 we started from discrete approximation of the path integrals in order to justify the definition and rules of calculation for truly *path* integrals in the continuous spacetime. After a suitable regularization (which can also be defined in *continuous* spacetime: an example is the dimensional regularization) we can obtain direct results for the continuous case. In contrast, in lattice (gauge) theory, all calculations are carried out within the discrete (lattice) approximation of the spacetime and of the corresponding multiple integrals. Only at the final step of the calculation do we recover results which are valid for the continuous spacetime limit. For this to be possible, the lattice system should satisfy certain conditions, in particular its parameters must be close to a critical point which corresponds to a second type of phase transition (see, e.g., Kogut (1979)). For the reader's convenience, we have schematically presented these two approaches (based on conventional path integrals and lattice field theory) to the calculation of field theoretical quantities in figure 4.7.

By now, lattice gauge theory has become a very extended subject with many specific problems, technical and theoretical achievements and restrictions, as well as computational methods. In this section, we shall discuss some basic ideas of this approach; for further details we refer the reader to the previously cited reviews and books.

◇ The pure gauge lattice action

Space and time are assumed to be discretized in a hypercubic lattice. It is usual to call the lattice spacing a. In quantum chromodynamics (QCD), a gauge-field configuration is described by a set of $SU(3)$ matrices attached to oriented links (bonds) of the lattice. The relation between these $SU(3)$ matrices and the usual continuum gauge field is:

$$\mathsf{U}_\mu(x) = \mathcal{P}\left\{\exp\left\{iag_0 \int_0^1 d\tau\, A^i_\mu(x + \tau a\bar{\mu})\frac{\lambda_i}{2}\right\}\right\} \qquad (4.7.36)$$

where $\mathsf{U}_\mu(x)$ is the $SU(3)$ matrix (see figure 4.8) attached to the link starting from the site x (x denotes a set of four integers labelling the site) and going to the site $x + a\hat{\mu}$ ($\mu = 0, \ldots, 3$); g_0 is the bare coupling constant, $\bar{\mu}$ is the unit vector in the positive μ direction, i is a color (i.e. $SU(3)$) index and λ_i ($i = 1, \ldots, 8$) are the Hermitian traceless 3×3 complex matrices (*Gell–Mann matrices*). $\mathcal{P}$ means a

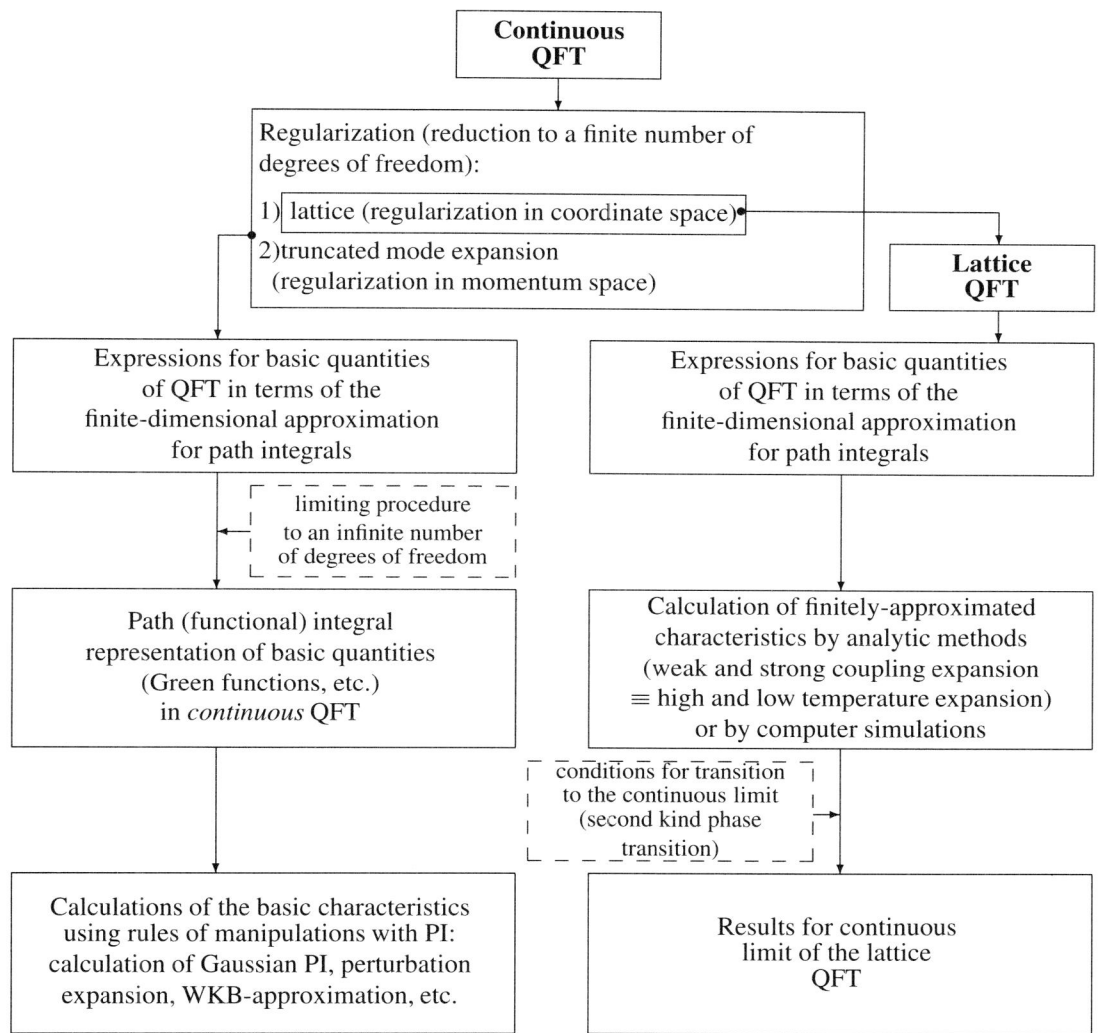

Figure 4.7. Schematic presentation of the relation between continuous quantum field theory, lattice field theory and path integrals.

path-ordered product. The same link, oriented in the opposite direction, corresponds to the inverse matrix

$$\mathsf{U}_{-\mu}(x + a\bar{\mu}) = \mathsf{U}_{\mu}^{-1}(x). \tag{4.7.37}$$

A gauge transformation is represented by an arbitrary set of $SU(3)$ matrices labeled by the sites of the lattice $\mathsf{g}(x)$ and acts as

$$\mathsf{U}_{\mu}(x) \to \mathsf{g}(x)\mathsf{U}_{\mu}(x)\mathsf{g}^{-1}(x + a\bar{\mu}) \tag{4.7.38}$$

A *plaquette* is an elementary square composed by four adjacent links, to which the following $SU(3)$ matrix is attached:

$$\mathsf{P}(x)_{\mu,\nu} = \mathsf{U}_{\mu}(x)\mathsf{U}_{\nu}(x + a\bar{\mu})\mathsf{U}_{\mu}^{-1}(x + a\bar{\nu})\mathsf{U}_{\nu}^{-1}(x) \tag{4.7.39}$$

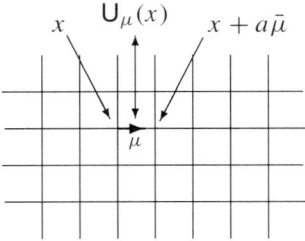

Figure 4.8. The attachment of lattice gauge fields.

the trace of which is gauge invariant which can be easily checked. The simplest pure gauge lattice action is:

$$S_{\text{lat}}[\mathsf{U}] = \sum_{x,\mu,\nu} \eta^{\mu\nu} \frac{2}{g^2} \operatorname{Re}\{\operatorname{Tr}[1 - \mathsf{P}(x)_{\mu,\nu}]\}$$

$$\xrightarrow[a\to 0]{} \frac{1}{4} \sum_{i,\mu,\nu} \int d^4x \, F^i_{\mu\nu}(x) F^{i\mu\nu}(x) \qquad (i = 1, \ldots, 8) \qquad (4.7.40)$$

for any gauge configuration defined by the set of $\mathsf{U}_\mu(x)$ for all links.

We restrict ourselves to the $SU(3)$ gauge group because of its physical importance and for concreteness. In fact, the construction can, quite straightforwardly, be generalized to any gauge group.

It is necessary to note that a large number of alternative gauge-invariant lattice actions which have the same limit when $a \to 0$ as the action in (4.7.40) exist.

◇ Starting point of lattice gauge theory: discrete approximation of the Euclidean path integral

The principal object in lattice field theory is the lattice (discrete) form of the path integral (in Euclidean space) for the corresponding gauge theory. In particular, the two-point correlation function for the local quantities $\mathcal{O}(x)$, $\mathcal{O}'(y)$ is given by

$$\frac{\int \prod_{x,\mu,\nu} d\mathsf{U}(x)_{\mu\nu} \, \exp\{-S_{\text{lat}}[\mathsf{U}]\} \mathcal{O}(x) \mathcal{O}'(y)}{\int \prod_{x,\mu,\nu} d\mathsf{U}(x)_{\mu\nu} \, \exp\{-S_{\text{lat}}[\mathsf{U}]\}} \qquad (4.7.41)$$

where $S_{\text{lat}}[\mathsf{U}]$ is given by (4.7.40).

Although the lattice represents a broad framework for the non-perturbative definition of a field theory, the subject is currently dominated by one approach, that of Monte Carlo simulation.

◇ Monte Carlo simulations

The crucial remark is that the positivity of $S_{\text{lat}}[\mathsf{U}]$ in (4.7.40) gives to $\exp\{-S_{\text{lat}}[\mathsf{U}]\}$ the meaning of a probability distribution, properly normalized by the denominator in (4.7.41), and (4.7.41) is simply the formula for a mean value in a probabilistic sense: thus we use the analogy of a field theory with statistical physics in the path-integral formalism, which we have already stressed many times. Although the probabilistic ensemble in (4.7.41) is the huge ensemble of Euclidean gauge-field configurations, most of them contribute for an exponentially suppressed amount. It was suggested that an algorithm called *Monte Carlo simulations* that selects gauge-field configurations at random according to the probability law $\exp\{-S_{\text{lat}}[U]\}/\mathcal{Z}$ should be used. A Monte Carlo program sweeps over a lattice stored in a computer memory and makes random changes biased by this 'Boltzmann weight'. Such an algorithm will automatically discard the configurations with negligible contributions. Once a large number N of uncorrelated configurations has been produced by the algorithm, the result is

$$\frac{\int \prod_{x,\mu,\nu} dU(x)_{\mu\nu} \, e^{-S_{\text{lat}}} \mathcal{O}(x) \mathcal{O}'(y)}{\int \prod_{x,\mu,\nu} dU(x)_{\mu\nu} \, e^{-S_{\text{lat}}}} = \frac{1}{N} \sum_{n=1}^{N} \mathcal{O}_n(x) \mathcal{O}'_n(y) + \mathcal{O}\left(\frac{1}{\sqrt{N}}\right) \quad (4.7.42)$$

where $\mathcal{O}_n(x)$ is the quantity $\mathcal{O}(x)$ with the values of the fields in the nth gauge configuration. Of course, many different algorithms may be used, improved, tested and intense activity continues on this issue.

One of the attractive features of the technique is that the entire lattice is available in the computer memory; so, in principle, we can measure anything. On the other hand, there are inherent statistical fluctuations which may make some things hard to extract. This represents a new aspect of theoretical physics, wherein theorists have statistical errors. In addition, these calculations have several sources of systematic errors, such as the effects of finite volume and finite lattice spacing. Quark fields, added to the pure Yang–Mills $SU(3)$-fields, introduce further sources of error. Furthermore, many calculations are made feasible by what is termed the valence or quenched approximation, wherein virtual quark loops are neglected.

◇ Quark actions and the problem of simulating fermions

Up to now, we have considered QCD with only gauge fields. Quark fields introduce serious unsolved problems. One problem is the issue of finding a reasonable computer algorithm for simulating fermions. In particular, since the quark fields are anticommuting, the full action is not an ordinary number, and the analogy with classical statistical mechanics breaks down. Algorithms in current practice begin by formally integrating out (at the stage of *continuous gauge theory*) the fermions, to give a determinant

$$\mathcal{Z} = \int \mathcal{D}A \, \mathcal{D}\bar{\psi} \, \mathcal{D}\psi \, \exp\{-[S_{\text{YM}} + \bar{\psi}(\gamma^\mu \partial_\mu - i\gamma^\mu A_\mu + m)\psi]\}$$

$$= \int \mathcal{D}A \, e^{-S_{\text{YM}}} \det(\gamma^\mu \partial_\mu - i\gamma^\mu A_\mu + m).$$

After the discretization, this determinant is, however, of a rather huge matrix, and is quite tedious to simulate. Over the years, many clever tricks have been used to simplify the problem.

The other old problem with quarks has to do with the issue of *fermion doubling* and chiral symmetry. Quark fields are located on lattice sites, and look quite like continuum quark fields, except for a different normalization: four spin and three color components. The naive action that we are tempted to use for quarks on the lattice is

$$S_{\text{quarks}} = \sum_x \left\{ \frac{1}{2a} [\bar{\psi}(x)(-\gamma_\mu) U_\mu(x) \psi(x + a\bar{\mu}) + (\bar{\mu} \to -\bar{\mu}, U_\mu \to U_\mu^{-1})] + \bar{\psi}(x) m \psi(x) \right\} \quad (4.7.43)$$

whose formal limit when $a \to 0$ is the standard continuum QCD quark action and which is gauge invariant as well as chiral invariant when $m = 0$. However, this naive action is not satisfactory because it encounters the 'doubling problem'. The point is that for one species of quarks in (4.7.43), the quark spectrum has 16 quarks in the continuum limit $a \to 0$.

Let us discuss the problem in somewhat more detail in one space dimension. A naive discretization of the Dirac Hamiltonian is

$$\widehat{H}_0 = K \sum_{\alpha=1}^{N} i(\widehat{a}^\dagger_\alpha \widehat{a}_{\alpha+1} - \widehat{b}^\dagger_\alpha \widehat{b}_{\alpha+1}) + m \sum_{\alpha=1}^{N} \widehat{a}^\dagger_\alpha \widehat{b}_\alpha + h.c.$$

where $\hat{a}_\alpha$ and $\hat{b}_\alpha$ are fermionic annihilation operators on the sites α located along a line and $h.c.$ denotes the Hermitian conjugate terms. The operators $(\hat{a}_\alpha^\dagger, \hat{a}_\alpha)$, $(\hat{b}_\alpha^\dagger, \hat{b}_\alpha)$ represent the upper and lower components of a two-component spinor (the index α labels the sites on a one-dimensional lattice). The spectrum of single-particle states for this Hamiltonian is easily found in momentum space:

$$E^2 = m^2 + 4K^2 \sin^2(q)$$

where q runs from 0 to 2π. Filling the negative energy states to form a Dirac sea, the physical excitations consist of particles as well as antiparticle 'holes'. The doubling problem manifests itself in the fact that there are low-energy excitations for the momenta q in the vicinity of π, as well as 0. In d spatial dimensions, this doubling increases to a factor of 2^d.

A simple solution to the doubling was presented by Wilson (1977), who added a term that created a momentum-dependent mass

$$\hat{H} = \hat{H}_0 - rK \sum_{\alpha=1}^{N} (\hat{a}_\alpha^\dagger \hat{b}_{\alpha+1} + \hat{b}_\alpha^\dagger \hat{a}_{\alpha+1} + h.c.)$$

where r is called the Wilson parameter. The energy spectrum is now

$$E^2 = 4K^2 \sin^2(q) + (m - 2Kr\cos(q))^2$$

and we see that the states at q near π have a different energy than those near zero. For the continuum limit, the parameters should be adjusted so that the extra states become infinitely heavy.

This Hamiltonian has a special behaviour when $2Kr = m$. In this case, one of the fermion species becomes massless. This provides a mechanism for obtaining light quarks and chiral symmetry. Unfortunately, when the gauge interactions are turned on, the parameters renormalize, and tuning becomes necessary to maintain the massless quarks. This is the basis of the conventional approach to chiral symmetry with Wilson fermions: we tune the 'hopping parameter' K until the pion is massless, it is then called the chiral limit.

In the four-dimensional spacetime, the *Wilson action* takes the form

$$S_{\text{Wilson}} = \sum_x \left\{ \frac{1}{2a} \sum_\mu [\bar{\psi}(x)(r - \gamma_\mu)\mathsf{U}(x)_\mu \psi(x + a\bar{\mu}) + (\bar{\mu} \to -\bar{\mu}, \mathsf{U}_\mu \to \mathsf{U}_\mu^{-1})] \right.$$
$$\left. + \bar{\psi}(x)\left(m + \frac{r}{a}\right)\psi(x) \right\}. \tag{4.7.44}$$

The effect of the additional terms proportional to r is to yield a mass $\mathcal{O}(r/a)$ to the 15 doublers, leaving only one quark with a finite mass when $a \to 0$, as desired.

Other quark actions that improve the $a \to 0$ limit also exist. All these aim at the same theory, QCD, and the lattice versions may be viewed as a class of QCD regularization scheme. Discretization cuts off the ultraviolet singularities. The different actions are different regularization schemes. Furthermore, any regularization procedure of a field theory has to be complemented by a renormalization scheme.

Clearly, all regularization and renormalization schemes must give the same physical quantities when $a \to 0$ and the volume goes to infinity. However, due to the numerical uncertainties and to the finite value used for a and the volume, these different methods do indeed differ somehow in their physical predictions. These differences are taken as a tool to estimate the systematic errors due to finite a, finite volume, etc. The statistical errors due to the finite number of gauge configurations used in the Monte Carlo may be estimated from the Monte Carlo itself, under some assumptions about the probability distribution of the quantities in question.

◇ **The quenched approximation**

The difficulties with simulating dynamical fermions have led to the majority of simulations being done in the 'valence' or 'quenched' approximation. Here the feedback of the fermion determinant on the dynamical gauge fields is ignored (i.e. it is replaced by unity). Hadrons are studied via quark propagators in a gauge field obtained in a simulation of gluon fields alone. In terms of Feynman diagrams, all gluonic exchanges are included between the constituent quarks, but the effects of virtual quark production ('quark loops' in the language of Feynman diagrams), beyond simple renormalization of the gauge coupling, are dropped. The primary motivation is to save orders of magnitude in computer time. These dynamical quark loops increase by several orders of magnitude the computation time and also the difficulty. This becomes even worse for larger lattices, since the computation time increases like $\propto N^{4/3}$ instead of $\propto N$ for a pure gauge theory (N being the number of lattice sites). While this may seem a drastic approximation, the fact that the naive quark model works so well hints that things might not be so bad.

◇ **An example: computation of a pseudoscalar meson mass and of f_π**

Let us consider in short how we compute a physical quantity on a simple example. The axial-current two-point correlation function in Euclidean spacetime and integrated over spatial coordinates reads as

$$\int d^3x \, \langle 0|(\hat{\bar{\psi}}(0)\gamma_0\gamma_5\hat{\psi}(0))(\hat{\bar{\psi}}(x_4,\boldsymbol{x})\gamma_0\gamma_5\hat{\psi}(x_4,\boldsymbol{x}))|0\rangle = \sum_n |\langle 0|\hat{\bar{\psi}}(0)\gamma_0\gamma_5\hat{\psi}(0)|n\rangle|^2 \frac{e^{-m_n x_4}}{2m_n}$$

$$\stackrel{x_4 \to \infty}{\simeq} |\langle 0|\hat{\bar{\psi}}(0)\gamma_0\gamma_5\hat{\psi}(0)|\pi\rangle|^2 \frac{e^{-m_\pi x_4}}{2m_\pi} \equiv f_\pi^2 \frac{m_\pi}{2} e^{-m_\pi x_4} \qquad (4.7.45)$$

where f_π is the so-called leptonic decay constant of the π meson. The sum $\sum_n$ is a sum over a complete set in the Hilbert space that is coupled to the vacuum by the axial current. At large time, due to the exponential decay, the lightest state dominates the sum. We denoted the latter state as π, but it can also be some other pseudoscalar meson, depending on the mass of the quarks.

The two-point correlation function in (4.7.45) may be computed by the Monte Carlo method described earlier. Once the calculation is performed, the large x_4 exponential slope gives the mass m_π, and once the latter is measured, the prefactor provides an estimate of the leptonic decay constant f_π. An obvious limitation of this method is that the signals produced by the excited states are difficult to extract from the statistical noise of the dominant ground state.

The practical limitations in the computing power lead to limitations for the domain of lattice QCD: the momenta are quantized to a few allowed values due to the finite volume, and also bounded to values below or equal to ~ 1 GeV. Generally speaking, only ground states and only systems with one hadron at a time are available.

Another area of extensive investigation in lattice gauge theory is the phase transition of the vacuum to a plasma of free quarks and gluons at a temperature of a few hundred MeV. Both theoretical analysis and numerical simulations have shown the existence of a high temperature regime wherein confinement is lost and chiral symmetry is manifestly restored.

◇ **Hamiltonian lattice gauge theory**

In 1975, Kogut and Susskind (see Kogut (1979) and references therein) derived a lattice Hamiltonian in which only the $(d-1)$-dimensional space is discretized while the time variable remains continuous. Theoretically, in the continuum limit this approach is equivalent to Wilson's Lagrangian formulation of lattice gauge theory where the gauge-field theory is discretized on the d-dimensional spacetime lattice.

The Hamiltonian formulation starts from the lattice Hamiltonian

$$H = \frac{g^2}{2a} \sum_{l,i} E_l^i E_l^i - \frac{1}{a} S_{\text{lat}}[\mathsf{U}] \qquad (4.7.46)$$

where E_l^i stands for the chromo-electric fields (cf (3.2.97)) on the three-dimensional links l. The mass spectrum can be obtained directly by solving the eigen-equation $\widehat{H}\Psi[U] = \epsilon_\Psi \Psi[U]$. Here ϵ_Ψ is the eigenvalue of $\widehat{H}$. When $a \to 0$, a huge number of gauge configurations are correlated, and it is very difficult to diagonalize the Hamiltonian with sufficient accuracy. In a feasible calculation, the gauge configuration space has to be truncated. An inappropriate truncation scheme often violates the continuum limit, and destroys the scaling behaviour for the physical quantities. Therefore, special care must be taken when choosing a truncation scheme. On the other hand, the Hamiltonian method does have some advantages over the Lagrangian method. In particular, it is relatively simple to obtain wavefunctions of the hadronic states.

Since the lattice provides a non-perturbative definition of a field theory, there have been numerous efforts in the methods are used on other models. A particularly active area has been developed towards understanding gravity. The general idea would be to discretize the points of spacetime and then do a sum over curvatures. So far, the results of this program have been limited, but no other non-perturbative approach to quantum gravity has yet been successful and the potential payoff is great.

There have been suggestions that there might indeed be some fundamental (*physical*) lattice or *fundamental minimal length* at a scale below current observations (e.g., at the scale of the Planck length) and this assumption opens up a vast number of variations. The main obstacles which appear in the path of a practical implementation of this suggestion are the necessity to match the requirements of the relativistic Poincaré invariance. The latter is obviously violated by a solid lattice structure introduced 'by hand'. During the last years, essentially new possibilities to achieve this goal based on the so-called *non-commutative geometry* have appeared (for a review see, e.g., Chaichian and Demichev (1996)).

4.7.3 Problems

Problem 4.7.1. Calculate the matrix element $\langle s_{i+1}|e^{-\varepsilon \widehat{H}}|s_i\rangle$ for Hamiltonian (4.7.10) which is the quantum-mechanical counterpart of the Ising Hamiltonian (4.7.7) and show that it can be presented in the form (4.7.14).

Hint. We start from the relations

$$\begin{aligned}\langle s_{i+1}|s_i\rangle &= \tfrac{1}{4}(s_i + s_{i+1})^2, \\ \langle s_{i+1}|\sigma_1|s_i\rangle &= \tfrac{1}{4}(s_i - s_{i+1})^2 \\ \langle s_{i+1}|\sigma_3|s_i\rangle &= \tfrac{1}{2}(s_i + s_{i+1}).\end{aligned} \qquad (4.7.47)$$

From the fact that for any i, $s_i = \pm 1$, we can easily deduce the set of algebraic identities:

$$\begin{aligned}(\tfrac{1}{2}(s_i + s_{i+1}))^{2n} &= (\tfrac{1}{2}(s_i + s_{i+1}))^2 \qquad \text{for } n \geq 1 \\ (\tfrac{1}{2}(s_i + s_{i+1}))^{2n+1} &= \tfrac{1}{2}(s_i + s_{i+1}) \qquad \text{for } n \geq 0 \\ (\tfrac{1}{2}(s_i - s_{i+1}))^{2n} &= (\tfrac{1}{2}(s_i + s_{i+1}))^2 \qquad \text{for } n \geq 1 \\ (\tfrac{1}{2}(s_i + s_{i+1}))^n (\tfrac{1}{2}(s_i - s_{i+1}))^m &= 0 \qquad \text{for } n, m \geq 1 \\ (\tfrac{1}{2}(s_i + s_{i+1}))^2 + (\tfrac{1}{2}(s_i - s_{i+1}))^2 &= 1\end{aligned} \qquad (4.7.48)$$

Path-integral formalism and lattice systems 309

which allows us to make the following transformations (Λ and λ are arbitrary constant parameters):

$$\exp\left\{\Lambda\left(\frac{1}{2}(s_i - s_{i+1})\right)^2 + \lambda\frac{1}{2}(s_i + s_{i+1})\right\}$$

$$= 1 + \sum_{n=1}^{\infty}\frac{1}{n!}\left[\Lambda\left(\frac{1}{2}(s_i - s_{i+1})\right)^2 + \lambda\frac{1}{2}(s_i + s_{i+1})\right]^n$$

$$= 1 + \sum_{n=1}^{\infty}\frac{1}{n!}\left[\Lambda^n\left(\frac{1}{2}(s_i - s_{i+1})\right)^{2n} + \lambda^n\left(\frac{1}{2}(s_i + s_{i+1})\right)^n\right]$$

$$= 1 + \sum_{n=1}^{\infty}\frac{\Lambda^n}{n!}\left(\frac{1}{2}(s_i - s_{i+1})\right)^2 + \sum_{n=1}^{\infty}\frac{\lambda^{2n+1}}{(2n+1)!}\left(\frac{1}{2}(s_i + s_{i+1})\right)^{2n+1}$$

$$+ \sum_{n=1}^{\infty}\frac{\lambda^{2n}}{(2n)!}\left(\frac{1}{2}(s_i + s_{i+1})\right)^2$$

$$= 1 + (e^\Lambda - 1)\left(\frac{1}{2}(s_i - s_{i+1})\right)^2 + \sum_{n=1}^{\infty}\frac{\lambda^{2n+1}}{(2n+1)!}\frac{1}{2}(s_i + s_{i+1})$$

$$+ \sum_{n=1}^{\infty}\frac{\lambda^{2n}}{(2n)!}\left(\frac{1}{2}(s_i + s_{i+1})\right)^2$$

$$= 1 + (e^\Lambda - 1)\left(\frac{1}{2}(s_i - s_{i+1})\right)^2 + \sinh\lambda\frac{1}{2}(s_i + s_{i+1})$$

$$+ (\cosh\lambda - 1)\left(\frac{1}{2}(s_i + s_{i+1})\right)^2$$

$$= \cosh\lambda\left(\frac{1}{2}(s_i + s_{i+1})\right)^2 + e^\Lambda\left(\frac{1}{2}(s_i - s_{i+1})\right)^2 + \sinh\lambda\frac{1}{2}(s_i + s_{i+1}). \quad (4.7.49)$$

Finally, neglecting the terms of the order $\mathcal{O}(\varepsilon^2)$, we can identify

$$\cosh\lambda = 1 \qquad \sinh\lambda = -\varepsilon\gamma \qquad e^\Lambda = \varepsilon\alpha \quad (4.7.50)$$

and this leads to the required result (in the required approximation).

Problem 4.7.2. Derive the partition function for the one-dimensional Ising model in the form (4.7.17) which is similar to the discrete approximation for the path integral representing the trace of the operator K given in (4.7.18). Calculate the eigenvalues of the matrix K (the answer is presented in (4.7.19)).

Hint. Making use of the identities (4.7.48), we derive

$$\exp\left\{\beta J s_i s_{i+1} - \frac{\beta h}{2}(s_i + s_{i+1})\right\} = e^{\beta J}\exp\left\{-2\beta J\left[\frac{1}{2}(s_i - s_{i+1})\right]^2 - \frac{\beta h}{2}(s_i + s_{i+1})\right\}$$

$$= e^{\beta J}\left[\cosh(\beta h)\left[\frac{1}{2}(s_i + s_{i+1})\right]^2 + e^{-2\beta J}\left[\frac{1}{2}(s_i - s_{i+1})\right]^2 - \sinh(\beta h)\frac{1}{2}(s_i + s_{i+1})\right].$$

If we define the matrix K as in (4.7.18), its matrix elements between the eigenstates $|s_i\rangle$ and $|s_{i+1}\rangle$ of the operator σ are calculated to be

$$\langle s_{i+1}|K|s_i\rangle = \exp\{\beta J s_i s_{i+1} - \beta h \tfrac{1}{2}(s_i + s_{i+1})\}$$

and this proves formulae (4.7.17) and (4.7.18). The eigenstates of the matrix K can easily be obtained from the equation: $\det(K - \xi \mathbb{I}) = 0$.

Note that this method of evaluating the partition function is known as the matrix method and K is called the *transfer matrix* (cf also next problem).

Problem 4.7.3. Find the form of the evolution operator $\widehat{U}_T$ (also called a *transfer matrix*) for the harmonic oscillator on a *discrete-time lattice* with the spacing a, postulating that it has the following matrix elements (for an *arbitrary* value of the spacing a):

$$\langle x_{i+1}|\widehat{U}_T(a)|x_i\rangle = \exp\left\{-\frac{1}{2\pi}\left[\frac{1}{a}(x_{i+1}-x_i)^2 + \frac{1}{2}\omega^2 a x_{i+1} + \omega^2 a x_i^2\right]\right\}.$$

Note that the usual evolution operator $\widehat{U}(t) = \exp\{-\widehat{H}_{\text{h.o.}} t\}$ (in imaginary time) has the same matrix elements in the limit of *infinitesimally small* time:

$$\langle x_{i+1}|\widehat{U}(\varepsilon)|x_i\rangle|_{\varepsilon \to 0} \approx \exp\left\{-\frac{1}{2\pi}\left[\frac{1}{\varepsilon}(x_{i+1}-x_i)^2 + \frac{1}{2}\omega^2 \varepsilon x_{i+1} + \omega^2 \varepsilon x_i^2\right]\right\}$$

(cf (2.2.31)).

Hint. Use the relation

$$\langle x'|\exp\{-\tfrac{1}{2}a\widehat{p}^2\}|x\rangle = \int dp\, dp'\, \langle x'|p'\rangle\langle p'|\exp\{-\tfrac{1}{2}a p^2\}|p\rangle\langle p|x\rangle$$
$$= \text{constant} \exp\{-\tfrac{1}{2}a\hbar^2(x'-x)^2\}$$

to obtain

$$\widehat{U}_T(a) = \exp\left\{-\frac{1}{4\hbar^2}a\omega^2\widehat{x}^2\right\}\exp\left\{-\frac{1}{2\hbar}a\omega^2\widehat{p}^2\right\}\exp\left\{-\frac{1}{4\hbar^2}a\omega^2\widehat{x}^2\right\}.$$

The operator ordering in this formula is important.

Problem 4.7.4. Show that in the limit of infinitesimally small evolution time, the logarithm of the transfer matrix $\widehat{U}_T(a)$ obtained in the preceding problem 4.7.3 becomes the usual Hamiltonian for the harmonic oscillator, that is

$$\widehat{H}_a \stackrel{\text{def}}{\equiv} -\left(\frac{\hbar}{a}\ln\widehat{U}_T(a)\right) \xrightarrow[a\to 0]{} \widehat{H}_{\text{h.o.}} = \frac{1}{2}(\widehat{p}^2 + \omega^2 \widehat{x}^2).$$

Hint. Use the Baker–Campbell–Hausdorff formula (2.2.6) to show that in the limit $a \to 0$,

$$\widehat{U}_T(a) \approx \exp\left\{-\frac{a}{\hbar}[\widehat{H}_{\text{h.o.}} + \mathcal{O}(a)]\right\}.$$

Supplements

I Finite-dimensional Gaussian integrals

- The basic one-dimensional Gaussian integral:

$$\boxed{\mathsf{G}_0(\alpha) \stackrel{\text{def}}{\equiv} \int_{-\infty}^{\infty} dx\, e^{-\alpha x^2} = \sqrt{\frac{\pi}{\alpha}} \qquad \alpha > 0.}\tag{I.1}$$

- The Gaussian-like integrals of the form

$$\mathsf{G}_{2n}(\alpha) \stackrel{\text{def}}{\equiv} \int_{-\infty}^{\infty} dx\, x^{2n} e^{-\alpha x^2} \qquad n = 1, 2, 3, \ldots \tag{I.2}$$

can be obtained recursively by differentiation:

$$\mathsf{G}_{2(n+1)}(\alpha) = -\frac{\partial \mathsf{G}_{2n}(\alpha)}{\partial \alpha}. \tag{I.3}$$

- The explicit form of some Gaussian-like integrals:

$$\int_{-\infty}^{\infty} dx\, x^2 e^{-\alpha x^2} = \sqrt{\frac{\pi}{\alpha}} \left(\frac{1}{2\alpha}\right) \tag{I.4}$$

$$\int_{-\infty}^{\infty} dx\, x^4 e^{-\alpha x^2} = \sqrt{\frac{\pi}{\alpha}} \left(\frac{3}{4\alpha^2}\right) \tag{I.5}$$

$$\int_{-\infty}^{\infty} dx\, x^6 e^{-\alpha x^2} = \sqrt{\frac{\pi}{\alpha}} \left(\frac{15}{8\alpha^3}\right). \tag{I.6}$$

- The one-dimensional Gaussian integral on a half-line:

$$\boxed{\mathsf{G}_1^{[0,\infty]}(\alpha) \stackrel{\text{def}}{\equiv} \int_0^{\infty} dx\, x e^{-\alpha x^2} = \frac{1}{2\alpha} \qquad \alpha > 0.}\tag{I.7}$$

- The Gaussian-like integrals of the form

$$\mathsf{G}_{2n+1}^{[0,\infty]}(\alpha) \stackrel{\text{def}}{\equiv} \int_0^{\infty} dx\, x^{2n+1} e^{-\alpha x^2} \qquad n = 0, 1, 2, \ldots \tag{I.8}$$

can be obtained recursively by differentiation:

$$\mathsf{G}_{2n+3}^{[0,\infty]}(\alpha) = -\frac{\partial \mathsf{G}_{2n+1}^{[0,\infty]}(\alpha)}{\partial \alpha}. \tag{I.9}$$

312   *Supplements*

- The explicit form of some Gaussian-like integrals on a half-line:

$$\int_0^\infty dx\, x^3 e^{-\alpha x^2} = \frac{1}{2\alpha^2} \tag{I.10}$$

$$\int_0^\infty dx\, x^5 e^{-\alpha x^2} = \frac{1}{\alpha^3} \tag{I.11}$$

$$\int_0^\infty dx\, x^7 e^{-\alpha x^2} = \frac{3}{\alpha^4}. \tag{I.12}$$

- The basic Gaussian integral with a linear term in the exponent:

$$\boxed{\mathcal{G}_0^{(\pm)}(\alpha, \beta) \stackrel{\text{def}}{=} \int_{-\infty}^\infty dx\, e^{-\alpha x^2 \pm \beta x} = \sqrt{\frac{\pi}{\alpha}} e^{\beta^2/4\alpha} \qquad \alpha > 0.} \tag{I.13}$$

- The Gaussian-like integrals of the form

$$\mathcal{G}_n^{(\pm)}(\alpha, \beta) \stackrel{\text{def}}{=} \int_{-\infty}^\infty dx\, x^n e^{-\alpha x^2 \pm \beta x} \qquad n = 1, 2, 3, \ldots \tag{I.14}$$

can be obtained recursively by differentiation:

$$\mathcal{G}_{n+1}^{(\pm)}(\alpha, \beta) = \pm \frac{\partial \mathcal{G}_n^{(\pm)}(\alpha, \beta)}{\partial \beta}. \tag{I.15}$$

- The explicit form of some Gaussian-like integrals with linear term in the exponent:

$$\int_{-\infty}^\infty dx\, x e^{-\alpha x^2 + \beta x} = \sqrt{\frac{\pi}{\alpha}} \left(\frac{\beta}{2\alpha}\right) e^{\beta^2/(4\alpha)} \tag{I.16}$$

$$\int_{-\infty}^\infty dx\, x^2 e^{-\alpha x^2 + \beta x} = \sqrt{\frac{\pi}{\alpha}} \left[\frac{1}{2\alpha} + \left(\frac{\beta}{2\alpha}\right)^2\right] e^{\beta^2/(4\alpha)}. \tag{I.17}$$

- The basic Gaussian integral with purely imaginary exponent:

$$\boxed{\mathcal{I}_0^{(\pm)}(\alpha, \beta) \stackrel{\text{def}}{=} \int_{-\infty}^\infty dx\, e^{-i(\alpha x^2 \pm \beta x)} \stackrel{\text{def}}{=} \lim_{\eta \to +0} \int_{-\infty}^\infty dx\, e^{-i(\alpha x^2 \pm \beta x) - \eta x^2} = \sqrt{\frac{\pi}{-i\alpha}} e^{i\beta^2/(4\alpha)}} \tag{I.18}$$

where $\alpha > 0$ or $\alpha < 0$.

- The Gaussian-like integrals of the form

$$\mathcal{I}_n^{(\pm)}(\alpha) \stackrel{\text{def}}{=} \int_{-\infty}^\infty dx\, x^n e^{-i(\alpha x^2 \pm \beta x)} \qquad n = 1, 2, 3, \ldots \tag{I.19}$$

can be obtained recursively by differentiation:

$$\mathcal{I}_{n+1}^{(\pm)}(\alpha, \beta) = \mp \frac{1}{i} \frac{\partial \mathcal{I}_n^{(\pm)}(\alpha, \beta)}{\partial \beta}. \tag{I.20}$$

- The basic multiple d-dimensional Gaussian integral:

$$\boxed{\mathcal{G}_0^{(d)}(A_{ij}, \beta_j) \stackrel{\text{def}}{=} \int_{-\infty}^\infty d^{(d)}x\, \exp\left\{-\sum_{i,j=1}^d x_i A_{ij} x_j \pm \sum_{i=1}^d \beta_i x_i\right\} = \frac{\pi^{d/2}}{\sqrt{\det \mathsf{A}}} \exp\left\{\sum_{i,j=1}^d \beta_i (A^{-1})_{ij} \beta_j\right\}} \tag{I.21}$$

$((A^{-1})_{ij}$ are the matrix elements of the inverse matrix A^{-1}).

II Table of some exactly solved Wiener path integrals

We list some practically useful Wiener path integrals which can be explicitly calculated and expressed in terms of elementary functions. Recall that $d_W x(\tau)$ is the Wiener path-integral measure

$$d_W x(\tau) \stackrel{\text{def}}{\equiv} \exp\left\{-\frac{1}{4D}\int_{t_0}^{t} d\tau\, \dot{x}^2(\tau)\right\} \prod_{\tau=t_0}^{t} \frac{dx(\tau)}{\sqrt{4\pi D\, d\tau}}$$

and the basic Wiener formula is the integral with this measure over the set of all continuous but non-differentiable functions:

$$\boxed{\int_{\mathcal{C}\{x_0,t_0;x,t\}} d_W x(\tau) = \frac{1}{\sqrt{4\pi D(t-t_0)}} \exp\left\{-\frac{(x-x_0)^2}{4D(t-t_0)}\right\}.} \quad \text{(II.1)}$$

The generating formula for the rest of the path integrals which we list below is the path integral

$$\int_{\mathcal{C}\{x_0,t_0;x,t\}} d_W x(\tau)\, \exp\left\{-\int_{t_0}^{t} d\tau\, [k^2 x^2(\tau) - \eta(\tau) x(\tau)]\right\}$$

$$= \left[\frac{k}{2\pi\sqrt{D}\sinh(2k\sqrt{D}(t-t_0))}\right]^{\frac{1}{2}} \exp\left\{-k\frac{(x^2+x_0^2)\cosh(2k\sqrt{D}(t-t_0)) - 2x_0 x}{2\sqrt{D}\sinh(2k\sqrt{D}(t-t_0))}\right\}$$

$$\times \exp\left\{\sqrt{D}\int_{t_0}^{t} d\tau \int_{t_0}^{\tau} d\tau'\, \eta(\tau)\eta(\tau')\frac{\sinh(2k\sqrt{D}(t-\tau))\sinh(2k\sqrt{D}(\tau'-t_0))}{k\sinh(2k\sqrt{D}(t-t_0))}\right\}$$

$$\times \exp\left\{\int_{t_0}^{t} d\tau\, \eta(\tau)\frac{x_0 \sinh(2k\sqrt{D}(t-\tau)) + x\sinh(2k\sqrt{D}(\tau-t_0))}{\sinh(2k\sqrt{D}(t-t_0))}\right\} \quad \text{(II.2)}$$

for the *driven oscillator* (an oscillator in a field of an external force) which was obtained in chapter 1 (see (1.2.262)). As we have learned, this path integral is, actually, the generating (characteristic) functional. Therefore, all other path integrals presented in this appendix can be derived from it via functional differentiation or as particular cases for specific forms of the external force. For the reader's convenience, we present them in explicit form. The corresponding Feynman path integral can be reduced to the Wiener one by the transition to the Euclidean time, as has been explained in section 2.1.

- Transition to the limit $k \to 0$ and an appropriate functional differentiation of (II.2) give the following sequence of path integrals:

$$\int_{\mathcal{C}\{x_0,t_0;x,t\}} d_W x(\tau)\, \exp\left\{\int_{t_0}^{t} d\tau\, \eta(\tau) x(\tau)\right\} = \frac{1}{\sqrt{4\pi D(t-t_0)}} \exp\left\{-\frac{(x-x_0)^2}{4D(t-t_0)}\right\}$$

$$\times \exp\left\{2D\int_{t_0}^{t} d\tau \int_{t_0}^{\tau} d\tau'\, \eta(\tau)\eta(\tau')\frac{(t-\tau)(\tau'-t_0)}{t-t_0}\right.$$

$$\left. + \int_{t_0}^{t} d\tau\, \eta(\tau)\frac{(t-\tau)x_0 + (\tau-t_0)x}{t-t_0}\right\} \quad \text{(II.3)}$$

$$\int_{\mathcal{C}\{x_0,t_0;x,t\}} d_W x(\tau)\, x^n(s) = \frac{1}{\sqrt{4\pi D(t-t_0)}} \exp\left\{-\frac{(x-x_0)^2}{4D(t-t_0)}\right\}$$

$$\times \sum_{k=0}^{[n/2]} \frac{n!}{k!(n-2k)!}\left[2D\frac{(t-s)(s-t_0)}{t-t_0}\right]^k \left[\frac{(t-s)x_0 + (s-t_0)x}{t-t_0}\right]^{n-2k} \quad \text{(II.4)}$$

where $[n/2]$ denotes the integral part of $n/2$;

$$\int_{\mathcal{C}\{x_0,t_0;x,t\}} d_W x(\tau) \int_{t_0}^t ds \, x^n(s) = \frac{1}{\sqrt{4\pi D(t-t_0)}} \exp\left\{-\frac{(x-x_0)^2}{4D(t-t_0)}\right\}$$
$$\times \sum_{k=0}^{[n/2]} \sum_{p=0}^{n-2k} \frac{D^k}{(n+1)} \frac{(p+k)!}{k!p!} \frac{(n-p-k)!}{(n-2k-p)!} (t-t_0)^{k+1} x^p x_0^{n-2k-p} \quad \text{(II.5)}$$

$$\int_{\mathcal{C}\{x_0,t_0;x,t\}} d_W x(\tau) \, x(s') x(s) = \frac{1}{\sqrt{4\pi D(t-t_0)}} \exp\left\{-\frac{(x-x_0)^2}{4D(t-t_0)}\right\}$$
$$\times \left[2D \frac{(t-\max(s',s))(\min(s',s)-t_0)}{t-t_0} + \frac{(t-s)x_0 + (s-t_0)x}{t-t_0} \frac{(t-s')x_0 + (s'-t_0)x}{t-t_0}\right] \quad \text{(II.6)}$$

$$\int_{\mathcal{C}\{x_0,t_0;x,t\}} d_W x(\tau) \int_{t_0}^t ds \, ds' \, [x(s)-x(s')]^2 = \frac{1}{\sqrt{4\pi D(t-t_0)}}$$
$$\times \exp\left\{-\frac{(x-x_0)^2}{4D(t-t_0)}\right\} \frac{(t-t_0)^2}{6} [D(t-t_0) + (x-x_0)^2] \quad \text{(II.7)}$$

$$\int_{\mathcal{C}\{x_0,t_0;x,t\}} d_W x(\tau) \exp\{\alpha x(s)\} = \frac{1}{\sqrt{4\pi D(t-t_0)}} \exp\left\{-\frac{(x-x_0)^2}{4D(t-t_0)}\right\}$$
$$\times \exp\left\{2\alpha^2 D \frac{(t-s)(s-t_0)}{t-t_0} + \alpha \frac{(t-s)x_0 + (s-t_0)x}{t-t_0}\right\} \quad \text{(II.8)}$$

$$\int_{\mathcal{C}\{x_0,t_0;x,t\}} d_W x(\tau) \exp\left\{\alpha \int_{t_0}^t ds \, x(s)\right\} = \frac{1}{\sqrt{4\pi D(t-t_0)}} \exp\left\{-\frac{(x-x_0)^2}{4D(t-t_0)}\right\}$$
$$\times \exp\left\{\alpha(t-t_0)(x+x_0) + \frac{\alpha^2 D}{12}(t-t_0)^3\right\}. \quad \text{(II.9)}$$

- The following integrals with non-zero k^2 are obtained by separating the time interval into two parts and by using the ESKC relation:

$$\int_{\mathcal{C}\{x_0,t_0;x,t\}} d_W x(\tau) \exp\left\{-k^2 x^2(s)\right\} = \frac{1}{\sqrt{4\pi D(t-t_0 + 4Dk^2(t-s)(s-t_0))}}$$
$$\times \exp\left\{-\frac{1}{4D} \frac{(x-x_0)^2 + 4Dk^2[(t-s)x_0^2 + (s-t_0)x^2]}{t-t_0 + 4Dk^2(t-s)(s-t_0)}\right\} \quad \text{(II.10)}$$

$$\int_{\mathcal{C}\{x_0,t_0;x,t\}} d_W x(\tau) \exp\left\{-k^2 \int_0^t d\tau \, [x(\tau)-x(s)]^2\right\} = \left[\frac{k}{2\pi\sqrt{D}\sinh(2k\sqrt{D}(t-t_0))}\right]^{\frac{1}{2}}$$
$$\times \exp\left\{-k \frac{(x^2-x_0^2)^2}{2\sqrt{D}\sinh(2k\sqrt{D}(t-t_0))} \cosh(2k\sqrt{D}(s-t_0)) \cosh(2k\sqrt{D}(t-s))\right\}. \quad \text{(II.11)}$$

- The following path integrals are obtained from (II.2) with non-zero k^2 by a specific choice of the external force and/or by functional differentiations:

$$\int_{\mathcal{C}\{x_0,t_0;x,t\}} d_W x(\tau) \exp\left\{-k^2 \int_0^t d\tau\, x^2(\tau)\right\}$$
$$= \left[\frac{k}{2\pi\sqrt{D}\sinh(2k\sqrt{D}(t-t_0))}\right]^{\frac{1}{2}} \exp\left\{-k\frac{(x^2+x_0^2)\cosh(2k\sqrt{D}(t-t_0)) - 2x_0 x}{2\sqrt{D}\sinh(2k\sqrt{D}(t-t_0))}\right\} \quad \text{(II.12)}$$

$$\int_{\mathcal{C}\{x_0,t_0;x,t\}} d_W x(\tau) \exp\left\{-k^2 \int_0^t d\tau\, x^2(\tau) + \alpha \int_{t_0}^t d\tau\, x(\tau)\right\}$$
$$= \left[\frac{k}{2\pi\sqrt{D}\sinh(2k\sqrt{D}(t-t_0))}\right]^{\frac{1}{2}} \exp\left\{-k\frac{(x^2+x_0^2)\cosh(2k\sqrt{D}(t-t_0)) - 2x_0 x}{2\sqrt{D}\sinh(2k\sqrt{D}(t-t_0))}\right\}$$
$$\times \exp\left\{\frac{\alpha}{2k\sqrt{D}}(x+x_0)\tanh(\sqrt{D}k(t-t_0)) + \frac{\alpha^2}{4k^2}\left[(t-t_0) - \frac{1}{k\sqrt{D}}\tanh(\sqrt{D}k(t-t_0))\right]\right\}$$
(II.13)

$$\int_{\mathcal{C}\{x_0,t_0;x,t\}} d_W x(\tau) \exp\left\{-k^2 \int_0^t d\tau\, x^2(\tau)\right\} x(s)$$
$$= \left[\frac{k}{2\pi\sqrt{D}\sinh(2k\sqrt{D}(t-t_0))}\right]^{\frac{1}{2}} \exp\left\{-k\frac{(x^2+x_0^2)\cosh(2k\sqrt{D}(t-t_0)) - 2x_0 x}{2\sqrt{D}\sinh(2k\sqrt{D}(t-t_0))}\right\}$$
$$\times \frac{x_0 \sinh(2k\sqrt{D}(t-s)) + x\sinh(2k\sqrt{D}(s-t_0))}{\sinh(2k\sqrt{D}(t-t_0))} \quad \text{(II.14)}$$

$$\int_{\mathcal{C}\{x_0,t_0;x,t\}} d_W x(\tau) \exp\left\{-k^2 \int_0^t d\tau\, x^2(\tau)\right\} x(s) x(s')$$
$$= \left[\frac{k}{2\pi\sqrt{D}\sinh(2k\sqrt{D}(t-t_0))}\right]^{\frac{1}{2}} \exp\left\{-k\frac{(x^2+x_0^2)\cosh(2k\sqrt{D}(t-t_0)) - 2x_0 x}{2\sqrt{D}\sinh(2k\sqrt{D}(t-t_0))}\right\}$$
$$\times \left[\frac{x_0 \sinh(2k\sqrt{D}(t-s)) + x\sinh(2k\sqrt{D}(s-t_0))}{\sinh(2k\sqrt{D}(t-t_0))}\right.$$
$$\times \frac{x_0 \sinh(2k\sqrt{D}(t-s')) + x\sinh(2k\sqrt{D}(s'-t_0))}{\sinh(2k\sqrt{D}(t-t_0))}$$
$$\left. + \frac{\sqrt{D}\sinh(2k\sqrt{D}(t-\max(s,s')))\sinh(2k\sqrt{D}(\min(s,s')-t_0))}{k\sinh(2k\sqrt{D}(t-t_0))}\right] \quad \text{(II.15)}$$

$$\int_{\mathcal{C}\{x_0,t_0;x,t\}} d_W x(\tau) \exp\left\{-k^2 \int_0^t d\tau\, x^2(\tau) + \alpha x(s)\right\}$$
$$= \left[\frac{k}{2\pi\sqrt{D}\sinh(2k\sqrt{D}(t-t_0))}\right]^{\frac{1}{2}} \exp\left\{-k\frac{(x^2+x_0^2)\cosh(2k\sqrt{D}(t-t_0)) - 2x_0 x}{2\sqrt{D}\sinh(2k\sqrt{D}(t-t_0))}\right\}$$
$$\times \exp\left\{\alpha \frac{x_0 \sinh(2k\sqrt{D}(t-s)) + x\sinh(2k\sqrt{D}(s-t_0))}{\sinh(2k\sqrt{D}(t-t_0))}\right\}$$
$$\times \exp\left\{\alpha^2 \frac{\sqrt{D}\sinh(2k\sqrt{D}(t-s))\sinh(2k\sqrt{D}(s-t_0))}{2k\sinh(2k\sqrt{D}(t-t_0))}\right\}. \quad \text{(II.16)}$$

III Feynman rules

The functional methods presented in sections (3.1.2)–(3.1.5) and (3.2.4)–(3.2.6) lead to the following general *Feynman rules* for the computation of (connected) Green functions:

(i) Draw all topologically distinct, connected diagrams of the desired order.
(ii) In each diagram, attach a propagator to each internal line according to tables 3.1, page 27; 3.2, page 38; 3.3, page 76; 3.4, page 78 and 3.5, page 79.
(iii) To each vertex, assign a vertex function given in the same tables, derived from the relevant term in the interaction Lagrangian.
(iv) For each internal momentum p not fixed by the momentum conservation at vertices, write a factor $\int d^4 p/(2\pi)^4$.
(v) Multiply the contribution for each diagram by:

  (a) a factor (-1) for each closed fermion loop;
  (b) a relative factor (-1) for graphs which differ from each other only by an interchange of two identical external fermion lines;
  (c) a symmetry factor $1/r$ with

$$r = N_p \prod_{n=2,3,\ldots} 2^\beta (n!)^{\alpha_n}$$

where α_n is the number of *pairs* of vertices connected by n identical self-conjugate lines, β is the number of lines connecting a vertex with itself and N_p is the number of permutations of vertices which leave the diagram unchanged with fixed external lines.

Examples:

$N_p = 1$, $\alpha_2 = 1$, $\beta = 0$, so that $r = 2$

$N_p = 1$, $\alpha_n = 0$, $\beta = 1$, so that $r = 2$

$N_p = 2$, $\alpha_2 = 1$, $\beta = 0$, so that $r = 4$.

(vi) The OPI Green functions $\Gamma^{(n)}(p_1, \ldots, p_n)$ come from the OPI diagrams.
(vii) For the scattering amplitude $T(p_1, \ldots, p_n)$, put all the external lines on their mass shell, i.e. $p_i^2 = m_i^2$ and provide external fermion lines with spinors: $u(p)$ (or $v(p)$) for fermions (or antifermions) entering with momentum p; $\bar{u}(p)$ (or $\bar{v}(p)$) for fermions (or antifermions) leaving with momentum p. Provide the external vector bosons with polarization vectors: $u_l^j(-k)$ (or $u_l^j(k)$) for the vector boson entering (or leaving) with momentum k.

IV Short glossary of selected notions from the theory of Lie groups and algebras

In this glossary, we recall some basic definitions from the theory of Lie groups and algebras which are used in the main text (the definitions are presented in alphabetical order). Of course, rigorous definitions can be given only in the appropriate context of a complete exposition of the theory and we refer the reader, e.g., to the books by Wigner (1959), Barut and Rączka (1977), Zhelobenko (1973), Wybourn (1974) and Vilenkin (1968) for further details and clarifications.

◇ Baker–Campbell–Hausdorff formula

The *Baker–Campbell–Hausdorff formula* (sometimes also called the Campbell-Hausdorff or Campbell–Baker–Hausdorff formula) is a formula for computing the Lie algebra element Z defined by the relation

$$e^Z = e^X e^Y$$

where X and Y are elements of some Lie algebra $\mathfrak{g}$. According to the Baker–Campbell–Hausdorff formula, the element Z is given by the following series:

$$Z = \sum_{n=0}^{\infty} C_n(X, Y)$$

where the terms $C_n(X, Y)$ are determined by the recursion formula

$$(n+1)C_{n+1}(X, Y) = \tfrac{1}{2}[X - Y, C_n(X, Y)] \\ + \sum_{\substack{p \geq 1, 2p \leq 2}} K_{2p} \sum_{\substack{k_1, \ldots, k_{2p} > 0 \\ k_1 + \cdots + k_{2p} = n}} [C_{k_1}(X, Y), [\cdots, [C_{k_{2p}}(X, Y), X + Y] \cdots]] \quad \text{(IV.1)}$$

($n \geq 1$, $X, Y \in \mathfrak{g}$) and by the condition $C_1(X, Y) = X + Y$. Here the coefficients K_{2p} are found from the power series expansion of the following auxiliary even function $f(t)$ of an ordinary variable t:

$$f(t) = \frac{t}{1 - e^{-t}} - \frac{t}{2} = 1 + \sum_{p=1}^{\infty} K_{2p} t^{2p}.$$

The terms $C_n(X, Y)$ may be calculated from (IV.1) in succession; unfortunately, the calculations become complicated very rapidly. However, the first few terms may be calculated without too much difficulty:

$$C_2(X, Y) = \tfrac{1}{2}[X, Y]$$
$$C_3(X, Y) = \tfrac{1}{12}[X, [X, Y]] + \tfrac{1}{12}[Y, [Y, X]]$$
$$C_4(X, Y) = -\tfrac{1}{48}[Y, [X, [X, Y]]] - \tfrac{1}{48}[X, [Y, [X, Y]]].$$

◇ Character of a group

The *character of a group* $\mathfrak{G}$ is a homomorphism of the given group into some standard Abelian group $\mathfrak{G}_A$. Usually, $\mathfrak{G}_A$ is taken to be the multiplicative group T of complex numbers with unit absolute values: $T = \{z \in \mathbb{C} \mid z^*z = 1\}$. The characters form a linearly independent system in the space of all T-valued functions on $\mathfrak{G}$. They are also one-dimensional linear representations of $\mathfrak{G}$. In the one-dimensional case, the concept of a character of a group coincides with that of a character of a representation of a group. Sometimes, the characters of a group are understood to mean characters of any of its finite-dimensional representations.

◇ Character of a representation of a group

In the case of a finite-dimensional representation π, this is the function χ_π on the group $\mathfrak{G}$ defined by the formula

$$\chi_\pi(g) = \operatorname{Tr} \pi(g) \qquad g \in \mathfrak{G}.$$

318   Supplements

For arbitrary continuous representations of a group $\mathfrak{G}$, this definition is generalized as follows:

$$\chi_\pi(g) = \chi[\pi(g)] \qquad g \in \mathfrak{G}$$

where $\chi[\pi(g)]$ is a linear functional defined on some ideal I of the algebra A generated by the family of operators $\pi(g)$, $g \in \mathfrak{G}$, which is invariant under inner automorphisms of A.

◇ **Coset in a group**

A *coset* in a group G by a subgroup G_0 (from the left) is a set of elements of G of the form

$$gG_0 = \{gh \mid h \in G_0\}$$

where g is some fixed element of G. This coset is also called the left coset by G_0 in G defined by g. Every left coset is determined by any of its elements. Since $gG_0 = G_0$ if and only if $g \in G_0$, for all g_1, g_2 the cosets $g_1 G_0$ and $g_2 G_0$ are either equal or disjoint. Thus, G decomposes into pairwise disjoint left cosets by G_0. Similarly, we define right cosets (as sets $G_0 g$, $g \in G$) and also right decomposition of G with respect to G_0. These decompositions consist of the same number of cosets. We can define the set G/G_0 as the set of right (left) cosets. For normal (invariant) subgroups, the left and right decompositions coincide. In this case, we simply speak of the decomposition of a group with respect to a normal subgroup and G/G_0 is a group.

For finite groups G, the number of elements in each coset is clearly given by the order of G_0. Since the cosets define the decomposition of G into disjoint subsets, for finite groups this implies Lagrange's theorem stating that the order d_G of G is divisible by the order d_{G_0} of G_0 (d_G/d_{G_0} gives the number of elements in G/G_0). For a Lie group $\mathfrak{G}$ of dimension $d(\mathfrak{G})$ and $\mathfrak{G}_0 \subset \mathfrak{G}$ of dimension $d(\mathfrak{G}_0)$, an analogous result holds: $d(\mathfrak{G}/\mathfrak{G}_0) = d(\mathfrak{G}) - d(\mathfrak{G}_0)$. For example, if $\mathfrak{G} = SO(3)$ is the three-dimensional rotation group and $\mathfrak{G}_0 = SO(2)$ is the one-dimensional group of rotations around a given axis, $\mathfrak{G}/\mathfrak{G}_0 = SO(3)/SO(2)$ is the two-dimensional sphere.

◇ **Groups: finite, infinite, continuous, Abelian, non-Abelian; subgroup of a group**

A set G of elements $g_1, g_2, g_3, \ldots$ is said to form a *group* if a law of *multiplication* of the elements is defined which satisfies certain conditions. The result of multiplying two elements g_a and g_b is called the *product* and is written $g_a g_b$. The conditions to be satisfied are the following:

(i) The product $g_a g_b$ of any two elements is itself an element of the group, i.e.

$$g_a g_b = g_d \qquad \text{for some } g_d \in G.$$

(ii) In multiplying three elements g_a, g_b and g_c together, it does not matter which product is made first:

$$g_a(g_b g_c) = (g_a g_b) g_c$$

where the product inside the brackets is carried out first. This implies that the use of such brackets is unnecessary and we may simply write $g_a g_b g_c$ for the triple product. This property is called the *associativity* of the group multiplication.

(iii) One element of the group, usually denoted by e and called the *identity* or *unity*, must have the property

$$e g_a = g_a e = g_a.$$

(iv) To each element g_a of the group there corresponds another element of the group, denoted by g_a^{-1} and called the *inverse* of g_a, which has the properties

$$g_a g_a^{-1} = g_a^{-1} g_a = e.$$

In general, $g_a g_b$ is not the same element as $g_b g_a$. A group for which $g_a g_b = g_b g_a$ for all elements g_a and g_b is called an *Abelian* group. Its elements are said to *commute*. If at least one pair of elements do not commute, i.e. we have $g_a g_b \neq g_b g_a$, then the group is called *non-Abelian*.

The number of elements in a group may be finite, in which case this number is called the *order* of the group, or it may be infinite. The groups are correspondingly called finite or infinite groups. Among the latter, the most important for physics are the *continuous* ones, for which the group elements, instead of being distinguished by a discrete label, are labeled by a set of continuous parameters (see below the definition of a continuous *Lie group* and related objects). The simplest continuous non-Abelian group is the rotation group in the three-dimensional space. The rotations in a two-dimensional space (on a plane), however, form an Abelian group.

Given a set of elements forming a group G, it is often possible to select a smaller number of these elements which satisfy the group definitions among themselves. They are said to form a *subgroup* of G. An *invariant* (also called the *normal*) subgroup is a subgroup H of G with the property that $gHg^{-1} = H$ for any $g \in G$ (that is, for each $h \in H$ and $g \in G$, we have $g^{-1}hg \in H$). For example, the translations and rotations of the three-dimensional space generate a group which has the translations as the normal subgroup.

◇ **Haar and bi-invariant measures on a Lie group**

The so-called *Haar measure* defines the invariant integration measure for Lie groups. This means that we can identify a volume element $d\mu(g)$ defining the integral of a function f over a Lie group $\mathfrak{G}$ as $\int_\mathfrak{G} d\mu(g) f(g)$ and with the property that the integral is both left and right invariant

$$\int_\mathfrak{G} d\mu(g) f(g_0^{-1} g) = \int_\mathfrak{G} d\mu(g) f(g g_0^{-1}) = \int_\mathfrak{G} d\mu(g) f(g).$$

The invariance follows from the invariance of the volume element $d\mu(g)$:

$$d\mu(g) = d\mu(g_0 g) = d\mu(g g_0)$$

which implies that the expression for $d\mu(g)$ at a neighborhood of the point g can be found by fixing the value of $d\mu(g)$ at $g = e$ (unit element) and by performing a left or right translation by g: $d\mu(g) = d\mu(e)$. Let the action of a map $x \to g(x)$ (left translation) be given by $x^i \to y^i(x^j)$, with x^i being the coordinates in the neighbourhood of the unit element e and denote by $dx^1, \ldots, dx^n$ the volume element spanned by the coordinate differentials $dx^1, dx^2, \ldots, dx^n$ at the point e. Then, the volume element spanned by the same coordinate differentials at the point g is given by

$$d\mu(g) = |J|^{-1} dx^1, dx^2, \ldots, dx^n$$

where J is the Jacobian for the map $x \to g(x)$ evaluated at the unit element e:

$$J = \frac{\partial(y^1, \ldots, y^n)}{\partial(x_1, \ldots, x_n)}.$$

In a right- or a left-translation, $dx^1, dx^2, \ldots, dx^n$ is multiplied by the same Jacobian determinant so that $d\mu(g)$ is indeed right and left invariant. A straightforward manner to derive the Haar measure is to consider a *faithful matrix representation* of the group and take some subset of matrix elements as the coordinates x^i. The Lie groups also allow an invariant metric and $d\mu(g)$ is just the volume element $\sqrt{g} \, dx^1, \ldots, dx^n$ associated with this metric.

Example. The volume element of the group $SU(2)$. The elements a of $SU(2)$ can be represented as 2×2 unitary matrices of the form

$$\mathsf{a} = \sum_\mu x^\mu \tilde{\sigma}^\mu \qquad \sum_\mu x^\mu x^\mu = 1$$

where the $\tilde{\sigma}$-matrices are defined by $\sigma_0 = 1$ and $\tilde{\sigma}_i = -i\sigma_i$, with σ_i being the Pauli spin matrices, so that we have $\tilde{\sigma}_i \tilde{\sigma}_j = -\delta_{ij}\tilde{\sigma}_0 + \epsilon_{ijk}\tilde{\sigma}_k$. The coordinates of $SU(2)$ can be taken as the coordinates x^i, $i = 1, 2, 3$, so that we have $x^0 = \pm\sqrt{1 - r^2}$, $r \equiv \sqrt{\sum_i x^i x^i}$. Clearly, the $SU(2)$ group manifold can be regarded as a three-dimensional sphere of unit radius ($\sum_{\mu=1}^4 x^\mu x^\mu = 1$) in Euclidean space E^4. The unit element e corresponds to the origin: $x^i = 0$, $x^0 = 1$. The left action of the element a on b can be written as $\mathsf{c} = \mathsf{ab} = \sum_\mu z^\mu \sigma^\mu$, where the coordinates z^μ are given by

$$z^i = (x^0 y^i + x^i y^0) + \epsilon_{ijk} x^j y^k$$
$$z^0 = \sqrt{1 - \sum_i z^i z^i}.$$

From this relation, the Jacobian matrix at $y^i = 0$ can be deduced: $\partial z^i/\partial y^j = x^0 \delta_{ij} + \epsilon_{ijk} x^k$ and its determinant is $J = \pm\sqrt{1-r^2}$, depending on the sign of x^0. The invariant integration measure reads as

$$d\mu = \frac{1}{\sqrt{1-r^2}} dx^1 dx^2 dx^3.$$

Note that the metric of $SU(2)$ can be deduced as the metric induced from the Euclidean space E^4 into which $SU(2)$ is embedded as a sphere.

◇ Isomorphisms, automorphisms, homomorphisms

Let $\mathcal{X}$ and $\mathcal{X}'$ be two sets with some relations among the elements of the each set.

For example, $\mathcal{X}$ and $\mathcal{X}'$ can be groups, and the corresponding relations can be the group multiplications: $g_a g_b = g_c$ for $g_a, g_b, g_c \in \mathcal{X}$ and $g'_a g'_b = g'_c$ for $g'_a, g'_b, g'_c \in \mathcal{X}'$. Another example is ordered sets with defined inequalities $a > b$, $a, b \in \mathcal{X}$ and $a' > b'$, $a', b' \in \mathcal{X}'$.

Let there be a one-to-one correspondence (map) $\rho : \mathcal{X} \longrightarrow \mathcal{X}'$ preserving the relations among the elements of $\mathcal{X}, \mathcal{X}'$, i.e. if some relation is fulfilled for $a, b \in \mathcal{X}$, then the corresponding relation is fulfilled for $\rho(a), \rho(b) \in \mathcal{X}'$ and vice versa. In this case, the sets $\mathcal{X}$ and $\mathcal{X}'$ are called *isomorphic*: $\mathcal{X} \cong \mathcal{X}'$, and the correspondence ρ is called *isomorphism*.

In particular, if the sets coincide, $\mathcal{X} = \mathcal{X}'$, a one-to-one correspondence ρ, preserving structure relations, is called *automorphism*.

If each element $a \in \mathcal{X}$ is mapped into a unique image, a single element $a' \in \mathcal{X}'$, but reverse is not in general true (e.g., a' may be the image of several elements of $\mathcal{X}$ or not be the image of any elements of $\mathcal{X}$) and the map preserves structure relations in $\mathcal{X}$ and $\mathcal{X}'$, then this map is called *homomorphism*.

◇ Lie groups, Abelian and semisimple Lie groups; Lie algebras

The elements $g(a_1, a_2, \ldots, a_r)$ of a continuous group depend on some real parameters a_i which are all essential in the sense that the group elements cannot be distinguished by any smaller number of parameters. The number r is called *the dimension* of the group. Each parameter has a well-defined range of values. For the elements to satisfy the group postulate, a multiplication law must be defined, and the product of two elements

$$g(a_1, a_2, \ldots, a_r) g(b_1, b_2, \ldots, b_r) = g(c_1, c_2, \ldots, c_r)$$

must be another group element. Thus the new parameters c_i must be expressible as functions of the as and bs

$$c_i = \phi_i(a_1, a_2, \ldots, a_r, b_1, b_2, \ldots, b_r) \qquad i = 1, \ldots, r.$$

It is customary to define the parameters in such a way that the identity element has all the parameters equal to zero. The r functions ϕ_i must satisfy several conditions in order for the group postulates to be satisfied. The groups with differentiable functions ϕ_i are called *Lie groups*. More precisely, a group $\mathfrak{G}$ having the structure of a smooth manifold such that the mapping $(g, h) \to gh^{-1}$ of the direct product $\mathfrak{G} \otimes \mathfrak{G}$ into $\mathfrak{G}$ is analytic is called a Lie group. The main method of research in the theory of Lie groups is the infinitesimal method, which reduces the study largely to the consideration of a purely algebraic object, a Lie algebra.

An abstract *Lie algebra* is a vector space $\mathfrak{g}$ together with a bilinear operation $[\cdot, \cdot]$ from $\mathfrak{g} \times \mathfrak{g}$ into $\mathfrak{g}$, satisfying

$$[X_1 + X_2, Y] = [X_1, Y] + [X_2, Y] \qquad X_1, X_2, Y \in \mathfrak{g}$$
$$[\alpha X, Y] = \alpha [X, Y] \qquad \alpha \in \mathbb{C} \text{ or } \mathbb{R}, \; X, Y \in \mathfrak{g}$$
$$[X, Y] = -[Y, X] \qquad X, Y \in \mathfrak{g}$$
$$[X, [Y, Z]] + [Y, [Z, X]] + [Z, [X, Y]] = 0 \qquad X, Y, Z \in \mathfrak{g} \qquad \text{(Jacobi identity)}.$$

In all cases of our interest, the bilinear operation $[\cdot, \cdot]$ can be understood as the commutator in the corresponding associative algebra

$$[X, Y] = XY - YX.$$

There exists a tight interrelation between Lie groups and Lie algebras. Recovering a Lie group $\mathfrak{G}$ from its Lie algebra $\mathfrak{g}$ is possible by the *exponential map*: $\exp : \mathfrak{g} \to \mathfrak{G}$. If $\mathfrak{G}$ is a *linear* group (that is, a subgroup of a general linear group $GL(n, \mathbb{R})$), the exponential mapping takes the form:

$$g = e^X \equiv \sum_{m=0}^{\infty} \frac{1}{m!} X^m \qquad g \in \mathfrak{G}, \; X \in \mathfrak{g}.$$

In order to obtain a Lie algebra from the group structure, let us consider a representation $T(a_1, \ldots, a_r)$ of the group $\mathfrak{G}$ in a space V. By convention, the parameters are chosen such that the identity element has all $a_i = 0$, so that

$$T(0, \ldots, 0) = 1.$$

If all parameters a_i are small, then, to the first order in these parameters,

$$T(a_i) \simeq 1 + \sum_i a_i X_i$$

where the X_i are some linear operators, independent of the parameters a_i. These operators are called the *infinitesimal operators* or *generators* of the group in a given representation and are expressed explicitly as partial derivatives

$$X_i = \left. \frac{\partial T(a_1, \ldots, a_r)}{\partial a_i} \right|_{a_1 = \cdots = a_r = 0}.$$

For any representation T of a group $\mathfrak{G}$, the set of infinitesimal operators X_i satisfy the commutation relations

$$[X_i, X_j] = \sum_k f_{ij}^k X_k$$

where the numbers f_{ij}^k, called *structure constants*, are the same for all representations T of $\mathfrak{G}$. Thus, the infinitesimal operators (generators) of a Lie group form the Lie algebra with the commutator playing the role of the bilinear operation in the abstract Lie algebra.

A certain combination of generators of a Lie algebra which commutes with all the generators is called the *invariant* or *Casimir* operator of the group. The maximal number of such independent operators is equal to the *rank* of the group, the latter being defined as the maximal number of generators of the Lie algebra commuting among themselves.

A Lie algebra L is *semisimple* if it has no non-zero Abelian ideals.

A Lie algebra L is *simple* if it has no ideals (except the zero ideal and the whole L).

A connected Lie group which does not contain non-trivial connected Abelian normal subgroups is called a *semisimple* Lie group. A connected Lie group is semisimple if and only if its Lie algebra is semisimple. A connected Lie group is said to be simple if its Lie algebra is simple, i.e. if $\mathfrak{G}$ does not contain non-trivial connected invariant subgroups other than $\mathfrak{G}$.

◇ **Maurer–Cartan form**

The *Maurer–Cartan form*
   is a left-invariant one-form on a Lie group $\mathfrak{G}$, i.e. a differential form ω of degree 1 on $\mathfrak{G}$, satisfying the condition $\ell_g^* \omega = \omega$ for any left translation $\ell_g : h \to gh, g, h \in \mathfrak{G}$. The differential $d\omega$ of a Maurer–Cartan form obeys the following *Maurer–Cartan equation*:

$$d\omega(X, Y) = -\omega([X, Y])$$

where X, Y are arbitrary left-invariant vector fields on $\mathfrak{G}$ and $[X, Y]$ is their commutator.

◇ **Real forms of Lie algebras and groups**

The *complex extension* $\mathfrak{g}^c$ of an arbitrary real Lie algebra $\mathfrak{g}$ (Lie algebra over $\mathbb{R}$) is the Lie algebra which consists of elements of the form $X = X + iY$; $X, Y \in \mathfrak{g}$ (as the vector space) and with the Lie multiplication

$$\begin{aligned} Z = [Z_1, Z_2] &= ([X_1, X_2] - [Y_1, Y_2]) + i([X_1, Y_2] + [Y_1, X_2]) \\ &= X + iY \\ Z_1 = X_1 + iY_1 &\quad Z_2 = X_2 + iY_2. \end{aligned}$$

The *real form* of a complex Lie algebra $\mathfrak{g}^c$ (Lie algebra over $\mathbb{C}$) is the real Lie algebra $\mathfrak{g}^r$, such that its complex extension coincides with $\mathfrak{g}^c$.

Let $\mathfrak{G}$ be the complex Lie group generated by a complex Lie algebra $\mathfrak{g}^c$. The subgroup $\mathfrak{G}^r$ which corresponds to (i.e. generated by) the real form $\mathfrak{g}^r$ of the Lie algebra $\mathfrak{g}^c$ is called the *real form* of the complex Lie group $\mathfrak{G}$.

◇ Representations: faithful, irreducible, reducible, completely reducible (decomposable), indecomposable, adjoint

A *representation* of an algebra A (group G) is a homomorphism of A (or G) into an algebra (group) of linear transformations of some vector space V.

A representation is termed *faithful* if the homomorphism is an isomorphism.

A subspace $V_1 \subset V$ of a representation space V is called an *invariant subspace* with respect to an algebra A (group G) if $Tv \in V_1$ for all $v \in V_1$ and all $T \in A$ (or $T \in G$).

A representation is called *irreducible* if the representation space V has no invariant subspaces (except the whole space V and the zero space $\{0\}$). Otherwise, the representation is called *reducible*.

A representation is called *completely reducible* or *decomposable* if all linear transformations of the representation can be presented in the form of block-*diagonal* matrices, each block acting in the corresponding invariant subspace. Otherwise, the representation is called *indecomposable*.

The representation of a Lie group $\mathfrak{G}$ (Lie algebra $\mathfrak{g}$) in the vector space of the Lie algebra $\mathfrak{g}$ itself is called the *adjoint representation* and the corresponding transformations are denoted by Ad_g, $g \in \mathfrak{G}$ (ad_X, $X \in \mathfrak{g}$). In the case of a Lie algebra, the adjoint representation is defined by the commutator in $\mathfrak{g}$:

$$\mathrm{ad}_X Y = [X, Y] \qquad X, Y \in \mathfrak{g}.$$

◇ Root system of a semisimple Lie algebra; positive and simple roots

The Cartan subalgebra $\mathfrak{g}_H$ of a semisimple Lie algebra $\mathfrak{g}$ is the maximal Abelian subalgebra in $\mathfrak{g}$ with completely reducible adjoint representation. Let $\mathfrak{g}$ be a semisimple Lie algebra with the Cartan subalgebra $\mathfrak{g}_H \subset \mathfrak{g}$ and α be a linear function on $\mathfrak{g}_H$. If the linear subspace $L^\alpha \subset \mathfrak{g}$, defined by the condition

$$L^\alpha := \{Y \in \mathfrak{g} \mid [X, Y] = \alpha(X) Y \; \forall X \in \mathfrak{g}_H\}$$

does exist (i.e. $L^\alpha \neq 0$), the function α is called *the root* of $\mathfrak{g}$, and L^α is called *the root subspace*. The system of non-zero roots is denoted by Δ. Actually, all root subspaces are one dimensional, so that L^α is a *root vector*.

The Cartan subalgebra and root vectors give a very convenient basis for an arbitrary semisimple Lie algebra $\mathfrak{g}$ and provide their classification. In particular, the *Cartan–Weyl basis* of a semisimple Lie algebra $\mathfrak{g}$ consists of a basis $\{H_i\}$ of the Cartan subalgebra $\mathfrak{g}_H$ and the root vectors $E_\alpha \in L^\alpha$. In this basis, for any $\alpha, \beta \in \Delta$, the defining commutation relations have the form

$$[H_i, E_\alpha] = \alpha(H_i) E_\alpha \qquad H_i \in \mathfrak{g}_H$$

$$[E_\alpha, E_\beta] = \begin{cases} 0 & \text{if } \alpha + \beta \neq 0 \text{ and } \alpha + \beta \notin \Delta \\ H_\alpha & \text{if } \alpha + \beta = 0 \\ N_{\alpha,\beta} E_{\alpha+\beta} & \text{if } \alpha + \beta \in \Delta \end{cases}$$

where the constants $N_{\alpha,\beta}$ satisfy the identity $N_{\alpha,\beta} = -N_{-\beta,-\alpha}$.

Thus the problem of the classification is reduced to the study of the possible sets of constants $N_{\alpha,\beta}$.

A root α is called *positive* if the first coordinate of the corresponding H_α is positive. The subsystem of positive roots is denoted by Δ_+.

A positive root is called *simple* if it cannot be expressed as the sum of two other positive roots.

There is a one-to-one correspondence between the roots $\alpha \in \Delta$ and the elements $H_\alpha \in \mathfrak{g}_H$ of the Cartan subalgebra defined by the equality

$$\langle X, H_\alpha \rangle = \alpha(X) \qquad \forall X \in \mathfrak{g}_H.$$

Here $\langle \cdot, \cdot \rangle$ denotes the *Killing form* (scalar product) on $\mathfrak{g}$

$$\langle X, Y \rangle = \text{Tr}(\text{ad}_X \, \text{ad}_Y) \qquad \forall X, Y \in \mathfrak{g}.$$

Usually, the scalar product $\langle H_\alpha, H_\beta \rangle$ is denoted simply as (α, β).

◇ **Schur's lemmas**

Schur's first lemma. Let $T(g)$ be an irreducible representation of a group G in a space V and let A be a given operator in V. Schur's first lemma states that if $T(g)A = AT(g)$ for all $g \in G$, then $A = \lambda \mathbb{I}$, where $\mathbb{I}$ is the identity (or unity) operator. In other words, any given operator which commutes with all the operators $T(g)$ of an irreducible representation of the group G, is a constant multiple of the unit operator.

Schur's second lemma. Let $T_1(g)$ and $T_2(g)$ be two irreducible representations of a group G in two spaces V_1 and V_2 respectively, of dimensions s_1 and s_2 and let A be an operator which transforms vectors from V_1 into V_2. Schur's second lemma states that, if $T_1(g)$ and $T_2(g)$ are inequivalent and $T_1(g)A = AT_2(g)$ for all $g \in G$, then $A = 0$, i.e. it is the null (or zero) operator.

◇ **Semidirect sum of Lie algebras and semidirect product of Lie groups (inhomogeneous Lie algebras and groups)**

Let $\mathfrak{g}_M$ and $\mathfrak{g}_T$ be Lie algebras and $D : X \to D(X)$, $X \in \mathfrak{g}_M$ be a homomorphism of $\mathfrak{g}_M$ into the set of linear operators in the vector space $\mathfrak{g}_T$, such that every D is a differentiation of $\mathfrak{g}_T$ (i.e. D satisfies the Leibniz rule). Define in the direct space of $\mathfrak{g}_M$ and $\mathfrak{g}_T$ the Lie algebra structure using the Lie brackets in $\mathfrak{g}_M$ and $\mathfrak{g}_T$ and

$$[X, Y] = D(X)Y \qquad X \in \mathfrak{g}_M, \ Y \in \mathfrak{g}_T.$$

One can check that all the Lie axioms are satisfied. The obtained Lie algebra $\mathfrak{g}$ is called the *semidirect sum* of $\mathfrak{g}_M$ and $\mathfrak{g}_T$:

$$\mathfrak{g} = \mathfrak{g}_T \uplus \mathfrak{g}_M.$$

Such an algebra generates the *semidirect product* of the Lie groups $\mathfrak{G}_M$ and $\mathfrak{G}_T$, which can be defined independently as follows. Let $\mathfrak{G}_T$ be an arbitrary group, $\mathfrak{G}_T^A$ be the group of all automorphisms of $\mathfrak{G}_T$, $\mathfrak{G}_M \subset \mathfrak{G}_T^A$ be some subgroup and $\Lambda(g)$ be the image of $g \in \mathfrak{G}_T$ under an automorphism $\Lambda \in \mathfrak{G}_M$.

The semidirect product $\mathfrak{G} = \mathfrak{G}_T \otimes \mathfrak{G}_M$ of the groups $\mathfrak{G}_T$ and $\mathfrak{G}_M$ is the group of all ordered pairs (g, Λ), where $g \in \mathfrak{G}_T$, $\Lambda \in \mathfrak{G}_M$, with the group multiplication

$$(g, \Lambda)(g', \Lambda') = (g\Lambda(g'), \Lambda\Lambda')$$

the unity element (e, id), e being the unity in G_T, and the inverse elements

$$(g, \Lambda)^{-1} = (\Lambda^{-1}(g^{-1}), \Lambda^{-1}).$$

Examples of semidirect product of Lie groups are the Poincaré group and the group of rotations and translations of a Euclidean space.

◇ **Tensor operators**

Let $g \to \mathsf{M}(g)$ be a matrix representation of a group G in a finite dimensional vector space V and $g \to U_g$ be a unitary representation of G in a Hilbert space $\mathcal{H}$. A set $\{T^a\}_{a=1}^{\dim V}$ of operators in $\mathcal{H}$ is called a *tensor operator* if

$$U_g^{-1} T^a U_g = M^a{}_b(g) T^b.$$

◇ **Universal enveloping algebra**

A *universal enveloping algebra* $\mathcal{U}(\mathfrak{g})$ of a Lie algebra $\mathfrak{g}$ is a quotient algebra

$$\mathcal{U}(\mathfrak{g}) \stackrel{\text{def}}{\equiv} A_{\mathfrak{g}}/J_{[\cdot,\cdot]}$$

where $A_{\mathfrak{g}}$ is a (free) associative algebra generated by all $X_i \in \mathfrak{g}$ ($i = 1, \ldots, \dim \mathfrak{g}$) and $J_{[\cdot,\cdot]}$ is the two-sided ideal generated by elements of the form $XY - YX - [X, Y], \forall X, Y \in \mathfrak{g}$.

V Some basic facts about differential Riemann geometry

This supplement contains basic facts about differential geometry of manifolds, mainly about Riemann spaces, needed for understanding some advanced applications of path integrals considered in this book. For further details, we refer the reader to the introductory books by Isham (1989), Visconti (1992) and to the classical monograph by Kobayashi and Nomizu (1969). In this supplement, we use the following condensed notation for the derivatives of a quantity $A^{\sigma\ldots\tau}_{\mu\nu\ldots\rho}(x)$ (which can be a tensor or non-tensor):

$$A^{\sigma\ldots\tau}_{\mu\nu\ldots\rho,\lambda}(x) \stackrel{\text{def}}{\equiv} \frac{\partial}{\partial x^{\lambda}} A^{\sigma\ldots\tau}_{\mu\nu\ldots\rho}(x).$$

◇ **Curved space: invariant distance, metric, parallel displacement and Christoffel symbols**

In a curved space, we are confined to curvilinear coordinates and the invariant distance ds between the point x^{μ} and a neighboring point $x^{\mu} + dx^{\mu}$ is given by

$$ds^2 = g_{\mu\nu}(x) dx^{\mu} dx^{\nu}.$$

With a network of curvilinear coordinates, the parameters $g_{\mu\nu} = g_{\nu\mu}$, given as functions of the coordinates, fix all the elements of distance; so, they fix the *metric*.

Suppose we have a vector X^{μ} located at some point P. If the space is curved, we cannot give a meaning to a parallel vector at a different point Q, as we can easily see if we think of the example of a curved two-dimensional space in a three-dimensional flat Euclidean space. However, if we take a point P' close to P, there is a parallel vector at P', with an uncertainty of the second order with respect to the distance between P and P'. We can transfer the vector continuously along a path by this process of *parallel displacement*. Taking a path from P to Q, we end up with a vector at Q which is parallel to the original vector at P with respect to this path. But a different path would give a different result.

We can obtain the formula for the parallel displacement of a vector by supposing that the curved space $\mathcal{M}$ under consideration is immersed in a flat space E of higher dimension with the coordinates y^n: $\mathcal{M} \to E: x^{\mu} \to y^n(x^{\mu})$. The metric $g_{\mu\nu}$ in this case has the form induced by this immersing:

$$g_{\mu\nu} = y^n_{,\mu} y_{n,\nu}. \tag{V.1}$$

Then shifting a vector X^n, tangent to the 'surface' $\mathcal{M}$, in the space E so as to keep it parallel to itself (which means, of course, keeping the components constant), to a neighboring point on the surface and projecting it down to this surface, we obtain the change dX^{μ} in X^{μ} (i.e. in the vector X^n written in terms of the x-coordinates: $X^n = y^n_{,\mu} X^{\mu}$):

$$g_{\nu\rho} dX^{\rho} = X^{\mu} y^n_{,\mu} y_{n,\nu,\sigma} dx^{\sigma}. \tag{V.2}$$

This can be written in the more compact and convenient form:

$$dX_{\nu} = X^{\mu} \Gamma_{\mu\nu\sigma} dx^{\sigma} \tag{V.3}$$

with the help of the *Christoffel symbols*

$$\Gamma_{\mu\nu\sigma} = \tfrac{1}{2}(g_{\mu\nu,\sigma} + g_{\mu\sigma,\nu} - g_{\nu\sigma,\mu}) \qquad (V.4)$$

(everywhere the indices are lowered and raised with the help of the metric tensor $g_{\mu\nu}$ and its inverse $g^{\mu\nu}$: $g_{\mu\nu}g^{\nu\sigma} = \delta_\mu{}^\sigma$). In this form, all reference to the flat space used for the derivation of the parallel displacement has disappeared, as the Christoffel symbol involves only the metric $g_{\mu\nu}$ of the curved space. Thus we can forget about the auxiliary flat space and define the parallel displacement in a curved space by equation (V.3). By differentiation and by using the matrix identity

$$g^{\mu\nu}_{,\sigma} = -g^{\mu\rho}g^{\nu\tau}g_{\rho\tau,\sigma} \qquad (V.5)$$

we can infer that the length of a vector is unchanged by the parallel displacement: $d(g_{\mu\nu}X^\mu X^\nu) = 0$.

It is frequently useful to raise the first index of the Christoffel symbol so as to form

$$\Gamma^\mu_{\nu\sigma} = g^{\mu\lambda}\Gamma_{\lambda\nu\sigma}.$$

This symbol is symmetrical in its two lower indices.

A curve in $\mathcal{M}$, the tangent vector of which is parallel to itself along the whole curve, is called a *geodesic*. This is defined by the equation

$$\frac{\partial^2 x^\mu}{\partial \tau^2} + \Gamma^\mu_{\nu\sigma}\frac{\partial x^\nu}{\partial \tau}\frac{\partial x^\sigma}{\partial \tau} = 0 \qquad (V.6)$$

where τ is a parameter on the curve.

◇ **Covariant differentiation, the curvature tensor and Bianci identities; Ricci and Weyl tensors**

The usual derivative of a vector field, $X_{\mu,\nu}$, is no longer a tensor because its transformations under a general change of coordinates, $x^\mu \to x'^\mu(x^\mu)$, contain an inhomogeneous part:

$$X_{\mu',\nu'} = X_{\rho,\sigma}x^\sigma_{,\nu'}x^\rho_{,\mu'} + X_\rho x^\rho_{,\mu'\nu'}.$$

We can, however, modify the process of differentiation so as to get a tensor. To this aim, we define the *covariant derivative*:

$$\nabla_\nu X_\mu \equiv X_{\mu;\nu} \stackrel{\text{def}}{\equiv} X_{\mu,\nu} - \Gamma^\sigma_{\mu\nu}X_\sigma. \qquad (V.7)$$

This derivative satisfies the usual Leibniz rule:

$$(X_\mu Y_\nu)_{;\sigma} = X_{\mu;\sigma}Y_\nu + X_\mu Y_{\nu;\sigma} \qquad (V.8)$$

and can be straightforwardly generalized to any tensor of a higher rank,

$$Z_{\mu\nu\ldots;\sigma} = Z_{\mu\nu\ldots,\sigma} - \Gamma\text{-term for each index}.$$

For example, for the second-rank tensor, the differentiation reads as

$$Z_{\mu\nu;\sigma} = Z_{\mu\nu,\sigma} - \Gamma^\rho_{\mu\sigma}Z_{\rho\nu} - \Gamma^\rho_{\nu\sigma}Z_{\mu\rho}. \qquad (V.9)$$

Formula (V.8) for the covariant derivative of a product

$$(Z_1 Z_2)_{;\sigma} = (Z_1)_{;\sigma}Z_2 + Z_1(Z_2)_{;\sigma} \qquad (V.10)$$

holds quite generally, with Z_1 and Z_2 being any kind of tensor quantity.

With product law (V.10), it is seen that the covariant differentiation is very similar to the ordinary differentiation. But there is an important distinction: the covariant differentiation is not commutative:

$$(\nabla_\mu \nabla_\nu - \nabla_\nu \nabla_\mu) X_\sigma \equiv X_{\sigma;\nu;\mu} - X_{\sigma;\mu;\nu} = X_\rho R^\rho_{\sigma\nu\mu} \qquad \text{(V.11)}$$

where

$$R^\rho_{\sigma\mu\nu} = \Gamma^\rho_{\sigma\mu,\nu} - \Gamma^\rho_{\sigma\nu,\mu} + \Gamma^\tau_{\sigma\mu} \Gamma^\rho_{\tau\nu} - \Gamma^\tau_{\sigma\nu} \Gamma^\rho_{\tau\mu}. \qquad \text{(V.12)}$$

This quantity proves to be a tensor and is called the Riemann–Christoffel tensor or the *curvature tensor*. It has the following symmetry properties ($R_{\lambda\sigma\nu\mu} = g_{\lambda\rho} R^\rho_{\sigma\nu\mu}$):

$$R^\rho_{\sigma\nu\mu} = -R^\rho_{\sigma\mu\nu}$$
$$R^\rho_{\sigma\nu\mu} + R^\rho_{\nu\mu\sigma} + R^\rho_{\mu\sigma\nu} = 0$$
$$R_{\mu\nu\rho\sigma} = -R_{\nu\mu\rho\sigma}$$
$$R_{\mu\nu\rho\sigma} = R_{\rho\sigma\mu\nu} = R_{\sigma\rho\nu\mu}$$

and satisfies the *Bianchi identities*:

$$R^\rho_{\mu\nu\sigma;\tau} + R^\rho_{\mu\sigma\tau;\nu} + R^\rho_{\mu\tau\nu;\sigma} = 0. \qquad \text{(V.13)}$$

If the space under consideration is flat, the curvature tensor $R^\rho_{\sigma\nu\mu}$ vanishes. Conversely, if $R^\rho_{\sigma\nu\mu} = 0$, we can prove that the space is flat.

In many applications, the *Ricci tensor* $R_{\mu\nu}$, which is obtained by contracting two indices, appears:

$$R_{\mu\nu} = R_{\nu\mu} \stackrel{\text{def}}{\equiv} R^\rho_{\mu\nu\rho}$$
$$= \Gamma^\rho_{\mu\nu,\rho} - \Gamma^\rho_{\mu\rho,\nu} + \Gamma^\tau_{\mu\nu} \Gamma^\rho_{\tau\rho} - \Gamma^\tau_{\mu\rho} \Gamma^\rho_{\nu\tau} \qquad \text{(V.14)}$$

the *scalar curvature*

$$R = g^{\mu\nu} R_{\mu\nu} \qquad \text{(V.15)}$$

and the *Weyl (conformal) tensor* which, in a four-dimensional space, has the form

$$C_{\mu\nu\rho\sigma} = R_{\mu\nu\rho\sigma} + \tfrac{1}{2}(g_{\mu\sigma} R_{\nu\rho} - g_{\mu\rho} R_{\nu\sigma} + g_{\nu\rho} R_{\mu\sigma} - g_{\nu\sigma} R_{\mu\rho}) + \tfrac{1}{6}(g_{\mu\rho} g_{\nu\sigma} - g_{\mu\sigma} g_{\nu\rho}) R. \qquad \text{(V.16)}$$

The vanishing of the Weyl (conformal) tensor is the necessary and sufficient condition for the spacetime to coincide locally with the Euclidean space after a suitable *conformal* transformation of the metric, i.e. after a substitution $g_{\mu\nu}(x) \to \Omega(x) g_{\mu\nu}(x)$ with some function $\Omega(x)$.

Note that all the formulae presented are correct under the assumption that the space under consideration has zero *torsion* and *non-metricity* (see the cited books at the beginning of supplement V for an explanation).

◇ Einstein's law of gravitation, Schwarzschild solution and black holes

The essence of *Einstein's law of gravitation* is expressed in his equations which, in the case of empty spacetime (i.e. spacetime without matter and other physical fields except the gravitational field), reads as

$$R_{\mu\nu} = 0. \qquad \text{(V.17)}$$

The flat spacetime obviously satisfies (V.17). The geodesics are then straight lines and so, the particles move along straight lines. If a spacetime is not flat, the Einstein equations put restrictions on the curvature and the metric.

Even for an empty spacetime, the Einstein equations are nonlinear and therefore very complicated. There is, however, one special case which can be solved without too much trouble, namely, the static spherically symmetric field produced by a spherically symmetric body at rest. The most general form for ds^2 in the four-dimensional spacetime, compatible with the spherical symmetry, is

$$ds^2 = U(r)\,dt^2 - V(r)\,dr^2 - W(r)r^2(d\theta^2 - \sin^2\theta\,d\phi^2)$$

where U, V and W are functions of r only. In a convenient coordinate system, the spherically symmetric static solution of equation (V.17) has the form

$$ds^2 = \left(1 - \frac{\rho_{\text{b.h.}}}{r}\right)dt^2 - \left(1 - \frac{\rho_{\text{b.h.}}}{r}\right)^{-1}dr^2 - r^2\,d\theta^2 - r^2\sin^2\theta\,d\phi^2 \qquad (\text{V.18})$$

and is known as the *Schwarzschild solution*. It holds outside the surface of the body producing the field, where there is no matter. The parameter $\rho_{\text{b.h.}}$ of the solution is related to the mass of this body. A physical analysis of this solution shows that the region $r < \rho_{\text{b.h.}}$ cannot communicate with the space for which $r > \rho_{\text{b.h.}}$. Any signal, even a light signal, would take an infinite time to cross the boundary $r = \rho_{\text{b.h.}}$. Thus, we cannot have a direct observational knowledge of the region $r < \rho_{\text{b.h.}}$. Such a region is called a *black hole*, because objects may fall into it but nothing can come out.

◇ **Extrinsic and intrinsic curvatures**

Let some manifold $\mathcal{X}$ be embedded into a higher-dimensional Riemann space $\mathcal{M}$. There are two approaches to the definition of the curvature for $\mathcal{X}$. On the one hand, we can consider $\mathcal{X}$ as a Riemannian space with a metric induced by that of $\mathcal{M}$ (in full analogy with (V.1)) and then use formula (V.12) to define its curvature. This yields what is called the *intrinsic curvature*. On the other hand, we can carry out the same construction that gives the definition of the curvature for surfaces in a usual flat Euclidean space and apply it to submanifolds in a Riemann space. The result is a different concept of the curvature, known as the *extrinsic curvature*. We have the following relationship:

$$K_{\text{i}} = K_{\text{e}} + K_{\text{t}}$$

where K_{t} is the curvature of $\mathcal{M}$ in the direction of the tangent plane to $\mathcal{X}$ and K_{i}, K_{e} are the intrinsic and extrinsic curvatures, respectively.

◇ **Lie derivative**

Besides the covariant derivative, we can define another useful derivative, called the *Lie derivative* acting on a tensor field Z defined on a manifold. This derivative produces a tensor field $\mathcal{L}_X Z$ of the same type as Z. The Lie derivative is constructed using a vector field $X = \{X^\mu\}$ defined on the same manifold and in local coordinates is given by the formula

$$(\mathcal{L}_X Z)^{\mu_1\cdots\mu_k}_{\nu_1\cdots\nu_p} = X^\mu \partial_\mu Z^{\mu_1\cdots\mu_k}_{\nu_1\cdots\nu_p} - \sum_{i=1}^{k}\partial_\rho X^{\mu_i} Z^{\mu_1\cdots\mu_{i-1}\rho\mu_{i+1}\cdots\mu_k}_{\nu_1\cdots\nu_p} + \sum_{j=1}^{p}\partial_{\nu_j} X^\rho Z^{\mu_1\cdots\mu_k}_{\nu_1\cdots\nu_{j-1}\rho\nu_{j+1}\cdots\nu_p}.$$

◇ **Differential forms**

An exterior *differential form* of degree p, or simply a p-form, on a differentiable manifold $\mathcal{M}$ is a p times covariant skew-symmetric tensor field on $\mathcal{M}$. If $\{x^1, \ldots, x^n\}$ is a local system of coordinates in a domain of the manifold $\mathcal{M}$, the one-forms $\{dx^1, \ldots, dx^n\}$ constitute a basis of the cotangent space $T_x^*(\mathcal{M})$ at a given point x. For this reason, any exterior p-form ω can be written in this domain as follows:

$$\omega = \sum_{\mu_1, \ldots, \mu_n} a_{\mu_1 \cdots \mu_n}(x) \, dx^{\mu_1} \wedge \cdots \wedge dx^{\mu_n}$$

where $a_{\mu_1 \cdots \mu_n}(x)$ is a skew-symmetric tensor field on $\mathcal{M}$. The wedge-product $\wedge$ produces from a p-form α and a k-form β the $(p+k)$-form $\alpha \wedge \beta$ and satisfies the condition of graded commutativity:

$$\alpha \wedge \beta = (-1)^{pk} \beta \wedge \alpha.$$

The concept of differentiation of a function is generalized to the concept of *exterior differential* which maps the space of p-forms into the space of $(p+1)$-forms and has the following properties:

$$d(d\omega) = 0 \qquad d(\alpha \wedge \beta) = d\alpha \wedge \beta + (-1)^p \alpha \wedge d\beta$$

where ω, α, β are arbitrary exterior forms (α has the degree p). In local coordinates, the exterior differential reads as

$$d\omega = \sum_{\rho, \mu_1, \ldots, \mu_n} \frac{\partial a_{\mu_1 \cdots \mu_n}(x)}{\partial x^\rho} dx^\rho \wedge dx^{\mu_1} \wedge \cdots \wedge dx^{\mu_n}.$$

Differential forms are an important component of the apparatus of differential geometry (see, e.g., Isham (1989), Visconti (1992) and Kobayashi and Nomizu (1969)).

VI Supersymmetry in quantum mechanics

The concept of *supersymmetry*, which relates bosonic and fermionic states in quantum mechanics, i.e. combines integer and half-integer spin states (particles) in one multiplet, has been playing an important role in the development of quantum field theory during the last two decades (see, e.g., Wess and Bagger (1983), Roček and Siegel (1983), West (1987), Weinberg (2000) and references therein). The supersymmetric models of unification of the fundamental interactions are the most promising candidates to extend the standard model of strong and electroweak interactions. Gravity was also generalized by incorporating supersymmetry (SUSY) into a theory called supergravity. In this theory, Einstein's general theory of relativity turns out to be a necessary consequence of a local gauged SUSY. Thus, local SUSY theories provide a natural framework for the unification of gravity with the other fundamental interactions of nature. Another theoretical motivation for studying supersymmetry is offered by the string theory (Green *et al* 1987).

In our book, we have used or, at least, mentioned several times the supersymmetry transformations (in particular, in sections 2.6.3, 3.2.7 and 3.4.5) which involve Grassmann (anticommuting) parameters. These are examples of SUSY Lie *group* transformations. In order to obtain an insight about the physical meaning and consequences of the supersymmetry, it is helpful to consider the corresponding Lie *superalgebras*. In this supplement, we shall consider the non-relativistic SUSY quantum mechanics (Witten 1981) as a simple realization of a superalgebra involving fermionic and bosonic operators.

A quantum-mechanical system characterized by a self-adjoint Hamiltonian $\widehat{H}$, acting on some Hilbert space $\mathcal{H}$, is called *supersymmetric* if there exists a *supersymmetry* (also called supercharge) operator $\widehat{Q}$, obeying the following anticommutation relations:

$$\{\widehat{Q}, \widehat{Q}\} = 0 = \{\widehat{Q}^\dagger, \widehat{Q}^\dagger\} \qquad \{\widehat{Q}, \widehat{Q}^\dagger\} = \widehat{H}. \tag{VI.1}$$

An immediate consequence of these relations is the conservation of the supercharge and the non-negativity of the Hamiltonian,

$$[\widehat{H}, \widehat{Q}] = 0 = [\widehat{H}, \widehat{Q}^\dagger] \qquad \widehat{H} \geq 0. \tag{VI.2}$$

A simple model of supersymmetric quantum mechanics is defined on the Hilbert space $\mathcal{H} = L^2(\mathbb{R}) \otimes \mathbb{C}^2$, that is, it characterizes a spin-$\frac{1}{2}$-like particle (with mass $m > 0$) moving along the one-dimensional line. In constructing a supersymmetric Hamiltonian on $\mathcal{H}$, let us first introduce the bosonic operators $\widehat{A}, \widehat{A}^\dagger$ and the fermionic operators $\widehat{f}, \widehat{f}^\dagger$:

$$\widehat{A} = \frac{\hbar}{\sqrt{2m}} \frac{d}{dx} + W(x) \qquad \widehat{A}^\dagger = \frac{-\hbar}{\sqrt{2m}} \frac{d}{dx} + W(x)$$
$$\widehat{f} = \sigma_+ = \begin{pmatrix} 0 & 1 \\ 0 & 0 \end{pmatrix} \qquad \widehat{f}^\dagger = \sigma_- = \begin{pmatrix} 0 & 0 \\ 1 & 0 \end{pmatrix} \tag{VI.3}$$

where the *superpotential* $W(x)$ is assumed to be continuously differentiable. Obviously, these operators obey the following commutation and anticommutation relations

$$[\widehat{A}, \widehat{A}^\dagger] = \frac{\sqrt{2}\hbar}{\sqrt{m}} W'(x) \qquad \{\widehat{f}, \widehat{f}^\dagger\} = 1 \tag{VI.4}$$

and allow us to define suitable supercharges

$$\widehat{Q} = \widehat{A} \otimes \widehat{f}^\dagger = \begin{pmatrix} 0 & 0 \\ \widehat{A} & 0 \end{pmatrix} \qquad \widehat{Q}^\dagger = \widehat{A}^\dagger \otimes \widehat{f} = \begin{pmatrix} 0 & \widehat{A}^\dagger \\ 0 & 0 \end{pmatrix} \tag{VI.5}$$

which satisfy the required relations $\{\widehat{Q}, \widehat{Q}\} = 0 = \{\widehat{Q}^\dagger, \widehat{Q}^\dagger\}$. Note that $\widehat{Q}$ is a combination of the generalized bosonic annihilation operator and the fermionic creation operator. Finally, we may construct a supersymmetric quantum system by defining the Hamiltonian in such a way that the second relation in (VI.1) also holds:

$$\widehat{H} = \{\widehat{Q}, \widehat{Q}^\dagger\} = \begin{pmatrix} \widehat{A}^\dagger \widehat{A} & 0 \\ 0 & \widehat{A}\widehat{A}^\dagger \end{pmatrix} = \begin{pmatrix} \widehat{H}_1 & 0 \\ 0 & \widehat{H}_2 \end{pmatrix} \tag{VI.6}$$

with

$$\widehat{H}_1 = -\frac{\hbar^2}{2m} \frac{d^2}{dx^2} + W^2(x) - \frac{\hbar}{\sqrt{2m}} W'(x) \tag{VI.7}$$

$$\widehat{H}_2 = -\frac{\hbar^2}{2m} \frac{d^2}{dx^2} + W^2(x) + \frac{\hbar}{\sqrt{2m}} W'(x) \tag{VI.8}$$

being the standard Schrödinger operators acting on $L^2(\mathbb{R})$.

Example: SUSY harmonic oscillator. We can introduce a Fock space of bosonic occupation numbers and the creation and annihilation operators a and $a^\dagger$ which, after a suitable normalization, obey the commutation relations

$$[\widehat{a}, \widehat{a}^\dagger] = 1 \qquad [\widehat{N}, \widehat{a}] = -\widehat{a} \qquad [\widehat{N}, \widehat{a}^\dagger] = \widehat{a}^\dagger \qquad \widehat{N} = \widehat{a}^\dagger \widehat{a} \qquad \widehat{H} = \widehat{N} + \tfrac{1}{2}. \tag{VI.9}$$

For the case of the SUSY harmonic oscillator, we can rewrite the operators $\widehat{Q}$ ($\widehat{Q}^\dagger$) as a product of the bosonic operator $\widehat{a}$ and the fermionic operator $\widehat{f}$. Namely, we write $\widehat{Q} = \widehat{a} \otimes \widehat{f}^\dagger$ and $\widehat{Q}^\dagger = \widehat{a}^\dagger \otimes \widehat{f}$, where the matrix fermionic creation and annihilation operators are defined in (VI.3) and obey the usual algebra of the fermionic creation and annihilation operators, namely

$$\{\widehat{f}^\dagger, \widehat{f}\} = 1 \qquad \{\widehat{f}^\dagger, \widehat{f}^\dagger\} = \{\widehat{f}, \widehat{f}\} = 0 \tag{VI.10}$$

as well as the commutation relation

$$[\widehat{f}, \widehat{f}^\dagger] = \sigma_3 = \begin{pmatrix} 1 & 0 \\ 0 & -1 \end{pmatrix}. \tag{VI.11}$$

The SUSY Hamiltonian can be rewritten in the form

$$\widehat{H} = \widehat{Q}\widehat{Q}^\dagger + \widehat{Q}^\dagger\widehat{Q} = \left(-\frac{d^2}{dx^2} + \frac{x^2}{4}\right)I - \frac{1}{2}[\widehat{f}, \widehat{f}^\dagger]. \tag{VI.12}$$

The effect of the last term is to remove the zero-point energy. This is a general property of SUSY systems: if the ground state is SUSY invariant, i.e. $Q|0\rangle = Q^\dagger|0\rangle = 0$, then, from the expression of the Hamiltonian, $\widehat{H} = \{\widehat{Q}, \widehat{Q}^\dagger\}$, we can immediately infer that the ground state has zero energy.

The state vector can be thought of as a matrix in the Schrödinger picture or as the state $|n_b, n_f\rangle$ in the Fock space picture. Since the fermionic creation and annihilation operators obey anticommutation relations, the fermion number is either zero or one. We will choose the ground state of H_1 to have zero fermion number. Then, we can introduce the fermion number operator

$$n_F = \frac{1-\sigma_3}{2} = \frac{1-[\widehat{f},\widehat{f}^\dagger]}{2}. \tag{VI.13}$$

The actions of the operators $a, a^\dagger, f, f^\dagger$ in this Fock space are then:

$$\begin{aligned} \widehat{a}|n_b, n_f\rangle &= |n_b - 1, n_f\rangle & \widehat{f}|n_b, n_f\rangle &= |n_b, n_f - 1\rangle \\ \widehat{a}^\dagger|n_b, n_f\rangle &= |n_b + 1, n_f\rangle & \widehat{f}^\dagger|n_b, n_f\rangle &= |n_b, n_f + 1\rangle. \end{aligned} \tag{VI.14}$$

Now we can see that the operator $\widehat{Q}^\dagger = -i\widehat{a}\widehat{f}^\dagger$ has the property of changing a boson into a fermion without changing the energy of the state. This is the boson–fermion degeneracy, characteristic of all SUSY theories.

As can be seen from (VI.3), for the general case of SUSY quantum mechanics, the operators $\widehat{a}, \widehat{a}^\dagger$ are replaced by $\widehat{A}, \widehat{A}^\dagger$ in the definition of $\widehat{Q}, \widehat{Q}^\dagger$, i.e. we write $\widehat{Q} = \widehat{A} \otimes \widehat{f}^\dagger$ and $\widehat{Q}^\dagger = \widehat{A}^\dagger \otimes \widehat{f}$. The effect of $\widehat{Q}$ and $\widehat{Q}^\dagger$ is now to relate the wavefunctions of $\widehat{H}_1$ and $\widehat{H}_2$ which have fermion number zero and one respectively, but now there is no simple Fock space description in the bosonic sector because the interactions are nonlinear. Again, as in the harmonic oscillator, in quantum theory with an exact symmetry the ground state must be invariant with respect to the group transformations. This means, in turn, that the ground state must be annihilated by the generators of the symmetry group and the ground-state energy *vanishes*. Otherwise, SUSY is said to be broken.

Bibliography

Abrikosov A A *et al* 1965 *Methods of Quantum Field Theory in Statistical Physics* (Oxford: Pergamon)
Adler S 1969 *Phys. Rev.* **177** 2426
Alfaro J and Damgaard P H 1990 *Ann. Phys.* **202** 398
Arnowitt R *et al* 1960 *Phys. Rev.* **117** 1595
Ashcroft N W and Mermin N D 1976 *Solid State Physics* (New York: Holt, Rinehart and Winston)
Ashtekar A 1991 *Lectures on Non-Perturbative Canonical Gravity* (Singapore: World Scientific)
Ashtekar A *et al* 1993 *Int. J. Mod. Phys.* D **2** 15
Bailin D and Love A 1993 *Introduction to Gauge Field Theory* (Bristol: Institute of Physics)
Balescu R 1975 *Equilibrium and Nonequilibrium Statistical Mechanics* (London: Wiley)
Bardeen J *et al* 1957 *Phys. Rev.* **108** 1175
Barut A O and Rączka R 1977 *Theory of Group Representations and Applications* (Warsaw: Polish Science)
Baxter R J 1982 *Exactly Solved Models in Statistical Mechanics* (London: Academic)
Becchi C *et al* 1975 *Commun. Math. Phys.* **42** 127
——1976 *Ann. Phys.* **98** 287
Bekenstein J D 1973 *Phys. Rev.* D **7** 2333
Belavin A *et al* 1975 *Phys. Lett.* B **59** 85
Bell J S and Jackiw R 1969 *Nuovo Cimento* A **60** 47
Berezin F A 1975 *Commun. Math. Phys.* **40** 153
Bertlmann R A 1996 *Anomalies in Quantum Field Theory* (Oxford: Clarendon)
Bjorken J D and Drell S D 1965 *Relativistic Quantum Fields* (New York: McGraw-Hill)
Bogoliubov N N and Shirkov D V 1959 *Introduction to the Theory of Quantized Fields* (New York: Wiley)
Bose S N 1924 *Z. Phys.* **26** 178
Brosens F *et al* 1997 *Phys. Rev.* E **55** 227
Brown J D and York J W 1994 *The Black Hole (Proc. Int. Conf. 'The Black Hole 25 Years After')* ed C Teitelboim *et al* (Santiago, CA: University Press)
Callan C G *et al* 1979 *Phys. Rev.* D **19** 1826
Carlip S 1995 Lectures on $(2+1)$-dimensional gravity *Preprint* UCD-95-6, gr-qc/9503024
Carlip S and Cosgrove R 1994 *J. Math. Phys.* **35** 5477
Chaichian M *et al* 1994 *Phys. Rev.* D **49** 1566
——2000 *Nucl. Phys.* B **567** 360; *J. Math. Phys.* **41** 1647
Chaichian M and Demichev A 1996 *Introduction to Quantum Groups* (Singapore: World Scientific)
Chaichian M and Hagedorn R 1998 *Symmetries in Quantum Mechanics. From Angular Momentum to Supersymmetry* (Bristol: Institute of Physics)
Chaichian M and Nelipa N F 1984 *Introduction to Gauge-Field Theories* (Berlin: Springer)
Chaichian M and Senda I 1993 *Nucl. Phys.* B **396** 737
Chandrasekhar S 1983 *The Mathematical Theory of Black Holes* (Oxford: Clarendon)
Chaturvedi S 1996 *Phys. Rev.* E **54** 1378
Cheng T-P and Li L-F 1984 *Gauge Theory of Elementary Particle Physics* (Oxford: Clarendon)
Cohen A *et al* 1986 *Nucl. Phys.* B **267** 143
Connes A 1994 *Noncommutative Geometry* (New York: Academic)

Connes A and Lott J 1990 *Nucl. Phys. Suppl.* B **11** 19
Cooper L N 1956 *Phys. Rev.* **104** 1189
Creutz M 1983 *Quarks, Gluons and Lattices* (Cambridge: Cambridge University Press)
Damgaard P H and Hüffel H 1987 *Phys. Rep.* **152** 227
Dashen R *et al* 1974 *Phys. Rev.* D **10** 4114
——1974 *Phys. Rev.* D **10** 4130
——1974 *Phys. Rev.* D **10** 4138
Davydov A S 1976 *Quantum Mechanics* (Oxford: Pergamon)
de Carvalho C A *et al* 1985 *Phys. Rev.* D **31** 1411
de Gennes P G 1989 *Superconductivity of Metals and Alloys* (New York: Addison-Wesley)
Dell' Antonio G and Zwanziger D 1989 *Cargese Proc., Probabilistic Methods in Quantum Field Theory and Quantum Gravity* vol 107 (New York: Gordon and Breach)
Deser S *et al* 1984 *Ann. Phys.* **152** 220
De Witt B S 1967 *Phys. Rev.* **160** 1113
——1967 *Phys. Rev.* **162** 1195
——1967 *Phys. Rev.* **162** 1239
Dirac P A M 1931 *Proc. R. Soc.* A **133** 60
——1950 *Can. J. Math.* **2** 129
——1958 *Proc. R. Soc.* A **246** 326
——1964 *Lectures on Quantum Mechanics* (New York: Belfer Graduate School of Science)
——1975 *General Theory of Relativity* (New York: Wiley)
Doplicher S *et al* 1994 *Phys. Lett.* B **331** 39
——1995 *Commun. Math. Phys.* **172** 187
Duff M J 1999 *The World in Eleven Dimensions: Supergravity, Supermembranes and M-Theory* (Bristol: IOP Publishing)
Dyson F J 1949 *Phys. Rev.* **75** 1736
Einstein A 1925 *Sitz. Kgl. Preuss. Akad. Wiss. (Berlin)* 3
Faddeev L D and Popov V N 1967 *Phys. Lett.* B **25** 29
Faddeev L D and Slavnov A A 1980 *Gauge Fields: Introduction to a Quantum Theory* (Reading, MA: Benjamin-Cummings)
Fermi E 1926 *Z. Phys.* **36** 902
Feynman R P 1948 *Rev. Mod. Phys.* **20** 367; reprinted in 1958 *Selected Papers on Quantum Electrodynamics* ed J Schwinger (New York: Dover)
——1950 *Phys. Rev.* **80** 440
——1955 *Phys. Rev.* **97** 660
——1972a *Statistical Mechanics: a Set of Lectures* (Reading, Mass.: Benjamin)
——1972b *Photon–Hadron Interactions* (Reading, MA: Benjamin)
Feynman R P and Hibbs A R 1965 *Quantum Mechanics and Path Integrals* (New York: McGraw-Hill)
Filk T 1996 *Phys. Lett.* B **376** 53
Floratos E G and Iliopoulos J 1983 *Nucl. Phys.* B **214** 392
Flügge S 1971 *Pratical Quantum Mechanics* (Berlin: Springer)
Fradkin E S and Shteingradt 1978 *Nuovo Cimento* A **47** 115
Fröhlich H 1937 *Proc. Phys. Soc.* A **160** 230
——1954 *Adv. Phys.* **3** 325
Fujikawa K 1979 *Phys. Rev. Lett.* **42** 1195
——1980 *Phys. Rev.* D **21** 2848
——1980 *Phys. Rev.* D **22** 1499(E)
——1984 *Phys. Rev.* D **29** 285
——1985 *Phys. Rev.* D **31** 341
Furry W 1937 *Phys. Rev.* **51** 125
Gervais J L 1977 Lectures given at the XVI Schladming School *Preprint* PTENS 77/3
Gervais J L *et al* 1975 *Phys. Rev.* D **12** 1038

——1976 *Phys. Rep.* **23** 281
Gibbons G W *et al* 1978 *Nucl. Phys.* B **138** 141
Gibbons G W and Hawking S W 1977 *Phys. Rev.* D **15** 2752
Gitman D M and Tyutin I V 1990 *Quantization of Fields with Constraints* (Berlin: Springer)
Glashow S L 1961 *Nucl. Phys.* **22** 579
Gozzi E 1983 *Phys. Rev.* D **28** 1922
Gradshteyn I S and Ryzhik I M 1980 *Table of Integrals, Series and Products* (New York: Academic)
Green H S 1953 *Phys. Rev.* **90** 270
Green M *et al* 1987 *Superstring Theory* vols 1 and 2 (Cambridge: Cambridge University Press)
Greiner W and Reinhardt J 1989 *Field Quantization* (Berlin: Springer)
Gribov V N 1978 *Nucl. Phys.* B **139** 1
Griffin A 1993 *Excitations in a Bose-Condensed Liquid* (New York: Cambridge University Press)
Griffin A *et al* (ed) 1995 *Bose–Einstein Condensation* (New York: Cambridge University Press)
Griffiths R B 1972 *Phase Transitions and Critical Phenomena* vol 1, ed C Domb and M S Green (London: Academic)
Gross D J and Wilczek F 1973 *Phys. Rev. Lett.* **30** 1343
——1973 *Phys. Rev.* D **8** 3633
Grosse H *et al* 1997 *Commun. Math. Phys.* **185** 155
Halpern M B and Koplik J 1978 *Nucl. Phys.* B **132** 239
Hamermesh M 1964 *Group Theory and its Application to Physical Problems* (Reading, MA: Addison-Wesley)
Hawking S W 1975 *Commun. Math. Phys.* **43** 199
Heeger A J 1988 *Rev. Mod. Phys.* **60** 781
Heisenberg W 1954 as quoted in: Dürr H P 1961 *Werner Heisenberg und die Physik unserer Zeit* (Braunschweig: Vieweg) and in Rampacher H *et al* 1965 *Fortsch. Phys.* **13** 385
Hepp K 1969 *Theorie de la Renormalisation* (Berlin: Springer)
Hoppe J 1989 *Elem. Part. Res. J.* **80** 145
Horowitz G 1996 *Black Holes and Relativistic Stars (Proc. Symp. on Black Holes and Relativistic Stars)* ed R M Wald (Chicago, IL: Chicago University Press); gr-qc/9704072
Hüffel H and Nakazato H 1994 *Mod. Phys. Lett.* A **9** 2953
Intriligator K and Seiberg N 1996 *Nucl. Phys. Proc. Suppl.* BC **45** 1
Isham C I 1989 *Modern Differential Geometry for Physicists* (Singapore: World Scientific)
Itzykson C and Drouffe J M 1989 *Statistical Field Theory* (Cambridge: Cambridge University Press)
Itzykson C and Zuber J-B 1980 *Quantum Field Theory* (New York: McGraw-Hill)
Izumov Yu A and Skryabin Yu N 1988 *Statistical Mechanics of Magnetically Ordered Sysytems* (New York: Consultants Bureau)
Keldysh L V 1965 *Sov. Phys.–JETP* **20** 1018
Kittel C 1987 *Quantum Theory of Solids* (New York: Wiley)
Klauder J R 1983 *Acta Phys. Aust. Suppl.* **25** 251
Klauder J R and Ezawa H 1983 *Prog. Theor. Phys* **63** 664
Kobayashi S and Nomizu K 1969 *Foundations of Differential Geometry*, vols 1 and 2 (New York: Interscience)
Kogut J B 1979 *Rev. Mod. Phys.* **51** 659
Kummer W and Vassilevich D V 1999 *Ann. Phys.* **8** 801
Landau L D *et al* 1954 *Dokl. Akad. Nauk SSSR* **95** 497
——1954 *Dokl. Akad. Nauk SSSR* **95** 773
Landau L D and Lifshitz E M 1981 *Quantum Mechanics* (New York: Pergamon)
——1987 *The Classical Theory of Fields* (Oxford: Pergamon)
Lehmann H *et al* 1955 *Nuovo Cimento* **1** 205
Lemmens L F *et al* 1996 *Phys. Rev.* E **53** 4467
Ma S 1976 *Modern Theory of Critical Phenomena* (Reading, MA: Benjamin-Cummings)
Maldacena J *et al* 1997 *JHEP* 9712:002
Marolf D 1996 *Phys. Rev.* D **53** 6979
Matsubara T 1955 *Prog. Theor. Phys.* **14** 351
Misner C, Thorne K and Wheeler J 1973 *Gravitation* (New York: Freeman)

Montonen C and Olive D 1977 *Phys. Lett.* B **72** 117
Montvay I and Münster G 1994 *Quantum Fields on a Lattice* (Cambridge: Cambridge University Press)
Mottola E 1995 *J. Math. Phys.* **36** 2470
Nakanishi N and Ojima I 1990 *Covariant Operator Formalism of Gauge Theories and Quantum Gravity* (Singapore: World Scientific)
Namiki M 1992 *Stochastic Quantization* (Heidelberg: Springer)
Niemi A J and Semenoff G W 1984 *Ann. Phys.* **152** 105
——1984 *Nucl. Phys.* B **220** 181
Novikov I D and Frolov V P 1989 *Physics of Black Holes* (Dordrecht: Kluwer)
Novozhilov Yu V 1975 *Introduction to Elementary Particle Theory* (London: Pergamon)
Ohba I 1987 *Prog. Theor. Phys.* **77** 1267
Okun L B 1982 *Leptons and Quarks* (Amsterdam: North-Holland)
Onsager L 1944 *Phys. Rev.* **65** 117
Osborn H 1979 *Phys. Lett.* B **83** 321
Parisi G and Wu Y 1981 *Sci. Sin.* **24** 483
Peskin M E and Schroeder D V 1995 *An Introduction to Quantum Field Theory* (Reading, MA: Addison-Wesley)
Plechko V N 1988 *Physica* A **152** 51
Polchinski J 1994 What is string theory? *Lectures Given at NATO Advanced Study Institute: Fluctuating Geometries in Statistical Mechanics and Field Theory (Proc. Les Houches, France)*; hep-th/9411028
——1996 TASI lectures on D-branes *Preprint* NSF-ITP-96-145; hep-th/9611050
Politzer H D 1973 *Phys. Rev. Lett.* **30** 1346
Polyakov A 1974 *JETP Lett.* **20** 194
——1977 *Nucl. Phys.* B **120** 429
——1981 *Phys. Lett.* B **103** 207
——1981 *Phys. Lett.* B **103** 211
——1987 *Gauge Fields and Strings* (Chur: Harwood)
Polychronakos A P 1996 *Nucl. Phys.* B **474** 529
Popov V N 1983 *Functional Integrals in Quantum Field Theory and Statistical Physics* (Dordrecht: Reidel)
Rajaraman R 1982 *Solitons and Instantons. An Introduction to Solitons and Instantons in Quantum Field Theory* (Amsterdam: North-Holland)
Rivers R J 1987 *Path-Integral Methods in Quantum Field Theory* (Cambridge: Cambridge University Press)
Roček M and Siegel W 1983 *Superspace: or One Thousand and One Lessons in Supersymmetry* (London: Benjamin-Cummings)
Ryser H J 1963 *Combinatorial Mathematics* (New York: Wiley)
Salam A 1968 in: *Elementary Particle Physics (Nobel. Symp. 8)* ed N Svartholm (Stockholm: Almqvist and Wilsell)
Schäfer T and Shuryak E V 1998 *Rev. Mod. Phys.* **70** 323
Schweber S S 1961 *An Introduction to Relativistic Quantum Field Theory* (New York: Row, Peterson)
Schwinger J 1951 *Proc. Natl Acad. Sci.* **37** 452
——1951 *Proc. Natl Acad. Sci.* **37** 455
——1957 *Ann. Phys.* **2** 407
——1961 *J. Math. Phys.* **2** 407
Seiberg N and Witten E 1994 *Nucl. Phys.* B **426** 19
——1994 *Nucl. Phys.* B **430** 485 (erratum)
——1994 *Nucl. Phys.* B **431** 484
——1999 *JHEP* 9909: 032
Stoof H T C 1999 *J. Low-Temp. Phys.* **114** 11
Strominger A and Vafa C 1996 *Phys. Lett.* B **379** 99
Suranyi 1990 *Phys. Rev. Lett.* **65** 2329
Symanzik K 1964 *Analysis in Function Space* ed W T Martin and I Segal (Cambridge, MA: MIT Press)
Taylor J R 1972 *Scattering Theory. The Quantum Theory of Nonrelativistic Scattering* (New York: Wiley)
ter Haar D 1971 *Elements of Hamiltonian Mechanics* (New York: Pergamon)
't Hooft G 1971 *Nucl. Phys.* B **33** 173

Bibliography

——1971 *Nucl. Phys.* B **35** 167
——1974 *Nucl. Phys.* B **79** 276
——1976 *Phys. Rev.* D **14** 3432
Treiman S B *et al* 1985 *Current Algebra and Anomalies* (Princeton, NJ: Princeton University Press)
Tyutin I V 1975 *Lebedev Inst. Preprint* N39
Umezawa H *et al* 1982 *Thermo Field Dynamics and Condensed States* (Amsterdam: North-Holland)
van Kampen N G 1981 *Stochastic Processes in Physics and Chemistry* (Amsterdam: North-Holland)
Vasiliev A N 1998 *Functional Methods in Quantum Field Theory and Statistical Physics* (Amsterdam: Gordon and Breach)
Vilenkin N Ya 1968 *Special Functions and the Theory of Group Representations (Transl. Math. Monographs)* (New York: American Mathematical Society)
Visconti A 1992 *Introductory Differential Geometry for Physicists* (Singapore: World Scientific)
Weinberg S 1967 *Phys. Rev. Lett.* **19** 1264
——1995 *The Quantum Theory of Fields. I. Foundations* (Cambridge: Cambridge University Press)
——1996 *The Quantum Theory of Fields. II. Modern Applications* (Cambridge: Cambridge University Press)
——2000 *The Quantum Theory of Fields. III. Supersymmetry* (Cambridge: Cambridge University Press)
Wentzel G 1949 *Quantum Theory of Field* (New York: Interscience)
Wess J and Bagger J 1983 *Supersymmetry and Supergravity* (Princeton, NJ: Princeton University Press)
West P 1986 *Introduction to Supersymmetry and Supergravity* (Singapore: World Scientific)
Wiegel F W 1986 *Introduction to Path-Integral Methods in Physics and Polymer Science* (Singapore: World Scientific)
Wigner E P 1959 *Group Theory and its Application to the Quantum Mechanics of Atomic Spectra* (New York: Academic)
Wilson K 1977 *New Phenomena in Sub-Nuclear Physics* ed A Zichichi (New York: Plenum)
Wilson K and Fisher M E 1972 *Phys. Rev. Lett.* **28** 240
Witten E 1981 *Nucl. Phys.* B **188** 513
——1988 *Nucl. Phys.* B **311** 46
Wybourn B G 1974 *Classical Groups for Physicists* (New York: Wiley)
Yang C N and Mills R L 1954 *Phys. Rev.* **54** 191
Zakrzewski W J 1989 *Low Dimensional Sigma Models* (Bristol: Hilger)
Zhelobenko D P 1973 *Compact Lie Groups and Their Applications (Transl. Math. Monographs 40)* (New York: American Mathematical Society)
Zinn-Justin J 1989 *Quantum Field Theory and Critical Phenomena* (Oxford: Clarendon)

Index

Page numbers in **bold** typeface indicate volume I.

action functional, **140**
anharmonic oscillator, 44
   perturbation theory expansion, 45
annihilation operators, 2, **132**
anomalous dimension, 239
anti-instanton, 121, 125
anti-kink, 112
anti-normal symbols, **207**
anticommutation relations, **286**
anticommutator, **286**
anticommuting variables, **286**
Arnowitt–Deser–Misner (ADM) formalism, 151
Arnowitt–Deser–Misner decomposition, 170
asymptotic freedom, 88
asymptotic series, **171**
asymptotic states, **222**
asymptotically flat spacetime, 150

Baker–Campbell–Hausdorff formula, **155**, 317
Bargmann–Fock realization of CCR, **206**
Belavin–Polyakov–Schwartz–Tyupkin (BPST) instanton, 128
Berezin integral, **292**
Bernoullian random walk, **49**
Bianci identities, 327
black hole, 166, 328
   entropy, 166, 174
   stabilization by a finite box, 168
Bloch equation, **65**, **72**, **76**
Bohr–Sommerfeld quantization condition, **123**, **145**, **146**, **176**, 0180
Borel set, **5**
Bose–Einstein condensation, 257, 263
bosons, 194
bra-vector, **127**
Brownian bridge, **82**
Brownian motion, **12**, **13**

and fractal theory, **28**
discrete version, **13**
in field of non-conservative force, **115**
independence of increments, **23**
of interacting particles, **67**
   under external forces, **68**
under an arbitrary external force, **66**
under an external harmonic force, **64**, **84**
with absorption, **73**
Brownian particle, **6**, **13**
   drift velocity, **49**
   time to reach a point, **118**
   under an external force, **76**
   with inertia, **71**
BRST symmetry, 1, 80
BRST transformations, 87

canonical anticommutation relations, 11
canonical commutation relations (CCRs), **132**
canonical loop space, **308**
canonical transformations, **305**
Cartan–Weyl basis, 323
Casimir operator, 322
Cauchy–Schwarz–Bunyakowskii inequality, **143**
caustics, **176**
central charge, 184
change of variables in path integrals, **45**
character of a group, 317
characteristic function, **101**
characteristic functional, **101**
chemical potential, 197, 223
Chern–Simons characteristic, 126
chiral symmetry, 130
chiral transformations, 130
   local, 131
Christoffel symbols, 326
chronological ordering, **74**

337

coherent-state path integrals, **200**, **218**
coherent states, **207**
    normalized, **218**
    on the group $SU(2)$, **280**
    overcompleteness, **219**
Coleman theorem, 89
collective coordinates (modes), 115
commutator of operators, **129**
compound event, **22**
Compton scattering, 79
conditional probability, **15**
configuration integral, 195, 200
conjugate points, **174**, **175**
connected vacuum loops, 30
constrained systems, 55
constraints, 54
    first-class, 56
    involution condition, 57
    primary, 55
    second-class, 56
    secondary, 56
continuity equation, **13**
continuous integral, **1**
contraction operator, **144**
contravariant symbol, **220**
coordinate representation, **129**
correlation length, 239
    divergence at critical temperature, 240
correspondence principle, **201**
coset space, 318
Coulomb problem, **122**
counter-terms, 84
coupling constant, 11, 24
covariance, **103**
covariant density, 151
covariant derivative, 51, 326
creation operators, 2, **132**
critical exponents, 238
critical temperature, 230
cross section, 80
curvature, 150, **246**
    scalar, **246**, 327
    tensor, 327
cyclicity condition, 297
cyclotron frequency, **196**

Debye–Hückel approximation, 205
deformed oscillator algebra, **298**

degenerate theory, 57
delta-functional (δ-functional), **71**
density operator, 197
    canonical, 197
    grand canonical, 198
    microcanonical, 197
differential forms, 328
diffusion constant, **13**
diffusion equation, **13**
    inhomogeneous, **20**
    solution, **18**, **19**
dimensional regularization, 83
Dirac bracket, 56
Dirac conjugate spinor, 12
Dirac equation, 12
Dirac matrices, 12
Dirac 'sea', 14
discrete random walks, **108**
discrete-time approximation, **37**
domain walls, 233
double-well potential, **91**
driven harmonic oscillator, **104**, 313
    classical (stochastic), **104**
        transition probability, **107**
    quantum
        propagator, **198**
duality, 120
Duistermaat–Heckman theorem, **307**
    loop-space generalization, **310**
Duru–Kleinert method, **267**
Dyson–Schwinger equations, 31–33
    for anomalous Green functions, 259
    for fermions, 262

effective action, 36, 228, **247**
effective mass, 137
effective potential, 228, 229
    convexity, 230
Einstein action, 150
Einstein's law of gravitation (Einstein equations), 327
electrodynamics (QED)
    Compton scattering, 79
    first-order formalism, 60
    second-order formalism, 60
electron self-energy, 81
entropy, 198
equilibrium state, 195

equivariant exterior derivative, **306**, **309**
ergodic property, 281
ESKC (semigroup) relation, **21**
Euler–Lagrange equations, **79**, **80**, **141**
evolution operator, **125**
    as a ratio of path integrals, **213**
    normal symbol, **217**
    Weyl symbol, **212**
excluded volume problem and the Feynman–Kac formula, **110**
exterior differential, 329
extrinsic curvature, 169, 328

Faddeev–Popov ghosts, 74
Faddeev–Popov trick, 69
Fermat principle, **188**
fermions, 194
Feynman diagrams, 1
    connected, 29
    disconnected, 29
    one-particle irreducible (OPI), 29
Feynman parametrization, 83
Feynman rules, 1, 316
Feynman variational method, 140, 141
Feynman–Kac formula, **65**, **73**, **137**
    in quantum mechanics, **123**
    proof for the Bloch equation, **76**
finite-difference operators, **95**
fluctuation factor, **80**, **82**
fluctuation–dissipation theorem, 272
focal points, **176**
Fokker–Planck equation, **61**, **111**
Fourier decomposition, **95**, **100**
    for Brownian trajectories, **95**
    independence of coefficients, **117**
free energy, 198
free Hamiltonian, **222**
functional derivatives, **86**
functional integral, **1**
functional space, **26**
functionals
    characteristic, **31**
    generating, **101**
    integrable, **34**
    simple, **33**
        Cauchy sequence, **34**
fundamental interactions, 88
fundamental solution, **20**

Furry theorem, 144

gas
    ideal, 199
    van der Waals, 201
gauge fields, 50
gauge group, 50
gauge invariance, **186**
gauge transformations, 57, **186**
gauge-field tensor, 51
gauge-field theory, 1, 50
    Lagrangian, 51
    lattice, 295
    non-Abelian, 1
gauge-fixing conditions, 57
    α-gauge, 73
    axial, 63
    Coulomb, 61, 63, **195**
    Lorentz, 61, 63
    unitary, 92
Gaussian distributions, **59**
Gaussian integral, **18**
    complex, **135**
    Grassmann case, **295**
    multidimensional, **38**
Gelfand–Yaglom method, **43**, **87**, **168**, **263**
general coordinate transformations, 150
generalized eigenfunctions, **130**
generating function, **101**
generating functional, **101**
    for Green functions in QFT, 20
        connected, 33
        on non-commutative spacetimes, 187
        one-particle irreducible (OPI), 35
        spinor, 22
    for thermal Green functions, 226
        non-relativistic, 225
        path-integral representation, 227
    in stochastically quantized QFT, 290
    in Yang–Mills theory, 66
        with matter, 77
    its Legendre transformation, 35
generators of a group (infinitesimal operators), 321
goldstone fields, 89, 91
Goldstone theorem, 89
grand canonical partition function, 198
    classical

path-integral representation, 203
in gravitation, 172
Grassmann algebra, **286**, **288**
infinite-dimensional, **298**
Grassmann variables, **289**
gravitons, 154
Green ansatz, 216
Green functions, **20**
anomalous, 259
causal, relativistic, 18
connected, 30
of Dirac equation, causal, 22
of the stationary Schrödinger equation, **177**
one-particle irreducible (OPI), 30
truncated (amputated), 30
Gribov ambiguity, 64
ground state, **132**
group, 318
Abelian, 319
invariant, 319
Lie, 321
semisimple, 322
non-Abelian, 319

Haar measure, 319
hadrons, 97
Hamiltonian vector field, **306**
harmonic oscillator, **103**
quantum, **131**
with time-dependent frequency, **168**
hedgehog ansatz, 126
Hermite function, **132**
Hermite polynomial, **132**
Higgs boson, 92
Higgs mechanism, 88, 92
high temperature expansion, 230
Hilbert–Schmidt theorem, **142**
hopping-path approximation, **91**
hopping-path solution, **92**

ideal gas, 199
index of a bilinear functional, **175**
infrared divergences, 106
instanton tunneling amplitude, 124
with account of fermions, 125
instantons, 2, **91**, 101, 110, 120, 121
tunnel transitions, 120
integral kernel of an operator, **78**, **103**

convolution, **134**
integral over histories, **1**
integral over trajectories, **1**
interaction representation, **223**
internal symmetry groups, 50
global, 50
local, 50
intrinsic curvature, 328
invariant subspace, 323
invariant torus, **178**
Ising model, 296
critical point, 301
free energy
Onsager's solution, 301
partition function, 299
as Grassmann path integral, 299
Ito stochastic integral, **63**, **120**, **183**
Stokes formula, **120**

Kac–Uhlenbeck–Hemmer model, 202
ket-vector, **127**
Killing form, 324
kink, 112
Klein–Gordon equation, 11
Kolmogorov second equation, **61**
Kolmogorov's theorem, **36**
Kubo–Martin–Schwinger propagator relation, 227
Kustaanheimo–Stiefel transformation, **245**, **266**, **269**

Lagrangian, **141**
Langevin equation, **61**, **62**
Laplace–Beltrami operator, **246**
lapse function, 156
large fluctuations, **91**
lattice derivative, **95**
lattice gauge theory, 302
fermion doubling, 305
lattice regularization, 82
Legendre transformation, 9, 35, **180**, 288
leptons, 96
Lie algebra, 321
real form, 322
semisimple, 322
simple, 322
Lie derivative, **306**, 328
Liouville measure, **306**

Lorentz group, 9, 10
Lyapunov exponent, **181**

M-theory, 175
magnetic monopole, 110
    of 't Hooft–Polyakov type, 110
magnetization, 238
Markov chain, **13**
Markov process, **23**
Maslov–Morse index, **174**, **181**
mass operator, 33
matter-field Lagrangian, 51
Maurer–Cartan equation, 322
Maurer–Cartan form, **273**, 322
Mayer expansion, 200
mean value, **58**
measure, **5**
    Feynman (formal), **138**
    Lebesgue, **5**
    Wiener, **25**
method of collective coordinates, 115
method of images, **231**
method of square completion, 28, **159**
metric in a curved space, 325
midpoint prescription, **47**, **186**, **210**
minisuperspace, 156
    matrix elements of the evolution operator, 160
Minkowski space, 9
Misner parametrization, 156
mixed states, **127**
mode expansion, **100**
Monte Carlo simulations, 304
Morse function, **307**
Morse theorem, **176**
Moyal bracket, 187
multiplicative renormalizability, 25

Nambu–Goto action, 180
normal coordinates, 4
normal symbol, **206**, **207**

observable, **123**
occupation numbers, 8
operator
    annihilation, **132**
    compact, **142**
    conjugate, **124**
    creation, **132**
    Hamiltonian, **125**
    self-adjoint (Hermitian), **124**
    symmetric, **124**
operator ordering problem, **129**, **183**, **187**, **200**
operator spectrum, **124**
    study by the path-integral technique, **141**
orbits of a gauge group, 61
ordering rules, **129**
Ornstein–Uhlenbeck process, **63**, **90**
overcompleted basis, **207**

parallel displacement in a curved space, 325
parastatistics, 194, 216
    propagator for particles, 219
Parisi–Sourlas integration formula, **314**
partial summation of the perturbation expansion, 101
    $1/N$-expansion, 102
    separate integration over high- and low-frequency modes, 106
partition function, **56**
    classical, 196
        path-integral representation, 205
    quantum, 197
        for identical particles in harmonic potential, 214
partons, 88
path integrals, **1**
    and singular potentials, **267**
    and transformations of states on non-commutative spacetimes, 191
    calculation by ESKC relation, **39**
    change of variables
        via Fredholm equation, **45**
        via Volterra equation, **46**
    coherent state PI on $SU(2)$ group, **281**
    definition via perturbation theory, 46
    discrete-time (time-sliced) approximation, **36**
    Feynman, **122**, **137**
    for topology-change transition amplitudes, 162
    holomorphic representation, 16
    in phase space, **122**, **139**, **155**
    in terms of coherent states, **218**
    Wiener, **25**
    with constraints, **122**
    with topological constraints, **122**

path length, **188**
periodic orbit theory, **154**, **181**
periodic orbits, **176**, **178**
perturbation expansion, **152**
perturbation theory, 24
    covariant, 67
phase transition
    first-order, 230
    second-order, 230
phase-space path integral, **122**
phonons, 137
    acoustic, 138
    optical, 138
physical–optical disturbance, **187**
Planck length, 185
Planck mass, 149
plaquette, 303
Poincaré group, 9, 10
Poincaré map, **181**
Poisson brackets, **129**
Poisson distribution, **50**
Poisson formula, **232**
Poisson stochastic process, **51**
polaron, 101, 137
Pontryagin index, 127
postulates of quantum mechanics, **123**
Potts model, 296
probability amplitude, **124**, **125**
probability density, **57**
probability distribution, **19**, **56**
    canonical, 196
    grand canonical, 197
    initial, **14**
    microcanonical, 196
    normal (Gaussian), **19**
probability space, **56**
propagator (transition amplitude), **135**
    and stochastic quantization, 286
    for a particle in a box, **232**
    for a particle in a curved space, **248**
    for a particle in a linear potential, **197**
    for a particle in a magnetic field, **195**
    for a particle on a circle, **240**
        and α-quantization, **241**
    for a particle on a half-line, **236**
    for a relativistic particle, 177, 179
    for a short time interval, **140**
    for a torus-like phase space, **242**
    for free identical particles, 208
    for the driven oscillator, **198**
    interrelation in different coordinate systems, **257**
    path-integral representation for QFT, 15
        via Feynman integral, 15
    radial part for a free particle, **265**
    radial part for the harmonic oscillator, **265**
pure states, **127**
px-symbol, **202**

quadratic approximation, **86**, **87**
quantization
    around a non-trivial classical configuration, 110
    canonical, **128**
    of constrained systems, 58
    of field theories, 4
    of non-Abelian gauge theories, 63
quantum anomalies, 2, 83, 87, 130
    chiral, 130
    covariant, 137
    non-Abelian, 136
    singlet, 130–132
quantum Boltzmann equation, 272
quantum chaos, **181**
quantum chromodynamics, 2, 88
quantum electrodynamics
    action in α-gauge, 79
    generating functional, 79
quantum field theory
    φ^4-model, 24
    at finite energy, 245
    at non-zero temperature, 223
        doubling of fields, 233
        path-integral representation in real time, 237
        real-time formulation, 233
    non-relativistic, 223
    non-renormalizable, 81
    on non-commutative spacetimes, 185
    regularized, 25
    renormalizable, 25, 81
    renormalization, 25
quantum fields, 7
    scalar, 7
    spinor, 11
    vector, 7

quantum fluctuations, **162**
quantum gravity, 149
    first-order formalism, 151
quarks, 88
quasi-geometric optics, **187**
quasi-periodic boundary conditions, **230**, **241**
quenched approximation, 307

R-operation, 82, 84
radial path integrals, **258**, **260**
random field, **102**
random force, **62**
random function, **57**
random variable, **56**
random walk model, **108**
rank of a group, 322
real form of a Lie group, 322
reduced partition function, 299
reduction formula, 27
regularization in quantum field theory, 82
    dimensional, 83
    lattice, 82
relaxation time, 195
renormalization in quantum field theory, 82, **212**
renormalization point, 84
representation of an algebra (group), 323
    adjoint, 323
    decomposable, 323
    faithful, 323
    indecomposable, 323
    irreducible, 323
    reducible, 323
resolvent of a Hamiltonian, **177**
Ricci tensor, 327
Riemann ζ-function, **100**
Riemann–Lebesgue lemma, **170**
root system, 323

S-matrix, **222**
    coefficient functions, 23
        generating functional, 24
    for fermionic fields, 21
        normal symbol, 22
    for gravitational fields
        path-integral representation, 153
    for scattering on an external source, 18
        normal symbol, 19
        path-integral representation, 17
    in quantum electrodynamics, 64
    in Yang–Mills theory, 65
        normal symbol in QFT, 23
        normal symbol in Yang–Mills theory
            in α-gauge, 73
            in Coulomb gauge, 66
            in Lorentz gauge, 69
            with ghost fields, 74
            with spontaneous symmetry-breaking, 94
    operator, 15
    path-integral representation, 17
saddle-point approximation, **170**
scalar curvature, **246**
scaling laws, 240
scaling limit, 240
scattering of elementary particles, 15
scattering operator, **222**, **223**
    adiabatic, **224**
Schrödinger equation, **125**
    stationary, **133**
Schwarz test functions, **130**
Schwarzschild solution, 328
Schwinger variational equation, **1**, 32
Schwinger–Keldysh contour, 266
Schwinger–Keldysh formalism, 266
semiclassical approximation, **80**, **144**
semidirect product of Lie groups, 324
semigroup property, **21**
Slavnov–Taylor–Ward–Takahashi identities, 35, 80, 85, 86
solitons, 2, 101, 110, 111
source functions, **102**
spacetime
    curved, 150
        asymptotically flat, 150
    Euclidean, 45, **139**
    Minkowski, **139**
    of the Bianchi type, 155
spacetime transformations in path integrals, **253**
spatially homogeneous cosmologies, 155
spontaneous symmetry-breaking, 43, 88
    of global symmetry, 91
    of local symmetry, 92
square completion method for path-integral calculation, 40
standard model, 2, 88, 96
star-product (star-operation), **201**

state vector, **123**
stationary Schrödinger equation, **126**
stationary state, **126**
stationary-phase approximation, **170**
steepest descent method, **170**
stochastic (random) field, **57**, **102**, 288
  Gaussian, **102**
stochastic chain, **57**
stochastic equations, **61**
stochastic function, **57**
stochastic integral, **63**
stochastic process, **17**, **23**, **57**
  Gaussian (normal), **58**
  Markov, **58**
  stationary, **58**
  white noise, **59**
  Wiener, **59**
stochastic quantization, 280
  of gauge theories, 291
stochastic sequence, **57**
stochastic time, 280
string tension, 180
strong CP problem, 129
summation by parts, **96**
superconductivity, 257
superficial divergence index, 81
  in non-Abelian YM theory, 82
  in QED, 81
superfluidity, 257
superposition principle, **126**
superselection rules, **126**
superselection sectors, **127**
superspace, **296**
supersymmetry, 125, 329
supersymmetry operator, **306**, 329
susceptibility
  magnetic, 239
Symanzik's theorem, 230
symbol of an operator, **200**
  px-symbol, **202**
  xp-symbol, **201**
  contravariant, **220**
  normal, **207**
  Weyl, **202**
symplectic two-form, **305**

tadpole, 30
Tauberian theorem, **149**, **326**

tensor operator, 324
thermal Green functions, 226
time-ordering operator, **126**
time-slicing, **28**
topological charge, 113
topological term, **234**
trace anomaly, 182
transfer matrix, 310
transformations of states on non-commutative
    spacetimes, 191
transition amplitude (propagator), **135**
transition matrix, **14**
transition probability, **14**
Trotter product formula, **156**, **157**

ultraviolet divergences, 25
uncertainty principle
  and path integrals, **159**, **216**
universal enveloping algebra, 325

vacuum polarization, 107
vacuum state, **132**
Van Vleck–Pauli–Morette determinant, **173**
variational methods, **80**
  Feynman's, 2
volume quantization condition, **243**

Ward–Takahashi identities, 35
  anomalous, 87, 134
wavefunction, **127**
Weyl anomaly, 181
Weyl invariance, 180
Weyl symbol, **202**, **203**
Weyl tensor, 161, 327
Wick rotation, 83
Wick theorem, **225**
Wiener measure, **25**
  conditional, **25**
  unconditional (full, absolute), **25**, **43**
Wiener path integral, 1, **25**
Wiener process, 24
  its derivative (white noise), **113**
Wiener theorem, **29**
  analog for phase-space path integrals, **214**
  and differential operators in path integrals,
      100
Wilson action, 306
winding (Pontryagin) number, 126

WKB approximation, **169**, **172**

xp-symbol, **201**

Yang–Mills fields, 51
    anti-self-dual, 127
    pure gauge, 126
    self-dual, 127
Yang–Mills theory, 51
    Abelian, 52
Yukawa coupling, 37
Yukawa model, 37

zero-mode problem, 113, 115